Alternative Energy Sources

Part A

INTERNATIONAL SYMPOSIUM OF THE KUWAIT FOUNDATION

Jamal T. Manassah, *editor*: ALTERNATIVE ENERGY SOURCES, Parts A and B, 1981

Jamal T. Manassah and Ernest J. Briskey, *editors*: ADVANCES IN FOOD-PRODUCING SYSTEMS FOR ARID AND SEMIARID LANDS, Parts A and B, 1981

ALTERNATIVE ENERGY SOURCES

Part A

Edited by

JAMAL T. MANASSAH

Sponsored by

ACADEMIC PRESS 1981
A Subsidiary of Harcourt Brace Jovanovich, Publishers
NEW YORK LONDON TORONTO SYDNEY SAN FRANCISCO

ACADEMIC PRESS RAPID MANUSCRIPT REPRODUCTION

Proceedings of a Symposium on Alternative Energy Sources, held in Kuwait, February, 1980.
International Symposium Series of the Kuwait Foundation for the Advancement of Sciences (KFAS).

ACADEMIC PRESS, INC.
111 Fifth Avenue, New York, New York 10003

United Kingdom Edition published by
ACADEMIC PRESS, INC. (LONDON) LTD.
24/28 Oval Road, London NW1 7DX

Library of Congress Cataloging in Publication Data

Main entry under title:

Alternative energy sources.

Proceedings of a symposium sponsored by the Kuwait Foundation for the Advancement of Sciences, and held in Kuwait Feb.9-13, 1980.

1. Renewable energy sources–Congresses. 2. Power resources–Congresses. I. Manassah, Jamal T. II. Kuwait foundation for the Advancement of Sciences.
TJ163 15 A46 662 6 80-27710
ISBN 0-12-467101-2 (Cloth)
ISBN 0-12-467121-7 (Paper) Part A

PRINTED IN THE UNITED STATES OF AMERICA

81 82 83 84 9 8 7 6 5 4 3 2 1

CONTENTS

CONTRIBUTORS FOR PARTS A AND B

Numbers in parentheses indicate the pages on which the authors' contributions begin.

George T. Abed (831), *International Monetary Fund, Washington, DC 20431*

Samir A. Ahmed (355), *Department of Electrical Engineering, City College, City University of New York, New York, NY 10031*

Carl Aspliden (405), *Battelle Memorial Institute, Richland, Washington 99352*

Edward J. Bentz, Jr. (733), *7915 Richfield Road, Springfield, Virginia 22153*

J. R. Bowden (147), *Conoco Coal Development Company, Research Division, Library, Pennsylvania 15129*

Samuel E. Bunker (895), *National Rural Electric Cooperative Association, Washington, DC 20036*

M. M. El-Wakil (519), *Department of Mechanical Engineering, University of Wisconsin, Madison, Wisconsin 53706*

E. Gorin (147), *Conoco Coal Development Company, Research Division, Library, Pennsylvania 15129*

Adel Hakki (797), *ICF Incorporated, New York, NY 10017*

Terry Healy (405), *Rocky Flats Plant, Rockwell International, Golden, Colorado 80401*

George F. Huff (331), *Gulf Science and Technology Company, Houston, Texas 77011*

Edward Johanson (405), *JBF Scientific Corporation, Wilmington, Massachusetts 01887*

David C. Junge (251), *Office of Energy Research and Development, Oregon State University, Corvallis, Oregon 97331*

T. Kammash (607), *Department of Nuclear Engineering, University of Michigan, Ann Arbor, Michigan 48109*

Theodore Kornreich (405), *JBF Scientific Corporation, Wilmington, Massachusetts 01887*

Richard Kottler (405), *JBF Scientific Corporation, Wilmington, Massachusetts 01887*

O. K. Mawardi (687), *Energy Research Office, Case Western Reserve University, Cleveland, Ohio 44106*

William Robins (405), *Wind Power Office, NASA-Lewis Research Center, Cleveland, Ohio 44106*

J. R. Thomas (99), *MHG International Ltd., Calgary, Alberta, Canada*

Ronald Thomas (405), *Wind Power Office, NASA-Lewis Research Center, Cleveland, Ohio 44106*

Khairy A. Tourk (831), *Department of Economics, Illinois Institute of Technology, Chicago, Illinois 60616*

Irwin Vas (405), *Solar Energy Research Institute, Golden, Colorado 80401*

Larry Wendell (405), *Battelle Memorial Institute, Richland, Washington 99352*

Richard Williams (405), *Rocky Flats Plant, Rockwell International, Golden, Colorado 80401*

B. T. Yocum (1), *YOCUM International Associates, Upper Black Eddy, Pennsylvania 18972*

PREFACE

This text has been assembled from the proceedings of the "Alternative Energy Sources Symposium of the International Symposium Series of the Kuwait Foundation for the Advancement of Sciences (KFAS)" that was held in Kuwait in February 1980.

The focus of this symposium was to review and assess those technologies that presently complement and will most likely substitute in the future for oil and gas extracted by conventional techniques. This text includes the state of the art of these technologies as seen by experts in their respective disciplines.

In the coverage of the technologies presented, an attempt has been made to include present developed technologies and those under development. As a consequence, the level of detail in each presentation is appropriate to the developmental stage of the technology under consideration, as assessed by the author. In general, the papers covering proven or nearly proven technologies mostly consist of detailed and or comparative assessments of the diverse engineering schemes without unduly dwelling on basics, while papers addressing technologies under development review the theoretical basis of these technologies in some details. In all instances where meaningful economics are available, numbers are included.

This text also includes review papers of electric storage technology and transportation and energy, topics that, along with conservation, affect most strategic energy planning for the foreseeable future. The text also includes economics methodology and economical development papers that will hopefully allow researchers in the energy field access to the more common tools and approaches of the economic and financial analysts and the international development economists.

During the symposium, participants were also invited to address the following questions in roundtable discussions:

- the role to be played by the Arab countries in the development of alternative energy sources technologies.
- the prioritization to be accorded to each such technology, i.e., to develop a strategy for deciding which technologies should be transferred, adapted, or developed;
- the infrastructure required for the execution of this strategy; and
- the techniques and operational steps to be adopted for implementing this strategy.

The summary of these discussions comprises the subject of a separate publication (Alternate Energy Sources Symposium, Summary Report, Jamal T. Manassah, KFAS).

This text and the symposium would not have been possible without the generous support of KFAS Board of Directors and the personal encouragement of H. H. Sheikh Jaber Al-Ahmed AL-SABAH, Chairman of the Board, Dr. Adnan Al-Aqeel, the Director General, and KFAS member companies. To all these, I am grateful.

Special appreciation is also directed to the KFAS staff for helping me complete this task.

CONTENTS OF PART B

ENHANCED OIL RECOVERY

BRYAN T. YOCUM

Yocum International Associates

ABSTRACT

The Enhanced Oil Recovery (EOR) processes are techniques for mobilizing the residual oil that cannot be recovered by water flood, gas injection, or primary production means.

This paper reviews the current status of laboratory experimental studies, theoretical research, and critical field pilot applications for the major types of EOR. Review and technical assessment are presented for the following types of EOR:

- Micellar surfactant polymer floods
- Alkaline floods
- Polymer floods
- Carbon dioxide floods
- Thermal methods, including:

  1. In situ combustion
  2. Combined thermal drive
  3. Steam drive
  4. Steam soaking
  5. Combined methods
  6. Miscible phase flood
  7. Stimulation method

The cost and economic studies that have been made on EOR processes are also reviewed.

This paper also estimates the long-range probable EOR production for the 1985-2000 period.

ISBN 0-12-467101-2

1.0 Introduction

This paper reviews the current status of laboratory experimental studies, theoretical research, and critical field pilot applications for the major types of Enhanced Oil Recovery.

The number of field applications of enhanced oil recovery (EOR) processes are increasing rapidly in response to higher oil prices and the improving technology that results from the extensive Government/Industry Research and Development programs begun in the 1974/75 period.

In 1975, the active Enhanced Oil Recovery projects in the U.S.A. were 156, made up of 106 Thermal Drive (21 Combustion, 54 Steam Soak, and 31 Steam Drive), 13 Micellar Surfactant, 13 Polymer, 1 Caustic, 9 Carbon Dioxide, and 13 Miscible Hydrocarbon.

In 1976, there were 24 Enhanced Oil Recovery projects active in Canada, 51 Thermal Drive projects in Venezuela, and 17 Enhanced Oil Recovery projects in other countries. The number of projects underway in 1979 is not known, but, in the U.S.A. at least, there has been a large increase in the level of activity in all types of Enhanced Oil Recovery compared with 1976.

Cost and economic studies that have been made on Enhanced Oil Recovery processes are also reviewed in this paper. The literature sometimes refers to Enhanced Oil Recovery as Tertiary Oil Recovery. We used the more generally accepted term Enhanced Oil Recovery.

2.0 Definition of Enhanced Oil Recovery (EOR) Processes

After discovery of an oil reservoir; production first takes place during the primary production, or natural depletion phase. Typically, 10 to 30 percent of the oil in situ in the reservoir rocks will be recovered. Natural depletion occurs under the forces of gas and rock expansion as pressure is reduced, gravity drainage, and natural water influx.

Beginning in the 1930s production methods were developed to increase oil recovery. The main technique was water flooding. Gas injection was also developed. A successful water flood may produce 50 to 60 percent of the oil in situ in the reservoir rocks under near ideal conditions. Because of the low mobility ratio of water relative to oil, water floods normally recover more oil than gas injection. The mobility ratio is the permeability of the reservoir to water divided by the viscosity of water, divided by the same term for the oil phase.

Because of non-uniformities in the reservoir sands, or rocks, the water often does not maintain a stable front. Water is lost into channels, into coning, and fingering. The efficiency with which the water flood enters the pores containing oil in every rock volume of the reservoir is called the sweep efficiency. When sweep efficiency approaches 100 percent, oil recovery will normally be in the 50 - 60 percent range. The water continues to flow through the porous rock volume; however, no more oil is produced from the volume. The residual oil is immobile. It may be located in small pores where interfacial tension forces are balanced against the driving force of the water flood. Also, even in channels where water is flowing, there is an equilibrium liquid oil held by the rock capillary forces, or that reaches equilibrium with the water velocities.

Gas injection is generally a more difficult operation because of the high mobility ratio of gas relative to oil or water. This leads to gas breakthrough and channeling with resulting poor sweep efficiency. Thus, many reservoirs cannot benefit from gas injection.

The enhanced oil recovery processes (EOR) are techniques for mobilizing the residual oil that cannot be recovered by water flood, gas injection, or primary production means. Intensive work on EOR began in the 1974/75 period when Research and Development funds were approved by the U.S. Congress, and DOE (Department of Energy) sponsored research programs were initiated. This work is beginning to bear fruit with over one hundred field pilot EOR projects underway or in design, as well as intensive laboratory and theoretical research programs. A recent economic analysis indicates that Government funding at ten times the present level would lead to 4 million BPD EOR production in 1985/90. This study will be reviewed in a later section.

EOR techniques and processes were experimented with starting twenty years ago. The concept of the micellar polymer floods were originated by Marathon and Union Oil companies in the 1960s. Approximately eighty patents have been taken out. The techniques of thermal recovery are part of EOR. Several thermal recovery projects using different processes have been in production since 1960. However, large scale research, both in thermal recovery fundamentals and field applications, is now underway with DOE support. A commercial carbon dioxide flood began in the early 1970s.

3.0 The Types of Enhanced Oil Recovery (EOR)

Current research and development, field pilot applications, and full scale projects, are going on in the following areas of EOR:

3.1 Micellar Surfactant Polymer Floods

These processes are based on a sequence of flooding operations: (1) the salinity of the reservoir in situ water (connate water) is adjusted by a fresh or low salinity preflush water flood. (2) This flooding is followed by the injection of a carefully designed and controlled chemical slug made up of surfactants and cosurfactants, such as sulfonates and long chain alcohols; as example, amyl alcohol. The action of this slug will be discussed in detail later on. Essentially, a microemulsion is formed between the oil, water, and chemical flood, in which the interfacial tension forces previously holding the oil immobile are reduced essentially to zero (ultralow). The velocity forces of the flood, plus buoyancy forces now able to act, force the oil to flow toward the producing well. (3) Following this slug, a large slug of polymers mixed in water with careful quality control to assure high viscosity are injected. The polymers increase the viscosity of the water flood and reduce the mobility ratio. This enables the flood to penetrate into smaller pores and increase sweep efficiency. It also may bridge larger pores where channeling would occur with water alone. The higher pressure drops required to flow the high viscosity fluid helps to mobilize the oil water emulsions. The polymer slug is tapered with the concentration of polymer normally decreasing with time in a logarithmic fashion. (4) Chase water flood then follows for a sufficient number of pore volumes to remove all residual oil mobilized.

3.2 Alkaline Flood

Because of the expense of the chemicals required for the micellar-polymer flood, a lot of research is devoted to finding cheaper chemical floods. The alkaline flood using sodium hydroxide and adjusted salinities is now under study in the laboratory and in the pilot flood stage.

3.3 Polymer Flood

Polymer floods are generally viewed as a way of improving water floods. The increased viscosity of the water/polymer mixture reduces the mobility ratio; thereby, increasing sweep efficiency and pressure drop in the reservoir. The polymers also serve to bridge channels and fissures; thus, reducing water loss and improving efficiency. Polymer flooding would not be expected to reduce residual oil saturation after conventional water flooding to the low levels that are hoped for in micellar surfactant polymer flooding.

3.4 Carbon Dioxide Flood

Extensive research studies are underway to determine how a CO_2 flood works and where it can be profitably applied. There are already commercial applications. When operating at reservoir temperatures where carbon dioxide is supercritical, the flood proceeds by miscible displacement. The oil and carbon dioxide mix to give a supercritical mixture which has very low surface tension and flows readily. This requires pressure levels in the reservoir that will create the supercritical gas. However, care must be taken to prevent parting of the formation. Recent experiments in low temperature reservoirs indicate that the carbon dioxide is effective in its liquid phase.

3.5 Thermal Methods

Several successful thermal methods are now in operation. The different techniques are being studied in pilot floods and fundamental research is underway. There is a need to determine where the thermal techniques are best applied. The main techniques now in use are described below.

3.5.1 In Situ Combustion

The air is injected under pressure into injection wells optimally located with respect to several producers. A down hole heater raises temperature at the sand face to approximately 600 — 700°F. The oil ignites in the immediate vicinity of the well. By controlling the rate of air injection and the pressures, the heat generated is transferred as temperature gradients throughout the surrounding area. Applications are normally in fields with low API gravity and very high viscosities. As the temperature is raised in the area, the viscosity decreases. The burning front moves oil under thermal gradient to the wells. Down hole heaters are often placed in the producing wells to maintain the temperature of the oil that arrives.

3.5.2 Combined Thermal Drive

Here water is injected periodically, while in situ combustion goes on. The water is converted to superheated steam in the burned rock section. It then passes through the fire front, and, after reaching the lower temperature areas in front of the fire front, it condenses to water forming a hot water flood which gives improved efficiency to the entire process.

3.5.3 Steam Drive

Steam generating plants are provided at the surface and high pressure supersaturated steam is injected. The producing wells are located nearby and benefit from the hot oil. Later, water breaks through and a cycling process combining thermal and water flood effects may be established.

3.5.4 Steam Soaking

The steam generator is moved to a producing oil well. Steam is injected in the oil well itself for a period of days. The well is shut in and allowed to stabilize. The well is then opened up and is produced for a period of time or until its production stops. The steam injection process is then repeated.

3.5.5 Combined Methods

There are many combinations of thermal drive; as example, in situ combustion in the lower part of the zone with water injection in the upper part. Other gases like nitrogen or carbon dioxide may be injected.

3.5.6 Miscible Phase Floods

Research is underway to lower the cost of enhanced oil recovery floods. A miscible phase theoretically can be established with a wide variety of gases and a given oil. The liquid banks of condensed gases will have miscible effects with the oil. Therefore, research is underway on the use of lower grade gases than carbon dioxide, such as, gas mixtures of hydrogen sulfide, carbon dioxide, methane, and others, such as flue gases. These are forms of enriched gas injection.

3.5.7 Stimulation Methods

Stimulation Methods enhance recovery of oil from the producing wells themselves and their surrounding regions. Examples are the foam fracturing techniques and the polymer gel treatment of wells in which water zones are bridged off by polymeric action.

4.0 Micellar Surfactant-Polymer Floods

4.1 Entrapment and Mobilization Correlating Parameters

Systematic research is underway on the entrapment and mobilization of residual oil by Morrow, et al. (1). The three problems under study are:

1. The mechanisms of entrapment and mobilization.

2. Development of correlations of the Capillary Number (the ratio of viscous to capillary forces) and the Bond Number (ratio of gravity to capillary forces).

3. Correction of well bore in situ residual oil saturations to the values actually in the reservoir.

Entrapment mechanisms determine the proportion of oil that is recovered from the wet zone. The fraction that can be produced is one main criteria of an EOR process. The conventional methods of determining residual oil by resistivity logs or laboratory core water flooding are not accurate enough for evaluating the formation for an EOR process. The region around the well bore normally has been stripped of residual oil because high pressure gradients during production enable the viscous forces to overcome the capillary retaining forces.

Entrapment and mobilization mechanisms control recovery by causing low flow rates, and preventing ultralow interfacial tension. The ratios of viscous to capillary forces are high when interfacial tensions are very low. Development of a continuously mobile oil bank means that entrapped oil must be mobilized to form a continuous bank which gathers low residual oil as it advances. Interfacial tension exists between the micellar bank and the oil, and between the micellar fluid and the aqueous polymer bank used to push the micellar fluid. Any entrapment of oil by the micellar bank, or of the micellar fluid by the polymer bank, would cause the process to fail. Figure 1 shows conditions of entrapment and mobilization.

An important finding about the displacement mechanisms when interfacial tensions are ultralow and flood flow rates are typically low is that buoyancy forces may have an important influence.

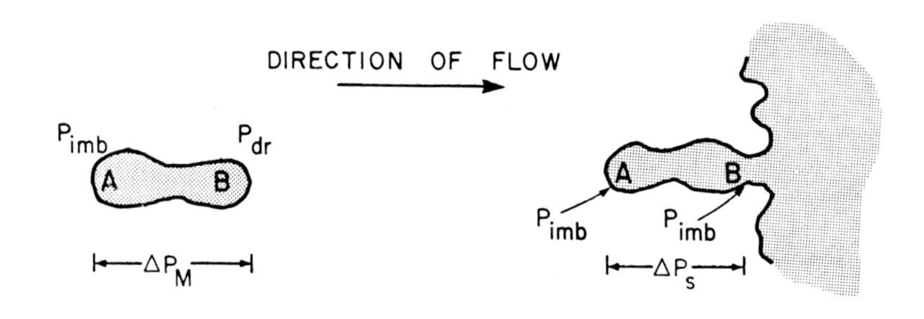

Illustration of change in trapping mechanism caused by hydrostatic contribution to imbibition capillary pressure

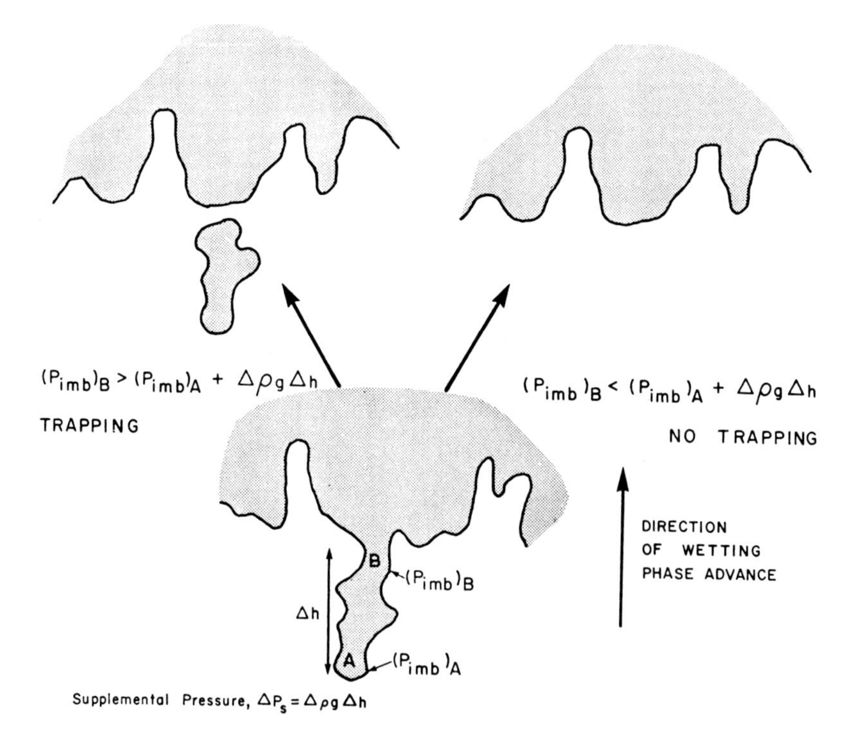

Conditions required for (a) mobilization of a trapped blob, (b) prevention of entrapment of a potential blob

Figure 1

A basic criteria of successful EOR processes is that residual oil can be recovered if the flowing phase causes viscous forces acting on the residual oil to exceed the capillary retaining forces. The ratio of viscous to capillary forces is, therefore, a key correlating parameter for evaluation of an EOR flooding process. The capillary number (N_{CA}) is expressed in the following equation:

$$N_{CA} = \frac{v\mu}{\sigma}$$

$v =$ fluid velocity normal to a unit area of rock

$\mu =$ viscosity of displacing fluid (poise)

$\sigma =$ interfacial tension (dyne/cm)

Capillary number is significant in trapping which occurs in the sequential banks of the micellar fluid process.

Taber established critical values of viscosities to capillary forces to produce residual oil from a given rock sample. This value appears to be an intrinsic property of the rock. For economic recovery of the oil, Taber's values must be exceeded by a factor of 10. However, because of practical and economic limitations on the pressure gradients that may be established across a rock unit, and the distance between injection and producing wells, it was realized that ultralow interfacial tensions, about 1/100 of a dyne/cm, are needed to mobilize the rock's residual oil.

The viscous forces in the flowing media can mobilize the residual oil freed by ultralow interfacial tension of the bank. They are proportional to the interstitial velocity of the flowing phase. To establish a given pressure drop, a given fluid velocity is required. Velocity is inversely proportional to the relative permeability of the flowing phase. Oil blocking the largest pore spaces will be displaced first. A large increase in permeability to the flowing phase occurs for a relatively small decrease in residual saturation. This effect is detrimental to oil recovery because local viscous forces are reduced while the oil remaining in smaller pores is harder to mobilize. This area of effective permeability under reduced residual oil conditions is yet unknown and is an important area of research for determining economic EOR floods. In one case when continuously displacing oil at a high capillary number, oil was produced as a clay stabilized emulsion with structural damage to the sandstone giving permeability loss and high residual oil saturations.

Since accurate values of residual oil saturation are essential for the economic evaluation of EOR processes, in situ saturations found by logging techniques in the well must be corrected for the high flow rates suffered in the well bore. The relationships of capillary number and permeabilities to the flowing phase for EOR target reservoir rocks must be determined. This research work is going on.

EOR mechanisms depend not only on the quantity of in situ residual oil, but also on its microscopic distribution within the pore spaces. There is very little qualitative information on the structural detail of residual oil.

With normal oil/water interfacial tensions, the capillary forces retaining residual oil far exceed the buoyancy forces. It is now believed that with ultralow interfacial tensions in surfactant flooding, the buoyancy forces become effective both on trapping and mobilization.

Experiments show that residual oil saturations remaining after water flooding range from less than 10 percent to more than 50 percent in rocks which may be similar in porosity and permeability. Thus, it becomes necessary to go beyond these two conventional reservoir engineering terms to develop appropriate correlations for estimating EOR project success. Therefore, a systematic investigation including particle shapes, size, size distribution, heterogeneities, change in fluid properties, and initial water saturation is underway.

The trapping mechanism is influenced by:

1. The geometry of the pore network.
2. Relative fluid properties, such as, interfacial tension, density difference, viscosity ratio, and phase behavior.
3. Fluid/rock interfacial properties which effect wetting, applied pressure gradient, and gravity.

The Bond Number and the Capillary Number are being proposed as dimensionless ratios for correlating the behavior of EOR fluid mechanical systems. The Bond number is given below (N_B):

$$N_B = \frac{\Delta\rho \; g \; R^2}{\sigma}$$

$\Delta\rho$ = fluid density difference (gm/cc)

g = acceleration due to gravity (cm/sec.^{2})

R = particle radius (cm)

σ = interfacial tension (dyne/cm)

The particle radius R is a characteristic length expressing pore dimensions. Laboratory experiments can be carried out with packings of various particles with varying R to give varying permeabilities, but similar geometries. The absolute permeability is expressed as a function of particle radius by the Kozeny Carman equation:

$$K = \frac{\phi^3}{K_z(1-\phi)^2 \, A_s^2}$$

A_s = specific surface area per unit solid volume

K_z = Kozeny constant (about five for uniform sands or sphere packings)

ϕ = porosity (about 0.38 for random packing of equal spheres)

Figures 2, 3, and 4 demonstrate the correlations developed for residual saturations as a function of Capillary and Bond Numbers and entrapment situations.

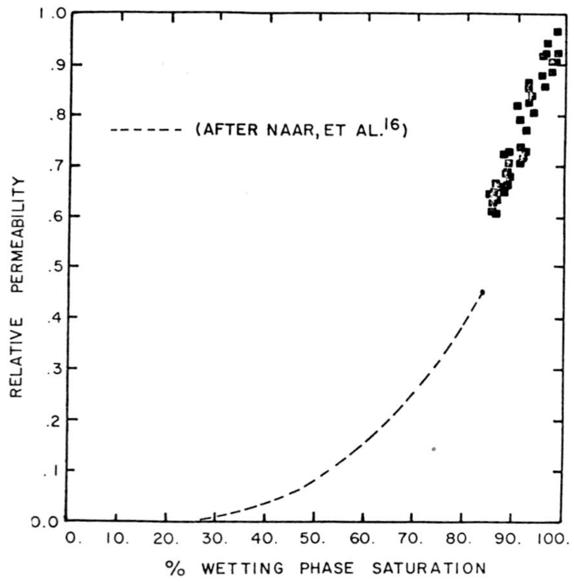

Relative permeability to the wetting phase at less than normal nonwetting phase residual saturations.

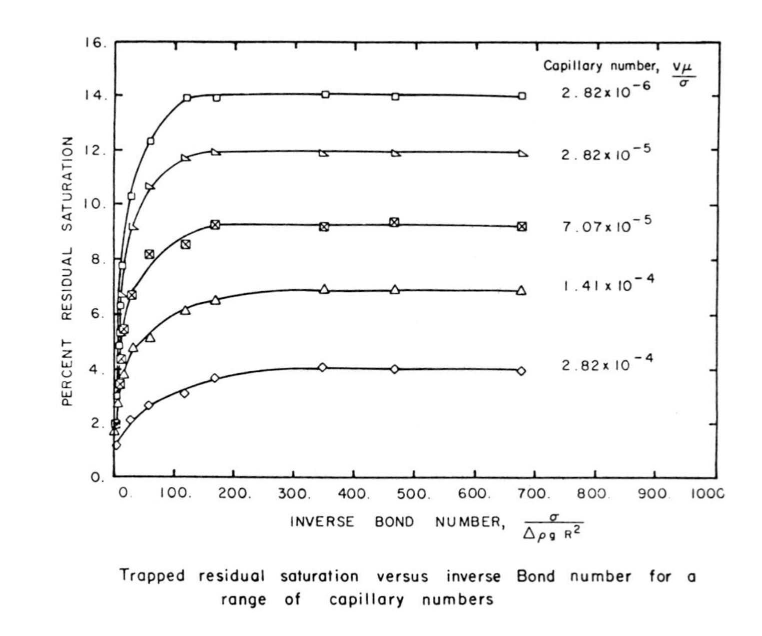

Trapped residual saturation versus inverse Bond number for a range of capillary numbers

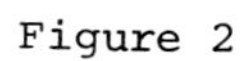

Figure 2

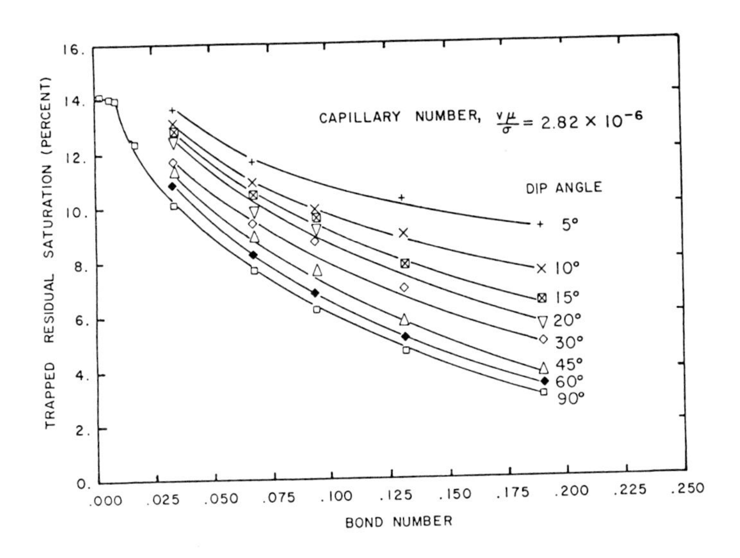

Effect of Bond number on trapping of residual saturation at different dip angles ($\frac{v\mu}{\sigma} = 2.82 \times 10^{-6}$)

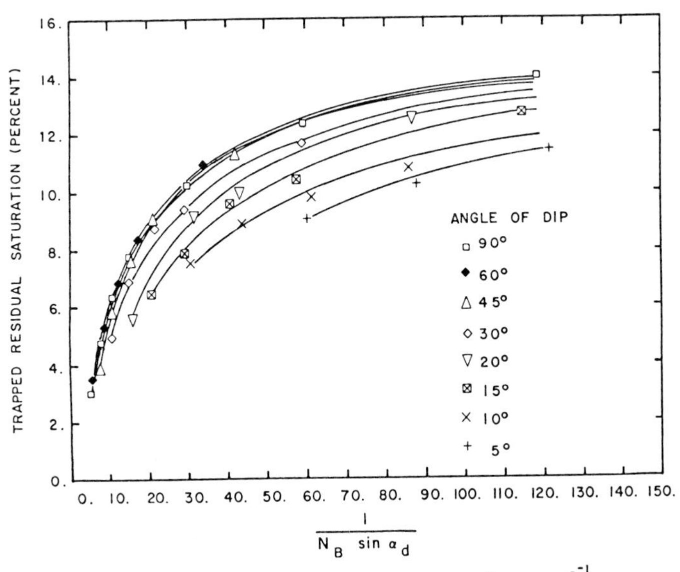

Trapped residual saturation versus $\left[N_B \sin \alpha_d\right]^{-1}$

Figure 3

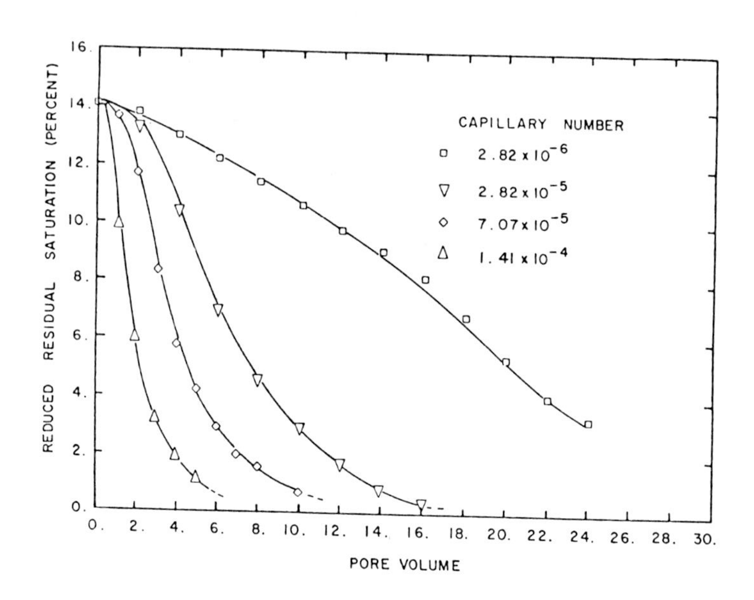

Effect of amount of injected fluid on reduced residual saturation at various capillary numbers

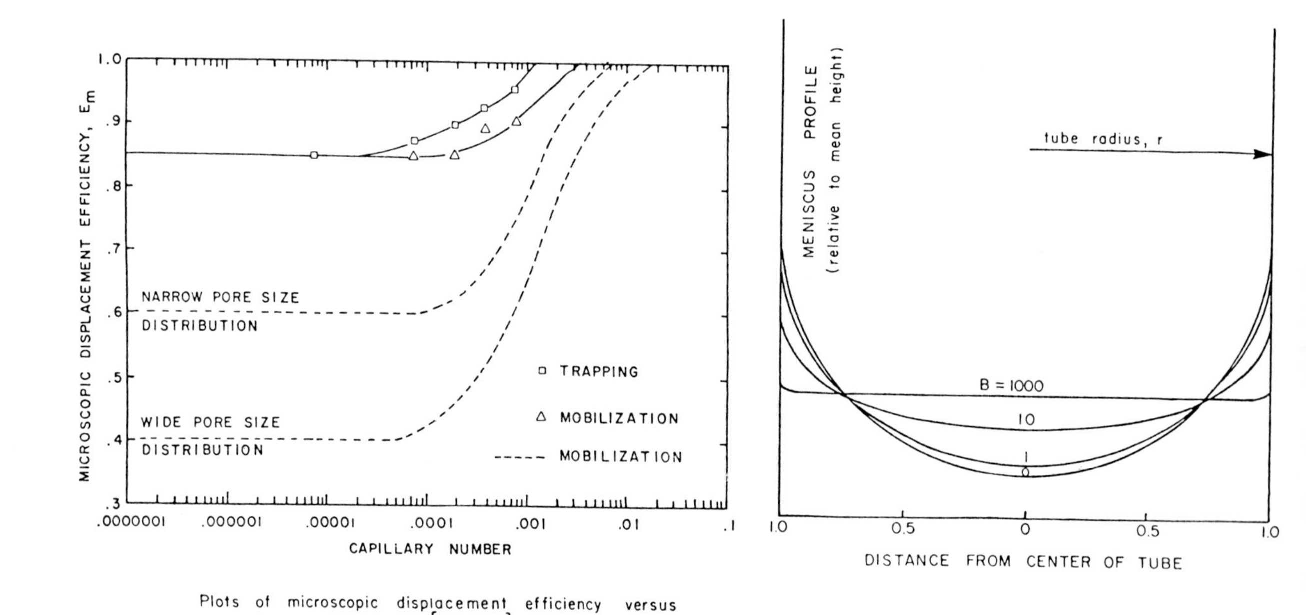

Plots of microscopic displacement efficiency versus capillary number $\left[v\mu/\phi\sigma\right]$

Interface profile at various Bond numbers (based on tube radius, r)

Figure 4

4.2 Reservoir Rocks and EOR Theory

The void volume in permeable rock is thought of as many intersecting pores of varying diameters, comparing pore networks with different mean radii that offer parallel paths for oil displacement. The residual oil will be trapped in the form of blobs that occupy a large set of neighboring interconnecting pores. Once a segment of oil is trapped in an irregular pore, it may or may not continue to be driven forward by the existing pressure gradient. If it does so, it will be by episodic motion. Therefore, the present qualitative theory for residual oil displacement is based on a momentum balance for oil and water in the pore during periods of slow deformation between jumps. See Slattery (2).

The conclusions are:

1. There is a critical value for the surface tension σ, or the dimensionless ratio $\frac{\sigma \phi}{|\vec{\nabla P}| K}$ above which no residual oil can be recovered by displacement in either oil-wet or water-wet rock.

   σ = surface tension (dyne/cm)
   ϕ = porosity
   $|\vec{\nabla P}|$ = pressure gradient in porous rock
   K = salt water permeability (millidarcies)

2. When the water-oil surface tension is less than the critical value, the efficiency of residual oil recovery will increase as the applied pressure gradient increases.

3. The efficiency of residual oil recovery increases as the surface tension decreases below its critical value.

4. With interfacial tension (IFT) less than critical, but the interfacial viscosities still large, both factors become equally important, and must be reduced by equal percentages to increase oil recovery.

4.3 Flow in Reservoir Porous Media

Fundamental research is also underway toward understanding reservoir rock and transport better, as example Davis and Scriven (3). They are studying pore connectivity which counts the ways in or out of a given pore body and, thus, the maximum number of transport paths through a pore system. This topology has been overlooked but is as significant as pore geometry. Pore and rock connectivity and wall curvature are interrelated and influence wettability, adhering film thickness, and mechanical properties. A unified picture of structure, strength, and transport is being developed to interpret sparse reservoir data, improve recovery processes, and stimulate innovations.

4.4 Ultralow Interfacial Tension Formation Theory

The key mechanism of ultralow interfacial tension formation is also under study by Franses, et al. (4). They find a dispersed liquid crystal between the oil rich and brine phases and believe it is responsible for the ultralow tensions. The interface between hydrocarbon and brine does not reach an ultralow value unless the third phase is present.

The state of dispersion on which low tension depends is sensitive to a large number of variables, many of which are difficult to control. Current studies are aimed toward deciphering the role of the third phase. The leading hypothesis is that deformation and coalescence of the viscous particles of the third phase are promoted by oil-brine interfaces. If the sum of the tensions of the third phase against the brine-hydrocarbon phases is less than the tension between these two phases themselves, it is possible that the third phase tends to spread over the hydrocarbon drops.

The effectiveness of surfactant formulations that create the dispersed third phase depend on the factors that effect its state of dispersion. Particle retention during flow through porous media can reduce the amount of third phase microcrystallites. This reduces effectiveness of ultralow tension formation, thus adversely affecting mobilization of residual oil.

4.5 Ultralow Interfacial Tension and the Middle Phase

Benton, et al. (5) are studying the ultralow interfacial tensions which must be achieved and maintained during EOR processes employing surfactants and forming micellar "middle phases". The mixture of surfactant, cosurfactant, brine, and oil form a characteristic middle phase, an oil-rich aqueous phase, and a water-and-surfactant-in-oil phase. The fluid phases formed are called the microemulsions and contain no liquid crystalline material. At intermediate salinities, the microemulsion phase called the surfactant, or middle phase, was seen in equilibrium with excess oil and brine phases. It contains crystalline material. The structure of the middle phase is not well understood. This phase forms as salinity increases and the oil and water microemulsion separates into two phases; one with a high concentration of oil drops, and the other with low concentration of oil drops in brine. Interfacial tension is ultralow between these phases.

4.6 Interfacial Tension (IFT) Correlations

Ultralow interfacial tension mixtures require "middle phases" in the surfactant/oil/brine system for a given reservoir crude oil/brine system. See Lipow, et al. (6). There are many different combinations of surfactant, cosurfactant, and concentration levels that will exhibit middle phases, in addition to the oil-rich and brine-rich phases. The question is to design the best surfactant-based chemical slug for the specific reservoir; and to maintain appropriate velocities in the reservoir. The critical "middle phase" contains microcrystalline emulsions called micelles that create ultralow IFT.

The optimum chemical slug must take into account surfactant retention on the rocks, mobility, compatibility between polymer and surfactant, and costs. Figures 5 and 6† show the effect of salinity, cosurfactants, oil composition, and concentrations on interfacial tension. The term ACN characterizes the oil (alkane carbon number).

Other factors of importance are: (1) the solubility of the surfactant, (2) the quantity needed to create miscibility between oil and water, (3) the required alcohol concentration. These are required to create the middle phase. It is necessary to use excess surfactant to make up the loss by absorption on the reservoir rock.

Clarification is being made of the problems in designing a surfactant slug for a specific reservoir oil and brine system; thus, laboratory experiments and a design study should be carried out on each reservoir candidate in order to determine the minimum cost-combination of chemicals, concentrations, fluid velocities, and pressure gradients that give minimum cost and maximal recovery.

† The optimal salinity S^* is given by:
$$\ln S^* = K(ACN) + f(A) - \sigma$$
where $f(A)$ is a function of the alcohol type and concentration,
σ is a parameter characteristic of the surfactant, and
K is .16 for all alkylaryl sulfonates.

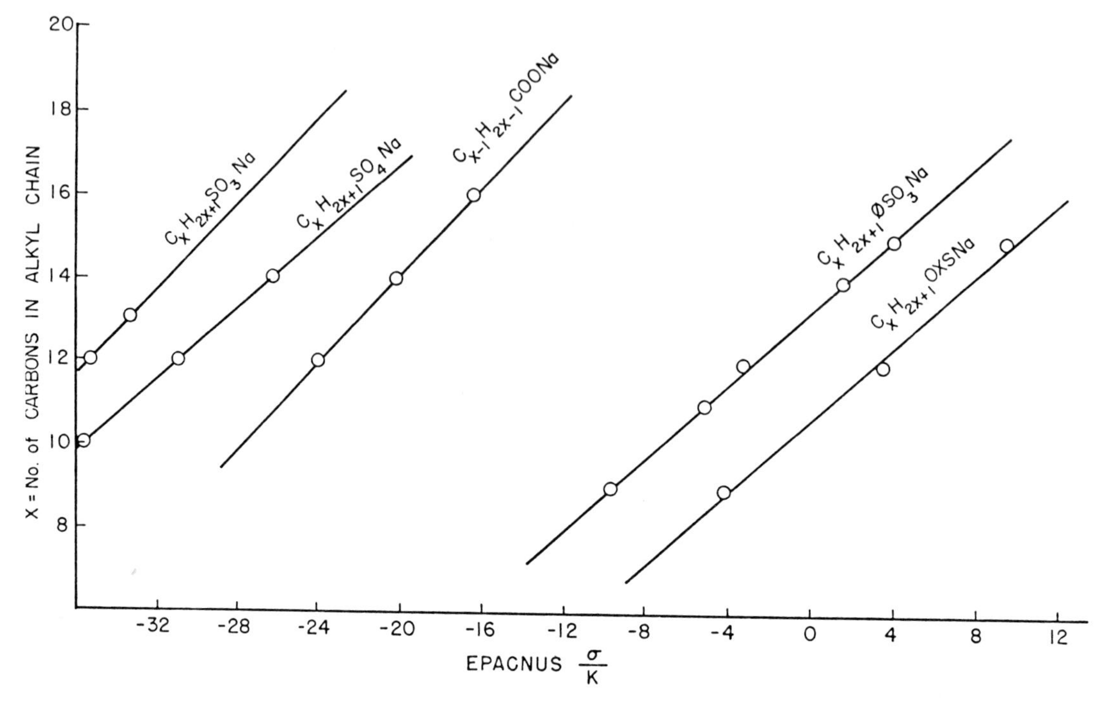

The effect of surfactant structure on EPACNUS.

Figure 5

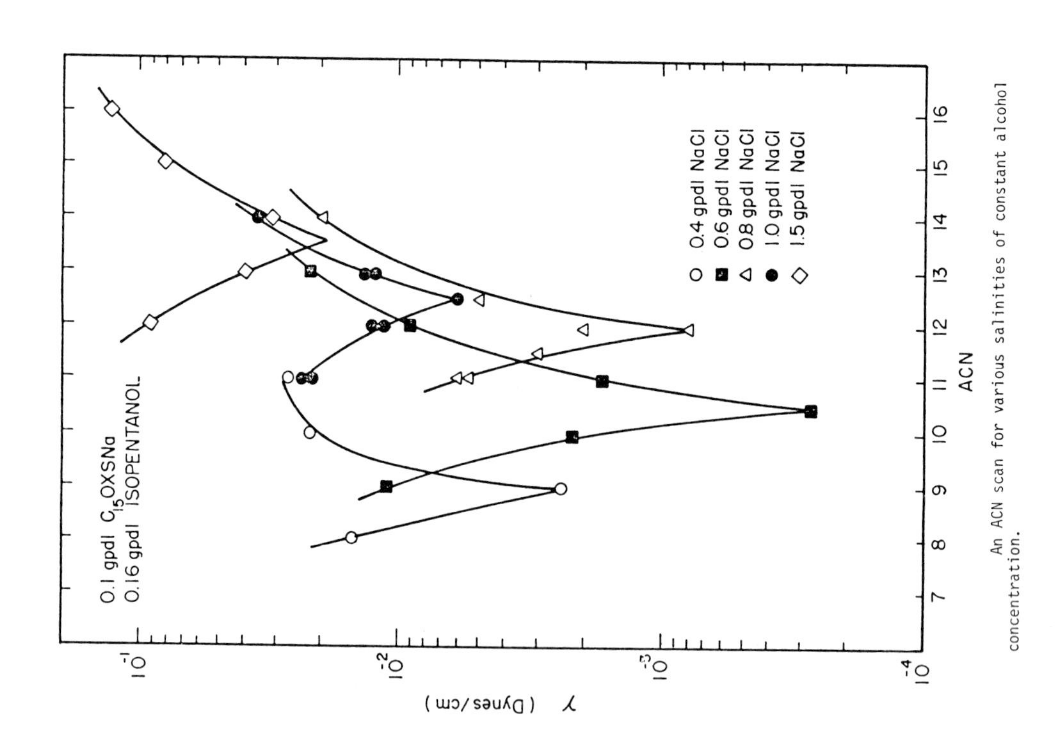

An ACN scan for various salinities of constant alcohol concentration.

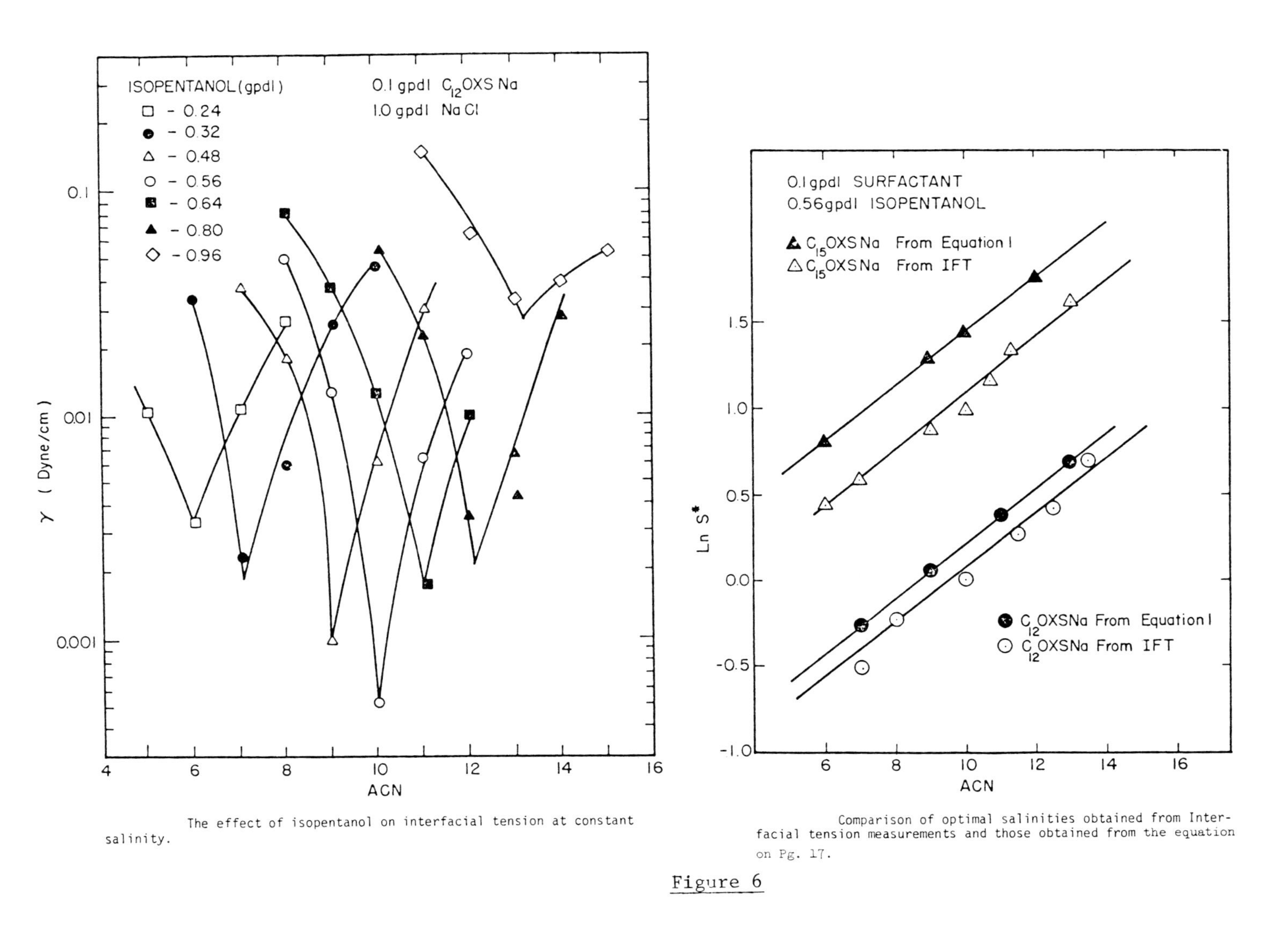

The effect of isopentanol on interfacial tension at constant salinity.

Comparison of optimal salinities obtained from Interfacial tension measurements and those obtained from the equation on Pg. 17.

Figure 6

4.7 Reservoir Clays and EOR

The recent research by Somerton, et al. (7) on the role of clays in enhanced oil recovery is important because many reservoirs contain significant amounts of clays. Because of the large effective surface areas of clay minerals in oil reservoir formations and because of the high degree of reactivity of such surfaces, clay minerals will play an important role in the success or failure of enhanced recovery techniques. Clays absorb many substances which may be injected into the reservoir to improve oil recovery. Surfactants and polymers may absorb on clay surfaces in sufficient amounts to make the recovery procedures uneconomic. The ion exchange capacity of clay minerals may cause surfactant precipitation loss, increasing of ultralow interfacial tension, and polymer degradation. Also, these reactions may cause structural damage to the formation, like clay swelling and migration. Flow rate, salinity, or temperature may cause plugging due to clay migration.

California reservoir formations, especially, contain clay minerals. Successful application of chemical flooding techniques will depend, to a large degree, on successful design to minimize reactivity and increase mobility in these formations. This is a major area of research.

4.8 Micellar Slug Formulation

Shah, et al. (8) have investigated interfacial phenomena in micellar systems. The oil-brine-surfactant systems have specific values for each variable; as example, salt concentration, oil chain length, and surfactant concentration to minimize interfacial tension (IFT). See Figure 7. Also, at the bottom of this figure, the makeup and combination of the six main variables to create ultralow IFT's are shown. These six variables influence two basic phenomena of the mixtures, the partition coefficient of the surfactant, and the effective CMC of the surfactant. The partition coefficient influences the mutual solubility of oil and brine. CMC represents the critical monomer concentration of surfactant in the aqueous phase. These two phenomena determine the three surface parameters that control interfacial tension, namely, the surface concentration of surfactant, surface charge density, and solubilization of oil and brine in each other. The magnitude of the effects on IFT is not known.

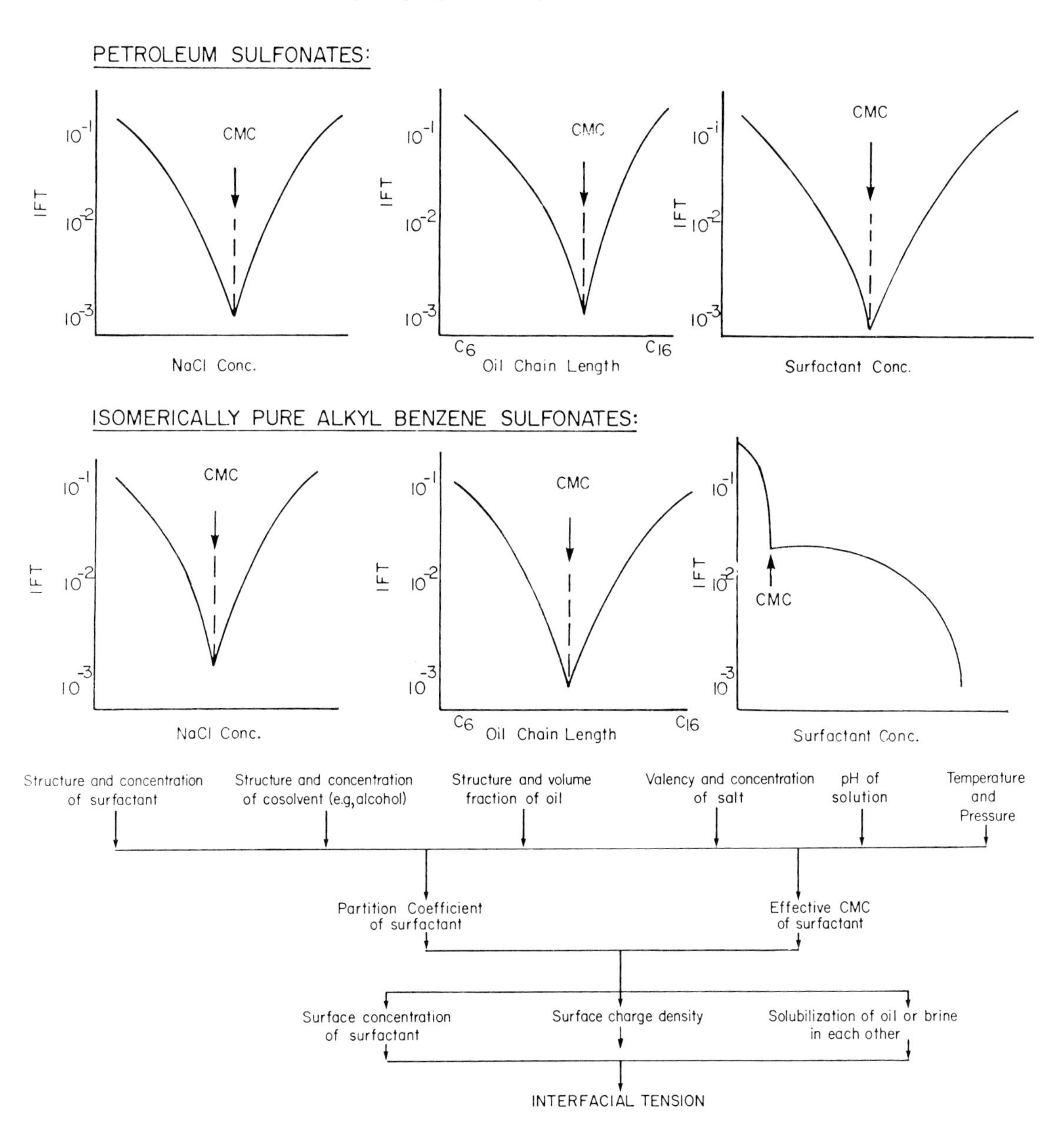
THE EFFECT OF SALT CONCENTRATION, OIL CHAIN LENGTH, AND SURFACTANT CONCENTRATION ON INTERFACIAL TENSION
PETROLEUM SULFONATES:
IFT
10^{-1}
10^{-2}
10^{-3}
CMC
NaCl Conc.
C_6
C_{16}
Oil Chain Length
Surfactant Conc.
ISOMERICALLY PURE ALKYL BENZENE SULFONATES:
Structure and concentration of surfactant
Structure and concentration of cosolvent (e.g, alcohol)
Structure and volume fraction of oil
Valency and concentration of salt
pH of solution
Temperature and Pressure
Partition Coefficient of surfactant
Effective CMC of surfactant
Surface concentration of surfactant
Surface charge density
Solubilization of oil or brine in each other
INTERFACIAL TENSION

Figure 7

4.9 Reduce Chemical Costs

Extensive research is going on to find ways of reducing the costs of the expensive chemicals used in micellar floods. Johnson (9) has been studying ways to reduce the absorption of the surfactants like petroleum sulfonate on the reservoir rocks. Studies to obtain cheaper raw materials from waste streams are underway. The polymers may be biologically degraded by microorganisms. Studies to eliminate this are underway. The production of mobility control polymers near the field where they will be injected, is an important cost minimization factor. This could involve the microbiological production of xanthan gums or similar polymers. Studies on the production of the higher alcohols from pulping wastes may result in lower cost processes.

4.10 Computer Models for Economic Evaluation of EOR Projects

In order to realistically estimate EOR project economics, computer models for comparative economic analysis are being developed. Validation with performance and economic data from several surfactant polymer projects have been made. The simple models are now becoming usefully complex with enhanced ability to analyze sensitivity to many types of injection sequences and selections of chemicals for polymer or surfactant polymer projects. The economic model can handle the present optimal injection sequence of preflush, chemical slug, polymer buffer with a tapered concentration, and post water-injection. Project performance is specified as cumulative fractional oil recovery versus cumulative pore volumes of material injected. Risk is handled by expected, good, and poor performance; each possibility is probability weighted. Economic parameters are entered as investment and operating costs. Sensitivity studies cover variations in barrels per acre/foot recovery, time recovery rate, cost of chemical slug, cost of polymer initial investment, and operating costs.

The economic model will be useful for screening of candidate leases or fields to determine where investments in micellar-polymer, or improved water flood with polymer are economically attractive with minimal risk to capital. They are also useful for optimizing the effects of new research findings on the project "state of the art" design basis.

4.11 North Burbank Pilot

Doe and Phillips Petroleum initiated a micellar polymer flood program in May 1975 in a 90-acre portion of the North Burbank Unit. See Kleinschmidt, et al. (10). The injection process is nearing completion now, with drive water injection going on. The preflush, surfactant slug, and polymer chemicals have already been injected. The tapered polymer slug had a final polyacrylamide concentration of 100 parts per million. Injection pressures were increased to increase production, but were limited to avoid extending fractures in the reservoir.

Selection of this unit for a pilot flood was due to its successful water flood. Production had declined to the stripper level. The WOR (water/oil ratio) was 100. However, cores showed high residual oil saturation in the sands with approximately 400 million barrels of sweet 39° API oil still, theoretically, recoverable. The average net pay was 43 feet, average permeability of 52 millidarcies, and oil saturation of 30 percent. The test pattern of 25 wells consists of nine inverted 10-acre five spots (refer to Figure 8). The injection schedule and the table of production are also shown in Figure 8.

As injection of chemical solutions proceeded, the pressures at the injection well heads rose reflecting the increased viscosity of the fluids moving through the reservoir and the building of higher oil saturations ahead of the chemical slugs. Pressure rise stabilized after completion of polymer injection. As viscous fluids moved out into the reservoir, declines in injection pressure were offset by increasing water injection rate.

Productivity limitations existed at several producing wells. Rates of injection were increased after successful fracturing of the producing wells and increase of their rates. A limit on the rates were fracture extensions that occur with excessive pressure. These occurred several times.

During the second year, oil production increased to 180 BPD with a WOR of 69. Water flood would have produced 59 BPD; so, the EOR process had increased production threefold. During the next year, oil production gradually increased to 281 BPD. The WOR had decreased to 49.5. Most of the increased production came from wells 6, 17, 28, and 29. The well pattern is shown in Figure 8. Figure 9 shows the chemical concentrations through time.

NORTH BURBANK PILOT PROGRAM

INJECTION SCHEDULE

START	CONCENTRATION, PPM NOM.	WT./VOL.	AVERAGE VISC., CP	INJ. RATE B/D
9-15-76	2500	2375	81	3600
10-14-76	2000	1954	53	4800
11-25-76	1500	1684	29	4800
1-17-77	1100	1170	13	5550
3- 8-77	800	850	8.4	5550
5- 1-77	600	657	4.9	5550
7- 8-77	460	481	4.1	6300
9-12-77	250	238	2.5	6300
11- 1-77	100	99	2.0	6300
1-16-78	Fresh Water Began			6300
3- 7-78	Fresh Water			6900
4- 4-78	Fresh Water			7500

TRACT 97 PRODUCTION

	Total Bbls.	B/D	Tertiary Oil Bbls.	B/D
Nov. '76	2113	70	275	9
Dec.	3018	97	1136	37
Jan. '77	3409	110	1543	50
Feb.	3823	137	2150	77
Mar.	4683	151	2844	92
Apr.	5515	184	3748	125
May	5548	179	3734	120
Jun.	5698	190	3953	132
Jul.	6044	195	4251	137
Aug.	6270	202	4487	145
Sep.	6631	221	4915	164
Oct.	6995	226	5230	169
Nov.	7022	234	5322	177
Dec.	7049	227	5301	171
Jan. '78	7208	233	5467	176
Feb.	6287	225	4721	169
Mar.	7290	235	5563	179
Apr.	7910	264	6245	208

Cumulative Tertiary Oil (5/12/78) - 73,569 Bbls.

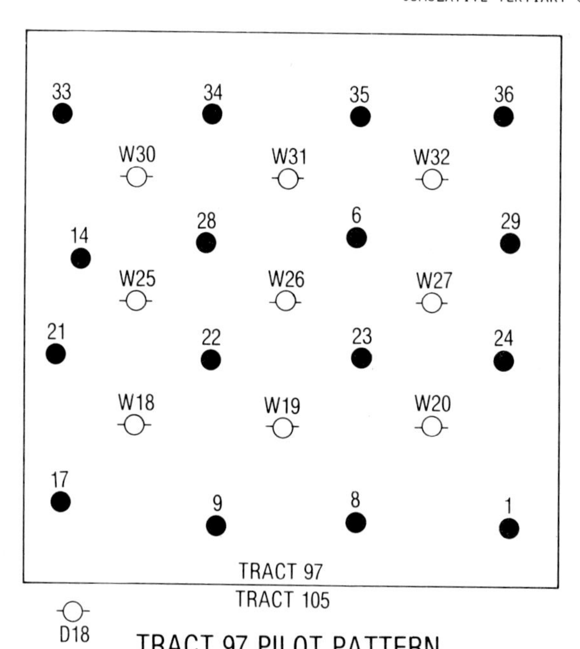

TRACT 97 PILOT PATTERN

Figure 8

NORTH BURBANK PILOT PROGRAM

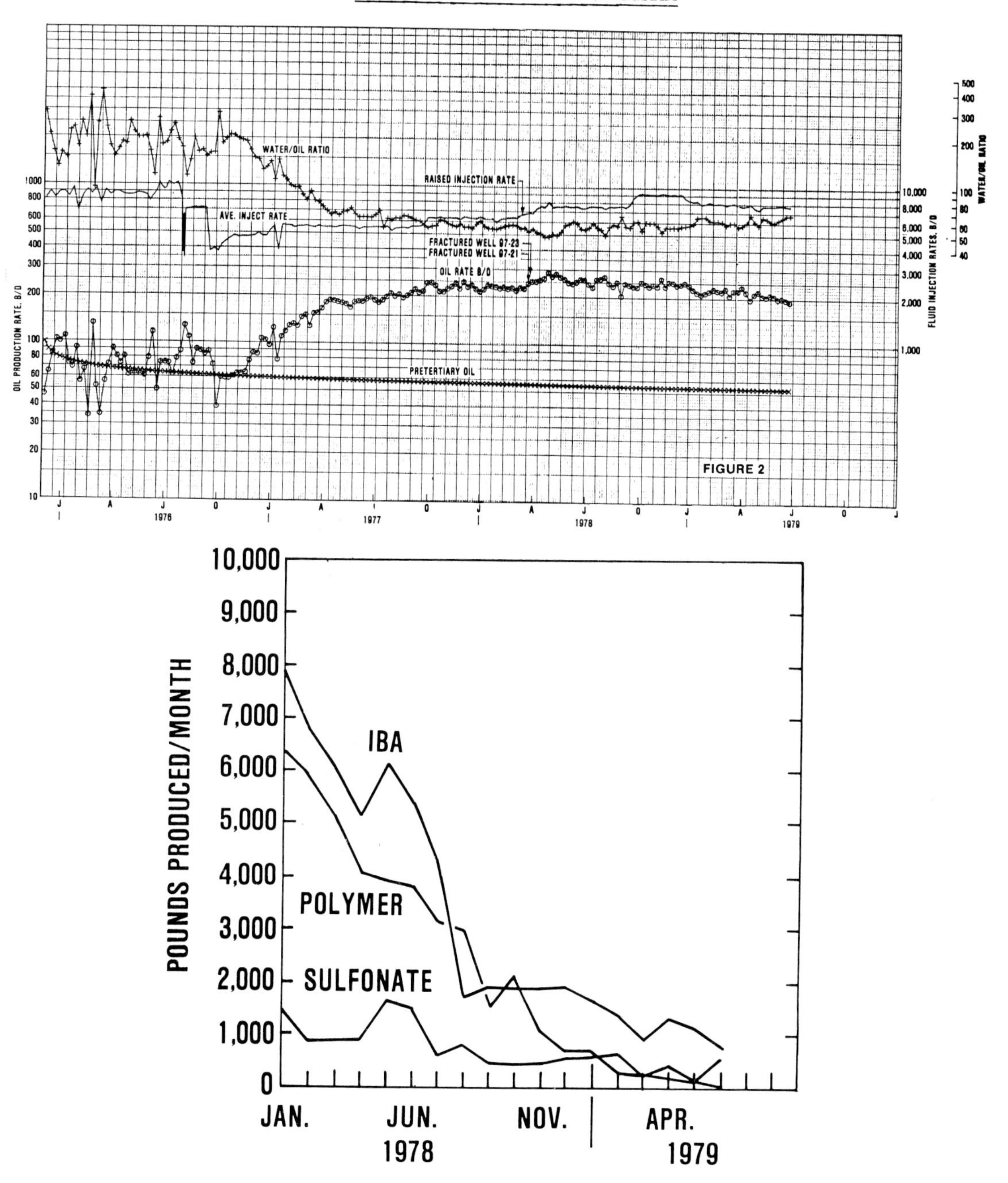

Figure 9

Two years after initiating the slug of sulfonate surfactant about 0.28% had been produced. Of the cosurfactant, isobutyl alcohol, about seven percent had been produced. Of the polymer slug, polyacrylamide production appeared about the same time as the sulfonate. By the third year, about 10.6 percent of the injected polymer had appeared in produced water. No correlations between oil production and the time of chemical breakthroughs are apparent.

The chemical solutions injected into the pilot pattern were confined by the production of offtract fluids at high WORs (about 200). Well fractures and channeling which reduce flood efficiency were corrected by injecting alternate stages of aluminum citrate and polymer solutions which result in the deposition of layers of polymer cross-linked together by aluminum with the first layer absorbed to the rock surfaces. Successive layers reduced the permeability in the zones where channeling is occurring.

Figure 10 shows the computer prediction of enhanced oil recovery versus the volume of fluid injected. Despite practical reservoir and well problems, actual production performance is better than predicted. The restriction on injection rates will cause a longer time to be necessary to recover the 600,000 barrels of additional oil. However, the project will be profitable. The lessons learned in the pilot flood plus the application of research findings to reduce the costs of the chemical slugs, and optimize the quantities and rates injected, will make the recovery of residual oil in many abandoned or high WOR stripper fields economically attractive in the 1980s. It is becomming clear that these projects must be well designed and engineered. Laboratory studies should be carefully made to determine the optimum process conditions. Considering the large number of abandoned stripper reservoirs in the United States, these results augur a good future for the micellar polymer process.

4.12 Crawford Robinson Sand Pilot Flood

A large commercial scale demonstration of the MARAFLOOD process is being carried out in the main consolidated Robinson Sand in Crawford County, Illinois. See Burdge (11). The M1 project is a micellar polymer flood, 407 acres in areal extent. This includes 248 acres of 2.5-acre five spot patterns and 159 acres of a 5.0-acre five spot pattern. The project has a total of 114 injection wells and 132 production wells. The project will

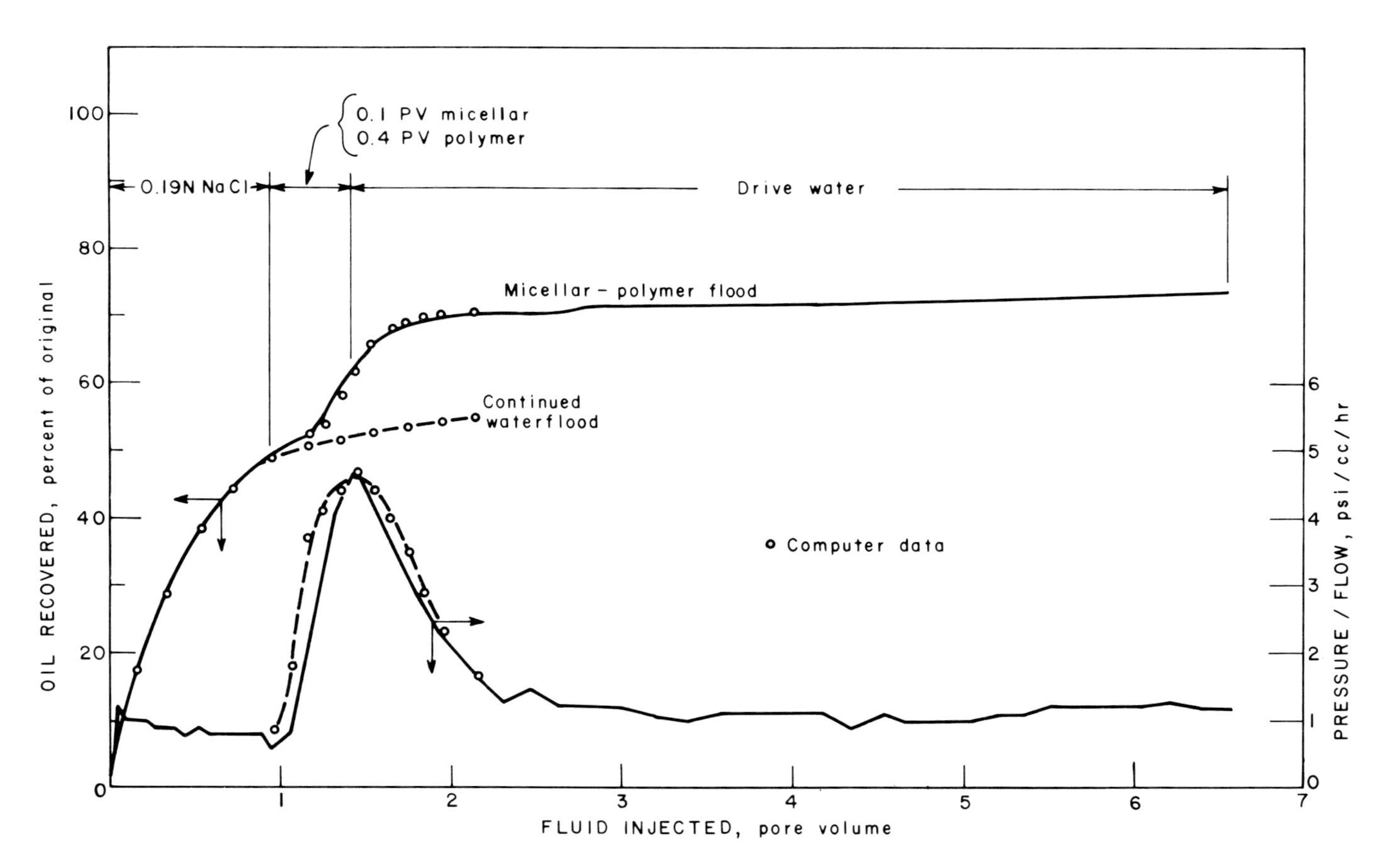

Figure 10. Cumulative oil recovery and pressure drop for Bartlesville core history match of fluid injected, pore volume.

take 10 - 15 years. The micellar slug was injected during 1977. Care was taken to work over the wells to improve injectivity and productivity and to eliminate bad flow distribution problems.

The characteristics of the sand selected are shown in Figure 11; net average thickness 27.8 feet, mean permeability 102.8 millidarcies (arithmetic), porosity 18.9 percent, reservoir pore volume 16,624,048 barrels. The composition of the original micellar slug was active sulfonate 10.2 percent, inorganic salt 3.1 percent, water 78.5 percent, unreactivated oil 7.6 percent, hexonol 0.75 percent. The alcohol content was increased to 0.9 - 1.2 percent in later slug formulations. The cost of the micellar solution slug was $5.53 per barrel of slug.

4.13 Big Muddy Wall Creek Test

A commercial size field test of low tension flooding is underway in the Wall Creek Reservoir of the Big Muddy field. See Ferrell, et al. (12). A successful low tension process here would be applicable to similar reservoirs in the Rocky Mountain Region. The reservoirs have low matrix permeability and fracture at pressures less than the hydrostatic. They are frequently faulted. The connate water, however, has low salinity and low hardness, an advantage in micellar floods. The demonstration site is 90 acres composed of nine 10-acre five spots. The approach is to correct fractures in both injection and producing wells, then to determine the feasibility of a low tension process in these low permeability sands.

The second Wall Creek Reservoir had been successfully water flooded. It has a thick net pay and high oil saturation. The chemical injection flow diagram is shown in Figure 12. A pilot flood of one acre established the validity of the low tension process. The criteria established for the surfactant/cosurfactant system are:

1. The system must be a stable single-phase aqueous solution under reservoir conditions.

2. It must displace essentially all of the oil from cores in laboratory testing.

3. It must have retention on the rock surfaces of less than 1 microequivalent per gram of rock.

4. It must be relatively insensitive to dilution by reservoir fluids.

CRAWFORD ROBINSON SAND PILOT FLOOD

RESERVOIR PARAMETERS AND RELATED DATA

Parameter	Mean Value		
	2.5-Acre	5.0-Acre	Total Project
Net Sand Thickness, Ft.	29.3	25.5	27.8
Arithmetic Mean Permeability, md.	100.3	107.0	102.8
Geometric Mean Permeability, md.	75.5	79.3	76.9
Porosity, %	18.8	18.9	18.9
Area, Acres	248	159	407
Reservoir Pore Volume, Bbls.	10,627,513	5,996,535	16,624,048
Wells (Total)	181	75	256
Producers	91	41	132
Injectors			
Slug	85	29	114
Water	8	6	14
Water Disposal	1	1	2

Figure 11a

CRAWFORD ROBINSON SAND PILOT FLOOD

MOBILITY BUFFER DESIGN

Volume (% P.V.)	Polymer Concentration		Total Lbs	Estimated Completion Date	
	P.P.M.	Lbs/Bbl		2.5-Acre	5.0-Acre
11%	1156	0.4030	736,944	6/79	12/79
19	800	0.2789	880,925	4/80	8/81
32	625	0.2179	1,159,162	6/81	12/83
12	411	0.1433	285,867	12/81	10/84
11	200	0.0697	127,457	5/82	8/85
10	100	0.0349	58,018	8/82	3/86
10	50	0.0174	28,927	11/82	9/86
35	0	0	0	11/83	8/88

Figure 11b

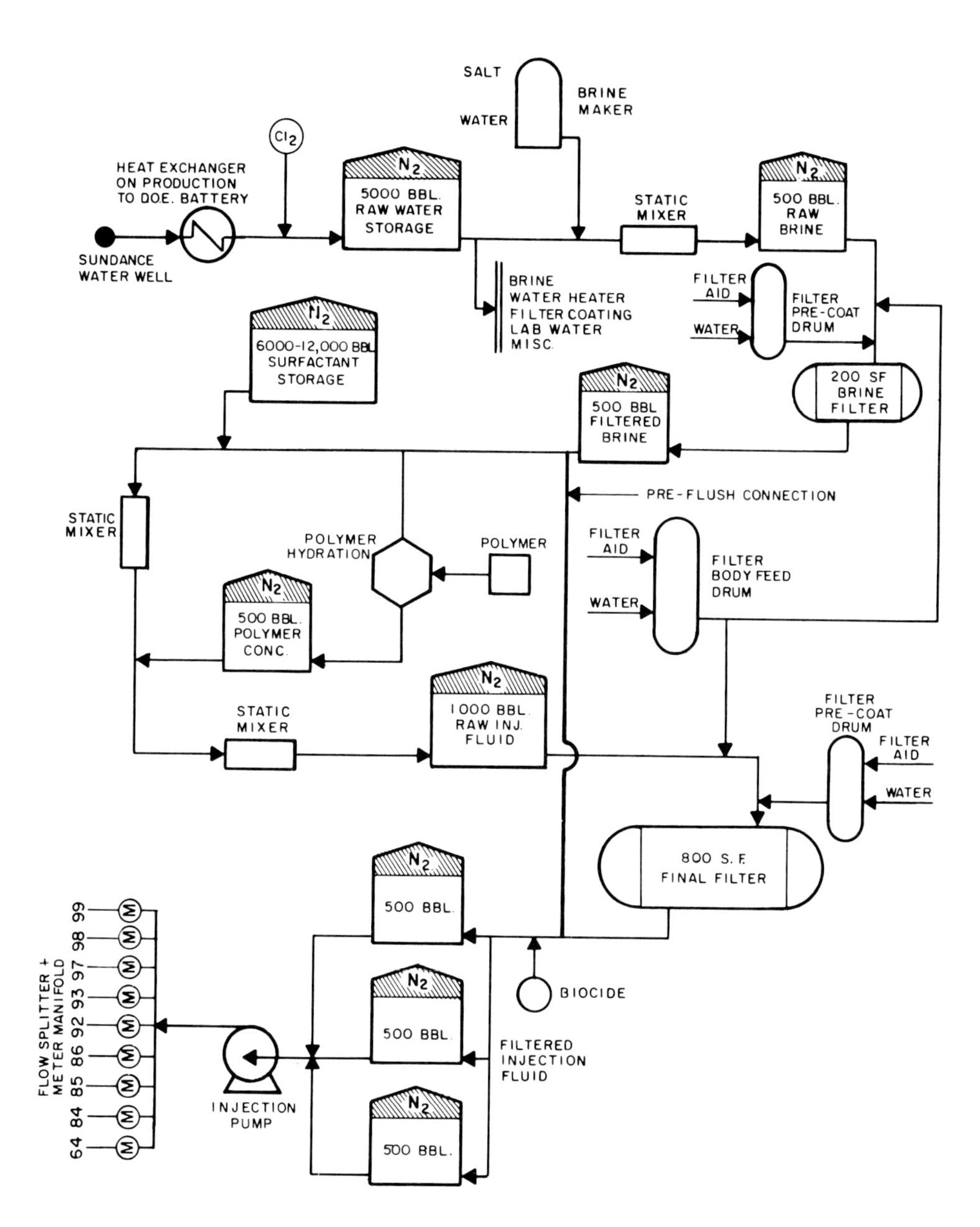
SALT
WATER
BRINE MAKER
Cl_2
HEAT EXCHANGER ON PRODUCTION TO DOE. BATTERY
SUNDANCE WATER WELL
N_2
5000 BBL. RAW WATER STORAGE
STATIC MIXER
500 BBL. RAW BRINE
BRINE
WATER HEATER
FILTER COATING
LAB WATER
MISC.
FILTER AID
WATER
FILTER PRE-COAT DRUM
6000-12,000 BBL SURFACTANT STORAGE
200 SF BRINE FILTER
500 BBL FILTERED BRINE
PRE-FLUSH CONNECTION
STATIC MIXER
POLYMER HYDRATION
POLYMER
FILTER AID
WATER
FILTER BODY FEED DRUM
500 BBL. POLYMER CONC.
STATIC MIXER
1000 BBL. RAW INJ. FLUID
FILTER PRE-COAT DRUM
FILTER AID
WATER
800 S.F. FINAL FILTER
500 BBL.
500 BBL.
500 BBL.
BIOCIDE
FILTERED INJECTION FLUID
INJECTION PUMP
FLOW SPLITTER + METER MANIFOLD
64 84 85 86 92 93 97 98 99
CHEMICAL INJECTION FLOW DIAGRAM
BIG MUDDY FLOOD

Figure 12

The optimum composition of the surfactant phase, including salinity and sulfonate, will be determined in the laboratory. The optimum slug size and viscosity can then be calculated by a computer model. The surfactant slug is followed by the main polymer slug of 20 percent pore volume and about 1.000 PPM concentration in water. A tapered polymer slug of about 25 percent pore volume will follow. This slug is divided into five batches with the viscosity progressively decreasing. Bacterial action would cause injectivity problems and destroy chemicals, so adding chlorine in caustic solution to adjust the pH to 12, was found cost effective. Water flooding will continue for several years after injection of the chemical slugs before all EOR oil is produced.

4.14 El Dorado Pilot Floods

Miller, et al. (13) describe two micellar-polymer pilot floods being carried out in the El Dorado 650-foot Sand (Kansas). The sand thickness is about 18 feet in the project areas. There are minifractures in the sand. Figure 13 shows the programs of the two floods and the reservoir parameters. Figure 14 shows the layout of producers, injectors, observation, test, and monitoring wells. Each 25.6-acre lease is drilled with four five spot patterns which were found to be optimum by detail model work. There are nine injectors, four producers, twelve monitoring, and two observation wells in each pattern.

Since the reservoir properties and well patterns are essentially the same, two alternative programs of micellar polymer flooding can be compared based on oil recovered and economies. Figures 15 and 16 show the design of fluid compositions and volumes to be injected for the North (Chesney) pattern and for the South (Hegberg) pattern.

In both floods, the original reservoir water is flushed out by preflush treatments to give the proper salinities and ionic concentrations for the micellar and polymer floods to maximize oil recovery.

The micellar slugs differ in that one is soluble oil based and the other is fresh water based. Surfactants and cosurfactants are chosen accordingly. Laboratory and experimental work were used in the design of these micellar slugs.

The polymer slugs are polyacrylamide following the soluble oil micellar slug and a polysaccharide following the water based micellar slug.

EL DORADO MICELLAR-POLYMER PROJECT

SUMMARY OF EVENTS

TASK	DATE
Site Evaluation	February, 1974
Contract Award	June, 1974
Pattern Development	December, 1974
NORTH PATTERN FLUID INJECTION	
Preflood I	November, 1975
Preflood II	December, 1976
Micellar	November, 1977
Polymer	November, 1978
SOUTH PATTERN FLUID INJECTION	
Pretreatment	November, 1975
Preflush	June, 1976
Micellar	March, 1977
Polymer	April, 1978

RESERVOIR PARAMETERS BY LEASE

	North Pattern (Chesney)	South Pattern (Hegberg)
POROSITY	0.243	0.245
PERMEABILITY, md (Air Log Mean)	265	208
DYKSTRA-PARSONS COEFFICIENT		
Upper Zone	0.71	0.85
Lower Zone	0.85	0.86
OIL SATURATION	0.333	0.307
THICKNESS, feet	18.4	17.5
CONFINED AREA, acres	25.6	25.6
TEMPERATURE, ^{o}F	69	69
OIL IN PLACE, bbl	303,200	261,200

Figure 13

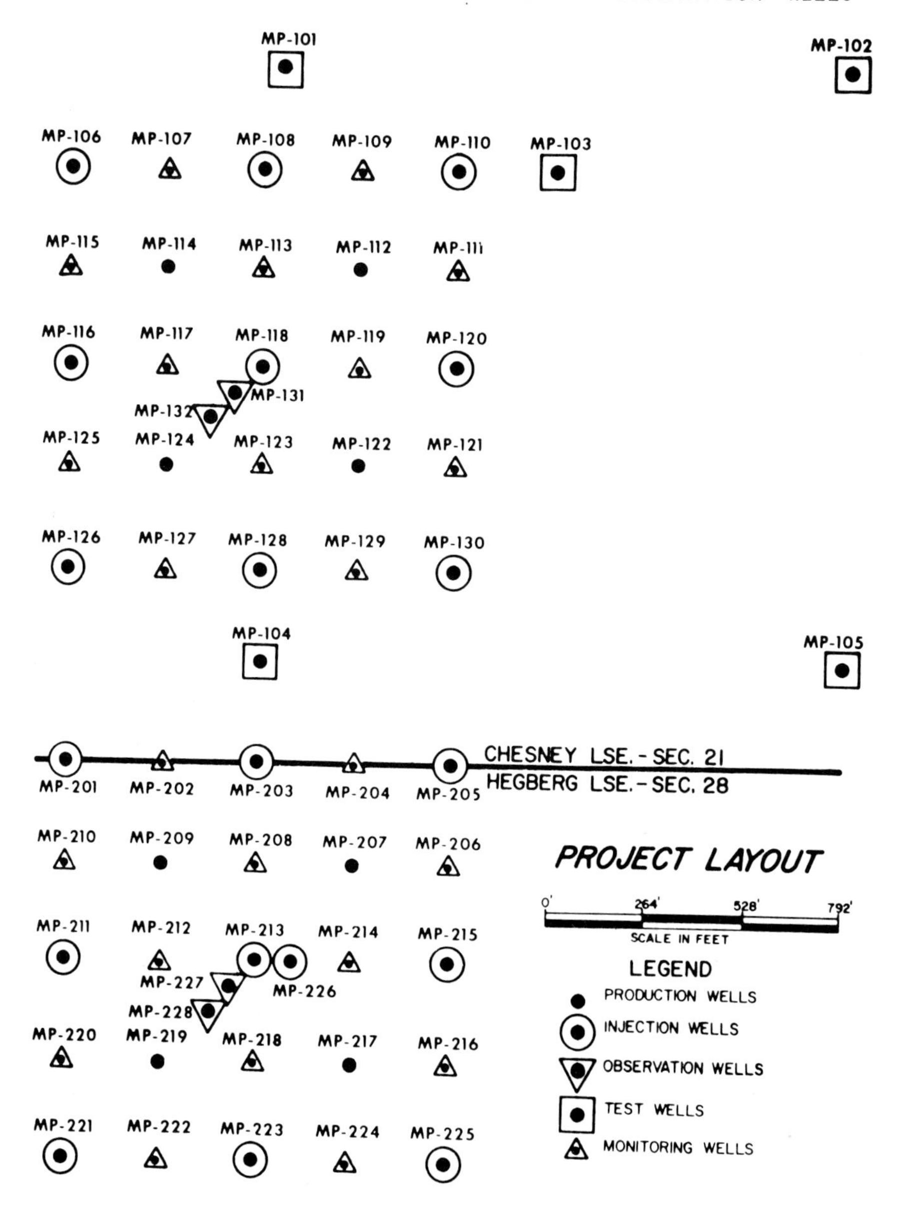
PROJECT LAYOUT SHOWING LOCATION OF THE OBSERVATION WELLS
MP-101
MP-102
MP-106
MP-107
MP-108
MP-109
MP-110
MP-103
MP-115
MP-114
MP-113
MP-112
MP-111
MP-116
MP-117
MP-118
MP-119
MP-120
MP-131
MP-132
MP-125
MP-124
MP-123
MP-122
MP-121
MP-126
MP-127
MP-128
MP-129
MP-130
MP-104
MP-105
CHESNEY LSE. - SEC. 21
HEGBERG LSE. - SEC. 28
MP-201
MP-202
MP-203
MP-204
MP-205
MP-210
MP-209
MP-208
MP-207
MP-206
PROJECT LAYOUT
0'
264'
528'
792'
SCALE IN FEET
MP-211
MP-212
MP-213
MP-214
MP-215
MP-227
MP-226
MP-228
MP-220
MP-219
MP-218
MP-217
MP-216
MP-221
MP-222
MP-223
MP-224
MP-225
LEGEND
PRODUCTION WELLS
INJECTION WELLS
OBSERVATION WELLS
TEST WELLS
MONITORING WELLS

Figure 14

SOUTH (HEGBERG) PATTERN DESIGNED FLUID COMPOSITIONS AND VOLUMES

Fluid	Description	Weight Percent	Volume, bbl
Pretreatment			118,500
	Sodium chloride	2.000[a]	
	Fresh water	98.000	
Preflush			134,776[b]
	Sodium silicate*	0.281	
	Sodium hydroxide	0.482	
	Softened fresh water	99.237	
Micellar Oil (Soluble Oil)			22,626
	Four sodium alkyl aryl sulfonates (Equivalent weight 250-650; Average equivalent weight 425)	23.85[c]	
	Ethylene glycol monobutyl ether	2.63	
	Crude oil	66.78	
	Fresh water added	6.74[d]	
Micellar Water			22,626
	Sodium chloride	0.25	
	Nitrilotriacetic acid trisodium salt	0.65	
	Fresh water	99.10	
Polymer			724,050
	Polyacrylamide	0.068[e]	
	Fresh water	99.932	

*Common name for a mixture of Na_2O and SiO_2.

[a] The last 8900 bbl contained one percent NaCl. The last 2800 bbl was softened.

[b] Volume actually injected. The calculated volume was 120,675 bbl.

[c] About 52 percent sodium sulfonate.

[d] Does not include the water in the petroleum sulfonates.

[e] Average of several steps.

Figure 15

NORTH (CHESNEY) PATTERN DESIGNED FLUID COMPOSITIONS AND VOLUMES

Fluid	Description	Weight Percent	Volume, bbl
Preflood I			352,700[a]
	Sodium chloride	1.4	
	Fresh water	98.6	
Preflood II			353,000
	Sodium chloride	2.900	
	Calcium chloride	0.102	
	Magnesium chloride	0.097	
	Fresh water	96.901	
Micellar Solution			100,660
	Two sodium alkyl aryl sulfonates (Equivalent weight 300-550; Average equivalent weight 430)	4.69[b]	
	C_{12}-C_{15} Alcohol ethoxysulfate sodium salt (about 60% active)	1.13	
	Secondary butyl alcohol	4.13	
	Polysaccharide	0.09	
	Sodium chloride	0.70	
	Fresh water added	89.26[c]	
Polymer			587,170
	Polysaccharide	0.076[d]	
	Sodium chloride	0.075	
	Calcium chloride	0.008	
	Magnesium chloride	0.007	
	Fresh water	99.834	

[a]Volume actually injected. The calculated volume was 335,525 bbls.

[b]About 56 percent sodium sulfonate.

[c]Does not include the water in the sulfonates.

[d]Average of several steps. The polymer design may change. The first 83,880 bbl contain 2.0 percent secondary butyl alcohol.

Figure 16

Operations were started in 1975 and much of the program is completed. The performance of two observation wells are shown in Figures 17 and 18. The characteristic rapid increase in oil production as the oil freed reaches the well, followed by a period of peak production, and then return to high water cut as the front moves past and mobile oil saturations drop, are shown. The floods will not be complete for quite a while, but these results are expected and augur success.

80 PPM of oxygen scavenger and 60 PPM of biocide were added to the polyacrylamide polymer flood to protect against degradation and viscosity loss by microorganisms.

Careful quality control was observed in making up and injecting the slugs. Problems were encountered with well plugging of lines and injection wells. Maintenance of constant viscosity has taken careful control.

Producing wells are operated at rates to maintain a production to injection ratio of approximately one. Mechanical and chemical stimulations have been required for both producers and injectors. These are required because of Barium sulfate scaling, the removal of paraffin waxes, and the removal of material deposited at the sandface during injection.

5.0 Alkaline Flooding

5.1 Current Research and Studies

5.1.1 General

Alkaline water flooding for enhanced oil recovery has been tested in selected oil reservoirs. An important feature of alkaline flooding is its low cost arising from the cheapness of sodium hydroxide. Also, alkaline flooding is a simple operation compared with the complex surfactant micellar polymer flooding, or polymer water flooding. There are processes where alkaline slugs alternate with polymer enriched water.

Research on alkaline flooding has not yet determined the operative mechanisms. The interaction with the oil phase is a complex process. Teh Fu Yen, et al. (14) have approached this problem by examining the chemical composition of the oil phase. Some petroleum constituents prove to have a high degree of interfacial activity when contacted with an alkaline environment.

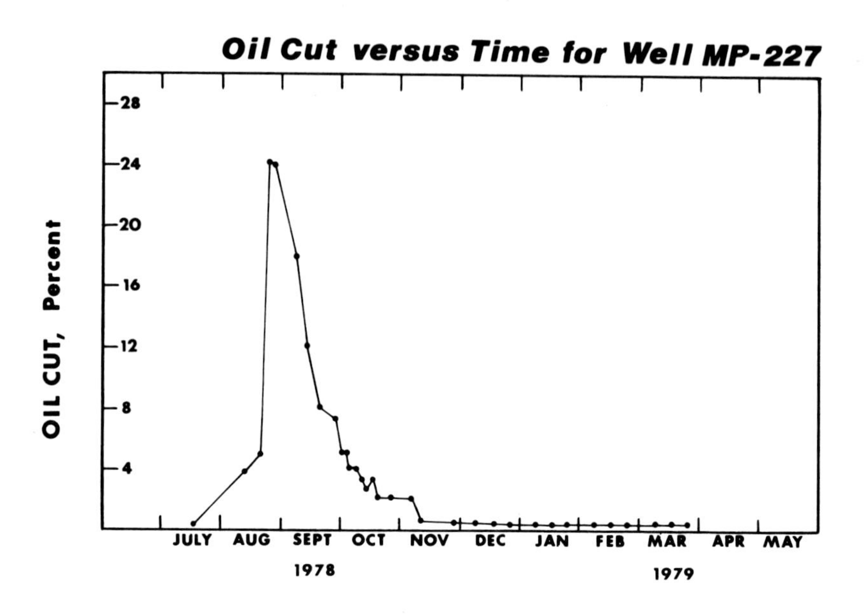
Oil Cut versus Time for Well MP-227
OIL CUT, Percent
28
24
20
16
12
8
4
JULY AUG SEPT OCT NOV DEC JAN FEB MAR APR MAY
1978
1979

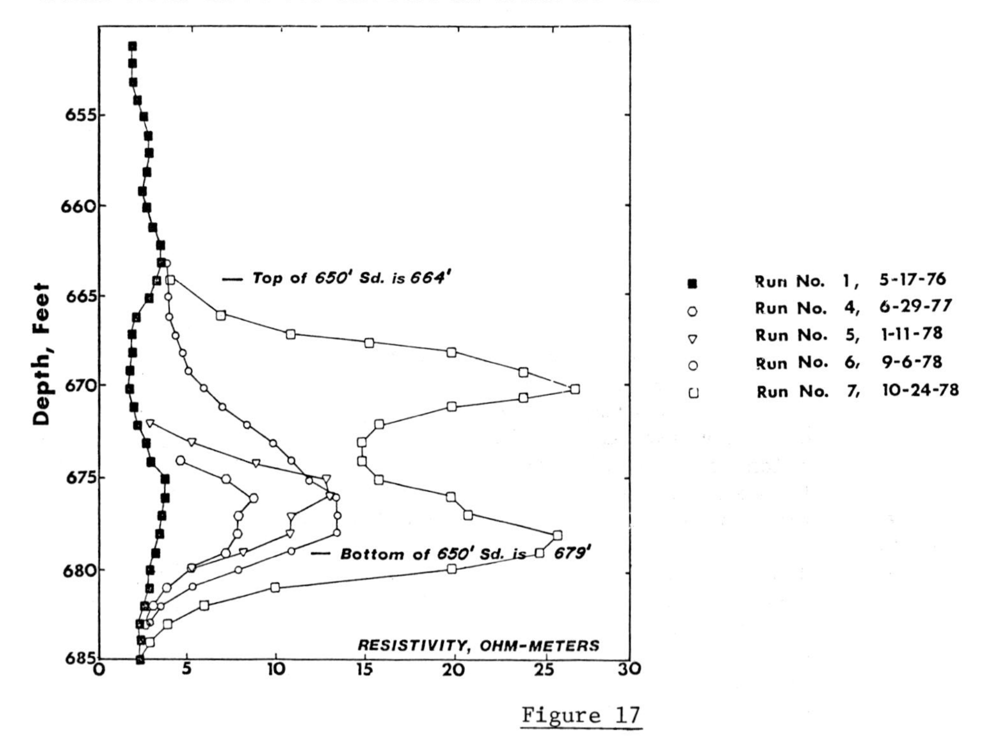
CASED-HOLE LOGGING RESULTS for WELL MP-227
Depth, Feet
655
660
665
670
675
680
685
Top of 650' Sd. is 664'
Bottom of 650' Sd. is 679'
RESISTIVITY, OHM-METERS
0
5
10
15
20
25
30
Run No. 1, 5-17-76
Run No. 4, 6-29-77
Run No. 5, 1-11-78
Run No. 6, 9-6-78
Run No. 7, 10-24-78

Figure 17

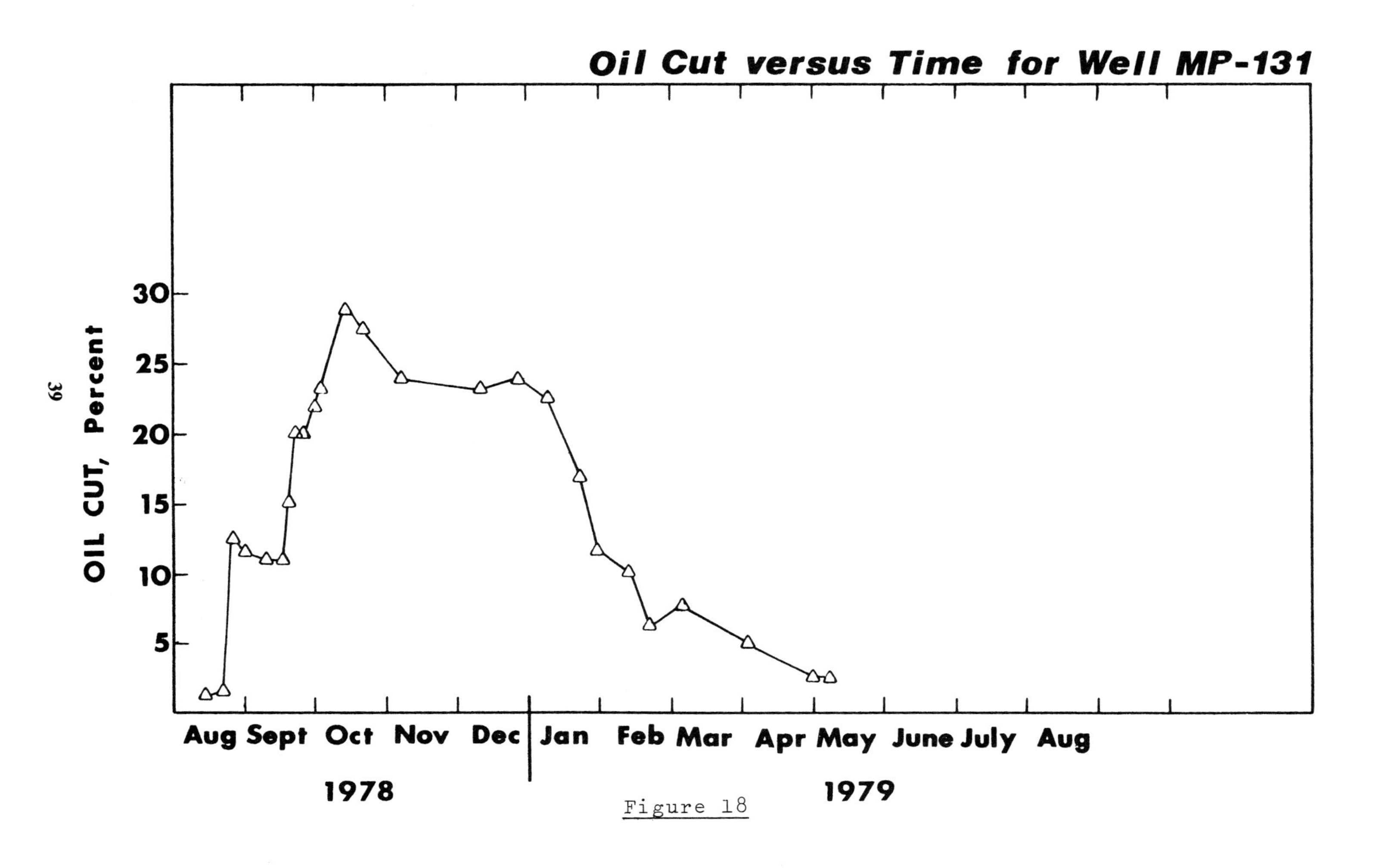
Oil Cut versus Time for Well MP-131
OIL CUT, Percent
30
25
20
15
10
5
Aug Sept Oct Nov Dec Jan Feb Mar Apr May June July Aug
1978
1979

Figure 18

5.1.2 Ultralow Tension Formation

The current belief is that a surfactant is formed in situ in the reservoir, and the ultralow interfacial tension is created by some mechanism after contact between the brine, oil, and surfactant. The lowering of interfacial tension has been attributed to substances, such as, carboxylic acids, phenols, or porphyrins that are active at the interface. Others have felt that a high acid number of a crude would be beneficial to oil recovery. However, this did not prove to be a good indicator in practice. In fact, candidate crude oils for alkaline flooding do not depend on a single compound, but on the collective properties of a number of compounds. Yen proceeded by fractionating crude oil and examining the role of petroleum fractions during alkaline flooding. This would enable him to identify active components. From this he could formulate a sound basis for determining which crude oils and reservoirs would respond actively to alkaline flooding.

Results show that the oil resin fraction is responsible for the reduction in interfacial tension when contacted with aqueous alkaline solution. The IFT values dropped to a minimum with increasing alkaline concentration and then increased gradually. This indicates that the active ingredients in the crude are not a single type of functional group. Yen suggests the possibility of amino acid groups. Figure 19 shows the behavior of interfacial tension with increasing alkaline content in a normal saline environment.

5.1.3 Viscosity (Mobility Ratios)

Viscosity is an important parameter in enhanced oil recovery because of its significant effect on the flowing properties of fluids in porous media. The effect of alkaline solution on the viscosity of the oil resin fraction has been studied. With alkaline concentrations less than 300 PPM, little effect on oil resin viscosities occurred. However, their viscosities started to increase as alkaline concentrations rose. At 3,000 PPM sodium hydroxide, oil resin viscosities reached values almost three times their normal. This increase in viscosity may arise from reactions between caustic and active ingredients of petroleum (carboxylic acids, thiols, and phenols) leading to the formation of salts of these compounds. Charged species would be produced which would increase interaction among petroleum molecules leading to higher viscosities. Figure 20 shows the viscosity study results.

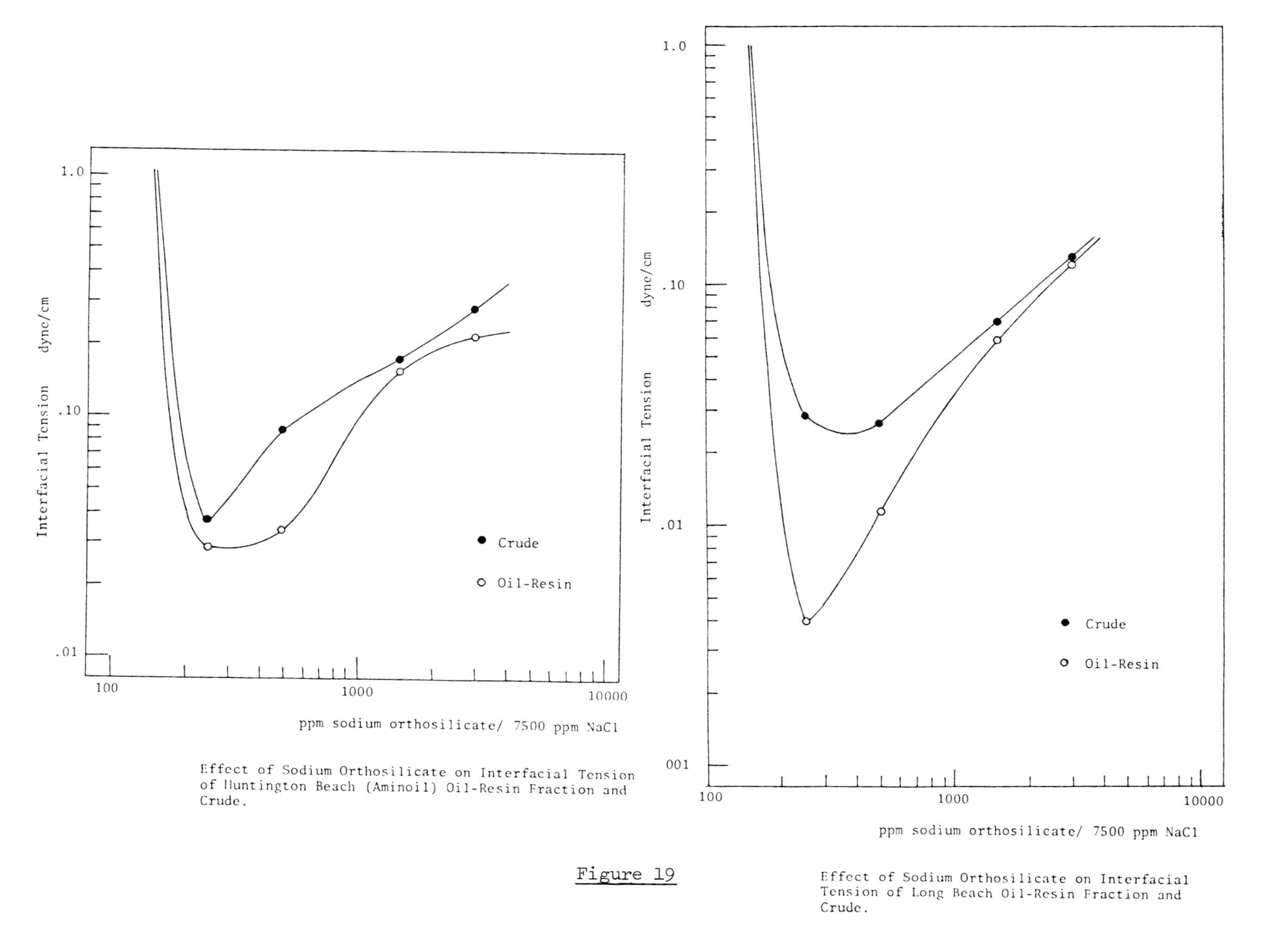

Effect of Sodium Orthosilicate on Interfacial Tension of Huntington Beach (Aminoil) Oil-Resin Fraction and Crude.

Effect of Sodium Orthosilicate on Interfacial Tension of Long Beach Oil-Resin Fraction and Crude.

Figure 19

Viscosity 10^{-3} ·centipoise

0 1 2 3 4 5 6 7 8 9

○ Huntington Beach

● Long Beach

0 1000 2000 3000

ppm NaOH/ 7500 ppm NaCl

Effect of Concentration of Alkaline Solution on Viscosity of Oil-Resin Fractions. Shear Rate: 19.2 sec^{-1}. Temp.: 25 C.

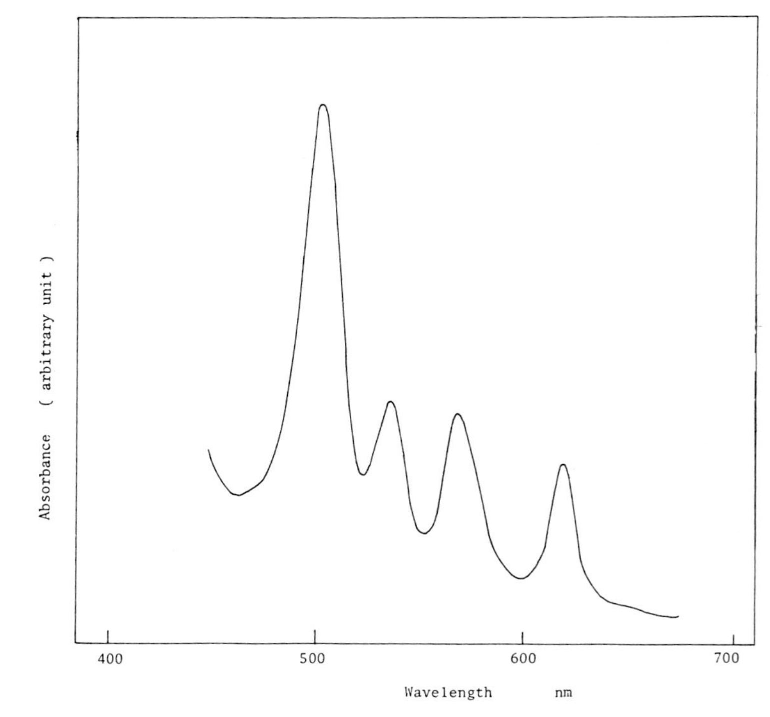

Absorption Spectrum of Porphyrins from Huntington Beach Oil-Resin Fraction

Figure 20

This study indicates that caustic concentrations around 300 PPM would be optimal in minimizing surface tensions. However, increases in oil viscosities would increase alkaline water/oil mobility ratios; thereby, decreasing enhanced oil recovery.

5.1.4 Oil Recovery Mechanisms

Radke, et al. (15) are studying the conditions for residual oil displacement of acidic oils with alkaline floods and are trying to discover the oil recovery mechanisms. They are studying displacement of oil, chemical transport, emulsion flow, and interfacial tensions in alkaline flooding. A major problem in alkaline flooding is the lack of mobility control. If polymer slugs are to be introduced behind caustic floods, they must be able to resist the high alkaline reservoir environment at elevated temperatures. Studies are underway to select the best polymer.

5.1.5 Emulsion Formation

Wasan, et al. (16) have found that a key factor in the success of enhanced oil recovery processes by chemical flood is the formation of an oil bank in the reservoir. The mechanism of oil bank formation and development are as yet unknown. However, experiments show that an emulsion zone is formed behind the oil bank by interaction between the alkaline flood and the reservoir oil. If these emulsions are severe, they cause large pressure drops and the immobilization and bypassing of oil in field tests. Special chemical treatments are needed to break these emulsions. Thus, studies are going on to understand the formation of these emulsions.

5.2 Field Pilot Floods and Applications

5.2.1 Wilmington Pilot Flood

A caustic water flooding demonstration project is underway in the Wilmington field. Extensive laboratory core flood tests have been made in preparation for the field tests which began in late 1979. It was found that salt concentration is essential in conjunction with the caustic to obtain high recoveries. Caustic with fresh water was ineffective.

5.2.2 Huntington Beach Field

A pilot test of the alkaline flooding process has also been started in the lower main zone of the Huntington Beach field. The area has already been water flooded to extremely high water/oil ratios. The average oil saturation is between 24 and 26 percent. Preflush injection of softened water has been underway to adjust the salinity of the water phase in the pilot flood area.

It will be two years before the efficiency of alkaline flooding in these two pilot fields will be known. The addition of further components, such as polymer slug, behind the alkaline may prove to be optimal.

6.0 Polymer Improved Water Floods

6.1 Storms Pool Pilot Flood

A pilot polymer water flood is underway in the Storms Pool near Carmi in Illinois. See Boghossian (17). The field is typical of a large number of oil fields which are candidates for enhanced oil recovery. The field has been water flooded for over 20 years and is now in stripper production. However, the residual oil saturation is estimated to be 40 percent. The polymer flood is designed to improve the sweep efficiency by reducing the mobility ratio of the flood; thus, allowing penetration of the water into smaller pores and the plugging of channels by the high viscosity aqueous solution. Also, the flood will be a prototype for determining the requirements to apply EOR to similar oil fields that would be abandoned because of lack of knowledge of EOR technology, or lack of investment.

6.1.1 Reservoir

The Storms Pool is a sandstone reservoir located 1800-1900 feet subsea. The sandstone exists as bar shaped lenses imbedded in shale. The sand thickness may be 70 feet and is interspersed with thin layers of shale and organic material. The well pattern is shown in Figure 21, and the reservoir properties in Figure 22. The formation shows a wide range in permeability which causes a poor sweep efficiency with conventional water flood. The oil is high gravity and low viscosity. The high viscosity polymer flood should extend oil recovery to low permeability sections of the reservoir.

Primary production was by solution gas drive with an original gas cap of 15 - 20 feet thick in the center. During primary production, water was channelled into the wells while the gas cap was blown down. When water flooding was initiated, large

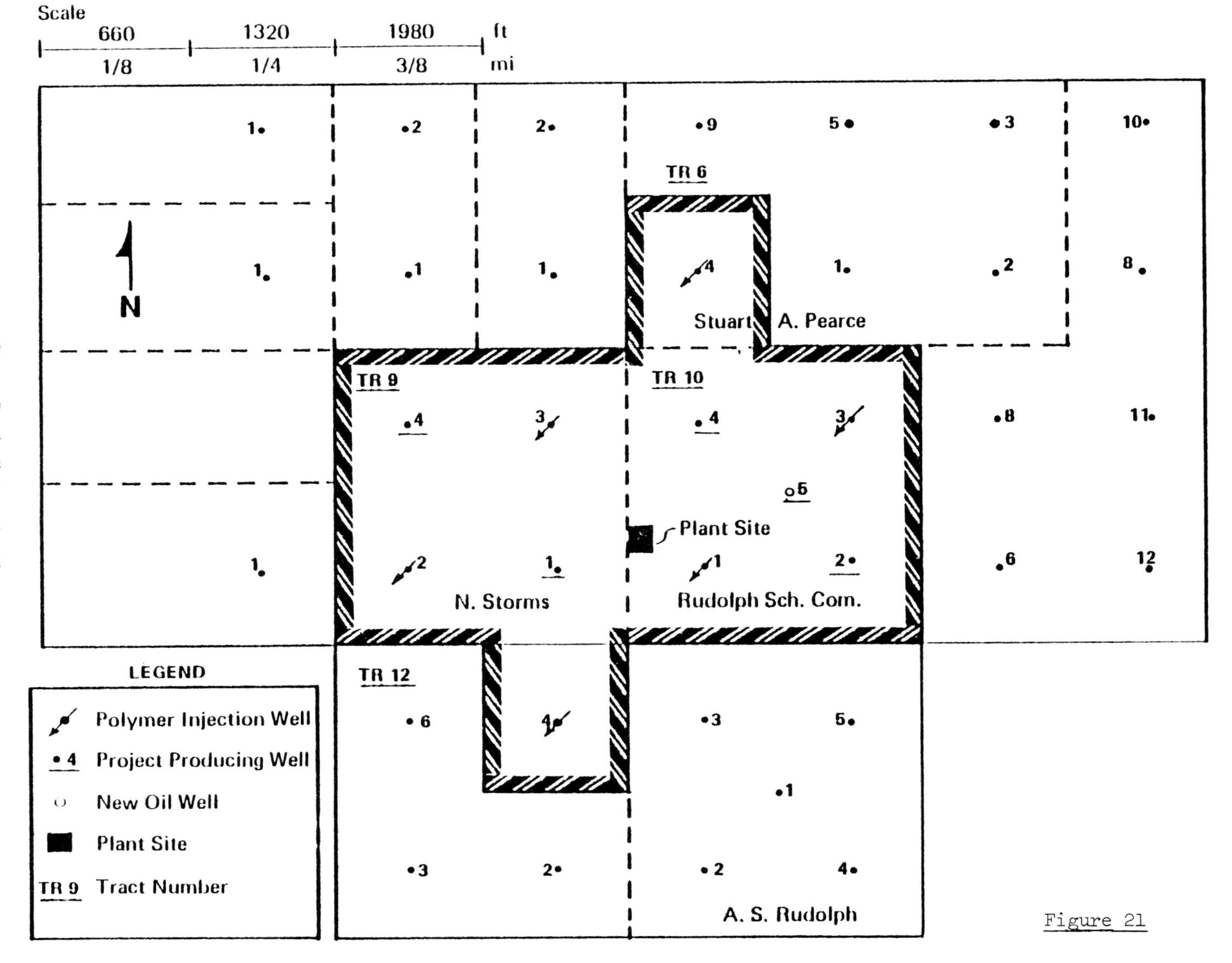

Storm Pool pilot project layout.

RESERVOIR PARAMETERS - STORMS POOL
POLYMER FLOOD PILOT PROJECT

Type of reservoir rock	Consolidated sandstone
Permeability range	50 - 900 md
Porosity range	18% - 22%
Wettability	Predominantly water wet
Reservoir temperature	95° F
Oil gravity @ 95° F	35.0° API
Oil viscosity @ 95° F	6.0 cp
Pilot area	60 acres
Acre feet of oil sand	2100
Average net sand thickness	60 ft
Average oil sand thickness	35 ft
Initial oil saturation	72% PV
Primary oil recovery	16% PV
Waterflood oil recovery	15% PV

DISTRIBUTION OF ORIGINAL OIL SATURATION AND VOLUMETRIC DATA
STORMS POOL PILOT AREA

	S_o=80%	S_o=75%	S_o=65%	
Tract 9	620	400	400	
Tract 10	640	400	400	acre-feet
Totals	1,260	800	800	

Average porosity - 20%
Correction for sand heterogeneity - 0.9
Formation volume factor - 1.1
Pore volume - 3,993,000 bbl
Original stock tank oil in place - 2,701,000 bbl
Cumulative production - 1,095,000 bbl
Present oil in place - 1,606,000 bbl
Present oil saturation - 40%

Figure 22

quantities of water were produced due to channeling through the gas cap area and water zone. By redrilling, and with the use of polymers, the channeling should be largely eliminated, and a stable front flood established. Core tests indicate that 17 percent of the 40 percent residual oil saturation is moveable by polymer water flood still residing in the top 25 feet of the pilot area. The polymer flood is designed to produce this target oil. Production had to be down structure because the former gas cap area would act as a thief zone for the polymer (water flood efficiency in the center of the field was low). The project incorporates six injectors and five producers on a 100-acre section.

6.1.2 Polymer Selection

The effectiveness of a polymer flood depends on the compatibility of the injected chemicals with the formation. Selection studies were made to choose the polymer. The choice appears to be whether to use polyacrylamides or polysacchrides. The following aspects of polymer behavior should be tested.

1. Response to shearing forces in the porous media. Figure 23 shows the effects of shear on decreasing viscosity and the effect of water salinities.

2. Response to electrochemical stress due to salts in the formation and injection waters.

3. Susceptibility to microbial attack.

4. Residual resistance factors and losses due to retention in the formation.

5. Ability to correct reservoir heterogeneities.

The tests concluded that polyacrylamides incur decreased resistance factor on shearing more than polysacchrides, although polyacrylamides have greater intrinsic viscosity. The degree of mobility control they achieve is likely to be reduced because of shear degradation at the well bore. The polysacchrides decreased viscosity also at high shear rates, but they recovered the loss on removal of the shear stress. The sensitivity of the polymers to salt were tested in 500 PPM solutions. The polyacrylamides were much more susceptible to electrochemical degradation than the polysacchrides.

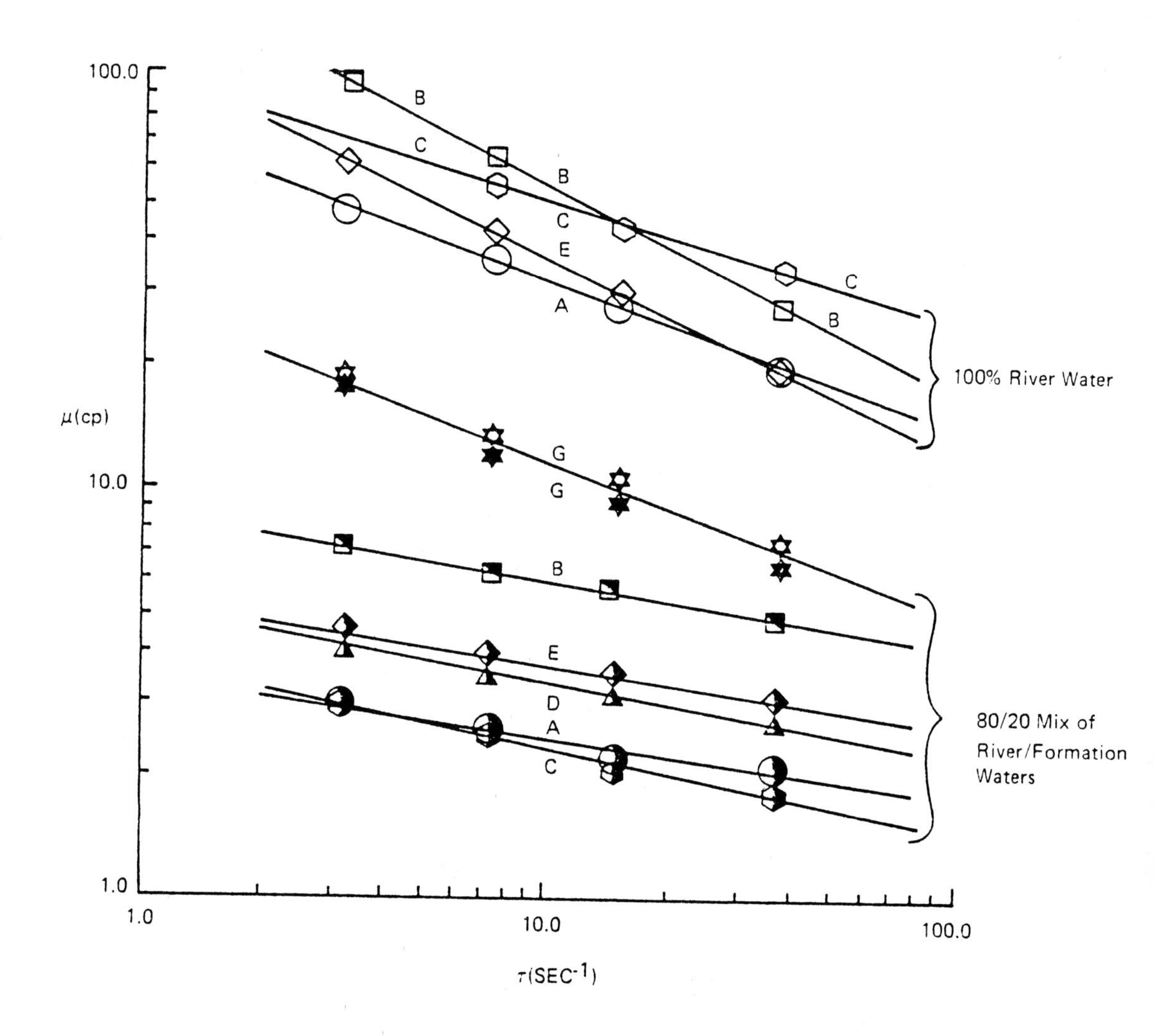

Figure 23. Viscosity of polymer solutions at 500 ppm as a function of shear with river water and with 80/20 mix of river/formation waters.

The reservoir formation supports significant populations of anarobic sulfate-reducing bacteria. A microbiological study is underway to simulate effects on the polymers. Polyacrylamides tend to absorb more on reservoir rock due to their higher charge density. This absorption results in a reduction in permeability. Experiments, however, found this effect also in polysacchride flooding in the presence of brine.

6.1.3 Heterogeneities

Minimizing reservoir heterogeneities in the Storms Pool could be aided by cross-linking polyacrylamides by using trivalent salts, such as aluminum citrate. They are, therefore, candidates for use in far-distant well plugging of permeability channels while there are indications that polysacchrides can also be used.

A polysacchride was chosen because of its low sensitivity to salts and mechanical degradation. Large zones of the reservoir were unswept by the water flood and will remain inaccessible to the preflush. The injected polymer slug should contact these zones; a loss in viscosity there is undesirable.

6.1.4 Water and Polymer Treatment Plants

Because of the poor quality of the injection water (river water), the injection system requires facilities for coagulation and sand pack filtration of the river water. Also, a biocide, a chelating agent, and an oxygen scavenger will be added prior to the addition of polymer. The fresh water and polymer quality will be continuously monitored for purity and viscosity. Figure 24 shows the water quality control plant. Figure 25 shows the polysacchride injection process plant.

6.2 North Stanley Flood

Screening criteria for polymer flooding have been studied by Pease, et al. (18). A polymer flooding project was undertaken in the Burbank Sand Reservoir of the North Stanley field, Osage County, Oklahoma. This reservoir is representative of mid-continent water flood properties. The Burbank sand is highly heterogeneous and consists of one high permeability zone on top of two lower permeability zones. There is an extensive system of natural fractures oriented at approximately right angles to the axis of the sand bar. This reservoir was in an advanced stage of depletion by conventional water flooding with a production water/oil ratio of about 70. Commencing mid-1976, a 17 percent pore volume slug of polymer solution was injected into the pilot area. In addition, to increase mobility control a cross-linked polymer was used as a selective plugging agent in wells.

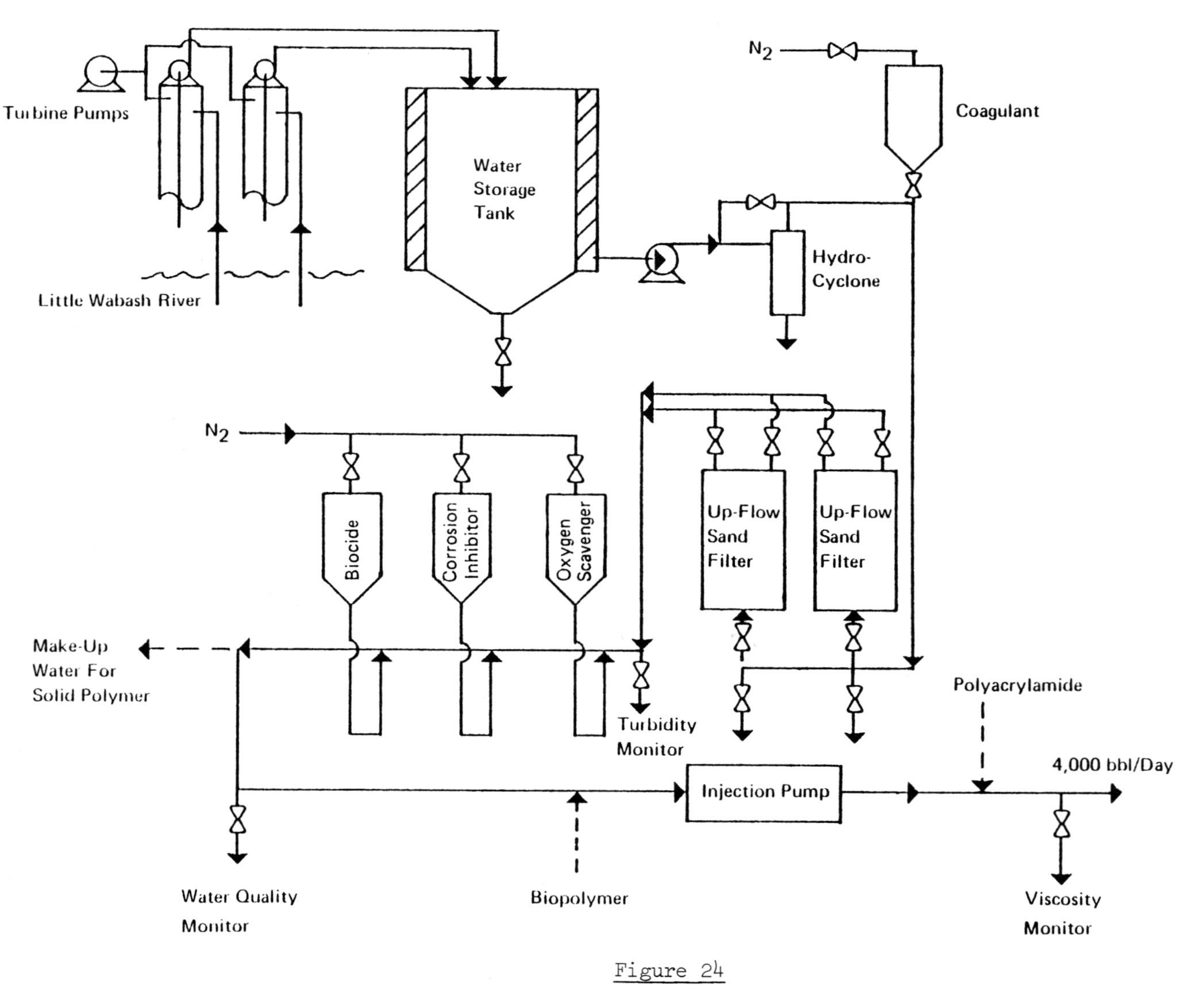

Figure 24

Water quality control system for the Storms Pool polymer injection facility.

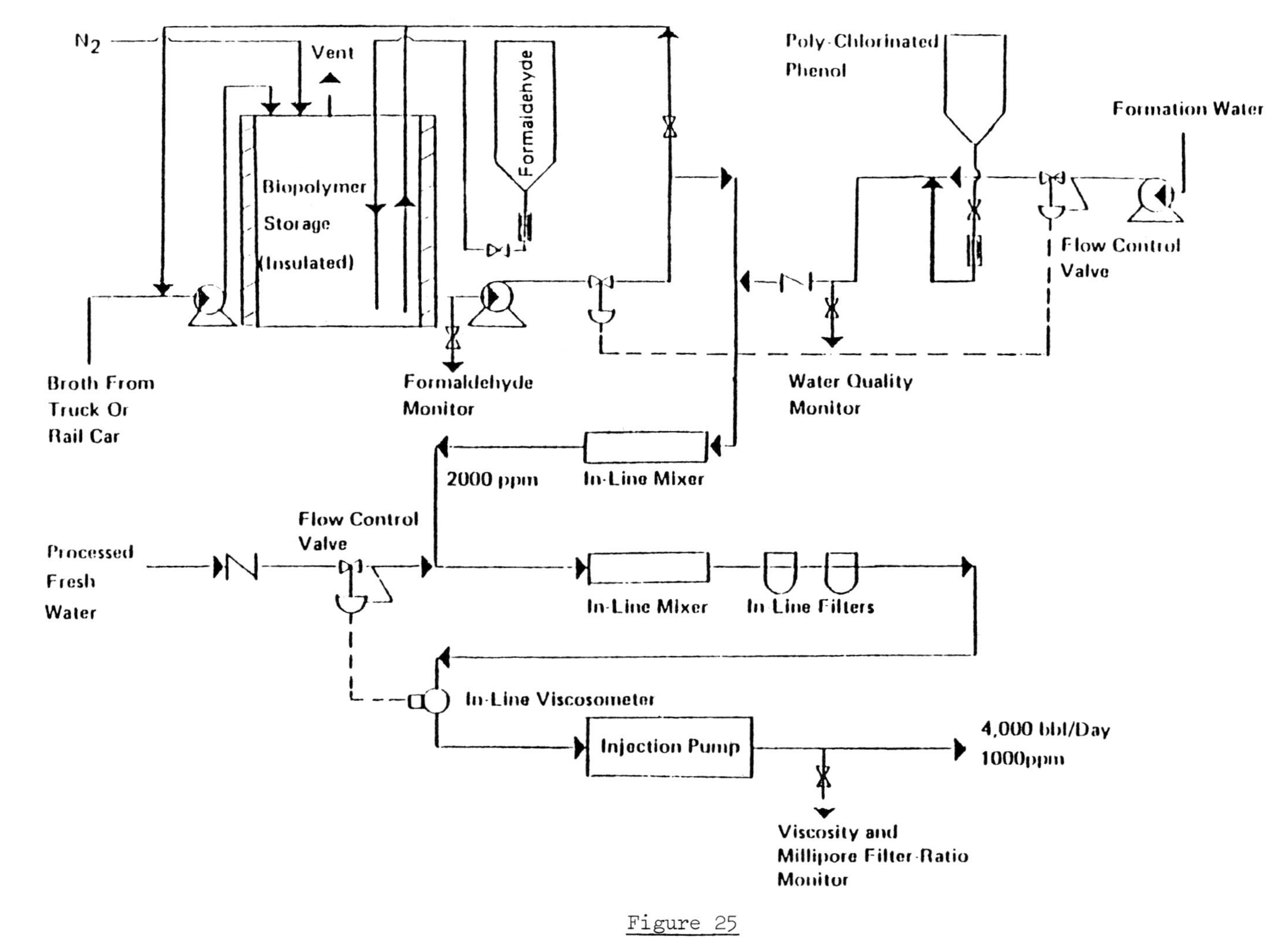

Figure 25

Polysaccharide injection process for the Storms Pool polymer injection facility.

6.2.1 Oil Recovery

There was a positive oil production response from this 1,000-acre project. The maximum EOR production rate of about 200 barrels per day was attained one and one-half years after starting polymer injection. Based on performance through January 1979, the ultimate EOR recovery is estimated at 500,000 barrels, or roughly 10 percent of the estimated mobile oil that remained after depletion by conventional flooding. The main adverse technical factors were the severe anisotropic reservoir condition (lack of homogeneity in the sands). The positive results indicate that current water flooding projects can be optimized through a program of well workovers, well pattern modifications, and balanced injection rates. These factors are felt to have contributed to the performance. The polymer slug alone was not solely responsible for recovery.

6.2.2 Screening Criteria

Studies are underway by DOE and its consultants to develop general screening criteria to eliminate those reservoirs having insufficient potential for polymer water flooding. Those remaining should be considered; however, the screening criteria are not adequate for priority ranking of candidates or for prospect evaluation. More information than is generally available is necessary for the ranking, while the evaluation requires technical measurements and analysis on each prospect.

Figure 26 shows a comparison of reservoir conditions in North Stanley before polymer injection and the most recently developed screening criteria. North Stanley was outside the recommended range for three parameters, the mobile oil saturation, producing water/oil ratio, and large quantity of channeling and fractures. The mobility ratio and oil viscosity were borderline. Yet incremental oil recovery indicates that the project was technically successful and will be economically successful with the oil prices of the 1980s (>$24.00 per barrel). This indicates that the parameters listed in the screening table have interactions and cannot be used as rejection criteria for a reservoir.

Pease has revised his screening criteria, as shown in Table 2 of Figure 26. The current water/oil ratio limit of 15 or less is too restrictive. Channeling and fractures can be mitigated by selective plugging. Thus, a candidate should be rejected only if conditions are too severe to permit conventional water flooding. Adequate permeability for conventional water flood means it's adequate for polymer flood. Mobile oil saturation

COMPARISON OF NORTH STANLEY RESERVOIR PARAMETERS
WITH CURRENT SCREENING CRITERIA

Parameter	Latest Published Screening Criteria	North Stanley
Oil viscosity, cp	less than 200	2.4
Mobility ratio, water/oil	greater than 1	1.1
Mobile oil saturation	greater than 0.10 PV	0.09 - 0.06 PV (*)
Current water-oil ratio	less than 15	67
Permeability, md	greater than 20	average 300, range 10 to 1500
Temperature, ^{o}F	less than 200	105
Lithology	sandstone	sandstone
Channeling-fractures	"no gross channeling or major natural fractures"	high permeability zones and natural fracture system within project area

* Based on irreducible oil saturation estimates ranging from 0.30 to 0.33

SCREENING CRITERIA FOR POLYMER APPLICATIONS
MID-CONTINENT REGION

Parameter	Range of Feasibility
Oil gravity	Between 16 and 42° API
Injectivity	Sufficient to support waterflood economics; usually 0.10 bbl/day per acre-foot minimum
Mobile oil saturation	0.10 PV minimum
Current water oil ratio	Less than 25:1
Channeling/fractures and Gross reservoir anomalies	Reject candidates if conditions are known to be too severe to permit conventional water flooding

Figure 26

of 0.10 pore volume and an API gravity of crude oil between 16 - 42° are within the feasible range. Other factors are a pay interval thickness of five feet.

6.2.3 Treatment of Anisotropy

Non-heterogeneties reduce oil recoveries. They impose operating conditions and require special treatments. Producing rates are often limited by restraints on injection pressures. Polymer flooding is a mobility control process that depends on reduced mobility of the displacing fluid in the reservoir environment. The fraction of displacing fluid entering the more highly swept portions is reduced by high viscosity fluid. This reduces the producing water/oil ratio. Increase in oil recovery, however, depends on the displacing fluids rate of entry into the less permeable and less swept portions of the reservoir. An increased pressure gradient in the reservoir is required. To accomplish this the injection well down hole pressures must be increased, or the bottom hole pressures of producers decreased. Injection pressures are limited by sand face breakdown pressures and mechanical facilities. Decreasing producing well pressures entails higher lifting costs.

Fracturing can reduce flow to wells and disrupt flooding patterns. Sealing off fractures by injecting polymerizing agents to block off the fractures will solve this problem.

6.2.4 Screening of Mid-Continent Fields

Utilizing the Petroleum Data System (PDS), the mid-continent fields were screened using the following parameters:

1. At least one sandstone reservoir.
2. Cumulative production of one million barrels of oil.
3. Five or more producing wells.
4. Pay interval thickness of at least five feet.
5. Oil gravity between 18 - 42° API.
6. No evidence of uncontrolled water flooding.

Access to PDS is gained through the General Electric Mark III computer network, an interactive time sharing system. The result of the computer screening, plus review of reservoir data, was that 730 fields containing 2,499 reservoirs passed through the screen. Further examination reduced these to 88 fields containing 632 reservoirs. It is estimated that 100 million barrels of EOR oil could be produced at a conservative minimum from these reservoirs by polymer flooding. This could be substantially increased by other uses of polymer, such as treatments of producing wells with selective plugging agents for fractures and channels, or in combination with surfactants.

It is important to note that polymers augmented water flooding will not, in general, produce more than 20 to 30 percent of the residual oil in situ after water flood. Thus, it does not have the inherent economics of using a surfactant micellar slug as the first chemical slug in order to essentially mobilize all residual oil it comes in contact with; and then follow this with a polymer slug for controlling mobility and increasing the sweep efficiency of the micellar phase. Water injection then follows to continue the fronts moving throughout the reservoir.

6.3 Mobility Control Agents

Whether for polymer water flooding or for surfactant-polymer flooding, the development of improved mobility control agents is important. Refer to the paper of Martin, et al. (19). Martin states that these processes will contribute to enhanced oil recovery during the 1980s . The processes are still developmental. The higher oil prices will make many hundreds, or even thousands, of potential EOR candidates commercial and economical. However, the second limit is process technology. Continuing research and development is required to maximize rate of return on investment and recover maximum in situ oil.

Approximately 250 technical references and 80 patents were cited during the 1968 to 1976 period. A large effort was directed toward developing improved surfactants. However, little effort went into improved polymers. Research work is required here to improve the efficiency of the surfactant polymer flooding processes. Experiments should be designed to simulate field conditions as much as possible.

The importance of mobility control can hardly be understated in determining the success of a surfactant polymer process. To insure process stability, the mobility of the surfactant slug should be equal to, or less than, the oil/water bank being displaced; and the mobility of the polymer solution should be equal to, or less than, the surfactant slug. Otherwise, surfactant will channel throughout the oil/water bank and the polymer will not have good sweep efficiency. Also, if adequate mobility control is not maintained, the polymer or chase water will invade the surfactant slug which would break up the slug and trap the oil that had been mobilized. Normally, the polymer slug will be three to six times larger than the surfactant slug. It will not be as expensive as the surfactant slug, but it is still a high cost.

Water soluble polymers have been applied in the mobility buffer solution. Practical applications have been limited to two types, a polysaccharide called xanthan gum (XG) and partially hydrolyzed polyacrylamides (HPAM). Xanthan gum is a natural biopolymer produced by a microbial fermentation action of the organism *xanthomonas campestris* on a carbohydrate. Partially hydrolized polyacrylamides are synthetic polymers made by hydrolysis of polyacrylamides with a caustic material, or by copolymerizing acrylamide with acrylic acid or acrylate. The degree of hydrolysis ranges from 15 to 35 mole percent. Both types of polymers are quite large macromolecules with molecular weights ranging from one to ten million.

The potential inadequacies of these mobility control agents are listed in Figure 27. These inadequacies cause process inefficiencies or excessive cost. The cheaper polyacrylamide is mainly used in field applications because of its lower cost. However, it is subject to reduced viscosity because of shear degradation in well bores and lacks brine tolerance. However, xanthan gum has degradation problems and is more difficult to handle in the field.

7.0 Carbon Dioxide Flooding

7.1 CO2 Flood Oil Recovery Mechanisms

7.1.1 Liquid Phase and Supercritical Drive

Kamath (20) points out that CO2 flooding has focused on oil displacement by the supercritical gas drive. However, at low temperatures of 88°F, or less, liquid CO2 has given good oil recoveries. Highest oil recoveries by either process should occur when the reservoir temperature is as close to the critical temperature as possible.

Kamath also points out that studies and tests on CO2 flooding have been misleading in that the displacement process is called a miscible supercritical gas drive. This concept has affected the screening of CO2 flood prospects by basing applicability on the ability of formations to withstand a "miscibility pressure". This is considered to be the minimum CO2 injection pressure required to insure a satisfactory level of oil recovery. However, the Appalachian oil reservoirs have low temperatures and favor a liquid CO2 drive.

7.1.2 CO2 - Hydrocarbon Phase Behavior

Figure 28 shows the pressure-temperature curves for saturated vapor pressures of various pore substances including Carbon Dioxide. Points above the CO2 curve denote liquid CO2. Those below the curve denote the vapor phase. The curve represents conditions at which liquid and vapor coexist in stable equilibrium. The lower terminus of the curve is the freezing point and the upper terminal point is its critical point (88°F and

POTENTIAL INADEQUACIES OF EXISTING

MOBILITY CONTROL AGENTS

Inadequacy	Polyacrylamide	Xanthan Gum
Economics	Moderate Cost	Higher Cost
Shear Degradation	Severe	Slight
Oxygen Degradation	Yes	Yes
Viscosity Loss in Brine	Severe	Slight
Thermal Degradation	$>250^{o}F$	$>160^{o}F$
Metal Ion Degradation	Yes	Yes
Hydrolysis Reaction	Yes	Yes
Microbial Degradation	Moderate	Severe
Filtration Required	No	Yes
Permeability Limitations	>10-20 md	>5 md
Surfactant/Polymer Interaction	Possible*	Possible*

* Extent of inadequacy would depend on formulation of surfactant slug as well as polymer type.

Other Potential Inadequacies:

- Reactions with Divalent Ions
- Polymer Retention or Adsorption
- Interactions with Porous Media
- Reduced Injectivity and Wellbore Impairment
- Oil Emulsions
- Producing Well Productivity Problems
- Premature Breakthrough of Polymer in Production Wells
- Handling Problems
- Product Quality Control
- Availability for Commercial Scale S/P Projects

Figure 27

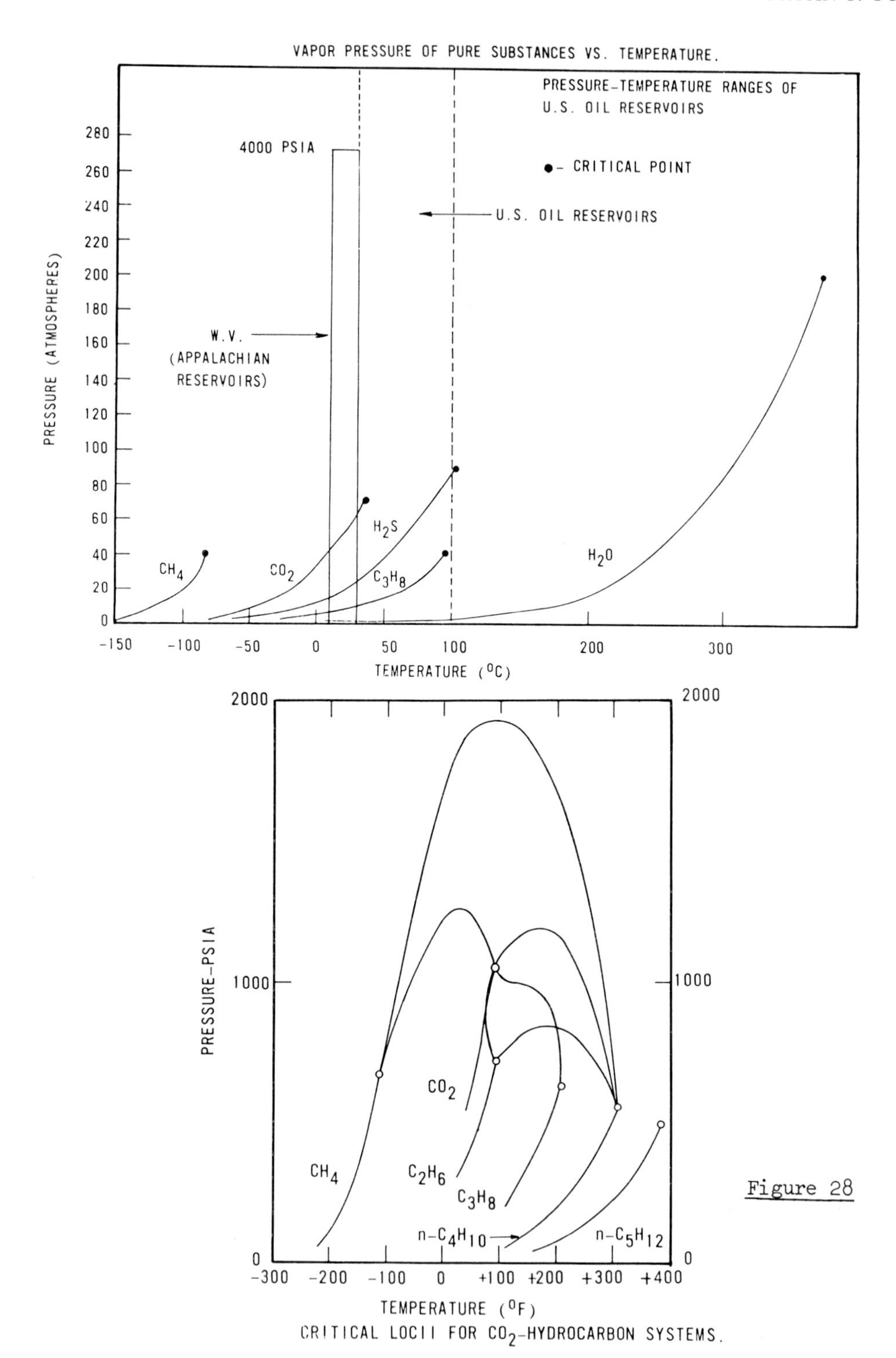
VAPOR PRESSURE OF PURE SUBSTANCES VS. TEMPERATURE.
PRESSURE–TEMPERATURE RANGES OF U.S. OIL RESERVOIRS
● – CRITICAL POINT
U.S. OIL RESERVOIRS
4000 PSIA
W.V. (APPALACHIAN RESERVOIRS)
PRESSURE (ATMOSPHERES)
280
260
240
220
200
180
160
140
120
100
80
60
40
20
0
CH_4
CO_2
H_2S
C_3H_8
H_2O
-150
-100
-50
0
50
100
200
300
TEMPERATURE (°C)
2000
1000
0
PRESSURE–PSIA
CO_2
CH_4
C_2H_6
C_3H_8
$n\text{-}C_4H_{10}$
$n\text{-}C_5H_{12}$
-300
-200
-100
0
+100
+200
+300
+400
TEMPERATURE (°F)
CRITICAL LOCII FOR CO_2-HYDROCARBON SYSTEMS.

Figure 28

1070 psia). Above the critical point, the CO_2 molecules are in the supercritical or true gas phase.

Figure 29 shows the behavior of CO_2 with hydrocarbon systems for the entire range of reservoir temperatures. There is a critical pressure above which the CO_2-hydrocarbon system becomes a gas phase, but normally they are two-phase or liquid phase. The CO_2-butane system is closest to the normal reservoir multicomponent fluids over the range of reservoir temperatures. This system would contain the liquid state for most reservoirs.

With CO_2 flooding, each hydrocarbon component forms a binary system on contact with CO_2. The lighter components of these systems have superior solvent properties with respect to the whole crude. This causes an extraction process of the heavier components that results in a more efficient miscible displacement. This would imply that CO_2 processes could recover heavy crudes as well as the lighter ones that are now considered its objective.

In deep hot reservoirs, the injected supercritical CO_2 gas forms an equilibrium with liquid oil in the two-phase region. Overall miscibility is not attained. As Figure 30 shows, in cold reservoirs below 88°F, both the oil and CO_2 are in the liquid state. Obviously, the concept of miscibility pressure, or the need to increase pressure in the reservoir to formation parting pressure, is not applicable operating with the liquid CO_2 phase.

7.1.3 Oil Recovery Mechanisms

With the supercritical gas drive, the displacement mechanism is essentially a Buckley Leverett piston type with mass transfer, and does not not differ from miscible methane or nitrogen displacements; except that very high pressures are required by the latter gases to achieve oil recoveries.

The recovery mechanism with liquid CO_2 is more complex. The mixing of CO_2 and oil, upon which oil recovery depends, is controlled by concentration gradients that are higher than in gas-liquid systems, but occur at lower diffusion rates. Laboratory tests show that more than one liquid phase is formed under some conditions. This may result in the trapping of CO_2-rich oil phase, or oil-rich CO_2 phase, behind the flood front. This could explain material balance losses observed in CO_2 floods.

The effectiveness of the CO_2 miscible process arises because CO_2 gas is several times more soluble in crude oil hydrocarbons

The Effect of Temperature on Secondary Oil Recovery by CO_2 Flooding

Run No.	Flow Rate (ml.hr)	Core Temp (°F)	CO_2 Slug (pore volume) Liquid	Gas	Oil So (pv)	Oil Recovered Displ. (pv)	Blow down (pv)	Total (pv)	Percent oil-in-place
E-19-7	20	70	.150	--	.620	.244	.112	.356	.574
E-59-2	20	86	.137	--	.660	.268	.170	.438	.664
E-19-6	20	100	.153	.395	.630	.308	.124	.388	.616
E-19-26	20	130	.177	.691	.640	.288	.068	.356	.556

Effect of Applied Back Pressure on Oil Recovery

Run No.	Back Pressure psig	So (PV)	CO_2 Slug (PV) (Liquid)	Oil Recovery (PV) Displ.	Blow Down	Total	Percent oil-in-place
E-89-6	1150	.696	.160	.226	.230	.456	65.5
E-89-4	1100	.696	.163	.235	.234	.469	67.4
E-89-1	850	.696	.169	.261	.144	.405	58.2
E-89-7	650	.696	.169	.278	.172	.450	64.7

Cores and Materials

	E-19	E-59	E-89
Sandstone Core	Berea	Berea	Berea
Diameter	3.81 cm	3.81 cm	3.81 cm
Length	20.0 cm	20.2 cm	20.3 cm
Pore Volume	50 ml	47 ml	46 ml
Porosity	21.9%	20.4%	19.9%
Permeability			
Ka (brime)	72 md	72md	97 md
Ko (oil)	62 md	25 md	60 md

Oil: Hilly Upland Crude Oil (3.64 cps).

Brine: Two percent NaCl in distilled water.

Alcohol: Commercial grade Isopropanol.

CO_2: Commercial grade cylinder with outlet tube reaching to bottom of inside cylinder for syphoning of liquid CO_2.

Figure 29

CO_2 DISPLACEMENT
TEST APPARATUS

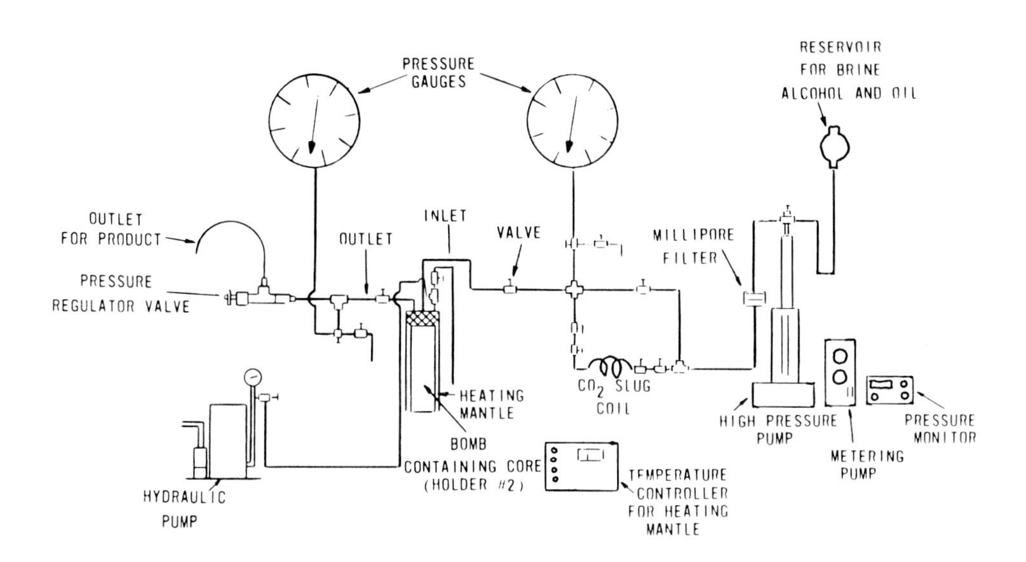

EFFECT OF TEMPERATURE ON SECONDARY OIL

RECOVERY BY CO_2

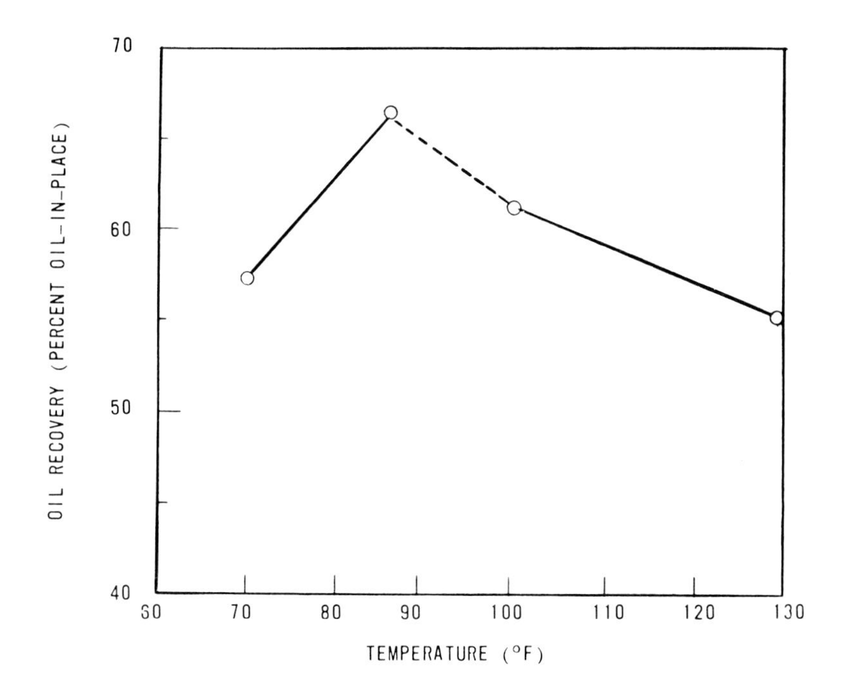

Figure 30

than it is in water. When dissolved in oil, it enhances oil mobility by swelling and viscosity reduction. Water slug behind the CO2 tends to form a bank to displace the CO2-oil mixture without excessive CO2 loss.

7.1.4 Liquid Flood Applications

The experimental runs show that liquid CO2 floods give higher oil recoveries than CO2 vapor floods under certain conditions; thus, applying back pressure to production wells should be considered. Oil recovery also increased with temperature in liquid CO2 floods, but decreased with temperature in supercritical CO2 floods. The usual screening parameter of miscibility pressure is not applicable to liquid CO2 floods. Figure 29 shows test data supporting these findings.

This kind of fundamental research on carbon dioxide flooding mechanisms is very valuable in increasing the scope of CO2 flooding and finding the most economical process combination for a given reservoir.

7.2 Scaled Physical Models of Reservoirs

Doscher and Gharib (21) are investigating carbon dioxide miscible displacement by use of scaled physical models. Their study of the literature shows few studies on miscible flooding employing scaled physical models. These models are required if laboratory observations are to be reliably extrapolated to actual reservoir operations. Disappointing results in miscible drive applications arise partially from this cause. The initial experiments revealed the significant influence of displacement velocity on the effectiveness of oil recovery. A model has been developed that simulates the performance of a low viscosity miscible fluid on residual crude oil displacement. Further experiments in progress will test the hypotheses formed to date.

7.2.1 Viscosity/Gravity Ratios

After breakthrough, miscible fluids recover considerable quantities of the residual oil because of the diffusion of the miscible fluids into each other behind the displacement front. Investigations have revealed the role of the ratio of viscous to gravity forces. Over a wide range of viscous/gravity forces, the recovery of high saturations of oil is independent of displacement rate and geometry. At higher ratios, numerous secondary fingers gave rise to higher oil recoveries. At low ratios, gravity tonguing dominated so that oil recovery was poor, unless the flow velocity was so decreased that molecular diffusion became controlled and stabilized or even increased the recovery. These findings point to the need for physical scaling models including viscosity/gravity ratios and proportioning of velocities in the upper and lower ranges of the ratio.

7.2.2 Capillary Trapping

Observations on core studies with solvent injection have concluded that residual oil recovery is limited by a shielding effect of the water which is related to the capillary trapping of residual oil.

7.2.3 Dimensionless Ratios

The principles of reservoir processes physical scaling from the laboratory model to the actual reservoir prototype requires that the ratios of all the dominant forces acting to control the behavior of the prototype, viscous to gravitational, gravitational to diffusion, etc., bear the same relation to each other in the model as in the reservoir for solvent displacement. The following equalities must be maintained:

$$\frac{\left(\frac{K\ \Delta\rho}{v\ \mu}\right)_P}{\left(\frac{K\ \Delta\rho}{v\ \mu}\right)_M} = \frac{\left(\frac{D}{v\ \ell}\right)_P}{\left(\frac{D}{v\ \ell}\right)_M} = \frac{\left(\frac{v\ \rho\sqrt{K}}{\mu}\right)_P}{\left(\frac{v\ \rho\sqrt{K}}{\mu}\right)_M}$$

K = permeability

$\Delta\rho$ = the density difference between the fluids

v = the linear velocity of the fluids

μ = a characteristic viscosity of the fluids

D = the diffusion constant

ℓ = a characteristic length of the system

M = model

P = prototype

Neglected in this equation are the capillary effects and the Leverett J function (distribution of pore sizes). The equation is a convenient assumption, since neither of the above terms can be scaled adequately. However, there is evidence that they do affect miscible displacement. These parameters will be treated as modifying influences in later studies after completing current scaling studies.

Flow within the model cores are laminar or viscous flow at velocities equivalent to reservoir displacement velocities of less than one foot per day. Reynolds Number scaling may be neglected. Experiments will be run varying Reynolds Number in the future.

7.2.4 Experimental Model

Thus, the experimental model consisting of a one-foot, 2.5-inch diameter tube packed with glass beads, saturated with water, flooded with oil, then with water, corresponds to a reservoir prototype of a 10-acre line drive pattern with a reservoir/producer spacing of 467 feet and an absolute permeability of 25 millidarcies. The porosity of the one-foot model averaged 0.37 and its permeability 13 darcies. Various rates of injection and solvent-oil pairs were studied.

7.2.5 Normal Velocity Experiments

A slightly less viscous solvent with a velocity corresponding to a reservoir velocity of 0.25 feet/day gave the following results. The water is displaced ahead of the solvent. The first oil production occurs after injection of 0.2 pore volume of solvent. The solvent breaks through after injection of 0.55 pore volume. Higher velocities did not change this behavior. The amount of water displaced corresponds to virtually all the mobile water, leaving a residual oil saturation of 24 percent. Practically 100 percent of the residual oil is recovered, but 75 percent of the recovered oil is produced after the first appearance of solvent in the effluent. Thus, at this velocity, a slightly less viscous solvent frontally displaces the water, but mixes with the bulk of the oil that is recovered. There is no evidence of significant viscous fingering or gravity overlay of the solvent while displacing water at this velocity.

7.2.6 Low Velocity Experiments

At a much lower reservoir velocity (0.017 foot/day), the water production begins immediately upon solvent injection; however, it reduces rapidly and only 0.25 pore volume of water is displaced, leaving a residual water saturation of 50 percent. The oil production starts after injection of only 0.08 pore volume of solvent and about half the residual oil is produced at the time solvent breaks through. This occurs after 0.23 pore volume of solvent has been injected (compared with 0.55 pore volume at the higher linear velocity). Only about 60 percent of the residual oil is recovered compared with virtually 100 percent at higher velocity. However, about half of the oil recovered at low velocity is prior to solvent breakthrough.

7.2.7 Effect of Viscosity

The viscosity ratio was changed with the solvent having the higher viscosity. At a rate of 0.05 feet/day, the displacement performance was almost identical to the previous experiment at a rate of 0.25 feet/day. The small improvement in viscosity ratio caused the viscous forces to dominate the system.

Extrapolating these experiments to carbon dioxide flooding where the solvent viscosity is only a fraction of water, it may not be possible to achieve stable displacement in the normal reservoir velocity range of a few tenths of a foot per day. Rather, the mode of displacement would be like the low velocity experiments. If carbon dioxide has low relative cost, the higher recovery at high injection rates and close spacing can be offset by the high costs involved in achieving the high fluid velocities in the reservoir possible with a solvent having a favorable viscosity ratio. The CO_2 would tend to be efficient in recovering oil relative to the pore volume injected; however, about half the recoverable oil would not be recovered.

7.2.8 Conclusions for Carbon Dioxide Flooding

Model scaling studies are just beginning, and ultimately, two-dimensional model studies will be developed using carbon dioxide crude oil systems.

The preliminary conclusions are that the unfavorable viscosity ratio of carbon dioxide to water and oil will not achieve piston-like displacement, unless reservoir velocities are greater than the normal few tenths of a foot per day. On the other hand, although perhaps only 50 - 60 percent of the residual oil is recovered at normal velocities, the oil recovery is efficient since the pore volumes of carbon dioxide injected would be small. The other conclusion is that very high residual oil recoveries are possible with viscosity ratios near one at normal reservoir velocities.

7.3 Carbon Dioxide Supply

An important factor in carbon dioxide flooding is the availability of sufficient quantities of CO_2. Supply studies have been made to determine the costs of producing CO_2 from CO_2 rich gas fields and pipelining it to the injection site. There are also studies on the generation of carbon dioxide. There are purification problems when CO_2 is obtained from industrial plants, but considerable quantities are available from industrial waste gases. This technology will be useful in carrying out the planned CO_2 flood programs.

7.4 Corrosion Problems

Deberry, et al. (22) have been studying surface and down hole corrosion problems associated with CO_2 miscible flooding. Their general conclusion is that corrosion can become severe if adequate prevention measurements are not planned in advance and implemented. Corrosion can be reduced by the proper choice of materials and coatings, the use of corrosion inhibitors and oxygen scavengers, as well as equipment design and operation. Metals are not attacked by CO_2 unless liquid water is also present in normal oil field situations. Stress corrosion cracking of plain steels occur with mixtures of CO_2, carbon monoxide, and water.

Since carbon dioxide flooding will be applied in fields where wells and much surface equipment are already in place, the use of protective coatings will become important.

Plastic coated steel must be properly selected. Thin film coatings have been used in high pressure CO_2 projects like SACROC injection well tubing. The SACROC project is converting to all 316 stainless steel well-head assemblies. Oxygen scavengers are added to the water streams to injection wells. A water soluble inhibitor is also used in the injection water. The produced fluids will contain CO_2 and will have increased corrosivity, raising problems in the oil handling equipment.

7.5 Carbonate Reservoirs and CO2 Floods

Pease, et al. (23) have studied carbonate target reservoirs for carbon dioxide miscible flooding. DOE has a current goal of increasing production by 124,000 BPD by 1985, utilizing carbon dioxide floods. The carbonate reservoirs located in the Permian, Williston, and other basins are being screened for this process.

7.5.1 Screening Guide

A preliminary screening guide for selecting reservoirs consisted of the following specifications:

1. Average permeability $\geq$ 5 millidarcies

2. Average oil gravity $\geq 36^{o}$ API

3. Average oil viscosity $\geq$ 10 CP

4. Average current oil saturation $\geq$ 38 percent

5. No high permeability fractures, stringers, or other thief zones.

7.5.2 CO2 Flood Candidates

The petroleum data system was accessed as a starting point for locating fields meeting these criteria. A total of 292 reservoirs in ten geologic provinces were found; 230 were in the Permian Basin.

It was noted that the selected list of fields missed some of the fields that major oil companies are testing for commercial carbon dioxide floods.

7.5.3 Existing Fields CO2 Flooding

Figure 31 gives a list of carbonate fields where CO2 flooding is underway or under study, showing the reservoir, rock, and fluid properties.

7.6 Granny's Creek Pilot

Conner (24) describes a 10-acre pilot flood carbon dioxide injection project in Granny's Creek field in Clay County, West Virginia. The purpose of the field tests is to determine the effectiveness of CO2 injection in recovering high gravity crude oil from shallow watered-out reservoirs. The reservoir was repressurized to 1000 psig with water prior to CO2 injection. Injection of CO2 began in June 1976, and a total of 9800 tons were injected in several stages, alternated periodically with water to improve mobility control and sweep efficiency. CO2 injection was completed one year later. Water injection then followed and continues.

7.6.1 Reservoir

Oil production from the central producer reached maximum shortly after cessation of CO2 injection and has continued for two years. The reservoir has three separate porous and permeable intervals in the Big Injun sand. Two horizons are coarse grained with low-to-medium porosities and medium-to-high permeabilities. The lower sand is fine grained sandstone with very high porosity and low permeability. There are six water injection wells and three oil producers.

RESERVOIR PARAMETER FROM CO_2 DISPLACEMENT PROJECTS IN "CARBONATE" ROCKS

	KELLY SNYDER FIELD SACROC UNIT	WASSON FIELD			CROSSETT FLD NORTH CROSS UNIT	SLAUGHTER FIELD SLAUGHTER ESTATE UNIT	LEVELLAND FLD LEVELLAND UNIT	ANETH FIELD	NORTH COWDEN FIELD	NORTH McELROY FIELD
			WILLARD UNIT	DENVER UNIT						
PHYSICAL RESERVOIR FEATURES										
FORMATION	CANYON	SAN ANDRES			DEVONIAN	SAN ANDRES	SAN ANDRES	PARADOX	GRAYBURG	GRAYBURG/ SAN ANDRES
LITHOLOGY	LIMESTONE	DOLOMITE			(C)	DOLOMITE	(E)	(E)		
STRUCTURE	ANTICLINAL	ANTICLINAL			STRATIGRAPHIC		(G)	(G)		
DEPTH, FT.	6700	5100			5300		4900	~ 5500	~ 5100	3000
WATER-OIL CONTACT, FT.	-4500				-3040				~ -1300	
AVG. GROSS THICKNESS, FT.	268		50-230	137	40-180 (88 EFFECTIVE)		GROSS 250 NET 100	140		150(NET)
LITHOLOGIC TYPE	(A)	(B)			CHALKY			(H)		
POROSITY TYPE	(J)	(D)			(D)			(I)		
FLUID AND ROCK PROPERTIES										
INITIAL RES. PRESS., PSIG	3122 @ 4300'		1805	1805	2400				1750 (EST)	755
RESERVOIR TEMP, °F	130° @ 4300'			105	106°					
OIL GRAVITY, °API	41		33	33	44 @ 60° F		30		34	32
BUBBLE POINT, PSIA	1850		2-65		2328					
SOLUTION GOR, CUFT/BBL	1000		4382		NOV .76 6000	JUN 1050				
OIL VOLUME FACTOR	1.5		1.312							
OIL VISCOSITY, CPS	0.35		1.18		0.38		1.9			
AVG. POROSITY OVER GROSS THICK., %	3.93		8.5	12	22		10.2	RANGE 10-12	.8	15
AVG. PERMEABILITY, MD	19.4		3	5	5.5		2.1	RANGE 10-50	5	10
AVG. SWI, %	21.9				35		23.4			
WATER-OIL MOBILITY RATIO	0.3			15			(ENRICHED GAS) 1100			
CO_2 OIL MOBILITY RATIO	8.0				5					
CO_2 MISCIBLE PRESSURE, PSI		1250	1250		1650					
AVG. SOR, %		30	>30							
DRIVE MECHANISM										
PRIMARY	(M)	SOLUTION GAS			SOLUTION GAS		SOLUTION GAS		SOLUTION GAS	
SECONDARY	(N)	(O)			(P)	WATERFLOOD	(Q)	WATERFLOOD CO_2 TERTIARY	WATERFLOOD	
CURRENT STATUS	COMMERCIAL		(R)		COMMERCIAL	TESTING	CONT TESTING			

(A) CALCARENITE WITH MINOR AMOUNTS OF CALCIRUDITE & CALCILUTITE
(B) FINELY CRYSTALLINE
(C) SILICEOUS DOLOMITE
(D) PRIMARY-INTERGRANULAR
(E) LIMESTONE/DOLOMITE
(G) STRATIGRAPHIC/STRUCTURAL
(H) CALCARENITES/CHALKY DOLOMITE
(I) VUGGY/FINELY INTERCRYSTALLINE
(J) SECONDARY VUGGY
(M) FLUID EXPAN. & SOL. GAS
(N) CENTERLINE WATERFLOOD PATTERN AREA CO_2 - WAG
(O) LINE DRIVE WATERFLOOD SINCE 1965
(P) CO_2 INVERTED 9 SPOT
(Q) ENRICHED HYDROCARBON CO_2 TERTIARY
(R) REPORTED TERMINATED SPE EOR FIELD REPORT V4 NO.2 9/78

Figure 31

7.6.2 Project Design Basis

Design of the project was based on:

1. Repressuring the reservoir to 1000 psi to guarantee the CO_2 being in the gas phase at reservoir temperature so the flood would be miscible with the contacted oil.

2. The residual oil saturation was determined by core analysis to be 35 percent.

3. The state of the art indicated that the CO_2 slug should be 200 SCF/BBL of oil in place or 5 percent of the pore volume. This indicated 4000 to 8000 tons of CO_2 required.

4. Previous field experience showed that the CO_2 slug should be broken up with alternate slugs of carbonated water to increase oil recovery.

7.6.3 Producing Problems

The producing problems were mainly due to paraffin blockage. Unexpectedly, low initial water volumes were produced when oil production increased. Oil production would markedly increase during the CO_2 slug injection, the water cut would gradually increase after water injection was resumed.

There were technical problems in handling the CO_2 which had a tendency to vapor lock the pump, especially in hot weather. Changes were made in insulation and piping design. Special packings and proper lubricating oil were critical. There has been no increase in leaks caused by corrosion due to CO_2.

A problem occurred in that only 3 - 6 percent of the injected CO_2 entered the pattern. However, the results are very encouraging and a more carefully designed pilot flood will be started.

7.7 Week's Island "S" Sand

An important pilot miscible CO_2 test has begun in the Week's Island "S" Sand reservoir B in Louisiana. The Deep Test is taking place at 12,800 feet depth. The CO_2 pilot is designed to field test downward CO_2 displacement in a steep dip, high temperature, high pressure Gulf Coast reservoir. These reservoirs are produced by natural water drives which leave a significant residual oil volume. There is a gas column above the residual oil and water below. These reservoirs are not suitable for surfactant flooding because the temperature and water salinities are too high for the currently available chemical processes. The great depth and good oil mobility preclude additional recovery by thermal stimulation.

The gravity forces should stabilize the downward CO_2 displacement and increase the sweep. The laboratory work shows the process will not be miscible; rather, the CO_2 will be in the liquid phase at reservoir contact. However, the core sample displacement studies show that substantial oil recovery will be obtained.

Injection of the 50,000 ton CO_2 slug will take one year. The project will displace 900-acre/feet of the reservoir. The CO_2 slug is being injected just above the producing gas/oil contact. The slug density will be reduced by methane absorbed from the oil and gas contacted in the reservoir. The CO_2 slug should spread between the less dense gas cap and the more dense oil column. The CO_2 slug density is only 5 percent less than the reservoir oil. Gravity forces should displace the CO_2 slug and the remaining oil column into the watered-out sand, as the water column is produced from a well completed in the water leg. A neutron log will monitor the movement of the CO_2 front down the reservoir and the oil bank ahead of the CO_2. Production of the oil will take place through the initial water producer. CO_2 is now being injected at an average rate of 107 tons per day. The results will be known in 1981.

8.0 Thermal Floods

8.1 Bodcau In Situ Combustion Flood

The Bodcau in situ combustion thermal flood is an example of an efficient and economical EOR application; see Garvey (25). Production is 19° API oil from the Bellevue field from the upper cretaceous Nacatoch sand at 400 feet depth. Primary production was by fluid expansion and gravity drainage. Residual oil saturation after primary production had decreased to an uneconomic level which was estimated at 95 percent. Various stimulation processes were attempted but none have been successful in producing the heavy crude, except the in situ combustion process thermal flood now underway.

8.1.1 Reservoir

The Nacatoch sand is very unconsolidated with strings of fossilized lime and sandy shale. Porosity averages 33.9 percent with a water saturation of 27.4 percent. Permeability is estimated at 700 millidarcies and average thickness was 56 feet. Calculations indicated 1909 barrels per acre foot. The reservoir temperature is 75°F and the pressure is 40 psig. The viscosity relationship with temperature, a key factor in thermal floods, is shown in Figure 32. At reservoir conditions the viscosity is 670 CP.

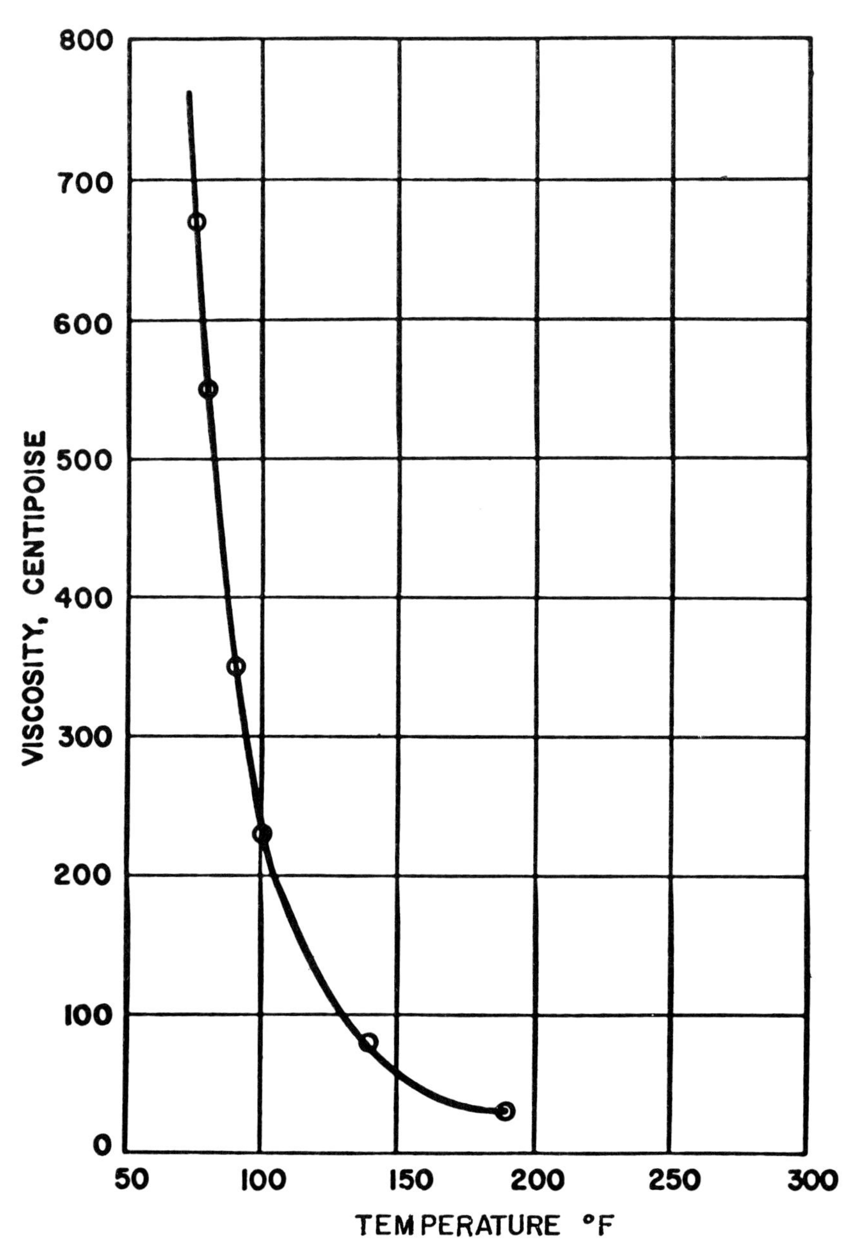

VISCOSITY - TEMPERATURE RELATIONSHIP
(19° API, BELLEVUE CRUDE)

Figure 32

8.1.2 Well Pattern and Completions

The Bodcau area has 33 producers and 5 injectors, as shown in Figure 33. The wells were completed with 7-inch casing cemented to the surface using high temperature resistant materials. The production wells were perforated selectively in the pay and equipped with beam pumping units. A down hole water cooling system is required in each producer to eliminate high temperature problems. The crude is produced through the tubing and flows via flow lines to a central header system. Any natural flow through the casing annulus flows to a pit. Liquids are extracted from the exhaust gases and sent to the central battery. The central battery consists of free water knockout tanks, an emulsion treater, oil storage, and automatic custody transfer metering.

8.1.3 Equipment

The air for the in situ combustion is supplied by three compressors operating at 250 psig supplying 20 MMCFD. The air flows through a trunk system with metering and regulation at each injector.

Ignition of the five inverted nine spot patterns was accomplished by injecting 600°F air into the formation. Down hole 30 kW electrical immersion heaters generated the ignition temperature at the sand face.

To improve volumetric sweep efficiency and to provide greater heat transfer ahead of the combustion zone, all five injectors now employ simultaneous air and water injection. This was started after the burning front had been established for approximately six months. The produced water is injected down the annulus at a 250 BBL/MMCF rate into the upper section of the zone, while air is injected through the tubing below a packer into the lower section of the reservoir (see Figure 34).

8.1.4 Operations

During the three years of operation so far, combustion front breakthrough has occurred at 19 of the 33 producing wells. If the combustion front approaches a producing well, the area surrounding the well becomes thermally stimulated; thereby, causing well producing rates to greatly increase as the mobility of the oil increases. Commercial producing rates cannot be obtained without this stimulation.

The combustion front breaks through, or channels, to the producer

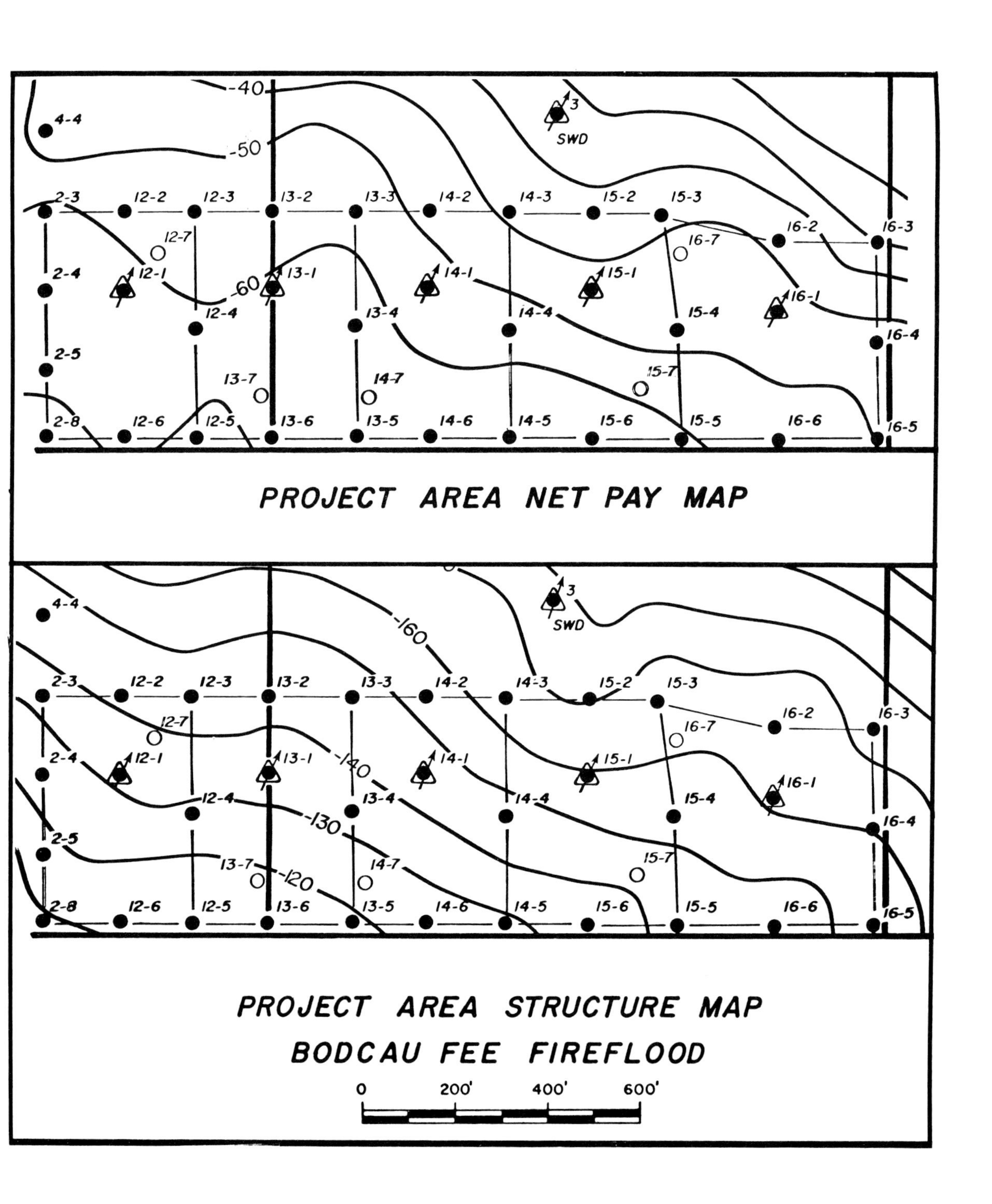
PROJECT AREA NET PAY MAP
PROJECT AREA STRUCTURE MAP
BODCAU FEE FIREFLOOD
0 200' 400' 600'
SWD
-40
-50
-60
-160
-140
-130
-120

Figure 33

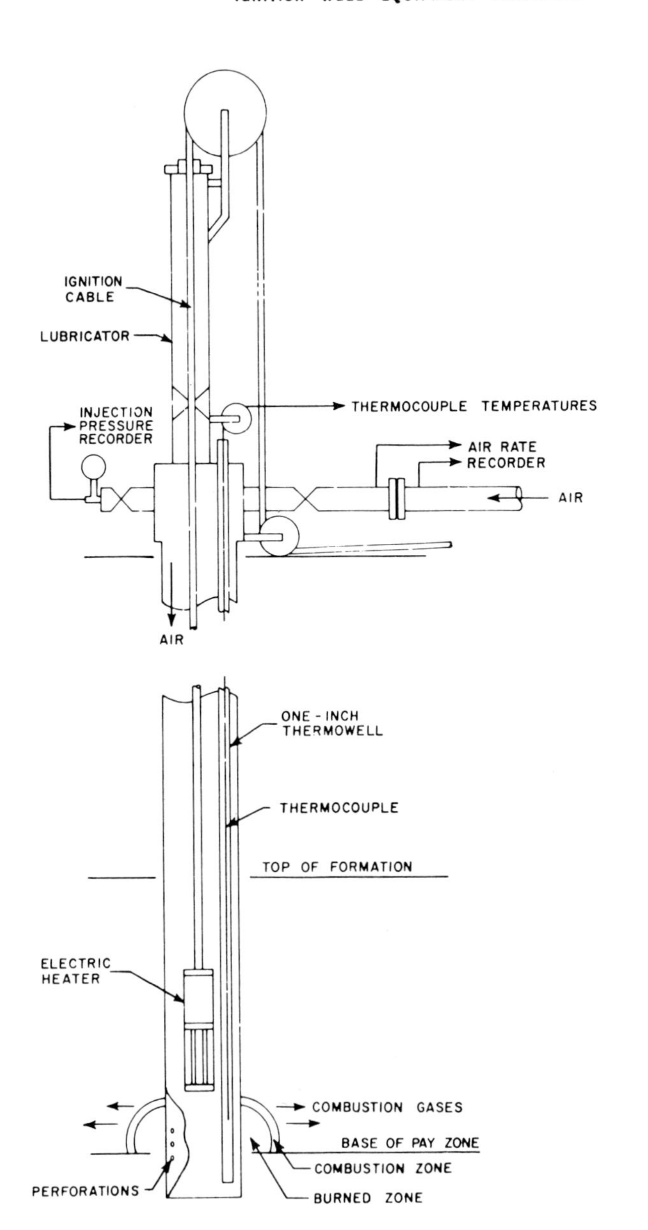
IGNITION WELL EQUIPMENT SCHEMATIC
IGNITION CABLE
LUBRICATOR
INJECTION PRESSURE RECORDER
THERMOCOUPLE TEMPERATURES
AIR RATE RECORDER
AIR
AIR
ONE - INCH THERMOWELL
THERMOCOUPLE
TOP OF FORMATION
ELECTRIC HEATER
COMBUSTION GASES
BASE OF PAY ZONE
COMBUSTION ZONE
PERFORATIONS
BURNED ZONE

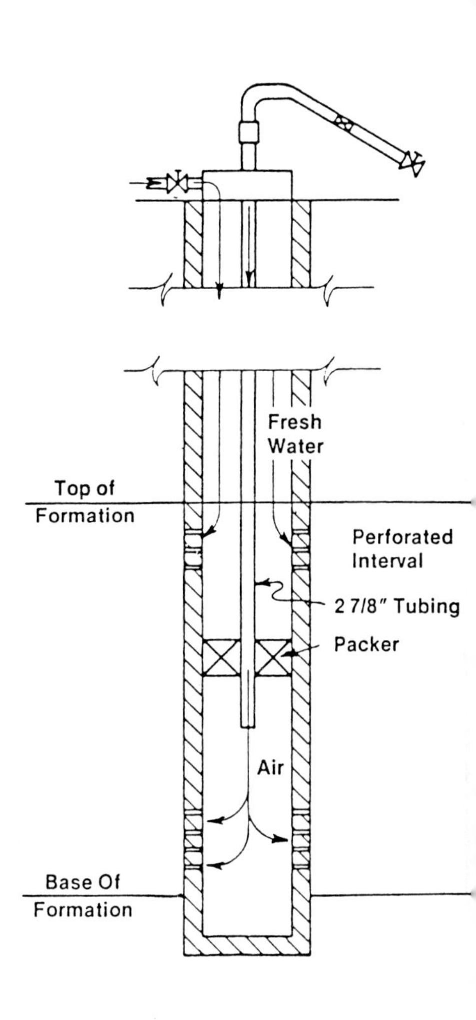
Injection Well Schematic
Simultaneous Injection
Fresh Water
Top of Formation
Perforated Interval
2 7/8″ Tubing
Packer
Air
Base Of Formation

Figure 34

before the entire reservoir can be thermally stimulated. The high flow of hot gases and sands into the producing wells cause extreme erosion of the well and pumping equipment. A quench water system was installed to combat the gases and sands, but this fails eventually. This problem is solved by squeeze cementing the portion of the zone that is burned. The well continues to produce following a breakthrough of the combustion front, but the continual operating problems and the reduced pay withdrawal capability reduce profitability.

Special problems arise with the treatment of the oil since the produced emulsions are hard to break because of the amount of sand produced with the oil. Studies are underway to minimize this problem.

In the three year period, 401,646 barrels of oil have been produced, while cumulative air injection is 6,381,364 MCF and water injected is 1,024,613 barrels. During dry combustion, the fuel deposit on the rocks was 2.05 lbs./cu. ft. with an air requirement of 17.0 MMSCF/acre foot. With combined air/water injection (225 barrels of water per MMSCF air), the fuel deposit was reduced to 1.65 lbs./cu. ft. and the air requirement was reduced to 10.6 MMSCF/acre foot.

8.1.5 Costs and Economics

The project expenditure has been approximately five million dollars over the three year period to April 1979. Total operating costs are 10.25 dollars per barrel of oil produced. However, only one-fifth of the producible oil has been produced, while most investment has been made, so future producing costs will be less.

8.2 Paris Valley Field, Cyclic Steaming and Wet Combustion Pilots

Shipley (26) reports on cyclic steaming, and a wet combustion pilot in the Paris Valley field, Monterey County, California. The test was to determine whether a very viscous crude could be produced from this type of reservoir. The project was terminated after four years because of adverse reservoir properties.

8.2.1 Cyclic Steaming

Fifty-six cyclic steam treatments were performed in 21 wells. Tests included varying volume of injected steam, heat loss reduction by use of packers, varying injection string configuration, adding compressed air with the steam. However, the thin net pay combined with the high oil viscosity resulted in only limited success.

Cyclic steaming requires the injection of high pressure steam for several days; the well is then shut in for a period. When opened up it produced at a peak initial rate which dropped to zero in 2-4 weeks. Adding air to the steam improved performance.

8.2.2 Wet Combustion

In trying the wet combustion process, poor air injectivity and severe channeling occurred in the upper interval of the Ansberry formation. Air injected into the lower interval was unsuccessful because of bottom water entry in adjacent producing wells. The depth of the formation is approximately 800 feet, net pay varies from 4 to 84 feet. Permeability to air was 3015 to 4454 millidarcies. The viscosity at reservoir temperature averaged 227,000 centipoises in the Upper lobe, and 23,000 centipoises in the Lower lobe.

Wet combustion is in situ combustion with compressed air injected and water added to the compressed air.

The wet combustion process was tried in a pattern of 5 air injectors and 18 producing wells in a staggered line drive pattern.

8.3 Lynch Canyon Pilot (CTD)

Scarborough, et al. (27) report on the Lynch Canyon field (San Ardo, California). The Lineagan sand of 500 acres contained 26 million barrels OOIP. Previous field development before abandonment in 1968 included drilling 11 wells, and utilization of production stimulation techniques; as example, flow line heaters, down hole electric heaters, and the cyclic system of steam stimulation (Steam and Soak). The field only produced 111,300 barrels.

8.3.1 Well Pattern

A pilot test will be made to provide productivity testing of an inverted five spot pattern to develop basic reservoir information. Then, a 60-acre line drive flood area with 33 wells will be developed. After ignition and stabilization of the first 6-well injection line using dry combustion with down hole heaters, the Combination Thermal Drive process will be applied.

Basic design, drawing on industry experience, is based on 2-1/2 acre well density, and, using six in-line wells to establish a line drive front. The injection rate will reach 6000 MSCFD and 1320 BPD of water. Project life will be five years. The large scale test will provide critical information on the operation of Combination Thermal Drive processes in a typical depleted heavy oil sand. Extensive and detailed records and measurements will be kept on all phases of the operation.

8.3.2 Previous Steam Treatments

During the 1960s, a steam stimulation program was carried out by treating wells with 1000 to 1500 BPD of steam for three to five days. Productivity indexes increased as much as eight times (0.5 BPD/foot to 4.6 BPD/foot). The high production rates, following a steam stimulation, lasted two to nine months. Subsequently, there were one or two larger steam treatments with three to six times as much steam injected. The wells were lost to high water production caused by damaged casing. The field was then abandoned. The Steam and Soak method of steam stimulation was only successful for a limited time.

8.3.3 Combination Thermal Drive Mechanism

Recent study of the improved in situ Combustion Thermal Drive process gave good reservoir model predictions of further oil recovery. The combination water/air process utilizes the steam which is superheated in the hot rocks behind the fire front, and then flows through the fire front, condensing in front of the fire front forming a hot water bank. This process rapidly extends the heating up of the reservoir because the steam carries the heat a long distance. The volumetric sweep efficiency is also increased by the water front. This process is often referred to as CTD (Combination Thermal Drive). The air to oil ratio often decreases as low as five; thereby, saving on compression costs.

After reaching stable burning conditions, with the heat pad existing around each injection well, the air will be injected for six days, while water is injected on the seventh day (220 BBL/MMSCF). The air injection distributes the water throughout the reservoir. Figure 35 shows a model study of steam flooding with correspondence between reservoir prototype and the model values for reservoir characteristics.

8.4 Steam Injection Cat Canyon Pilot

A 20 acre continuous steam injection thermal flood is underway in the Cat Canyon field, Santa Barbara, California. See Hanzlik, et al. (28). Four inverted five spot patterns are employed (four injection wells in the center). The S1-B sand is found at an average depth of 2500 feet. It produces a 9° API crude oil with an oil viscosity of 25,000 centipoises at gas saturated reservoir conditions. Figure 36 shows the

PROTOTYPE PROPERTIES OF STEAM FLOODING
AND THE CORRESPONDING MODEL VALUES

	Prototype *	Model
Pay thickness	73 ft	4.594 inch
Well spacing	361.5 ft	22.744 inch
Porosity	0.35	0.35
Residula oil saturation to steam	0.04	0.12
Irreducible water saturation	0.05	0.20
Initial movable saturation	0.91	0.68
Permeability (darcy)	7.6	3817
Reservoir temperature (°F)	80	38
Steam injection pressure (psia)	400	7.4
Initial reservoir pressure (psia)	240	6.6
API gravity	14	11.8
Oil viscosity (cp)	3651 @ 80°F 2.35 @ 100 psi	8325 @ 38°F 3.9 @ 76°F
Water viscosity (cp)	0.111 @ 100 psia	0.26 @ 5.8 psia
Thermal conductivity of reservoir (BTU/hr/ft °F)	1.2	0.5
Thermal conductivity of adjacent formations (BTU/hr/ft °F)	1.2	1.2
Heat capacity of reservoir $\left(\frac{BTU}{ft^3\ °F}\right)$	35	29
Heat capacity of adjacent formations (BTU/ft^3 °F)	35	35

*Based on properties of reservoirs in the Kern River Field, California

Figure 35

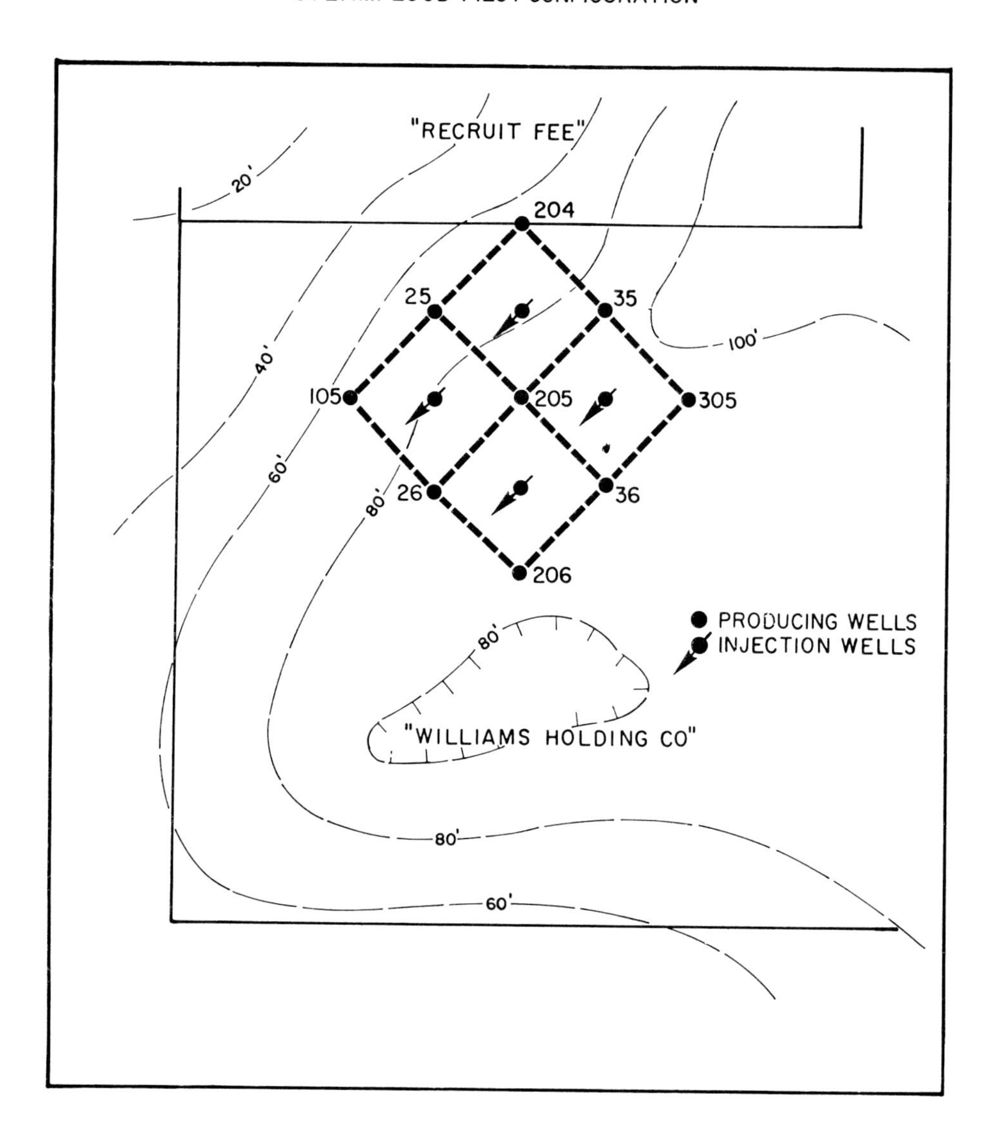
WILLIAMS HOLDING
STEAMFLOOD PILOT CONFIGURATION
"RECRUIT FEE"
20'
40'
60'
80'
100'
204
25
35
105
205
305
26
36
206
80'
"WILLIAMS HOLDING CO"
80'
60'
PRODUCING WELLS
INJECTION WELLS

Figure 36

well pattern.

8.4.1 Operating Problems

The steam displacement operations caused repeated failures of packers in the steam injection wells. There were also flow line failures. Thus, well workover programs were required. These initial problems have been solved and steam injection is going on. Figure 37 shows the well design.

Preliminary findings are that steam channeling occurred in four of the nine producing wells that are now having temperature and production response. These wells were shut in to force steam to move to the other five producing wells. When all producing wells have thermal response, oil production should increase because of the better areal sweep of the steam and the ability to keep low fluid levels in the producing wells.

Flow line temperatures have reached 320°F. The high temperatures are causing poor pump efficiency in the down hole pumps which results in high fluid levels in the wells and limiting production.

Although the initial operating costs have been high due to the well and pumping equipment problems, it is felt that operating costs will decrease and production increase in the future.

8.5 Midway Sunset Steam Flood

Alford (29) reports on a continuous steam flooding pilot project in the 200 Sand Pool in the Midway Sunset field, Kern County, California. This sand made no primary production and would not respond to steam stimulation. There are approximately 50 million barrels of 12° API oil in place. There is no gas, viscosity is about 6500 CP, and the reservoir pressure is low in the 20 - 60 psig range. There is no effective mechanism for producing the shallow dip reservoir. Therefore, steam flooding would not only reduce oil viscosity, but would also provide a displacement drive mechanism.

8.5.1 Optimum Steam Injection Rate

The response time of the flood is estimated at 18 - 24 months. The higher the steam injection rate, the higher the oil production will be because of the flat formation. An optimum injection rate was found which balances the highest oil production per dollar invested for steam injection. Early steam requirements do not represent an optimum steam injection rate. This can only be found after the formation is hot and the maximum well producing rates are established. There are no tech-

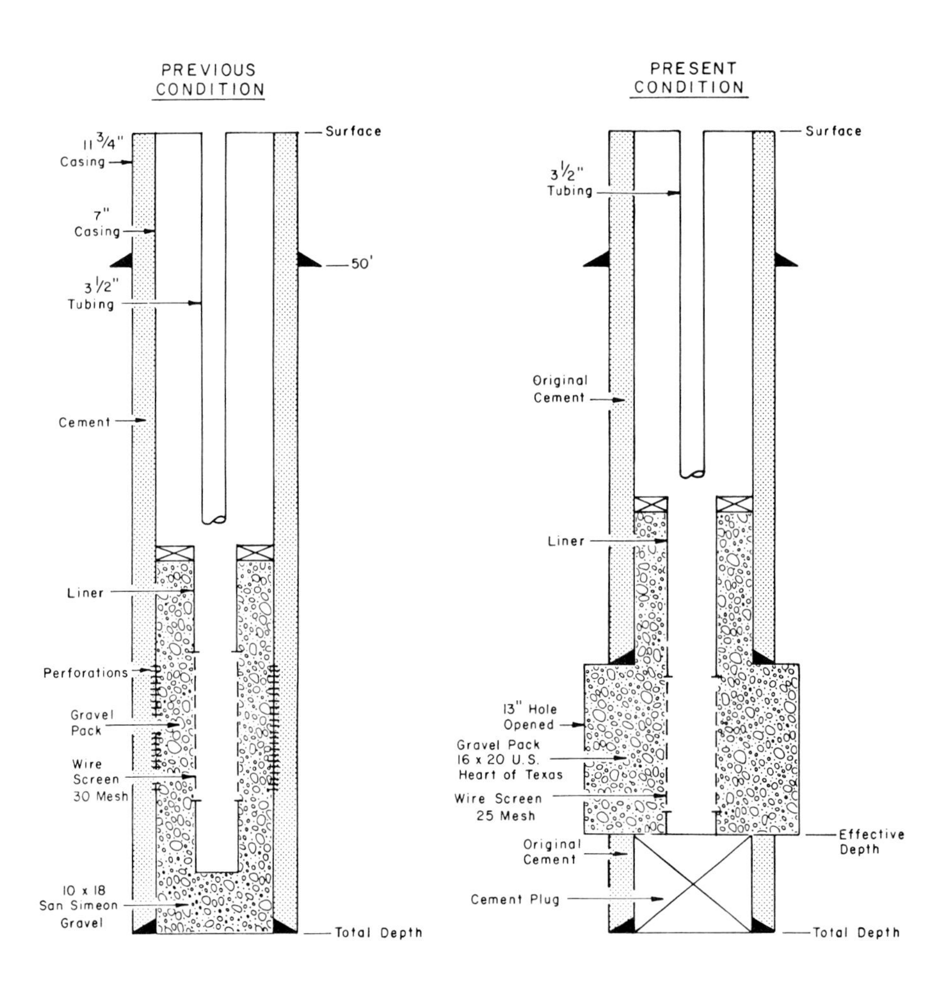
GENERAL SCHEMATICS OF THE STEAM
INJECTOR WELL COMPLETIONS
PREVIOUS
CONDITION
PRESENT
CONDITION
11 3/4"
Casing
7"
Casing
3 1/2"
Tubing
Cement
Liner
Perforations
Gravel
Pack
Wire
Screen
30 Mesh
10 x 18
San Simeon
Gravel
Surface
50'
Total Depth
3 1/2"
Tubing
Original
Cement
Liner
13" Hole
Opened
Gravel Pack
16 x 20 U.S.
Heart of Texas
Wire Screen
25 Mesh
Original
Cement
Cement Plug
Surface
Effective
Depth
Total Depth

Figure 37

nical limits on the steam injection rates, since the fractures in the formation have not been a problem. The reservoir is thick, about 150 - 200 feet, highly permeable, and unconsolidated. Tests indicate that a high rate of steam injection will cause a plastic deformation of the sands, rather than fracturing.

8.5.2 Operations and Completions

Operations began in May 1975. The ten pilot area producers now have temperature and fluid producing response. Five wells have steam breakthrough. Producers are given steam stimulation treatments as their well bore temperatures decline. Two key wells that are surrounded by injectors have produced at the highest rate. The wells are completed so that steam is injected in the lowest 50 feet of the formation. Experience shows that steam overriding will occur with larger intervals and result in poor vertical sweep efficiency. When the bottom interval is depleted, the upper sands can be steam flooded in increments. Experience shows that a slotted liner completion set across the entire sand is better than a cased through completion with jet perforations confined to the bottom pay steam drive interval.

8.5.3 Steam Channeling

A limitation on the steam drive recovery is that the well patterns are not fully developed with producers. A fully developed pattern improves control of frontal movement and volumetric recovery. It is observed that steam channeling from injector to producer is necessary to efficiently produce the highly viscous oil; therefore, high steam injection rates are required for commercial producing rates. Excessive steam channeling is controlled by restricting the producing wells' drawdown. Shutting in the casing vents creates a back pressure which reduces the rate of frontal advance in that channel and encourages channeling to other producing wells.

8.6 Steam Distillation Drive With Light Crudes

The economic feasibility of steam distillation drive in light oil reservoirs has been studied by Aydelotte, et al. (30). The model includes viscous and gravitational forces, compositional and thermal effects, multiphase flow of steam, water, oil and gas in porous media. The presence of free gas or high water saturation in a reservoir is not detrimental to steam distillation drive, provided enough distillable oil is present in the reservoir. Model studies indicate good vertical sweep efficiency as well as horizontal sweep.

8.6.1 Steam Distillation Process

In the steam distillation process, volatile components are stripped from the residual oil left from the steam displacement. The light components flow to the steam front where they condense to form a solvent slug. As the steam zone advances, the solvent slug also advances. Thus, residual oil saturations below those obtained by steam driving nonvolatile oils are possible. Some other mechanisms of the process are the reduction in oil viscosity which improves oil mobility and the thermal expansion of the oil and its residual gases that provide additional gas drive to the steam vapor phase.

8.6.2 Reservoir Type

The investigation is directed toward thin shallow reservoirs. Often depleted sandstone reservoirs have light distillable oil with free gas saturation and solution gas. The conclusion of the study is that these reservoirs could be economically produced by steam distillation drive if a 50 percent saturation of light distillable oil were present. High formation permeability and a high yield oil were desirable. Cost factors were entered into an economic model, and the relationships of Figures 38 and 39 were developed.

8.6.3 Economic Analysis

These include:

1. The fraction of oil recovered as a function of pore volumes of steam injected for three well spacings. Oil recovery is much better with 2.5 acre well spacing than 5 acres. This shows the importance of having optimum well spacing.

2. The present value per acre for high density well spacing calculated for the direct line drive pattern. Both high yield oil and high formation permeability are important to profitability. Profitability generally increases with shallower formation depths and tighter well spacings. An exception occurs at low permeabilities and shallow depths where not enough steam could be injected to effectively flood the reservoir.

3. The profitability of the five spot pattern was found to be greater than the direct line drive pattern. This may be artificial because an areal sweep efficiency of only 50 percent was assumed for the direct line drive pattern. The question of the best pattern can be resolved after field

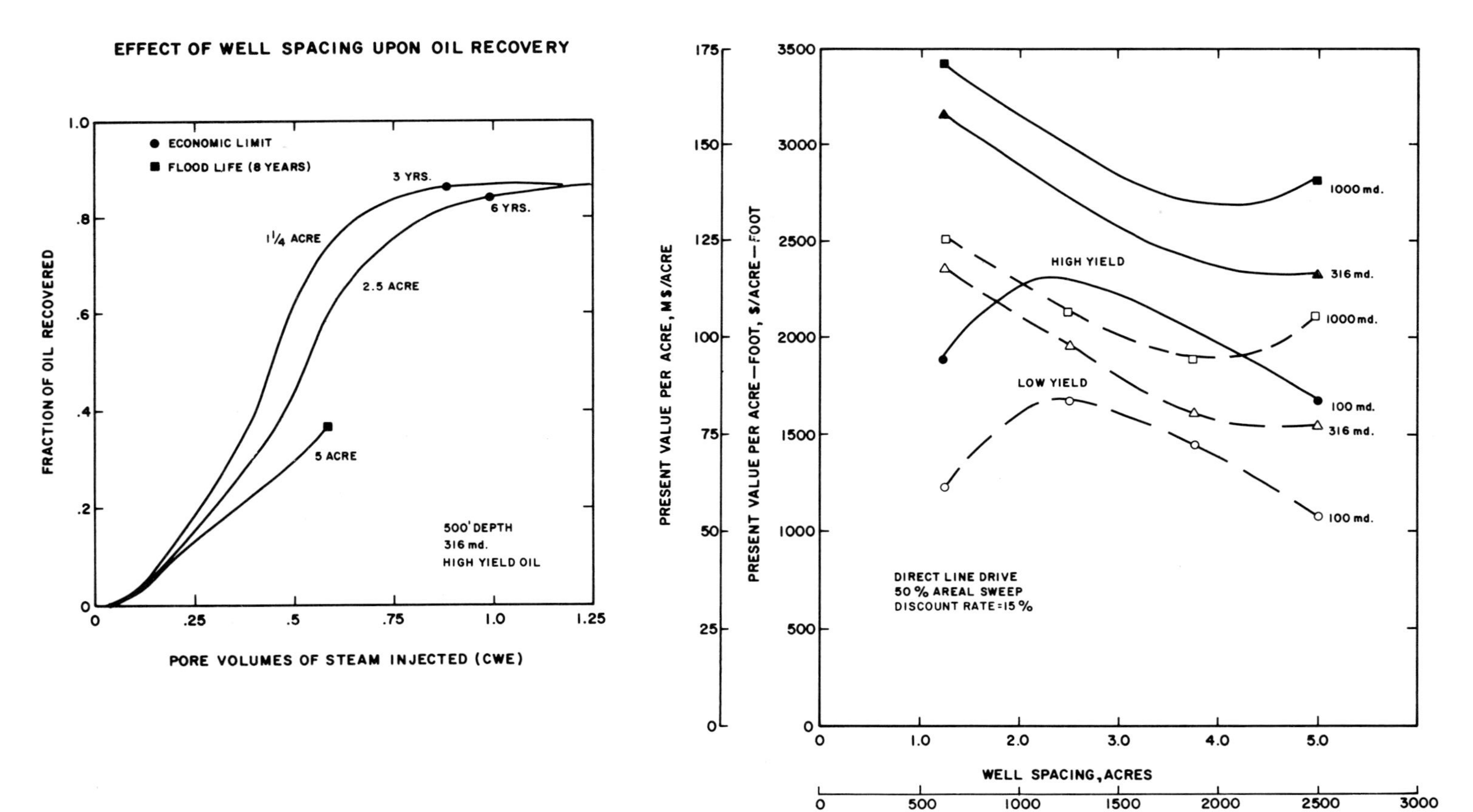
EFFECT OF WELL SPACING UPON OIL RECOVERY
ECONOMIC LIMIT
FLOOD LIFE (8 YEARS)
3 YRS.
6 YRS.
1 1/4 ACRE
2.5 ACRE
5 ACRE
500' DEPTH
316 md.
HIGH YIELD OIL
FRACTION OF OIL RECOVERED
PORE VOLUMES OF STEAM INJECTED (CWE)
0
.2
.4
.6
.8
1.0
.25
.5
.75
1.25
PRESENT VALUE PER PATTERN ACRE FOR HIGH DENSITY SPACING
PRESENT VALUE PER ACRE, M$/ACRE
PRESENT VALUE PER ACRE—FOOT, $/ACRE—FOOT
1000 md.
316 md.
1000 md.
100 md.
316 md.
100 md.
HIGH YIELD
LOW YIELD
DIRECT LINE DRIVE
50 % AREAL SWEEP
DISCOUNT RATE = 15 %
WELL SPACING, ACRES
DEPTH, FEET
25
50
75
100
125
150
175
500
1000
1500
2000
2500
3000
3500
2.0
3.0
4.0
5.0

Figure 38

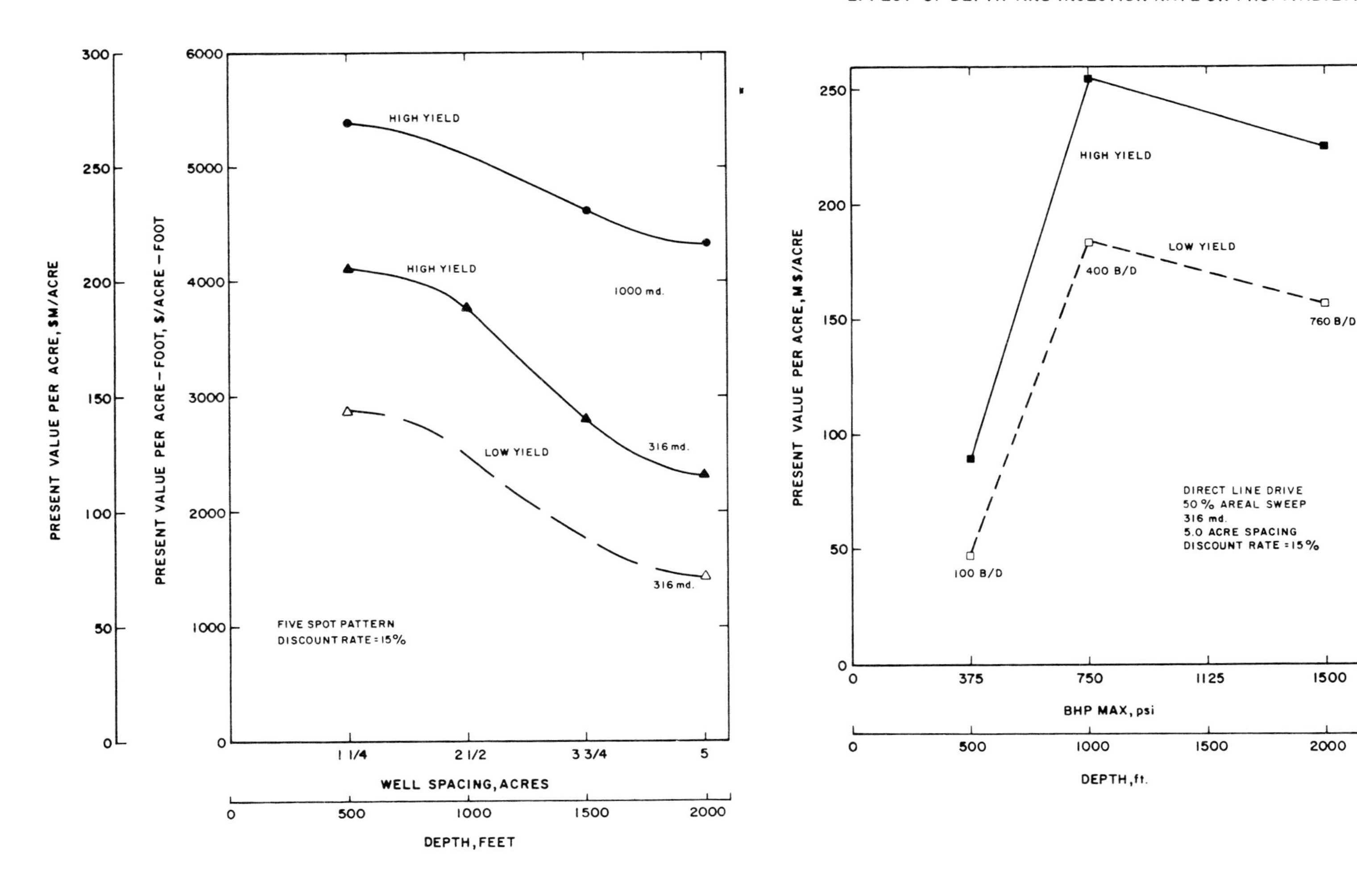
PROFITABILITY OF THE FIVE SPOT PATTERN
PRESENT VALUE PER ACRE, $M/ACRE
300
250
200
150
100
50
0
PRESENT VALUE PER ACRE – FOOT, $/ACRE – FOOT
6000
5000
4000
3000
2000
1000
0
HIGH YIELD
HIGH YIELD
LOW YIELD
1000 md.
316 md.
316 md.
FIVE SPOT PATTERN
DISCOUNT RATE = 15%
1 1/4
2 1/2
3 3/4
5
WELL SPACING, ACRES
0
500
1000
1500
2000
DEPTH, FEET
EFFECT OF DEPTH AND INJECTION RATE ON PROFITABILITY
PRESENT VALUE PER ACRE, M $/ACRE
250
200
150
100
50
0
HIGH YIELD
LOW YIELD
400 B/D
760 B/D
100 B/D
DIRECT LINE DRIVE
50 % AREAL SWEEP
316 md.
5.0 ACRE SPACING
DISCOUNT RATE = 15%
0
375
750
1125
1500
BHP MAX, psi
0
500
1000
1500
2000
DEPTH, ft.

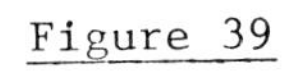

Figure 39

test data is available.

4. The effect of varying depth and injection rates on profitability for a 5 acre direct line drive pattern shows a maximum profitability at an injection rate less than the maximum. Thus, an optimal injection rate is indicated for a given spacing. Also, where low injection rates are only possible, tighter well spacings are more profitable; whereas, if higher injection rates are possible, wider well spacings are desirable.

The reservoir and hydrocarbon properties for this project are given in Figure 40.

8.7 Vapor Therm Process

8.7.1 Process Description

Sperry, et al. (31) report on a Thermal Flood process called Vapor Therm. A field test is underway in the Cherokee field in Missouri. A mixture of nitrogen, carbon dioxide, and superheated steam are injected into producing wells. There are a large number of shallow, highly viscous oil reservoirs in the mid-continent. Steam injection and in situ combustion had been tried with apparently a lack of success (this may be due to incorrect design of the methods). The Vapor Therm process would stimulate well production, and recover oil at higher rates after each stimulation. The process operates (1) by reducing oil viscosity around the well bore, (2) energizing the reservoir by repressuring it with carbon dioxide and nitrogen gas, (3) the absorption of the carbon dioxide results in swelling of the oil phase with consequent increased drive potential, as well as cutting viscosity. No damage is done to permeability.

8.7.2 Field Application

The injection/production cycle was a commercial success in the pilot wells. Now, a 30 well development drilling program is underway in the Cherokee field. The new wells have shown peak production rates of 290 BPD with an average of 100 MBD. This process can be used to recover heavy oils from "dirty" sandstone reservoirs. These reservoirs contain clays and shales, as well as fractures, and should be treated on a well-by-well basis, as is done in the Vapor Therm process.

8.8 In Situ Combustion of Tar Sands

Abou-Sayed, et al. (32) report on in situ recovery experiments that are beginning at Asphalt Ridge in Northeastern Utah. A hydraulic fracture

RESERVOIR PROPERTIES

	Class 1	Class 2
WATER SATURATION	50%	30%
OIL SATURATION	50%	50%
GAS SATURATION	0%	20%
RESIDUAL WATER SATURATION	25%	25%
RESIDUAL OIL SATURATION	50%	50%
RESIDUAL GAS SATURATION	0%	5%
RESERVOIR PRESSURE, psi	250	100
SOLUTION GOR, SCF/STB	0	50
RESERVOIR TEMPERATURE, ^oF	175	175
k_v/k_h RATIO	0.2	0.2
POROSITY, fraction	0.23	0.23
STRUCTURAL DIP, DIRECT LINE DRIVE	15°	15°
STRUCTURAL DIP, FIVE SPOT	0°	0°
ROCK COMPRESSIBILITY, psi^{-1}	$2x10^{-4}$	$2x10^{-4}$
NET THICKNESS, ft.	50	50
RESERVOIR THERMAL CONDUCTIVITY, BTU/(FT-DAY-^oF)	28.8	28.8
RESERVOIR HEAT CAPACITY, BTU/(FT^3-^oF)	35	35
OVERBURDEN AND UNDERBURDEN THERMAL CONDUCTIVITY, BTU/(FT-DAY-^oF)	23	23
OVERBURDEN AND UNDERBURDEN HEAT CAPACITY, BTU/(FT^3-^oF)	32	32

HYDROCARBON PROPERTIES

	C_1-C_5	C_6-C_{17}	C_{18+}
MOLECULAR WEIGHT	30	156	400
CRITICAL PRESSURE, psia	708	321	196
CRITICAL TEMPERATURE, ^oR	550.1	1160	1730
COMPRESSIBILITY, psi^{-1}	$2x10^{-3}$	$5x10^{-5}$	$1x10^{-5}$
THERMAL EXPANSION, $^oF^{-1}$	0.0	$5x10^{-5}$	$34x10^{-5}$
HEAT OF VAPORIZATION, BTU/lb	210.7	110.0	0.0
STOCK TANK DENSITY, lb/ft^3	0.0	48.87	66.14
HEAT CAPACITY BTU/(lb-^oF)	0.41	0.41	0.44
VISCOSITY @ 175°F, cp	0.013	1.0	35.0
VISCOSITY @ 600°F, cp	0.018	0.045	1.6

Figure 40

is planned within the tar sand layer in order to create the necessary flow path between the formation and the wells. In situ combustion would then begin at central points. The mobilized bitumen would flow through the paths created, and the control of the burn would be improved. Tar sands in Eastern Utah contain more than 28 billion barrels of oil in place. Only 10 percent can be surface-mined; the remainder requires development by in situ methods.

8.9 Deep Steam Kern River

Fox, et al. (33) report on Deep Steam Injection. One purpose of the program is to improve the design of thermal wells and completion methods. Two designs for generating the steam down hole are under study. There is a low pressure combustion generator system and a high pressure generator system. Hydrocarbon fuel is burnt to supply the down hole steam. Tests are underway in the Kern River field.

8.10 Add Surfactants to Improve Steam Flood Performance

8.10.1 High Temperature Surfactants

Handy, et al. (34) are evaluating surfactants as possible additives for steam flooding to reduce steam-oil ratios. The effect of high temperatures on the absorption of surfactants on reservoir rocks is a significant factor. Experimental findings, so far, show that the nonionic Igepal CO-850 did not precipitate at high temperatures. Higher oil recoveries were observed in some tests.

8.10.2 Steam in Reservoir

Handy points out that although steam injection improves oil recovery from high viscosity reservoirs, there is the major problem that the steam moves through the upper portion of the reservoir, while the lower portion is flooded only with hot or cold water. Residual oil saturation in the steam swept portion has been observed at near zero, but the normal residual oil saturations are retained in the water swept part of the reservoir.

In addition to mechanical ways of limiting thickness of the steam front, the possibility of adding suitable surfactants to increase recovery from the hot water flood is also possible. One class of surfactants is showing better oil recovery performance at the high steam flood reservoir temperatures, so this area of research combining steam and chemical flooding is promising. Further fundamental research is required followed by pilot plant testing before a combination commercial process can be available.

8.11 Willow Draw Attic Air Injection

Wilson (35) reports on the Attic Air Injection process in the Willow Draw field in Wyoming. Approximately 20 billion barrels OOIP are available in 30 similar fields in Wyoming's Big Horn Basin. The reservoir is a small anticlinal trap of heavily fractured carbonate with low matrix permeability. The pool is filled with 13 - 16° API heavy crude and is underlain by an active infinite aquifer. Primary recovery is essentially nil, since available pressure in the reservoir is soon blown down and productivity indicies are low.

Air injection was carried out from May 1975 to January 1978 with encouraging results. The objectives of this test were (1) to increase oil production and reduce water production by offsetting production with injected air, (2) to improve recovery of matrix oil by the gravity powered gas/oil interchange along the face of gas-filled fractures. Further testing and development is underway.

9.0 Enhanced Oil Recovery Information

Because of the large DOE/Industry sponsored research and development efforts, projects both in the field and the laboratory are generating an immense quantity of data, and technical knowhow, which should be implemented quickly in production practice. An effort is being made to organize this growing volume of information in retrieval systems, data banks, and technology transfer systems on computers which can be accessed from local computer terminals.

9.1 Technical Information Needs

Wilson and Scott (36) report on their studies of the technical information needs of potential EOR users. They found that EOR activities and interests are:

1. Greatest in major oil companies.

2. Independent oil companies do not have the technical and financial resources for pilot projects, and must be supplemented by DOE cost sharing.

3. Noted a general belief in the petroleum industry that economic constraints are more significant than technical shortcomings (the opposite of the actual situation).

4. Oil service companies are aware of the EOR state of the art, but have not undertaken innovative research programs for field application of EOR techniques.

5. Banks, under their current loan procedures, consider EOR too risky to lend money on the basis of projected production. Banks currently loan money only to finance proven production. Banks will require education in EOR and must develop policies and methods appropriate to its financing.

6. Engineering consultant firms keep up with the state of the art, but are only interested in application of proven techniques to commercially profitable projects operated by independent producers.

7. Major oil companies are anxious to receive the latest research findings and technical information. This indicates a high level of EOR activity.

9.2 Computer Information Retrieval

Mohr (37) reports on DOE's development of on—line computer systems for bibliographic and patent information retrieval. A terminal, telephone, and pass word are sufficient to obtain the information. Sixteen major sources of information are now available on the computer systems. The reasons for using the computer-based information retrieval systems for EOR evaluations and projects are summarized below:

1. To avoid duplication of research effort in this very active area.

2. Completeness of information, since the computer does not get tired.

3. Expansion of the information resource base.

4. Low cost relative to personal search methods.

5. Speed-Information is available immediately at terminals in two days in printed form.

6. Boolean logical combinations. There is no limit on input term combinations, so searches can be done on the computer that cannot be done by people.

7. Productivity-On-Line operations are accomplished in a few minutes that would take many hours by manual methods.

9.3 Petroleum Data System (Field Data Retrieval)

Tracy (38) reports on the Petroleum Data System (PDS) which is a comprehensive computer base of petroleum information containing data banks on over 80,000 oil and gas fields in the United States and Canada. Information indicates field and reservoir name, location, cumulative production, lithology, trap types, reservoir temperature, porosity, permeability, and pressure. The PDS system is being used extensively on various types of enhanced recovery projects. Effective information retrieval systems have been devised for accessing the immense data base in PDS. Further reservoir and geological data are being added into PDS in a continuing program. One of the main uses of the PDS system is in the screening of potential enhanced oil recovery candidates. The several methods of EOR may be compared and the best process selected.

9.4 Reservoir Screening Criteria All EOR Methods

Hicks, et al. (39) describe an interesting screening study on the fifty major reservoirs in the Texas Gulf Coast area evaluating them as enhanced oil recovery prospects. Their findings were that none of the reservoirs had an outstanding potential for thermal recovery; however, seven reservoirs have carbon dioxide miscible flood potential, seven have surfactant-polymer flood potential, and nine have polymer flood potential. None of these commercial processes were considered suitable for the remaining twenty-seven reservoirs.

The importance of investigations like this is great. By identifying the highest priority EOR candidates in a given region, the pilot floods can be developed. Profitable full scale floods can then be justified and implemented. Figure 41 shows the field summaries for this screening.

10.0 Long Range Probable EOR Production Rate (1985-2000)

A study was made by Korn, et al. (40) of Arthur D. Little, Inc. on the United States national benefits, costs, and future profitability of EOR Research and Development. The results are summarized in Figure 42, which shows the expected economic levels of EOR production versus time, based on programs of extensive Government support, or by industry only doing the job. Based on world oil prices in the first quarter of 1980, the chart shows that a high level of EOR recovery in the United States is economically attractive. If DOE continues to stimulate research and participate in investment, a production of 2.5 million BPD could be realized by 1985. This production would grow to slightly over 4 million BPD in 1990 and continue at that rate until the year 2000. By 2000, some alternative economic energy processes from nuclear fission and solar energy should be materializing.

Figure 43 shows a study of the cumulative probability of proposed EOR programs and the exploration and development of the Outer Continental Shelf as a function of minimum required price. With the first quarter 1980 world prices, the probability of both EOR and OCS programs approach certainty.

11.0 Conclusions

1. At the first quarter 1980 world price of $26-29 per barrel for oil, the latest long-range studies show U.S.A. enhanced oil recovery up to 2.5 million BPD by 1985, rising to 4.0 million BPD in the 1990-2000 period.

Screening criteria standards for enhanced oil recovery methods

Screening parameters	Enhanced oil recovery (EOR) method				
	Steam injection	In situ combustion	CO_2 miscible flood	Surfactant flood	Polymer flood
1. Oil gravity, °API.....	≤ 25	≤ 25	≥ 27	≥ 25	--
2. Permeability, md......	--	--	--	≥ 20	≥ 20
3. Oil net pay thickness, ft......	≥ 20	≥ 10	--	--	--
4. Oil viscosity, cp.....	≥ 20	≥ 20	≤ 10	≤ 30	≤200
5. Transmissibility, md-ft/cp	≥100	≥ 20	--	--	--
6. Initial reservoir pressure, psig........	--	--	>1,200	--	--
7. Reservoir temperature, °F....	--	--	< 250	≤250	≤200
8. Depth, ft............	>200 ≤5,000	≥5,000	>2,300	--	--
9. Target oil, STB/ac-ft.	>500	>500	--	--	--
10. Percent of initial oil in place before EOR begins................	≥ 50	≥ 50	≥ 25	≥ 25	≥ 50
11. Lithology............	Sandstone or Carbonate	Sandstone	Sandstone or Carbonate	Sandstone	--
12. Fractured............	0 to minor	0 to minor	no natural fractures	0 to minor	0 to minor
13. Water drive..........	0 to weak	0 to weak	0 to weak	0 to weak	0 to weak
14. Gas cap..............	0 to minor	0 to minor	0 to minor	0 to minor	0 to minor
15. Salinity (TDS), ppm....	--	--	--	≤200,000	--
16. Hardness, (calcium and magnesium) ppm........	--	--	--	<1,000	--

50 major Texas Gulf Coast reservoirs categorized by EOR method

EOR method	Number of reservoirs
Steam injection......................	0
In situ combustion..................	0
CO_2 miscible flood..................	7
Surfactant flood.....................	7
Polymer flood........................	9
None.................................	27

Figure 41

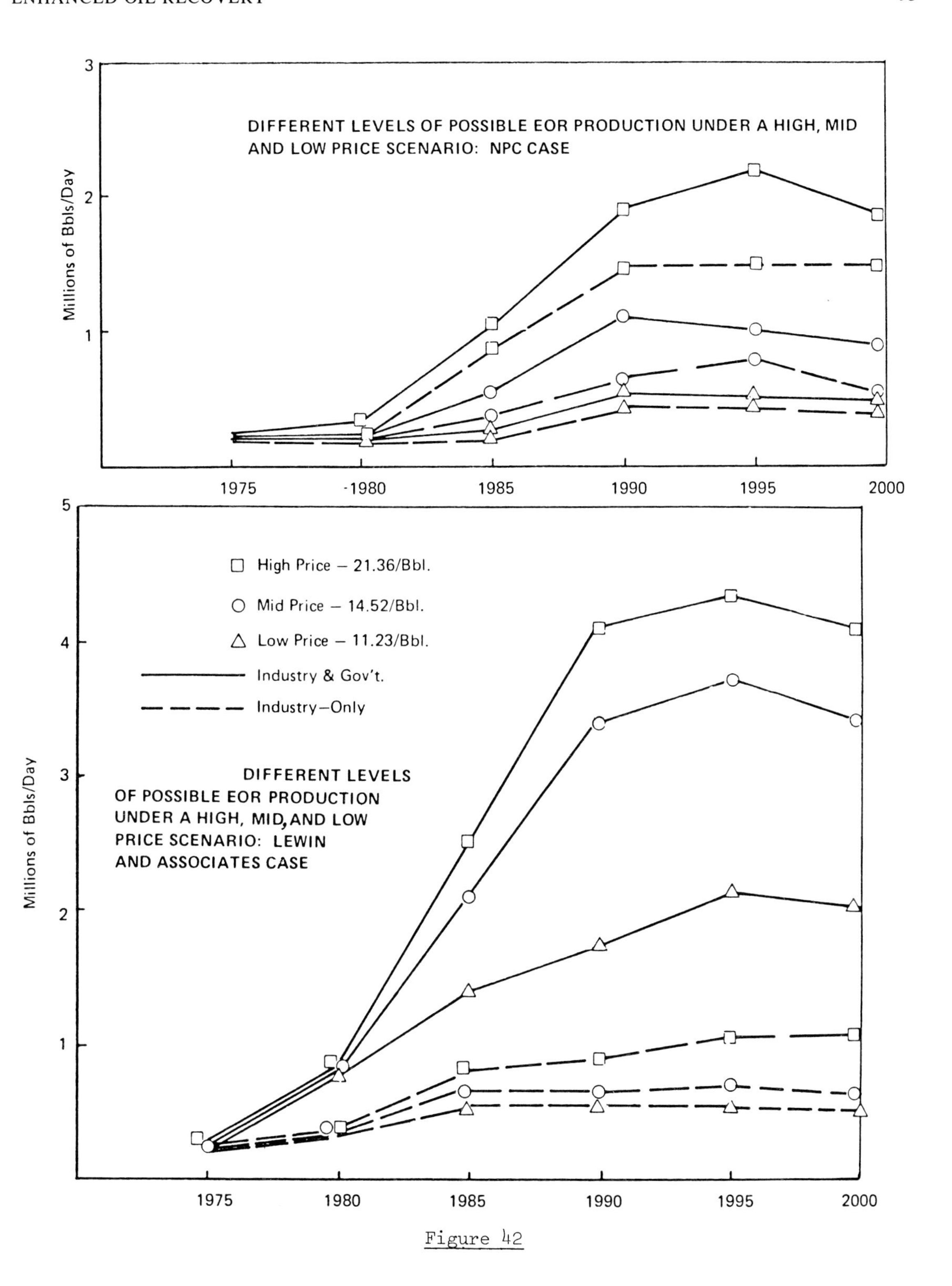
DIFFERENT LEVELS OF POSSIBLE EOR PRODUCTION UNDER A HIGH, MID AND LOW PRICE SCENARIO: NPC CASE
Millions of Bbls/Day
3
2
1
1975
1980
1985
1990
1995
2000
High Price – 21.36/Bbl.
Mid Price – 14.52/Bbl.
Low Price – 11.23/Bbl.
Industry & Gov't.
Industry–Only
DIFFERENT LEVELS OF POSSIBLE EOR PRODUCTION UNDER A HIGH, MID, AND LOW PRICE SCENARIO: LEWIN AND ASSOCIATES CASE
Millions of Bbls/Day
5
4
3
2
1
1975
1980
1985
1990
1995
2000

Figure 42

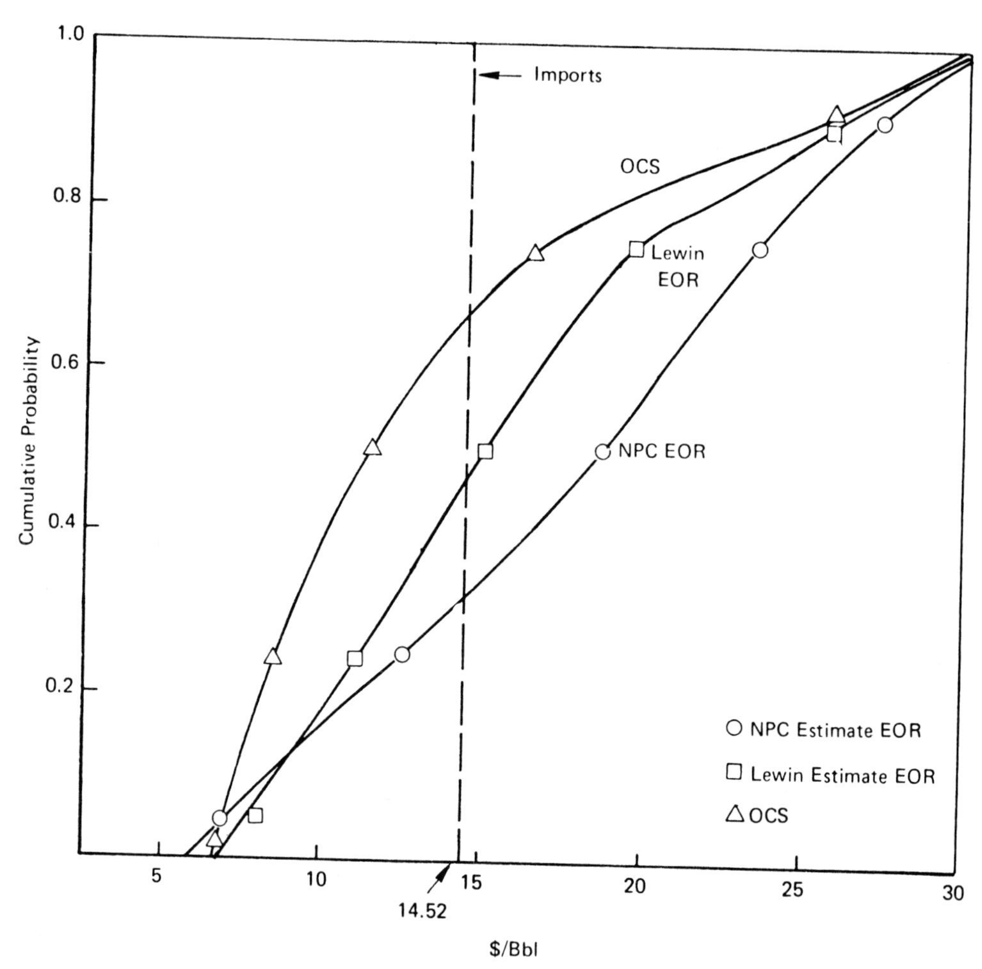

Source: Arthur D. Little, Inc.

Figure 43. CUMULATIVE DISTRIBUTION OF MINIMUM REQUIRED PRICE [1] OF EOR OIL AND OCS OIL

(1) Minimum required price,as used here,reflects average,after tax costs, allowing for a required rate of return to find,develop,and produce the next field.

No worldwide study of the economics of enhanced oil recovery has been made. One is required to determine the total world supply-demand picture.

2. Several hundred field pilot floods, and larger reservoir floods, have begun in the 1975-1979 period. Considerable progress has been made in determining field operating problems and solving them. As examples, successful solutions have been found to well burning, correct completions, viscosity control of polymers, and avoidance of shear degradation, corrosion, equipment, and well design. A field technology is rapidly being created to deal with problems formerly considered insoluble.

3. Based primarily on U.S. Government sponsorship, extensive laboratory experimental research and theoretical studies are underway on Enhanced Oil Recovery processes. The fundamental mechanisms are being elucidated, and the mathematical procedures and data bases are being developed to formulate physical scaling and computer models applicable to actual reservoirs. Innovations have already resulted, as well as progress towards lower cost, higher recovery floods.

4. The enhanced oil recovery processes are becoming more effective in specific reservoirs. There is a proliferation of process detail to cover different situations. Also, combinations of different flooding processes are being experimented with to determine the most economical solution for a specific reservoir type and matrix of problems.

5. Screening criteria for determining the best EOR process and field development plan to apply in specific reservoirs have been developed. These methods are being improved as field results from pilot floods come in. Information retrieval systems of great power have been developed under U.S. Government auspices to assist in the screening procedure.

6. In the U.S.A. alone, there are 80,000 reservoirs. Many of these are abandoned with 30-50 percent of residual oil in place. Many are now economical to produce using existing EOR processes. With a price of $16.55/BBL with only 30 percent windfall excise tax expected between world price and $16.55/BBL, there is a large profit potential for well engineered EOR processes. Most pertinent information for design and screening is available in the Public Domain since DOE has played a part in sponsoring the fundamental research and pilot floods.

REFERENCES

REFER TO THE 4TH AND 5TH ANNUAL
DOE SYMPOSIUMS FOR PAPERS.

(1) Morrow, N. R. and Songhran, B. "Measurement and Correlation of Conditions for Entrapment and Mobilization of Residual Oil".

(2) Slattery, J. C., "Interfacial Effects in the Recovery of Residual Oil by Displacement".

(3) Davis, H. T., Scriven, L. E., "Toward Understanding Reservoir Rocks and Transport Therein".

(4) Franses, E. I., Puig, J. E., Tolman, Y., Miller, W. G., Scriven, L. E., Davis, H. T., "Mechanism of Ultralow Interfacial Tension in Surfactant Flooding".

(5) Benton, W. J., Notoli, J., Mukher,Jr., S.,Qutuluddin, U., Miller, C. A., Fort,Jr.,T.,"Aqueous Surfactant Solutions and Microemulsions of Petroleum Sulfonates and Well Characterized Surfactants".

(6) Lipow, A., Bounel, M., Schechter, R. S., Wade, W. H., "Ultralow Interfacial Tension, Phase Behavior, and Oil Recovery".

(7) Somerton, W. H., Radke, C., "Role of Clays for Enhanced Recovery of Petroleum".

(8) Shah, D. O., Walker, Jr., R.D., O'Connell, J.P., "University of Florida Research Program on Surfactant-Polymer Oil Recovery Systems".

(9) Johnson, J.S., "Micellar Flood Research: Chemicals, Phase Behavior, and Absorption on Minerals".

(10) Kleinschmidt, R.F., Trantham, J.C., Gleinsmann, G.R., Dickinson, W.R., "North Burbank Tertiary Recovery Pilot Test - Fourth Year Status Report".

(11) Burdge, D.N.,"Commercial Scale Demonstration of the Maraflood Process, M1 Project, Crawford County, Illinois, 1978-79".

(12) Ferrell, H.H., Davis, J.G., "Big Muddy Low Tension Flood Process Demonstration Project".

(13) Miller, R.I., Rosenwald, B.W., "The El Dorado Micellar-Polymer Flood Project", El Dorado, Kansas.

(14) Teh Fu Yen, Rong J. Hwang, Manken Chan, Ping-Feng Lin, "Characterization of Alkaline Sensitive Fraction of California Crudes".

(15) Radke, C. J., Somerton, W.H., "Enhanced Recovery With Mobility and Reactive Tension Agents".

(16) Wasan, D.T., Perl, J., Miles, F., Brauer, P., Chang, M., McNamara, J.J., "The Mechanism of Oil Bank Formation, Coalescence in Porous Media and Emulsion Stability".

(17) Boghossian, D.M., "Enhanced Oil Recovery by Improved Waterflooding".

(18) Pease, R.W., Durham, E.N.,"Assessments of Potential Increased Oil Production by Polymer Waterflood in Mid-Continent Fields".

(19) Martin, F.O., Donaruma, L.G., Hatch, M.J., "Development of Improved Mobility Control Agents for Surfactant/Polymer Flooding".

(20) Kamath, K.I., "The Critical Role of Reservoir Temperature in CO2 Flooding".

(21) Gharib, S., Doscher, T. M., "Solvent Displacement of Residual Oil".

(22) Deberry, D.W., Clark, W.S., "Survey and Analysis of Corrosion Problems Caused by CO2 Injection for Enhanced Oil Recovery".

(23) Pease, R.W., Hartsock, J.H., Lokse, E.A., "Carbonate Target Reservoirs for CO2 Miscible Flooding".

(24) Conner, W.D., "Granny's Creek CO2 Injection Project, Clay County, West Virginia".

(25) Garvey, J., "Dodcau In Situ Combustion Project, Bossier Parish, Louisiana".

(26) Shipley, Jr., R., Patek, J.W., "Wet Combustion Pilot, Paris Valley Field, Monterey County, California".

(27) Scarborough, R.M., Cady, G.V., Stair, J.R., "The Lynch Canyon Thermal Drive Oil Recovery Project".

(28) Hanzlik, E.J., Adamson, G.R., "Williams Holding Steamflood Demonstration Project".

(29) Alford, W.O., "The '200' Sand Steamflood Demonstration Project".

(30) Aydelotte, S.R., Ramesh, A.B., "Economic Feasibility of Steam Drive in Light Oil Reservoirs".

(31) Sperry, J., Young, F.S., Poston, R.S., "Field Testing of the Vapor Therm Process in Eastburn (Cherokee) Field, Vernon County, Missouri".

(32) Abou-Sayed, A.S., McCain, R., Wolgemuth, K.M., Jones, A.H., "In Situ Thermal Processing".

(33) Fox, R.L., "Deep Steam: Development of Downhole Steam Generation Systems and Injection String Modifications for Deep Steam Injection".

(34) Handy, L.L., Ershaghi, I., Amaheoku, M., Amaefule, J., Elgassier, M., Ziegler, V., "The Use of Chemical Additives with Steam Injection to Increase Oil Recovery".

(35) Wilson, Q.T., Oswald, Jr., D.D., "Willow Draw Field, Attic Air Injection Project, Park County, Wyoming".

(36) Wilson, T.D., Scott, J.P., "Demonstration of EOR Technology Transfer Requirements".

(37) Mohr, E., "Bibliographic Data Bases: Key to EOR Information".

(38) Tracy, P., "Petroleum Data System: Scope and Applications".

(39) Hicks, J.N., Foster, R.S., "Evaluation of Target Oil in 50 Major Reservoirs in the Texas Gulf Coast for Enhanced Oil Recovery".

(40) Korn, D.H., Rothermel, T.W., Mansvelt-Beck, F., Calvert, M.O., Perry, C.W., "The National Benefits/Costs of Enhanced Oil Recovery Research".

TAR SANDS TECHNOLOGY

J.R. THOMAS

MHG International, Ltd.

ABSTRACT

Tar sands have lain dormant in prodigious quantities unexploited and unharnessed to the needs of society until relatively recently. The birth of an industry to develop this considerable resource has, to date, occurred only in Canada.

Nearly 1,950 billion barrels of heavy oil, that is, 93% of the total known volume, is concentrated in two geographic locations: the Province of Alberta in Western Canada (900 billion) and Eastern Venezuela (1,050 billion).

Current commercial operations in Canada consist of two separate complexes, each of which is an integrated tar sand surface mining, extraction, and bitumen upgrading facility. The construction of a third such plant will commence in early 1980, while another major world-scale production complex, but employing in situ techniques, is about to start activities at about the same time.

In the larger of the two operating plants a mining output of 235,000 tons/day is required in order to process and upgrade 130,000 barrels/day of synthetic crude.

A new industry inevitably has its own problems regardless of established experience as well as due to a real tendency to not seek specialized experience in disciplines often foreign to the promoter.

The only commercially proven process to extract bitumen is the Hot Water Process attributed to Dr. K. A. Clark and the early research pioneered by the Alberta Research Council.

Crude bitumen, when recovered from tar sands, has little value except as a raw material for refining into conventional petroleum products.

The research and development net is being cast far and wide in an endeavor to both improve and expand the industry, while foreign investors are showing significant interest in Canada's unconventional resource and to its ultimate availability as an alternative source of energy.

ISBN 0-12-467101-2

1. INTRODUCTION

The progressive depletion of conventional energy resources in general and oil reserves in particular — the threat imposed by such factors upon our standard of living, our established or accepted social practices, habits, and attitudes — focuses increasing attention upon alternative sources, often realized but more often hitherto ignored.

Whether ignored for reasons of complacency, lack of economic or financial justification, the lack of technology to facilitate efficient recovery or application, or whether due to reasons of location in remote or problematical or climatically inhospitable geographic areas — the urgency of the energy situation, as in any situation of crisis, is bringing forth a renewed interest in the need for innovation, inventiveness, research, development, and exploration in the widest possible sense, to harness whatever available alternatives may be around us to our everyday lives.

Tar sands have lain dormant in prodigious quantities, unexploited and unharnessed to the needs of society until relatively recently. The birth of an industry to develop this considerable resource has, to date, occured only in Canada, where significant investment is already evident in the establishment of operating plants, further research and development to improve the state of the art, and in the preliminaries of a program for the rapid growth of an industry contributing to a position of ultimate energy self-sufficiency.

This presentation addresses itself to the tar sand industry in Canada, dealing with its background, its status today, its problems, and its future potential.

2. DEFINITIONS

To avoid confusion in the use and interpretation of terminology associated with this discussion, the following explanations and qualifications are given:

2.1 Crude Bitumen (or Heavy Oil):

Crude bitumen is defined as a naturally occurring, viscous mixture mainly of hydrocarbons heavier than pentane that may contain sulphur compounds and that, in its naturally occurring viscous state, is not recoverable at an economic rate through a well. The API gravity would be generally in the range of 6° to 20°.

2.2 Tar Sands (or Oil Sands):

These may be described as sands and other rock materials which contain crude bitumen and other associated mineral substances.

More specifically, the tar sand in the Athabasca deposits is composed of sand, heavy oil, mineral-rich clays, and water — the sand making up about 84% of tar sand by weight and consisting predominantly of quartz, with traces of mica, rutile, zircon, tourmaline, and pyrite.

The bitumen, or heavy oil, averages about 11% by weight, made up of 50-60% oil, 30-35% resins, and 15-25% asphaltenes.

Water makes up some 4% of the tar sand by weight and is present in the form of a film, enveloping each grain of the sand, keeping it separate from the oil.

2.3 In Place Reserves:

These are defined as the total resource in a deposit, LESS any portion of it deemed unrecoverable by any forseeable technology or under any forseeable economic circumstances.

2.4 In Situ Recovery:

The in place recovery of heavy oils located in the deeper deposits where surface mining techniques would be deemed uneconomic or impractical.

A typical example would be the employment of "steam flooding", in association with wells drilled into the ore body such that the injection of steam would reduce the viscosity of the crude bitumen and thereby facilitate its expulsion from the deposit.

2.5 Synthetic Crude:

A blend of naphtha and gas oil produced by the upgrading of the recovered bitumen compatible with providing a suitable refinery feedstock.

3. IN PLACE RESERVES

Sixteen very large tar sand fields are currently known to exist worldwide, representing some 2,100 billion barrels of oil in place. On such a scale, this heavy oil almost equals the in place quantity of the total discovered conventional deposits of medium and light-gravity crudes.

Seven of the deposits contain 98% of the world's heavy oil and, as such, equate to as much oil in place as is known to exist in the world's 264 giant oil fields.

Nearly 1,950 billion barrles of heavy oil, that is 93% of the total known volume, is concentrated in two geographic locations: the Province of Alberta in Western Canada and Eastern Venezuela.

The "tar" deposits of these two locations, estimated at some 900 billion barrels in Alberta and 1,050 billion barrels in Venezuela, are individually comparable in size to the proven oil in place of the whole Middle East. These estimates are reputedly conservative.

In the Province of Alberta the Athabasca deposit alone, at some 650 billion barrels, has to be the world's largest, self-contained accumulation of hydrocarbons—four times as large as the Ghawar oil field in Saudi Arabia which, itself, is the largest of all the mammoth conventional oil fields.

Four major tar sand deposits occur in Alberta, in the regions of Athabasca (650 billion barrels), Cold Lake (160 billion barrels), Peace River (65 billion barrels), and Wabasca (25 billion barrels). All are generally in the northern half of the province.

Athabasca

This deposit varies in depth from surface outcrops to beds with overburden of more than 76 m (250 ft.) of muskeg, glacial till, sandstones and shales. Some 5% of the deposit is overlain by 45 m (150 ft.) or less of overburden. This represents the area in which surface mining recovery techniques are currently being applied.

Cold Lake

These resources lie under overburden extending from 305 m (1,050 ft.) to a depth of 610 m (2,000 ft.). In situ techniques for crude bitumen recovery are projected for this area.

Peace River

These deposits are buried by overburden ranging from 150 m (500 ft.) to over 760 m (2,500 ft.).

Wabasca

This field possesses the most viscous of the Alberta deposits and lies under an overburden ranging from 150 m (500 ft.) to 455 m (1,500 ft.).

4. HISTORICAL BACKGROUND TO A NEW CANADIAN INDUSTRY

Interest in the commercial exploitation of Alberta tar sands now spans a period from almost the very beginning of this century, while knowledge of

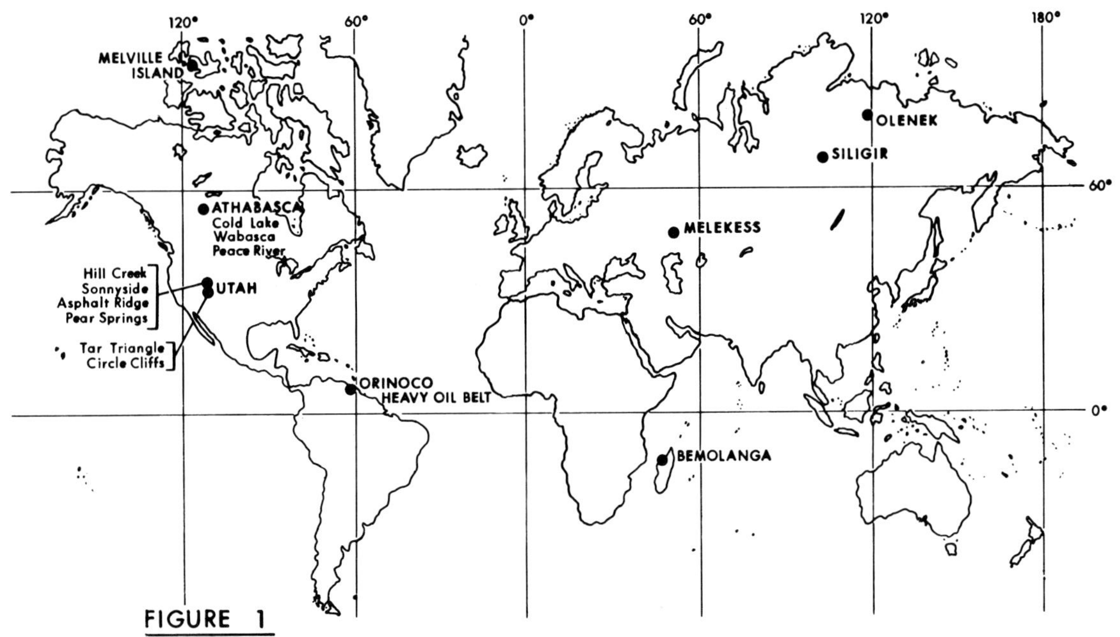

FIGURE 1

Very large heavy oil deposits of the world – Location map. Note concentration of occurrences in Northern Hemisphere, as for conventional crude oil reserves.

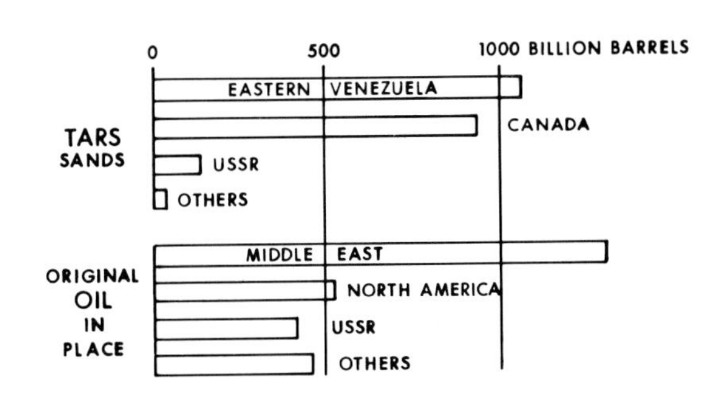

FIGURE 2

Compared orders of magnitude for world's tar sand reserves and the world's discovered, conventionally producible, crude oil reserves.

	VOLUME IN PLACE BILLION BARRELS
VENEZUELA	
ORINOCO HEAVY OIL BELT	1050
W. CANADA	
ATHABASCA	625
COLD LAKE	164
WABASCA	53
PEACE RIVER	50
CANADIAN ARCTIC	
MELVILLE ISLAND	0·1
USSR	
MELEKESS	123
SILIGAR	13
OLENEK	8
USA	
TAR TRIANGLE	16
CIRCLE CLIFFS	1
SUNNYSIDE	4
PEAR SPRINGS	4
HILL CREEK	1
ASPHALT RIDGE	1
MALAGASY	
BEMOLANGA	2

FIGURE 3

Very large tar sands/heavy oil deposits of the world – Statistical Tables.

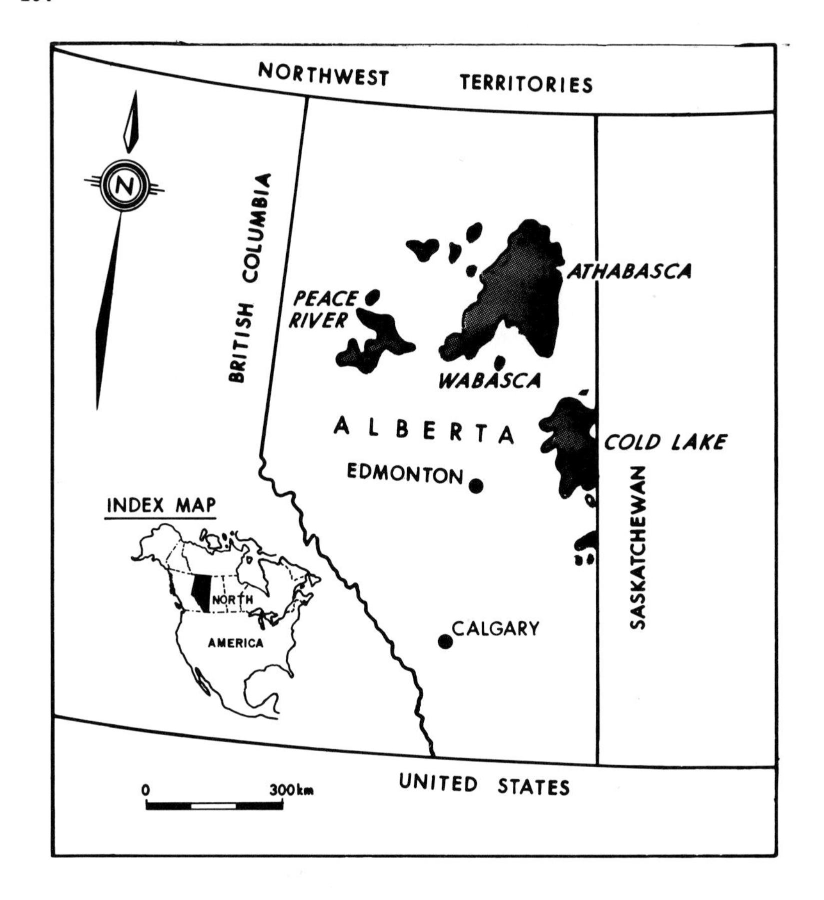

FIGURE 4

Location and extent of the Athabasca oil sands.

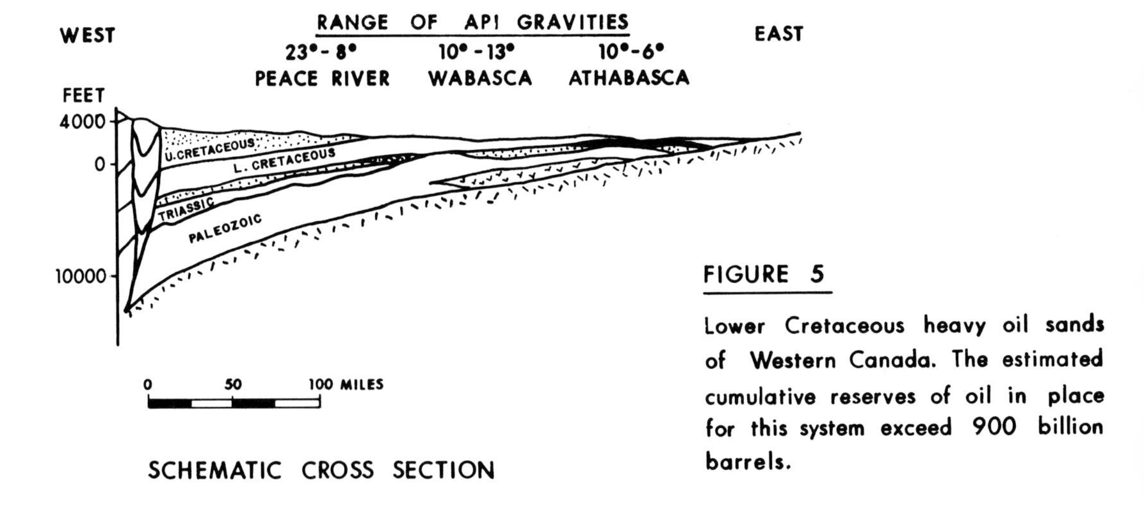

FIGURE 5

Lower Cretaceous heavy oil sands of Western Canada. The estimated cumulative reserves of oil in place for this system exceed 900 billion barrels.

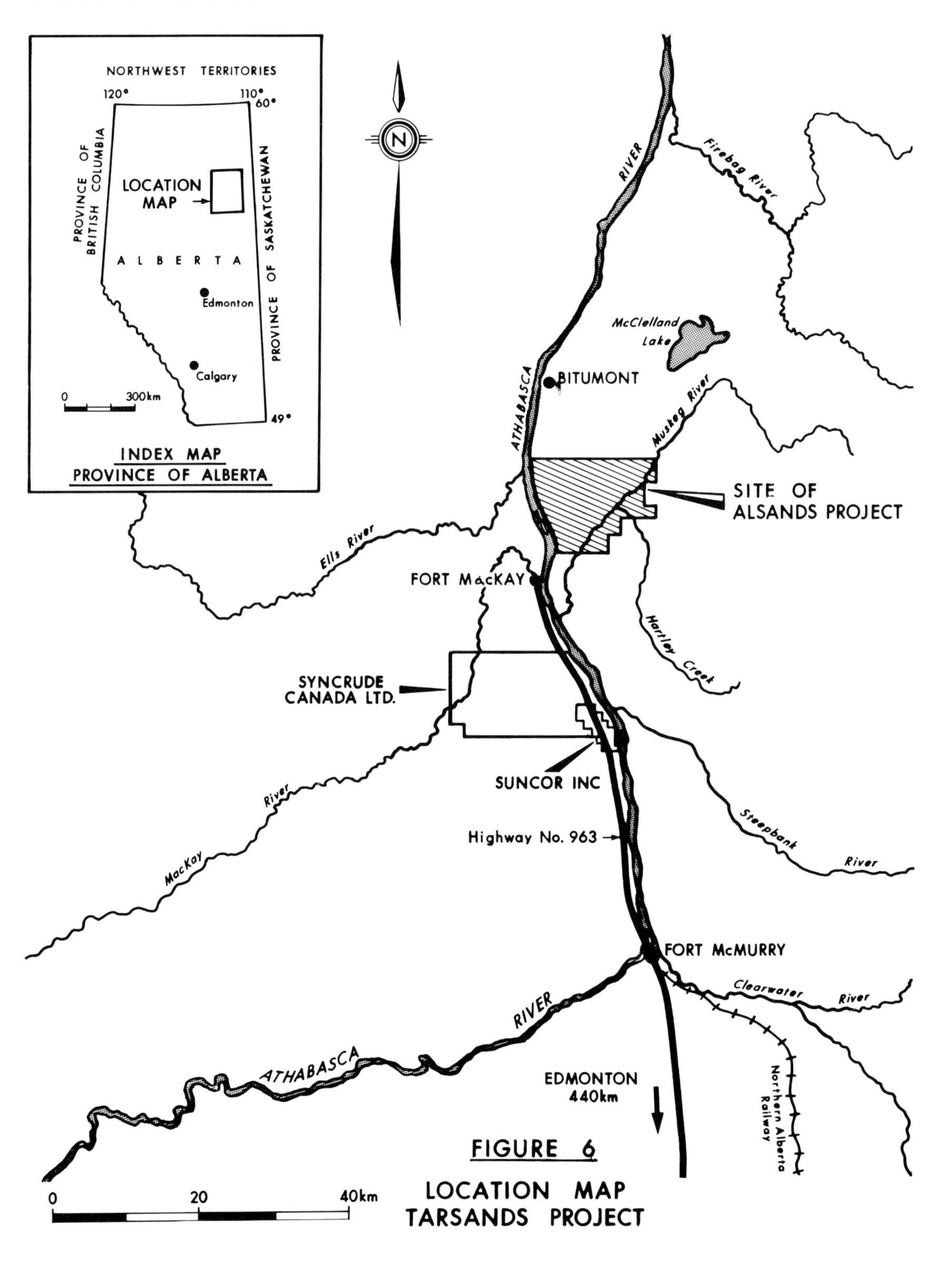

FIGURE 6

LOCATION MAP
TARSANDS PROJECT

the existence of the material, observed from outcrops in river banks, dates back to around the year 1800.

The first "white man" to see the tar sands was fur trader Peter Pond, who was lured to the area by tales of the rich fur harvests that were possible in the far north of the North American continent. Whether or not those tales were accurate, it would appear that our Mr. Pond had little alternative but to travel to the remote northwest as a fugitive from justice in the U.S.A.

A decade later another trader recorded in his journal that "bituminous fountains" were to be seen in the area and that local Indian tribes used the material, mixed with a gum from spruce trees, to caulk their canoes.

Other explorers were equally fascinated by the tar sands, including map-maker David Thompson and Arctic explorers Franklin, Richardson, and Simpson.

It was not until 1875 that the first government-sponsored geological study was initiated. During a subsequent expedition in 1889, it was chronicled "that this region is stored with a substance of great economic value is beyond all doubt, and, when the hour of development comes, it will prove to be one of the wonders of northern Canada".

These early adventurers' explorations familiarized Canadian officials with the tar sands and their potential and caught the imagination of entrepreneurs from around the world. They set the stage for the development that was soon to follow.

The first attempts to develop the Athabasca Oil Sands commercially were made under the illusion that the bitumen in the area must be coming from a pool of oil deep beneath the surface. In an attempt to locate this source, Alfred von Hammerstein drilled the first wells in the region, north of Fort McMurray.

Altogether, between 1906 and 1917, about 24 wells were sunk in the search for the mother lode of oil. None were successful.

In 1913, Sidney Ells, a young engineer employed by the federal government's Department of Mines, began his work in the oil sands, which was to last until 1945. Ells was an early advocate of the hot water flotation method of separating bitumen from sand, conducting a number of experiments to test this technique. He was the first to bring out samples from the area for laboratory testing. As a result, quantities of oil sand were shipped to Edmonton to be tested as road-paving material. While the paving was successful, oil sand could not compete economically with imported asphalt because of transportation difficulties; the project was dropped.

In the 1920s an independent entrepreneur named R.C. Fitzsimmons, using the same hot water flotation process, produced bitumen for roofing and road surfacing at a plant near Bitumount, 80 kilometers north of Fort McMurray. By 1942, however, he had encountered financial difficulties and was forced to sell. The new owners, Oil Sands Limited, also ran into money problems and, in 1948, the small plant was taken over by the Alberta Government to investigate Alberta Research Council extraction methods with large-scale equipment.

By 1949, the plant was processing 450 tons of tar sand a day, but it was decided to close the operation as the government was not interested in launching a commercial venture. Data from the experiments were used as the basis for a major study of the viability of commercial production.

The Alberta Research Council, a provincial government agency, had been conducting research methods of extracting oil from the tar sands since the 1920s . One of its scientists, Dr. Karl Clark, pioneered experiments with a hot water flotation process which involved mixing the tar sand with hot water and aerating the resultant slurry. This would then separate into a floating froth of bitumen and a clean layer of sand which would settle to the bottom of the tank.

While many other techniques were tried to extract the oil (including radiation, combustion, solvent extraction, and centrifuging) the hot water flotation method pioneered by Ells, Fitzsimmons, and Clark proved, over the years, to be the most viable.

In 1936, another developer, Max Ball, founded Abasand Oils Ltd. His plant, west of Fort McMurray, was a limited success in that it produced diesel oil from the oil sands. There was a brief flurry of interest in his project, especially during World War II but, when the plant burned down after being purchased by the federal government, the project died with the buildings.

The 1950s saw another upsurge of interest in the tar sands when the publiction of an Alberta government report indicated that production of oil from the sand could be a profitable venture. One result was the establishment of a 159 cubic meters (1,000 barrels) a day pilot plant at Mildred Lake by Cities Service Athabasca Inc., the forerunner of Syncrude Canada Ltd.

The first major producer of oil from the tar sands, Great Canadian Oil Sands Ltd. (now renamed Suncor Inc.), began plant construction in 1964 and started to produce oil in 1967. Permitted production capacity of the plant was 7,155 cubic meters (45,000 barrels) of synthetic crude oil per day. This permit level was subsequently raised to 10,336 cubic meters (65,000 barrels) per day. Even after starting production, the company had to overcome numerous technical and financial difficulties; but present production has reached as high as 11,130 cubic meters (70,000 barrels) of oil per day, with a sustained daily output in the range of 7,950 cubic meters (50,000 barrels).

In the mid-1950s , Royalite Oil Company Limited was conducting research work at the Bitumont plant. In June of 1958, a 90 percent interest in this project was purchased by Cities Service Co. and the Mildred Lake pilot plant was constructed, with operations and research under the direction of Cities Service Athabasca, Inc.

The following year Richfield Oil Corporation (later called Atlantic Richfield Canada Ltd.) acquired one third of the Cities Service interest in the project. Then Imperial Oil Limited joined the group. The working interest at that time was Royalite (now Gulf Oil Canada Limited), 10 percent; and Imperial, Richfield, and Cities Service each with 30 percent.

In 1962 the four-company group applied to the Alberta Oil and Gas Conservation Board for a license to produce 15,900 cubic meters of synthetic crude oil per day. The application was deferred five years and the activities of the organization remained confined to research and development.

On December 18, 1964, Syncrude Canada Ltd. was incorporated and, on January 1, 1965, assumed operation of the project in place of Cities Service Athabasca.

Another application, presented in 1968, was rejected when the discovery of oil reserves in Prudhoe Bay, Alaska, created what appeared to be a surplus of conventional oil, leaving no potential market for the tar sand product. The following year, Syncrude was authorized to build a 12,720 cubic meters (80,000 barrels) a day plant, provided it would not go on stream before July 1, 1976. This was amended to 19,875 cubic meters (125,000 barrels) per day and, in 1971, was recommended for approval by the Alberta Energy Resources Conservation Board. Permission to proceed was given by Order-in-Council in 1972. Just before the plant was completed, the rated capacity was increased to 20,511 cubic meters (129,000 barrels) per day.

In September, 1973, the Government of Alberta and the participants in the Syncrude Project reached an agreement on royalties. Clearing of the construction site commenced in December of that year, with actual construction beginning in the following spring.

In December, 1974, increased capital project costs, combined with other commitments, forced Atlantic Richfield Canada Ltd. to withdraw from the joint venture. New financial support was found in February, 1975, when the governments of Canada, Alberta, and Ontario purchased 15, 10, and five percent of the project respectively, leaving Imperial Oil Limited with 31.25 percent; Canada Cities Service with 22 percent, and Gulf Oil Canada Limited with 16.75 percent.

In December, 1978, the Ontario government's five percent share was purchased by Pan Canadian Petroleum Limited, at a cost of $160 million.

Site preparation for the construction of the Syncrude Project was started with the clearing of the future sites of the extraction and upgrading facilities, the tailings pond, and mine. The first of 3.8 million cubic meters of muskeg were removed and stockpiled. The area was then backfilled with sand and gravel to support the work crews, equipment, and materials which began moving onto the site in the spring of 1974. Construction started with the foundations of the fluid cokers.

In the plans were a mine which would eventually cover about 26 square kilometers; an extraction complex housed in a building averaging nine stories high, 151 meters by 64 meters; a utility plant which can provide 260 megawatts of electrical power, 472.5 kilograms per second of steam, and water treatment facilities capable of producing 20,511 cubic meters (130,000 barrels) of synthetic crude oil per day.

Product and intermediate tankage with a total volume of 731,400 cubic meters had to be constructed, as did all the warehousing, storage, maintenance shops, and administration buildings required to support the production process.

A work force of over 7,500 swarmed over the site during the peak construction periods in 1976 and 1977; 6,600 of whom lived in the construction camp on site. This small town was equipped with a variety of recreational facilities and three kitchens which served up to 27,000 meals a day.

Completing the complex required an estimated five million engineering and 39 million construction man-hours, with another four million construction hours for the utility plant. Among the materials required were 248,000 cubic meters of concrete, 853 kilometers of piping,and 2,920 kilometers of wire and cable.

Two new bridges had to be built, one over the mighty Athabasca River and one over Poplar Creek, a small watercourse between Fort McMurray and the construction site. Each had to have a load capacity of 450 tons to handle the immense weight of the equipment being trucked to the project. It is estimated that nearly 450,000 tons of materials, equipment, vessels, and plant components travelled the highway to the construction site.

Actual cost of the project, exclusive of the utility plant, was approximately $2.26 billion, making it one of the largest single construction projects in Canadian history.

The prediction of the chronicler of the 1889 expedition has certainly materialized and site work on a third major complex, that of the Alsands Project Group of Companies, is scheduled to commence in the spring of this year (1980) to ultimately cater to a production capability of 140,000 barrels per day of marketable synthetic crude oil. The facilities, both in magnitude and operation, are comparable with the currently operating Syncrude plant and are scheduled for completion in mid-1986.

From this background, development has obviously been predominant in the mineable tar sand areas and confined solely to the Athabasca deposits. However, another major world-scale production complex, but employing in situ recovery techniques, is about to start activities in early 1980 in the Cold Lake area where Esso Resources Canada Ltd. propose a 140,000 barrels per day (marketable crude) operation. Completion is again planned for the mid-1980s.

In this expanding environment the federal government's plan, relating to long-term energy needs, states that Canada may have to build as many as fifteen oil sands and heavy oil plants by the year 2000.

Clearly, considerable attention is being focused upon Canada's unconventional resources by governments and industry alike. Commensurate with this, significantly increasing research and development effort is being progressively applied to improve the state of the art in all its aspects.

To date, marked success has been achieved in the coupling of the vast and unusual earthbound commodities to the necessities of the age. The industry is young and there remains a great deal to be learned and even more to be accomplished if ultimate self-sufficiency and economic independence is to be assured.

5. THE INDUSTRY—A TECHNICAL OVERVIEW

Current commercial operations consist of two separate complexes, each of which is an integrated tar sand surface mining, extraction, and bitumen upgrading facility.

The two companies involved, Suncor Inc. (previously Great Canadian Oil Sands Ltd.) and Syncrude Canada Ltd., share adjacent sites located on the Athabasca deposits.

The oil sands mining operations of Suncor Inc. are located on Bituminous Sands Lease No. 86, which covers an area of about 1,830 hectares and is situated on the west bank of the Athabasca River about 34 km north of Fort McMurray, Alberta. An additional 618 hectares of land has been subleased by agreement with Syncrude Canada Ltd. (Lease 17A).

The Suncor operation was the first major scheme to produce a synthetic crude oil from the Athabasca tar sands on a commercial scale. It was commissioned in September, 1967.

Syncrude Canada Ltd., through its participants, controls mining rights on Bituminous Sands Leases 17 and 22, in the Mildred Lake area north of Fort McMurray. These leases cover an area of about 39,593 hectares.

5.1 Economic Geology

On both sites the overburden is soil and muskeg at the surface, underlain by glacial drift consisting of boulder sands, sands, and boulder clays, followed by grey to green glauconitic sands and shales of the Clearwater Formation, and top reject resting on the ore zone.

In the Suncor location, cut-off in relation to unproductive material (top reject) is set at sands with less than 8% bitumen by weight. Overburden averages 16 m but varies in thickness from 0 to 46 m.

Syncrude has established its cut-off at 6% bitumen and, in general, a minimum 1.5 m zone thickness. Within the Syncrude mining area the overburden averages 23 m but varies from 1.5 m to 43 m.

In both locations, the cumulative thickness of the economic oil sands average about 40 m in a range from 0 to 70 m. The average in situ bitumen saturation is about 11.6% with an average fines content in the area of 15%.

5.2 Muskeg Removal

Muskeg exposed by site clearing is drained by a systematic array of ditches approximately 1.5 m wide and 1.5 m deep. In the absence of actual measurements, the muskeg is estimated to average 1 m in thickness. Actual thickness ranges from zero to in excess of 4.5 m. After drainage, the muskeg is removed by a truck and loader operation. This work is generally carried out during the winter months, when the muskeg is frozen and trafficability is improved. Muskeg is stored in piles located around the site for use in reclaiming land disturbed by construction and mining activities.

5.3 Overburden Stripping

Overburden removed from the mining areas is used in the construction of dykes to contain extraction plant tailings, the latter being a feature of the hot water extraction process, which will be discussed at length later in this chapter.

5.4 Mining Operation — Suncor Inc.

The Suncor mining operation utilizes bucketwheel excavators for mining, and belt conveyors for transporting tar sands plant feed to the extraction plant. The main mining units are two Orenstein and Koppel LMG bucketwheel excavators with a rated output of 3,990 tons/hr, each operating on parallel benches; one bench leading the other. The average thickness of the ore body is about 40 m. The lead bench is usually 23 m high and the trailing bench varies in height. At an ore grade of 12%, approximately 104,300 tons of oil sands must be mined in one day to produce 45,000 BPCD of synthetic crude oil. The mineable tar sands are preblasted to facilitate BWE excavation, and slopes are cut to an angle of 40° to 45°. The Suncor pit is essentially dry with no water problems caused by high-pressure aquifers at the base of the McMurray Formation, as encountered in other areas of the Athabasca oil sands.

The tar sands excavated by the BEW are deposited on a face conveyor via the conveyor system installed on the BEW and a connecting belt wagon. These 1,524 mm wide conveyors run parallel to the mining face and are about 1,525 m long. Each BWE makes two passes of 1,220 to 1,525 m long and 43 to 46 m wide, taking two to three weeks for each pass. The face conveyors must be moved

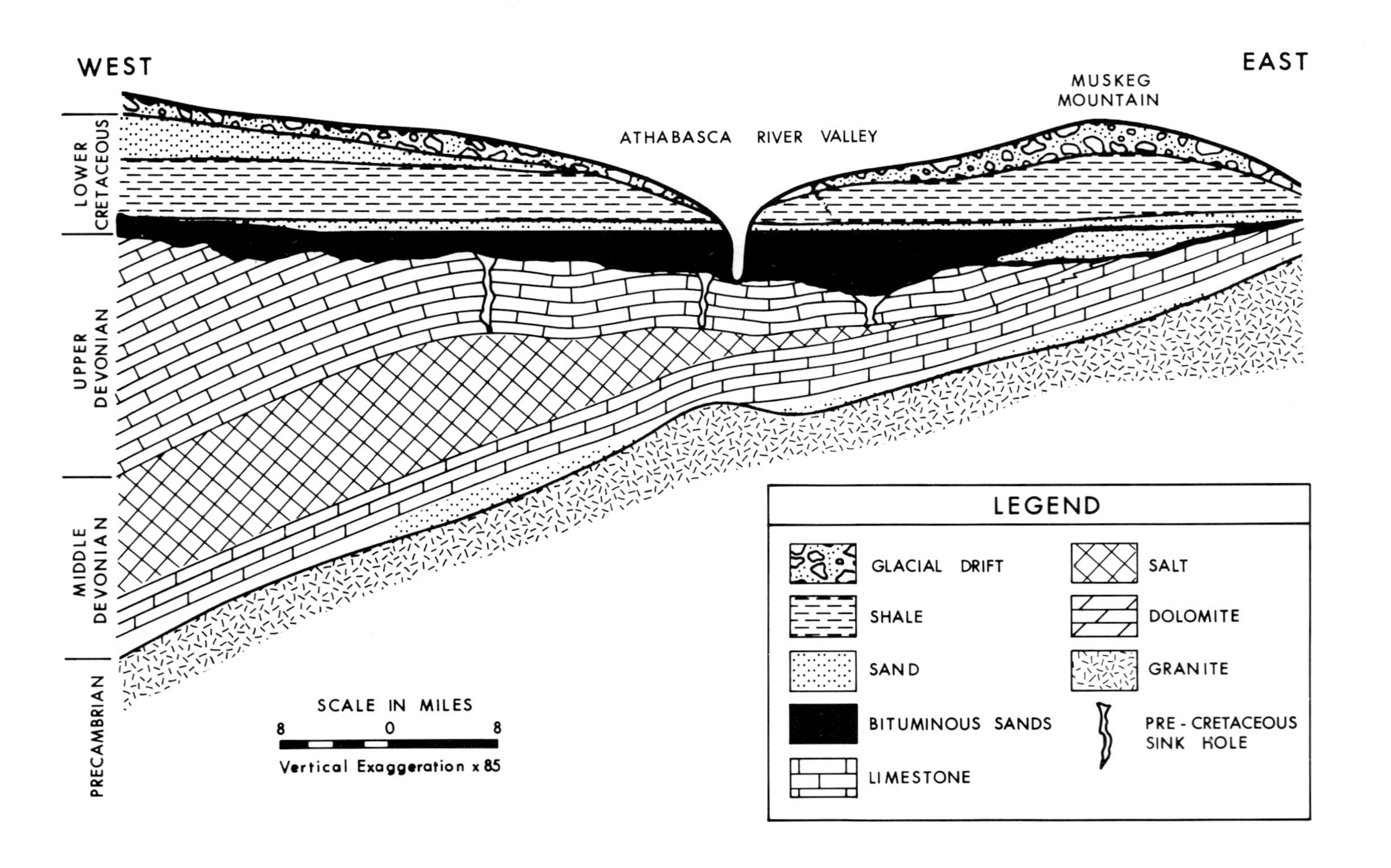

FIGURE 7

SIMPLIFIED GEOLOGICAL CROSS SECTION THROUGH THE ATHABASCA OIL SANDS DEPOSIT

every three months to keep up with the mine advance. The two face conveyors feed two trunk conveyor systems, which carry the mined oil sands to feed bins at the extraction plant. One trunk conveyor extension must be made for every two face conveyor moves.

5.5 Mining Operation—Syncrude Canada Ltd.

The Syncrude operation uses draglines for mining, bucketwheel reclaimers for loading, and belt conveyors for transporting mined tar sands to the extraction plant. For about the first five years of operation, the overburden and the tar sands will be excavated with four 61 m^3 draglines (2 Marion model 8750 and 2 Bucyrus-Erie model 2570 W). It is estimated that each machine is capable of producing about 14,000,000 m^3 of bank (based on 230,000 per year per m^3 of bucket capacity) annually.

The draglines will operate from a prepared working bench along each side of the initial opening boxcut. The present mine design requires the draglines to 'chop cut' overburden above the working bench. The draglines mine 24 wide strips in a north-south direction outward from the opening cut.

The feed-grade tar sands excavated from below the working bench are placed in windrows paralled to the pit wall, and the waste material is cast directly onto the pit floor of the mined-out area. The highwall is designed to be excavated at an overall angle of 50° from toe to crest.

Water sands, which underlie the oil sands throughout much of the mining area, contain saline aquifers that are under considerable piezometric pressure. Hydrological studies have shown that it is necessary to depressurize the basal aquifers well in advance of mining. Depressurization wells have been installed around the perimeter of the mining area. At an ore grade of 11.5% approximately 235,000 tons of tar sands must be mined daily to produce 130,000 BPCD of synthetic crude oil.

The material placed in windrows by each dragline are reclaimed by the bucketwheel that is paired with the dragline. The four reclaimers, with a rated output of 4,990 tons/hr each, were manufactured in Germany by Orenstein and Koppel LMG. They discharge, via bridge conveyor, onto a conveyor system that consists of four 1,829 mm wide collecting conveyors that transfer the plant feed to a radial stacker arrangement immediately adjacent to the extraction plant. The face conveyors are about 4,270 m long, and are skid-mounted so that they can be shifted in conjunction with the mining advances.

5.6 Bitumen Extraction and Upgrading

In both plants, crude bitumen is extracted from the tar sand plant feed using the Clark Hot Water Extraction process and dilution centrifuging. The bitumen produced is not suitable for market as a refinery feedstock, and must be upgraded before it can be shipped in a conventional pipeline.

In the Suncor operation the raw bitumen is subjected to a delayed coking operation following diluent recovery. Vapors from the coking drums are fractionated into gas, naphtha, kerosene, and gas oil components. After further processing, the latter three products are blended to form a synthetic crude oil, which is transshipped to Edmonton through a 428 km pipeline for eventual shipment, again by pipeline, to eastern Canadian and U.S. refineries.

In the Syncrude operation the upgrading plants receive diluted bitumen from the extraction plant and upgrade it to a mineral-free crude oil with reduced nitrogen and sulphur content. Upgrading is accomplished using a fluid coking process to produce gas, butanes, naphtha, gas oil, and coke. The naphtha and gas oil are hydrotreated and blended to form a synthetic crude oil which is suitable as a conventional refinery feedstock. A 559 mm diameter oil pipeline links the plant with facilities in Edmonton.

In both complexes the materials handling and process unit operations contained in the extraction plants are similar except for magnitude. The Syncrude extraction operation was designed to process approximately twice the Suncor tonnage of oil sands.

5.7 Tailings Disposal — Suncor Inc.

Initially, extraction plant tailings were disposed in a tailings pond constructed immediately adjacent to the Athabasca River. A starter dyke was constructed along an island in the river (Tar Island) and across a small channel. The dyke was subsequently raised with the coarse fraction of the tailings stream to form a structure that would contain sludge and fines. This pond was designed to store tailings until sufficient space became available in the mined-out area to allow for in-pit disposal; at present, its dyke reaches a height of over 82 m.

Recently, Suncor has commenced disposal of tailings in the pit. A large dyke composed of overburden material was constructed across the pit to form the first in-pit pond. Continued in-pit disposal will be accomplished by construction of a series of overburden dykes as mining progresses to the northwest. Current Suncor plans propose the development of various tailings ponds, some topped with sand and others filled with sludge, and a final pit which will not be back-filled.

Approximately 1,500 liters per second of tailings in the form of water (50%), bitumen, sand, silt, and clay are pumped by multi-stage centrifugal pumps to the disposal area. Clarified water is recycled to the extraction plant for use as make-up water so the quality can be carefully regulated.

5.8 Tailings Disposal — Syncrude Canada Ltd.

A large area within the Beaver Creek valley north of the plant site was selected for a tailings disposal area. The site covers an area of approximately 2,800 hectares and has a volume sufficient to contain at least eight years of tailings. Three starter dykes were constructed in the pre-production phase across the valley. The main dykes will be raised by the upstream 'step-over' method to an ultimate maximum crest height of 80 m with sand tailings.

When sufficient space becomes available in the pit, tailings will be diverted to the mined-out area. Sludge and water from the in-pit disposal area will be pumped to the main tailings pond in the Beaver Creek valley. Prior to in-pit disposal of tailings, a portion of the mined-out pit will be used to store overburden. At this time, it is not clear whether the out-of-pit tailings pond will remain or if it will, in due time, be removed to allow mining of tar sands covered by the pond.

At full production, the Syncrude extraction plant will generate approximately 103 million m^3 of tailings per year of which approximately 48 million m^3 is water to be recycled through the extraction plant.

6. SPECIFICS

6.1 Mining

Separate mining techniques are employed in the two established tar sand operating plants of Alberta and, between which, the major differences relate only to their lease areas and plant capacities.

In the Suncor environment, the first generation of tar sand processing plants, an average daily output of 104,300 tons/day is required to be mined and processed to meet the average daily upgrading output of 45,000 barrels of synthetic crude. In this,the mining capability is provided by two bucket-wheel excavators (BWE), each able to produce 3,990 tons/hour. Atheore-tical capability of 191,500 tons/day is indicated by this equipment, , reflecting a working efficiency demand of some 55%.

In the Syncrude operation a mining output of 235,000 tons/day (in order to process and upgrade 130,000 barrels/day of synthetic crude) is provided by four mammoth walking dragline units, each attended by a BWE. Each BWE has a capability of 4,990 tons/hour, on which basis the overall equipment configuration would appear to reflect a working efficiency demand of some 49%.

In each instance, downtime would be a significant factor — whether due to mechanical failure or high maintenance needs, or whether limited by the severity of the northern Alberta climate or, indeed, the impositions of the individual mining plans and techniques. All these components materially affect operational efficiency.

Significant cost differences, in both capital and operating environments, would appear evident between the two philosophies involved — especially in the area of prime mining equipment.

From a capital equipment point of view, cost comparisons based on a per ton of gross daily mining capability (of Syncrude and Suncor) is in the ratio of 2.5:1 respectively, while the same comparison in terms of power usage would be of the order of 3:1.

In another expensive exercise, conveyance of tar sands from mine to extraction unit represents a comparable difference. The Syncrude layout currently extends to an in-mine configuration of over 17 km of conveyor layout compared with some 3 km of Suncor facilities. On this basis, the comparative ratios are of the order of 2.9:1, in relation to the number of mining production lines or, alternatively, of the order of 2.2:1 when related to their respective mining outputs.

New plants, certainly benefitting from the hard-earned experiences of "pathfinder" operations, should reflect such an advantage. Surprisingly, in the proposals of the Alsands group of companies, a mining operation basically identical to that of Syncrude is contemplated. The arguments promoted to justify the Alsands decision are noteworthy.

The selected Alsands mining area was considered to be suitable for mining by large draglines containing reserves necessary to supply a tar sand extraction plant at a throughput rate of 25,040 m^3/day (157,500 b/d) of bitumen for 25 years.

In order to supply feed with an overall average grade of 12.1 percent bitumen to the extraction plant at that rate, an average mine output of 109,000 bank m^3/day (143,000 yd^3/d) of tar sands would be necessary. In this the movement of an average of some 170,000 bank m^3/d (230,000 yd^3/d) of total material within the mine area is required

The mining technique selected for the development of the orebody could incorporate four large draglines in the 61 to 76 m^3 (80 to 100 yd^3) capacity class, as primary excavating equipment, with four bucketwheel reclaimers, one per dragline, for transferring material stockpiled by the draglines onto a conveyor transportation system.

The draglines would selectively mine beds of ore and waste material to an average minimum bed thickness of 4.5 m (15 ft.). Ore would be transported either directly into the extraction plant or to a stockpile system ahead of the plant, depending on plant requirements. Waste material would either be cast directly back into the mined areas by the dragline, or if insufficient storage volume was accessible, rehandled by the bucketwheel reclaimers and transferred via the conveyor system to the mined-out areas.

Single bench and double-bench mining configurations have been developed to enable the draglines to mine to full orebody depths in the outlined mine area while applying the geometric constraints on highwall depth and slope angle imposed by geotechnical consideration for safe mining practice. These criteria were developed during the excavation of a test pit using a small dragline.

Extensive and continuous single-bench and double-bench mining areas have been identified within the mine area. An orebody mining sequence, initiating in a single-bench mining area, would be implemented. The mining sequence recognized the need to provide feed at a consistent bitumen grade to the extraction plant, accomplished in the later stages of the project by blending material from two separate mining areas.

Two basic systems were considered in developing a suitable economic method of excavating tar sand for feed to the proposed extraction plant, the chosen method having to fulfill the following requirements:

- Equipment of current technology;

- Proven capabilities of mining to the full depth of the orebody;

- Allow a safe working environment.

The two mining systems considered were:

- A bucketwheel mining scheme utilizing bucketwheel excavators as primary excavating equipment with conveyors for ore and waste transporation;

- A dragline mining scheme utilizing large draglines as primary excavating equipment, bucketwheel reclaimers for reclaiming mined tar sand and excess waste materials, with conveyors for material transporation.

The dragline mining scheme was the preferred system for the following reasons:

- The dragline system appeared to be the more economic mining scheme and therefore one which allowed more tar to be economically recovered.

- The dragline was considered better able to mine selectively interbedded waste and tar sand materials.

- Due to its ability to be selective, the dragline scheme could reduce extraction plant feed for the same bitumen production rate, thereby requiring an extraction plant and supporting facilities with a lower capacity and less area for external tailings disposal.

- Following the excavation of a test pit using a small dragline and the in-depth geotechnical studies associated with this test, confidence was expressed that, with necessary monitoring of high-wall slope performance, a dragline operation could be safely carried out to the depths anticipated with the constraints on the mining configuration.

- The proposed dragline mining scheme would not require equipment to work on a mining bench formed on a dewatered basal aquifer sands and clays as would be necessary on the lowest bench of a bucketwheel mining scheme. These materials, particularly during and directly after rains, would provide both difficult and hazardous operating conditions.

Clearly, many factors require to be taken into consideration in order to define the most appropriately efficient method for any situation.

Equally, over-indulgence in arguments used in isolation, or out of the total context of the circumstances, must also be obviated.

A new industry inevitably has its own problems regardless of established experience as well as due to a real tendency to not seek specialized experience in disciplines often foreign to the promoter.

A special study commissioned to review current tailings disposal and reclamation techniques, indicated that principles of the economy of scale for mining equipment did affect mining costs:

- The need to double handle material as in the Syncrude operation, obviously added to basic costs.

- The application of dragline philosophy appeared to be more attractive with the smaller mine size below a mining capacity of 150,000 tons/day of mined material. It also appeared likely that the costs of dragline systems increased rapidly for larger mines primarily due to the disadvantageous capital and operating costs of the entire materials handling system.

- At the 125,000 tons/day level of output, comparative configurations of dragline versus bucketwheel indicated the former to be preferred where a complement of one BWE and two draglines (of reduced proportions) could operate more efficiently than a complement of three BWE's.

- At a 250,000 tons/day requirement, the indications were that a dragline scheme reflected higher capital and operating costs and that this tendency increased with the larger operations due to the more extensive conveyor systems required.

Optimal bucketwheel mine size appeared to be in the range of 275,000 to 350,000 tons/day of mined tar sands.

- It is sufficient that in a dragline mining operation, the separation of reject material from acceptable quality sands is primarily dependent on the ability of the dragline operator to see the bucket while digging, and on the operator's ability to distinguish the reject material from the pay zone on the basis of color or perhaps diggability. A considerable amount of training and skill would be required before operators could make this distinction with precision. Under less than ideal conditions, such as as digging at depth and during night operation when the boom lights created shadows, it would be questionable whether the visual distinction between lower grade tar sands and reject material could be possible. To maximize tar sands recovery, a non-selective mode of operation might be necessary during periods of marginal visibility.

  In a bucketwheel excavator mining operation, the operator is much closer to the mining face, and has a much greater opportunity to distinguish between ore and waste. As a result, mining selectivity and plant feed quality should be improved over those attainable in a dragline mining operation. Selective mining with any equipment would have a negative effect on overall productivity and would require to be balanced against possible benefits.

The study referred to in the above text could be interpreted as suggesting that the mining techniques applied in both the existing plants in Alberta are in error - that the Suncor operation could be better undertaken using a dragline/bucketwheel concept and that the Syncrude facilities could be better suited to a bucketwheel operation.

In its recent report (Dec.1979) on the Alsands submission for a projected integrated plant, the Energy Resources Conservation Board of the Province of Alberta has made some significant comments in the mining area.

The Board's response to the Alsands submission was as described below:

The Board believed that Alsands failed to fully evaluate mining equipment capabilities and that it was therefore unable to consider aspects that could cause serious difficulties. Although the Shell test pit information, on which Alsands relied, provided valuable information relating to mine system design, several key questions remained unresolved — among them, the capability of draglines to mine to the base of the plant feed material; and their ability to cast back overburden material to the pit bottom after mining without interfering with adjoining ore-grade mining. Selective mining of plant feed and reject material to the proposed specifications had also to be established with reasonable engineering certainty before equipment choices and mining programs could be finalized.

Possible instability of in-pit spoil could critically affect space for in-pit overburden disposal, and could dilute plant feed by encorachment of in-pit spoil on the highwall. Also, seepage from intrabody aquifers and surface water run-off could aggravate stability problems.

Since the test-pit operation employed a two-bench system, the applicant had no opportunity to test whether the dragline could selectively mine to the full depth of ore-grade material, especially in the deeper section of the face. Neither had Alsands, at the time of the hearing, obtained actual operating information from Syncrude to substantiate its belief that selective mining of plant feed and reject material to an average minimum bed thickness of 4.5 m could be achieved.

The Board agreed that placement of bucketwheel excacators on basal aquifer sands could be impractical unless a dry pit bottom could be ensured by an effective mine dewatering program. However, the Board believed that a dragline operation would also require an essentially dry pit bottom in order to ensure stable slopes on pit walls and inpit spoil piles, and draglines could therefore only have a modest advantage over bucketwheels.

The Board noted that Alsands had not prepared its mining plan in sufficient detail to demonstrate that efficient mining recovery could be achieved from its box-cut and subsequent cuts. In particular, the Board believed that Alsands's proposed sequencing of draglines and bucketwheel reclaimers (necessarily a closely coupled operation) could result in lower productivity than estimated by Alsands, and also potential problems due to congestion and lack of equipment maneuvrability. Further, investigation would be similarly needed to determine the magnitude of problems created by frozen ore and attendant loss of production. Such information could change Alsands' initial estimates of mining equipment efficiencies and, in turn, alter comparisons of capital costs of dragline versus bucketwheel excavator systems.

In view of the above uncertainties, the Board regarded the applicant's proposed mining system, the initial (5-year) mine plan, and the long-range mine plans as conceptual rather than definitive. In the approval, the Board would require submission of equipment design details and operating plans, supported by commercial data that would ensure a practical plan for initial operation and maximum recovery of mineable bitumen over the long-range.

The author believes that a significant contribution to problems of production efficiency being experienced both in the mining area and further downstream, enroute to the extraction facilities (as well as in the presentation of suitably sized material to that plant), relates to the effects of the need to stockpile material as in the Syncrude-type operation.

The problem resulting from this technique is particularly troublesome where the material, stockpiled in the windrows above the mine, becomes frozen under the influences of severe winter conditions with temperatures frequently in the -30°C to - 40°C range. Exposure to temperature, wind, and snow in these aboveground heaps, provides significantly less protection to the material than when the material is in its more compacted natural state. From the windrows, which could lie untouched for weeks on end, oversize material (frozen to that size by the elements) adversely affect BWE production rates, cause machine damage, cause costly conveyor belt damage, and can increase the extent of oversize reject material dumped from the conditioning phase of the extraction plant.

As has been previously indicated, double handling is an expensive operation, especially where equipment costs are at an almost prohibitive level — even for the oil industry. Where double handling also compounds the problem it is certain that it should be avoided wherever possible.

Unquestionably, the more subjective and objective attentions being focused onto the mining components of the industry, the conclusions drawn by observation and study, and the experience being progressively gained by existing operations, will contribute to the next generation of integrated tar sand plants being supportably more efficient in concept and operation.

6.2 Bitumen Extraction

The only commercially proven process to extract bitumen is the HOT WATER PROCESS atrributed to Dr. K. A. Clark and the early research pioneered by the Alberta Research Council.

Progressively, since its first large-scale commercial application in the Suncor Inc. facilities at Fort McMurray, technical improvements have been incorporated such that its capability to handle such vast throughputs of material is reasonably efficient.

The process is used at Syncrude and is further proposed for the new Alsands complex currently in the design stage of development.

Clark's Hot Water Process is relatively simple in its concept where, through the medium of the addition of hot water, caustic, and steam, the tar sand is conditioned for gravity separation. The techniques of froth flotation and dilution centrifuging are used to provide a suitable upgrading plant feedstock.

The following processing steps describe the major components of this extraction process:

- Plant stockpile and feed systems

- Conditioning

- Primary separation

- Secondary separation or scavenging

- Froth treatment

Stockpile and Feed System

A stockpile and reclaim system is provided upstream of the extraction plant primarily as a surge facility providing a buffer between the mining and extraction operations. In the Syncrude facilities a 3-hour stockpile is catered for, representing some 35,000 tons of material.

Tar sand feedstock delivered from the mine by a conveying system is either stockpiled by stackers located at the entrance to the extraction plant or is delivered directly to the plant itself.

Apron feeders progress the material which is weighed and then presented to the extraction lines. Syncrude has four such lines, providing a throughput of 11,800 tons/hour. The Alsands proposal will be almost identical to this.

Conditioning

In the conditioning step, tar sand is combined with hot water at 85°C (185°F) and is agitated in rotary drums or tumblers located in each production line. These vessels would be of the order of 5.49 m (18 ft.) in diameter and 30.5 m (100 ft.) long, rotating at 2-3 revolutions per minute. Caustic is added at this stage to control the alkalinity of the "pulp" at a pH of 8-8.5 and steam is further added to maintain the temperature of the material within the vessel.

The conditioned slurry is screened at the discharge-end in order to separate out the oversize material comprising mainly of clay lumps and rocks (and not uncommonly, teeth from both dragline buckets and BWEs). The oversize reject material would be transported for disposal in a mined-out area or at the tailings pond.

Primary Separation

Separation of the dispersed bitumen from the slurry is accomplished in a separation cell by diluting the material (to 50% solids) with additional hot water (160°C) and recycled "middlings" (the central layer in the primary separation cell) and then holding it in acquiescent state. The cells or PSV's (primary separation vessels) would basically be thickeners, each, in the major plants, being some 18.9 m (62 ft.) in diameter. The sand particles rapidly settle to the conical bottom of the PSV while the bitumen rises, forming a frothy layer, is skimmed off, and pumped to the next process stage.

The sand in the bottom of the PSV is continuously raked to a central discharge point from where, after further dilution, it is pumped to the tailings pond.

The fines, consisting essentially of silt and clays from the tar sand, do not settle out and are held in suspension in the "middlings". This accumulation adversely affects the rate of rise of the bitumen particles and therefore requires to be continuously withdrawn and conveyed to the secondary separation unit for further bitumen separation.

Secondary Separation or Scavenging

In this operation conventional flotation cells are employed. Air is injected into the "middlings" which experience further agitation to promote bituminous froth. The recovered froth is sent to a settler for quality improvement after which it is combined with froth from the primary stage.

Froth Treatment

Recovered froth from primary and secondary circuits are treated in this section of the process by the application of heat, followed by deaeration, dehydration, and final demineralization by dilution centrifuging.

Dilution centrifuging uses naphtha, a recycle material, as a diluent to reduce the specific gravity and the viscosity of the recovered bitumen, with the result that entrained solids and water (heavier than diluted bitumen) can be segregated.

Typically, froth from the extraction plant can consist of 60 to 65 weight percent bitumen, 25 to 30 weight percent water and 8 to 10 weight percent mineral solids. Before the bitumen can be upgraded into useful petroleum products the majority of the water and solids must be removed from the froth. To accomplish this, the froth is diluted with naphtha so that its viscosity and specific gravity make the water and solids amenable to separation from the diluted froth by centrifuging.

First Stage Centrifuges effect removal of the coarse +325 mesh solids. Second Stage Centrifuges effect removal of the finer -325 mesh solids, and the majority of the water.

The diluted froth flows through a basket screen ahead of a bank of Solid Bowl-Scroll Centrifuges, the centrifuges removing roughly 50 per cent of the solids and almost all of the +325 mesh material.

Product from the First Stage Centrifuges gravitates into a horizontal surge drum before it is pumped through Cuno filters to the Second Stage Centrifuges. The Second Stage Centrifuges produce a product containing less than 1/2 weight percent solids and 4 to 7 weight percent water. These centrifuges are three-phase separators that produce a diluted bitumen product normally called the light phase, and two tailings streams, one called nozzel water which contains 25 to 27 weight percent solids with traces of hydrocarbon, and a heavy phase or water phase which contains traces of solids and hydrocarbons. The two tailings streams are combined and used to slurry the First Stage Centrifuge tailings.

In the Syncrude plant, the tailings disposal system consist of five separate 24-in. lines leading to the tailings pond. Each line has a pump box and three Canadian Allis Chalmers SRL pumps in series. The first two pumps are constant speed and the third pump is vari-speed through a Nelson fluid drive and a Falk gear reducer. There are magnetic flowmetres and gamma gauges on the pump discharge to control the system. The three essential controlling modes are density, line velocity, and sump level control. As the tailings lines are up to seven miles long, there are booster pumps in series on the lines at about every 10,000 ft.

Recovery

The recovery efficiency of the Clark Hot Water Process is directly related to the fines content of the plant feed.

Average recovery efficiency would be in the region of 91-92%. The effect of diluent losses would reduce efficiency to about 90.5%

6.3 Upgrading

Crude bitumen, when recovered from tar sands, has little value except as raw material for refining into conventional petroleum products.

From the Alberta tar sands locations, the bitumen or its derivatives must be transported to Edmonton, the terminus of the crude pipeline system through Canada and North America. However, because of the high viscosity and pour point of the raw bitumen, pipeline transportation of this material is impractical. The material must instead be processed at the plant to produce a product which can be pumped at ambient conditions.

In addition, the electrical power and steam requirements for recovery of bitumen are high. For example, surface mining and hot water extraction require about 10 percent of the energy in the bitumen mined. Much of this energy is provided by an associated thermal power plant. The initial source of fuel, of course, is the bitumen. However, with the relatively high sulphur content (four to five percent), it would appear that it is necessary to remove sulphur from either the fuel or from fuel gas. Raw bitumen contains considerably more carbon, aromatics, nitrogen, as well as sulphur than conventional Alberta crudes. Therefore, in order to feed Athabasca crude into existing conventional refineries, the bitumen requires upgrading.

The bitumen analysis is shown in Table 1. Fortunately, the chemical properties of the bitumen throughout the tar sand deposits are fairly constant. About 50 percent of the bitumen can be distilled without cracking. Viscosity is reported as having the widest variation with a range indicated in Table 1.

Many alternative schemes for upgrading of bitumen have been investigated and developed. Most follow the same general pattern. Where the first upgrading step is to convert the non distillable residue (approximately 50 percent) to a distillate, (1000 degrees F. minus), which can be further processed. The secondary conversion step, generally hydrogen treating, converts the distillate to an acceptable product mix and removes objectionable sulphur and nitrogen compounds. The final products, naphtha, light and heavy gas oils, are then recombined as synthetic crude oil.

Ancillary facilities include the hydrogen production unit which converts off-gas from conversion units and/or natural gas to hydrogen, a sulphur recovery unit, including an H_2S absorption system, and a residual handling system. The residual from the primary conversion steps is generally used for fuel in the power plant.

Past upgrading development work has ranged from direct conversion of tar sands, to conversion of bitumen after its extraction from tar sands by various methods. This discussion is confined to those processes which have been considered for application following the hot water extraction and bitumen recovery process. One of the earlier engineering studies on the production of synthetic crude oil from bitumen was by S. M. Blair in 1950. He concluded that such an operation was technically and economically feasible and favored a fluidized bed coking plus distillate hydrogen treating route. (See Fig. 8)

Many novel process schemes have been patented since that time. The processes of significance are those used or proposed for commercial operations and a summary of these operations is given in Table 2.

Delayed Coking

Delayed Coking is the process used by Suncor. The yields and product inspections shown in Table 3 are from this operation. A typical flowsheet is given in Figure 9. Preheated raw bitumen is rapidly heated in the feed furnace, and then given a lengthy residence time in coke drums where it cracks to lighter components in a vapor phase and yields coke as a by-product. Conventional delayed cokers run at 900 to 975 degrees F. and 10 to 100 psig.

The overhead products from the cokers are condensed and separated in a fractionator to gas, naphtha, kerosene, and gas oil. Heavy gas oil is recycled to feed. The three liquid products are fed to hydrodesulphurization (HDS) units. Naphtha produced from hydrogenation and mild hydrocracking in the kerosene and gas oil HDS units is combined with coker naphtha and kerosene from the gas oil HDS unit. The coker gas is combined with HDS unit off-gas for amine treating, and is used to produce 95 percent hydrogen for use in the HDS units. The H_2S removed is then converted to sulphur.

Because coke is produced and accumulated in the coke drums, it must be removed. Therefore three parallel sets of coke drums are used, one vessel in each set for the on-stream coking process, the others for coke removal once filled. Coke is removed hydraulically. High-speed, high-pressure water jets bore a hole vertically through the center of the drum. Horizontal jets, operating from a revolving drilling tool are then slowly raised through the open core to break up the coke. The coke is crushed at the coker and conveyed to the thermal power plant for preparation as primary fuel.

Delayed coking is a well-established, relatively simple process which offers a suitable product mix. However, it is fairly limited in flexibility; its product qualities, particularly gas, vary over the coking cycle, and the coke water/slurry systems present difficult clean-up problems.

Fluid Coking

Fluid Coking is the process selected by Syncrude and is that proposed for the Alsands project. The yields and product inspections shown in Table 3 are pilot plant data. Figure 10 is a schematic flowsheet of the process. Here the coke is maintained in a fluidized state at 900 to 1000 degrees F. The coke particles formed in the reactor are fluidized mostly by steam and bitumen feed which vaporizes on contact with the coke; they are transferred to a second "burner" vessel where steam and air maintain fluidization while burning some coke and reducing particle size; most "hot coke" is returned to the reactor to provide process heat while a net surplus is quenched and withdrawn.

TABLE 1

Typical Bitumen Analysis

Gravity	6.5° API	**Elemental Analysis / Wt%**	
Distillation		Carbon	83.2
IBP, °F	505	Hydrogen	10.4
5%	544	Oxygen	0.94
10%	610	Nitrogen	0.36
30%	795	Sulphur	4.2 (3.8-5.5)
50%	981		
EP	1030	Metals, ppm	
% Recovery	50	Vanadium	250
		Nickel	100
Viscosity cs/100°F, 7,700 (up to 500,000)		Copper	5
cs/210°F, 118 (up to 800)		Ash, Wt%	0.65
Pour Point, °F +50			
Ramsbottom Carbon, Wt% 10		Hydrocarbon type, Wt%	
Sp. Ht., Btu/lb/°F .35		Asphaltenes	19
		Polar Aromatics	32
		Aromatics	30
		Saturates	19

TABLE 2

Summary of Commercial Operations

	Suncor	Syncrude
Parent	Sun Oil	Gulf, Imperial and Others
Status	In operation since 1968	In operation since 1979
Synthetic Crude Production, MBPD	45	129
Primary Conversion	Delayed Coker	Fluid Coker
Secondary Conversion	Naphtha HDS Kerosene HDS Gas Oil HDS	Naphtha HDS Gas Oil HDS
Residual-Type Disposition	Coke Fuel + Excess stockpile	Coke Stockpile
Sulphur Treatment	None	None

TABLE 3

Comparison of Primary Conversion Yields

YIELDS	Delayed Coking Wt.%	Fluid Coking Wt.%
C_4-Gas	7.9	11.3
Naphtha, C_5-380°F	12.7(1)	19.1LV
Kerosene, 380-650°F	15.0	26.6LV
Gas Oil, 650-975°F	42.2(2)	35.4LV
Resid, Material Yield	Coke	Coke
Conversion	22.2	10.0

INSPECTIONS

Product	Delayed Coking Nap.	Delayed Coking Ker.	Delayed Coking G.O.	Fluid Coking Nap.	Fluid Coking Ker.	Fluid Coking G.O.
Gravity °API	16.8	32.9	18.3	52.1	25.8	10.7
Sulphur Wt.%	2.2	2.7	3.8	1.4	4.1	5.4
Nitrogen, ppm	150	400	2000	160	800	4100
Bromine No.	61	36	20	130	90	30
Aniline Pt, °F				50	65	85
CCR, Wt%						3.2

NOTES:

(1) FBP = 400°F
(2) G.O. FBP = 850°F

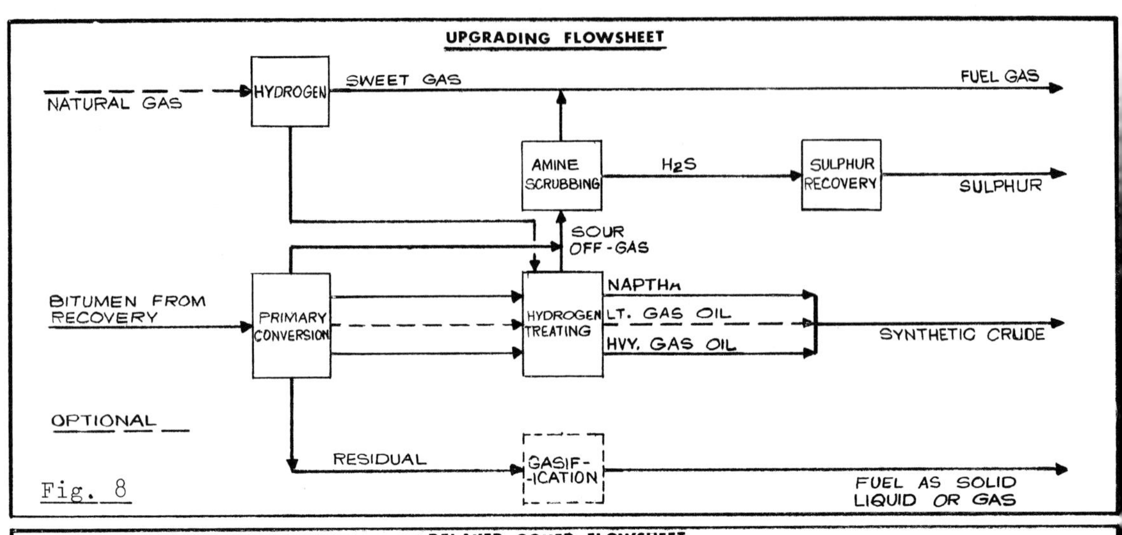

Fig. 8

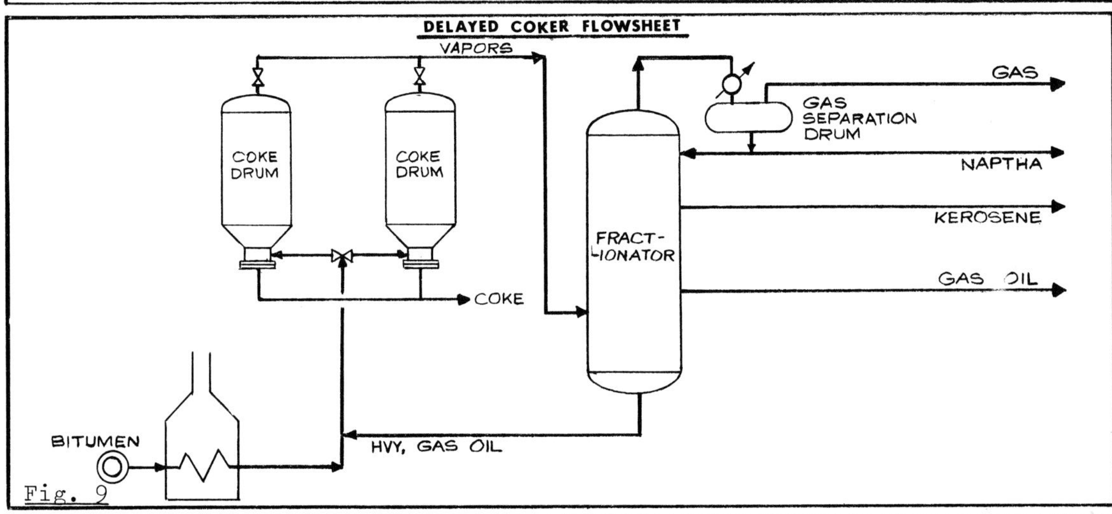

Fig. 9

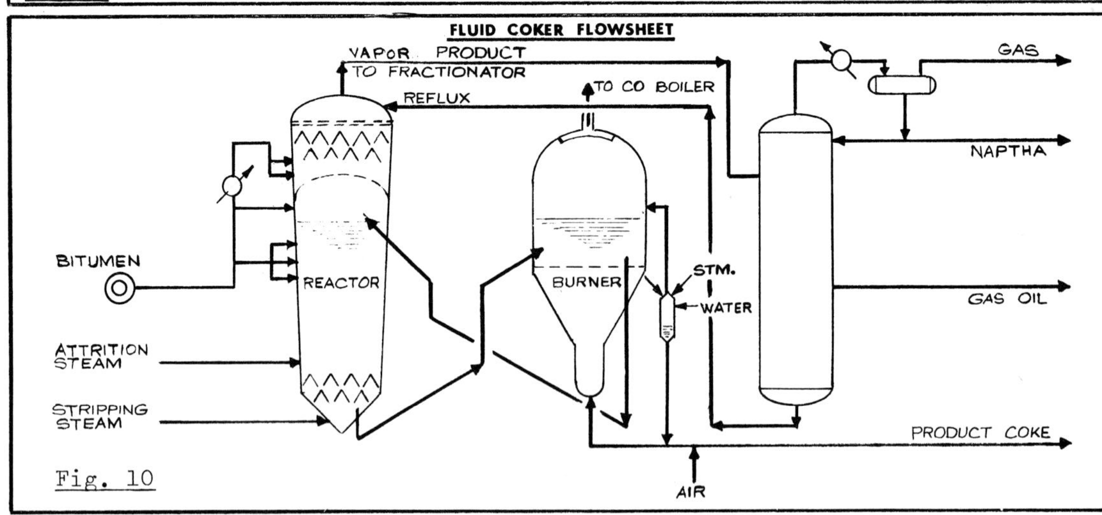

Fig. 10

Particle size of coke is controlled by steam elutriation before withdrawing coke, and with high-pressure attrition steam injected below the reactor feed inlet. Stripping steam is injected in the bottom of the reactor to displace hydrocarbons which could migrate to the burner. Coke transfer is accomplished with steam injection to the transfer lines. The system is pressure balanced at 10 to 30 psig.

The vapor products are a combination of lighter components of feed, and the material cracked from heavier components. These vapors pass through cyclones in the top of the reactor to a quench scrubber tower, where heavy material is condensed and the temperature is reduced to below cracking temperatures. Also, entrained coke fines are washed from the vapors and the bottom slurry is returned to reactor feed. Remaining vapors are further condensed and fractionated to gas, naphtha, and middle distillate streams.

The Syncrude operation will yield only two distillate products, naphtha and gas oil; so only two HDS units are required. The coker off-gas is amine treated and used for plant fuel. CO off-gas from the burner will be used with supplemental fuel in a CO boiler for steam generation. Net coke production, with a sulphur content of nine percent, will be stockpiled as Syncrude did not feel that there was a proven commercial process which would reduce the sulphur emission to atmosphere.

The fluid coking process requires less capital investment than delayed coking. Being a continuous process, with constant coke inventory, the reaction vessel itself is smaller. Although the burner vessel is additional, no "spare" coker reactor is required. Like delayed coking the process is limited in flexibility; however, the product qualities remain more constant. The net product coke yield is less than delayed coking because some coke has been used as the source of process heat; however, this operation concentrates the sulphur so that coke sulphur content is higher, and is more difficult to use.

Fluidized coke transfer systems have a very difficult service due to the abrasive qualities of fluid coke, and maintenance is high. The CO gas from the burner must be converted to CO_2 for heat recovery and for acceptable atmospheric emission. However, the gas still contains SO_2 which has to be removed to meet environmental standards.

Hydrotreating

Hydrotreating is the standard hydrogenation and desulphurization process used or proposed for secondary conversion in all commercial ventures. The sulphur compounds are distributed throughout the boiling range of all primary conversion distillates. While the Suncor operation includes hydrotreating naphtha, kerosene, and gas oil, subsequent proposals indicate only two units for light and heavy feeds. Product data from the Suncor operation are reported in Table 4. A simplified flowsheet is given in Figure 13.

TABLE 4

Suncor Commercial Hydrodesulphurization

INSPECTION	Coker Naphtha		Coker Kerosene		Coker Gas Oil	
	Feed	Product	Feed	Product	Feed	Product
Wt.% of Coker Distillate	16.5		13.7		51.8	
Vol.% of Syn.Crude		30.8		27.2		42.0
Gravity, °API	46.8	55.3	32.9	38.6	18.3	27.5
Sulfur, ppm	22000	15	27000	50	38000	410
Nitrogen, Total, ppm	150	2*	400	50*	2000	500*
FIA, Vol.%						
Aromatics	19		39.2	12.7	62.1	25.3
Olefins	32		14.4		2.0	
Paraffins & Naphthenes	49		46.4		35.9	
Distillation, °F						
IBP	180	162	380	358	515	498
5%	202	194	396	385	530	526
10%	220	206	409	398	550	540
30%	268	238	428	418	600	568
50%	295	278	441	438	645	588
70%	314	316	458	460	697	615
90%	347	369	477	496	780	675
95%	360	396	490	513	807	706
EP	400	462	535	533	850	715

*Typical Properties

In general, distillates are vaporized and passed down over a cobalt-molybdenum catalyst along with hydrogen at 800 to 1500 psig and 600 to 800 degrees F. Heavier feeds require more severe conditions. Hydrogen consumption is a function of the amount of sulphur and nitrogen compounds to be removed as well as the degree of required saturation of olefins and aromatics in the feed. Effluent from the reactor is stripped of H_2S, NH_3, and C_4, minus hydrocarbons. Light hydrocarbon liquids are fractionated, and then combined in their appropriate product pools, eg., naphtha from gas oil hydrotreater to the naphtha pool.

Ancillary Units

Ancillary units within upgrading include H_2S removal by amine-scrubbing the tail gas from primary conversion units and from hydrogen processing units. The H_2S is fed to standard Claus type units for greater than 95 percent sulphur recovery. The scrubbed tail gas may be used as fuel gas or used as feedstock in the hydrogen plant. Depending on the balance between process tail gas, fuel gas, and hydrogen required, some operations will import natural gas to supplement hydrogen production. This route will allow simpler processing with less capital cost. It will also allow higher overall synthetic crude yield from bitumen. However, the energy input per barrel of bitumen feed is greater with natural gas input,so that the overall thermal efficiency is about the same.

The residual from upgrading can be in the form of asphaltic liquid or coke. Because of its fuel value, it is desirable to use this material. Burning as fuel is difficult because of its high sulphur and metals content. In order to meet sulphur emission standards, it will be necessary to reduce sulphur content, but technology is not available to desulphurize this in liquid or solid state. The choice is: (1) to desulphurize stack flue gas, or (2) to gasify the residual to low Btu gas, remove H_2S and SO_2, and utilize the fuel gas. As reported, there are conflicting ideas on the commercial viability of flue gas clean-up. Also the high metals of the residual still present some firing problems.

Gasification is a process applicable to liquid hydracarbons, petroleum coke, and coal. In general, carbon and steam are converted to CO and H_2 gas at temperatures exceeding 1700 degrees F. The gasification is followed by shift conversion (to desired CO/H_2 mix) and purification (CO_2 and sulphur removal). Gasification systems could be installed with any conversion process discussed, as "add-on" units.

The fluid coking process has been advanced by Exxon Research and Engineering to a "Flexicoker", which integrates gasification of coke within the fluid solids system. This concept should result in some heat savings. A simplified flowsheet is presented in Figure 14. All gasification units will leave an ash residue which contains essentially all feed metals, which may be recovered at some later date when technology and economics improve.

As stated initially, the components of synthetic crude are fed to downstream refiniries for further processing. While some material may be directly marketable, the naphtha requires reforming to a suitable gasoline component and the heavy gas oil can achieve increased value by further con-

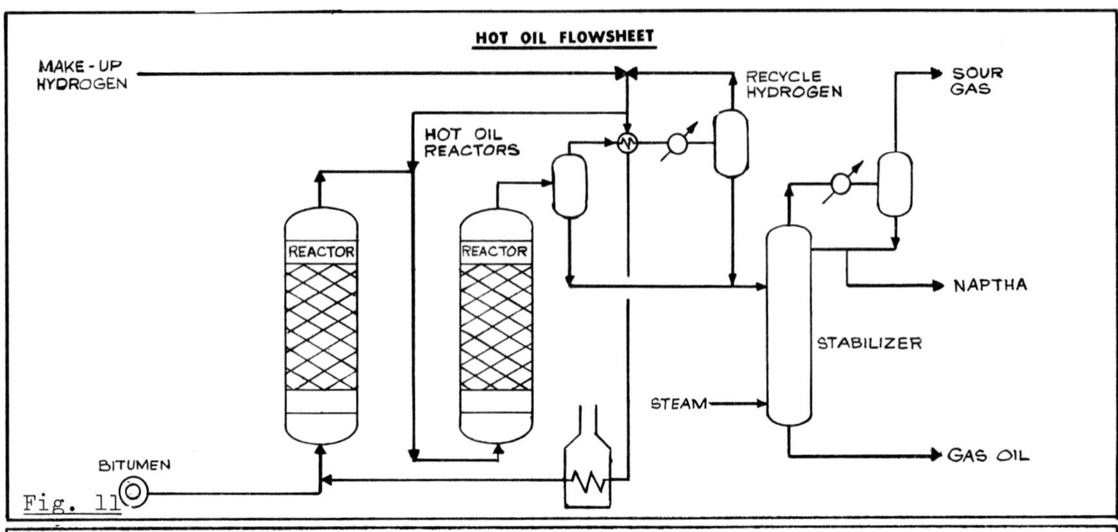

Fig. 11

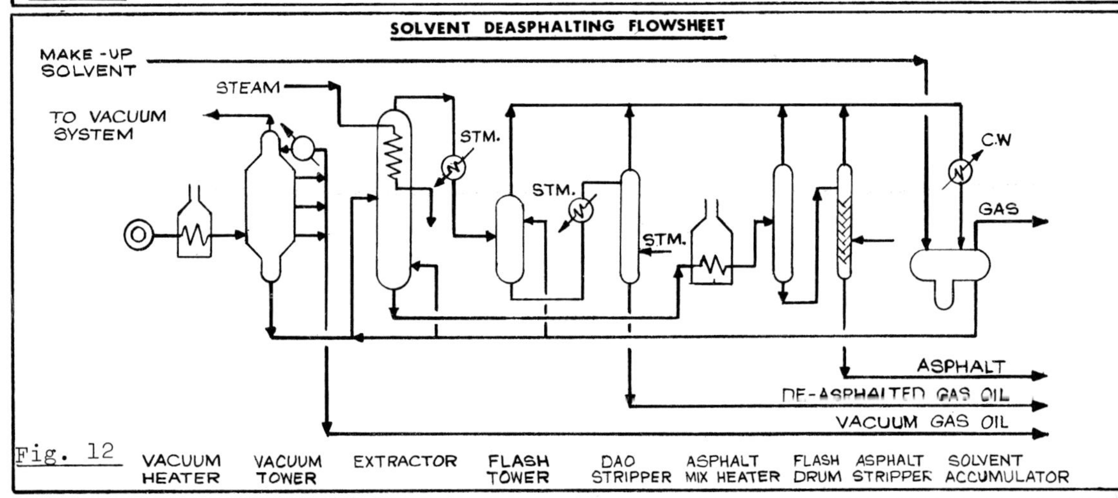

Fig. 12

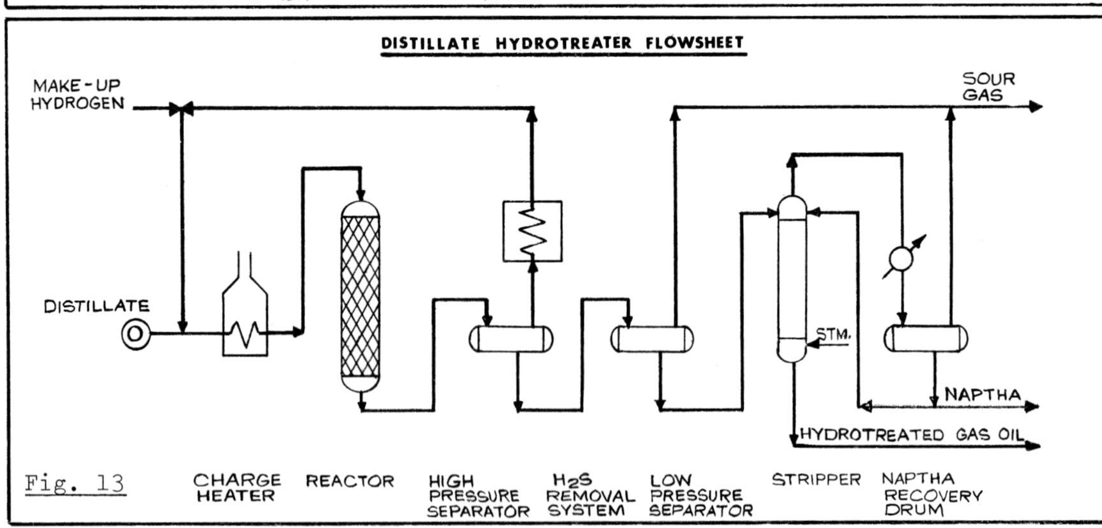

Fig. 13

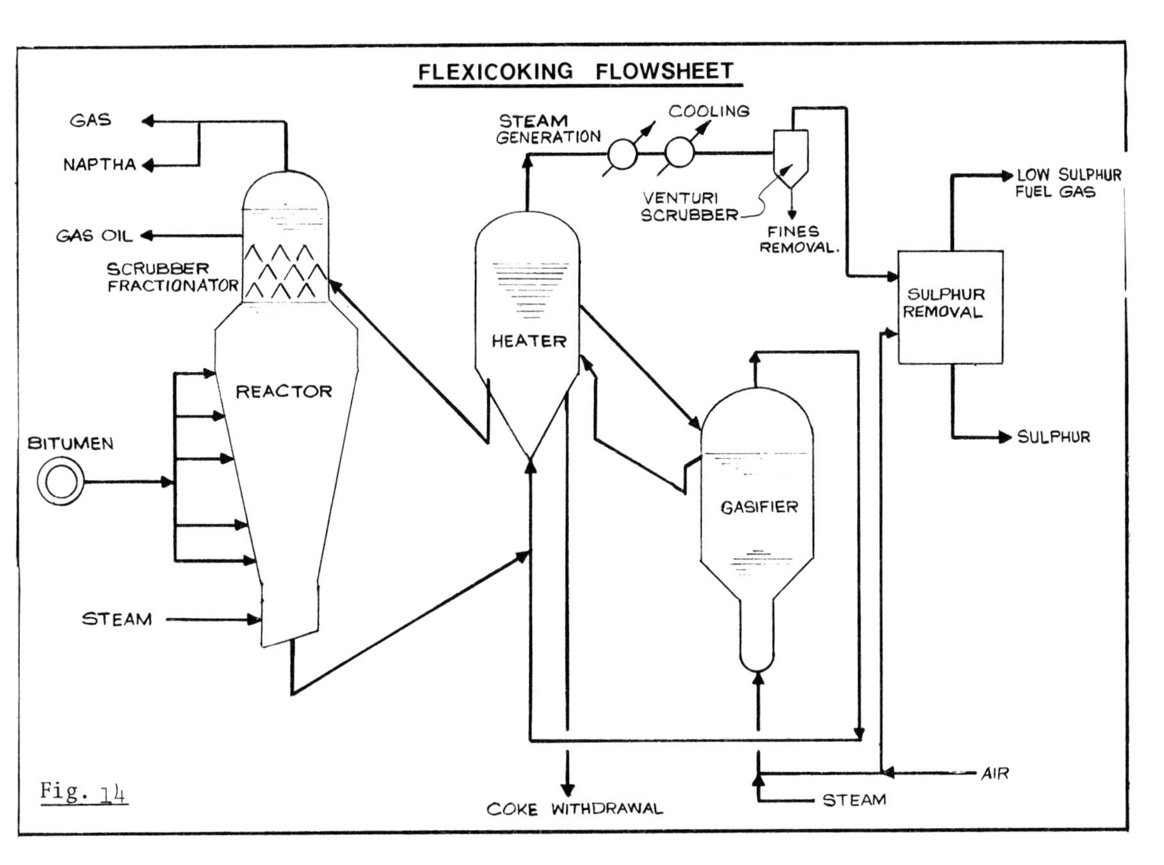

Fig. 14

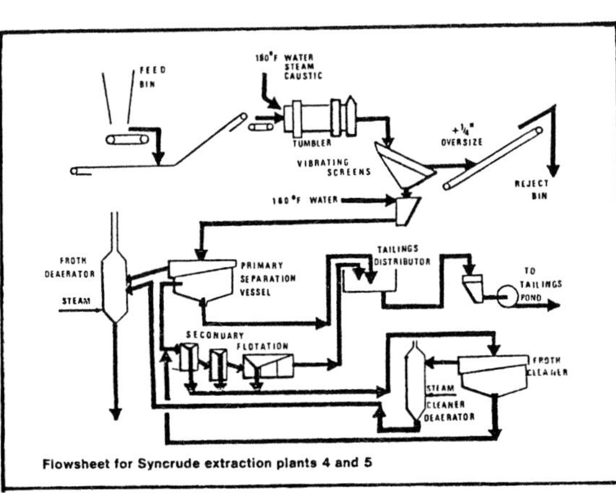

Flowsheet for Syncrude extraction plants 4 and 5

Fig. 15

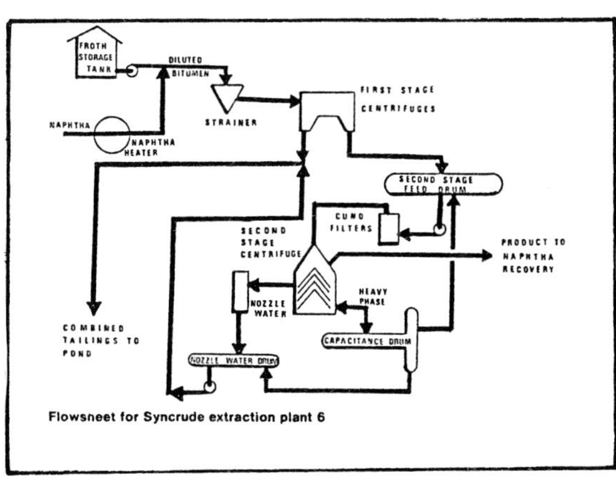

Flowsneet for Syncrude extraction plant 6

Fig. 16

version, eg., catalytic cracking. Pilot plant data shows that catalytic cracking of Suncor heavy gas oil (500 to 850 degrees) yields about 16 percent more C5+ gasoline than gas oil (500 to 980) from conventional crude, with less coke made. Suncor naphtha was reformed in a pilot plant with slightly higher yield than conventional naphtha below 95 RON-CI severity and slightly lower yield above 95. However, data is limited and the synthetic/conventional results are within about 1.5 percent.

THIRD-GENERATION PHILOSOPHIES

In the Alsands proposals for their new integrated complex many factors were considered in the selection process for the upgrading as well as the utilities configurations. Amongst these, primary components were the efficiency of resource utilization and the avoidance of the stockpiling of energy by-products.

The final selected scheme was based on Fluid Coking with Coke Gasification for hydrogen manufacture. Although coke gasification technology would not be available for initial plant design it was anticipated that incorporation could be effected after start up. Initial plant design therefore included gas oil or natural gas gasification.

In this way, the proposals by Alsands came closest to meeting efficiently any environmental criteria.

Representing, therefore, a third generation plant, the Alsands proposals probably provide the most advanced technologies for the upgrading phase of operations.

The process unit descriptions and objectives are described in the following text.

Diluent Recovery Units (DRU)

The purpose of these facilities is to produce a dry bitumen from the diluent bitumen/water mixture that is fed to them from the extraction plant, and to recycle the recovered diluent back to the extraction unit. The bitumen enters the unit diluted with an equal volume of naphtha and carries residual clay and water. The diluent is produced on the Naphtha Hydrotreater; its make-up rate to offset extraction plant losses is about 475 m^3/d (3000 b/d).

The feed rate to each Diluent Recovery Unit is approximately 28,620 m^3/d (180,000 b/d) and the separation of water, naphtha, and bitumen in a distillation process is done in a carefully staged heating and flashing operation. Recovered water is collected and routed to the sour water treating facilities and the dry bitumen product flows directly to the Fluid Cokers, which balance on intermediate bitumen storage.

Fluid Coking

First commercialized in 1954, this process now has a total capacity exceeding 47,700 m^3/d (300,000 b/d) in at least 11 units. This includes the two 11,600 m^3/d (73,000 b/d) units at Syncrude. Conceptually, the process is similar to the universally employed Fluid Catalytic Cracking Process; the difference is that a fluidized bed of coke granules formed in the coking process is used, rather than catalyst.

Bitumen feedstock from the Diluent Recovery Unit is fed directly into the reactor. The subsequent coking or cracking reaction produces a wide range of hydrocarbon products and causes additional coke to form in the fluidized bed.

The Fluid Coker is limited to a maximum product endpoint dictated by the maximum operating temperature on the overhead scrubber. Higher temperatures would result in unacceptable coke build-up. Coker gas and naphtha are routed to the Gas Recovery Unit while light and heavy gas oil streams are sent to downstream Hydrotreaters. Because of the unstable nature of the Coker product streams, there is a minimum of intermediate storage.

The CO Boiler acts as the incinerator for both Fluid Coker burner gas and the tail gas from the Sulphur Plant and generates some 3800 t/d (350,000 lbs. hr.) of steam on each unit. Fines entrained through the burner cyclones to the CO Boiler are removed by electrostatic precipitators on the CO Boiler flue gas. The fines from the electrostatic precipitators are collected and transferred to the coke storage hoppers.

The net product coke from the burner is cooled in the quench elutriator by a stream of water and steam. The vapor produced tends to sweep the small coke particles back into the burner, resulting in a larger particle size net coke. This is transported pneumatically to storage hoppers from where it can be transferred to storage, returned to the unit, or sent to Coke Gasification in the ultimate phase of this plant.

Hydrogen Manufacture

The $4.29 \times 10^6 m^3/d$ (152×10^6 SCF/D) of hydrogen demand for the hydrotreating units will be met initially by Partial Oxidation (POX) of Coker heavy gas oil and, in the ultimate configuration when 4.81×10^6 m^3/d (170×10^6 SCF/D) is required, by the Partial Oxidation of fluid coke. Also under consideration for the initial phase, while the Coke Gasification technology is being proven, is the use of natural gas.

Initial plant design will be predicated on ultimate operation using coke, although equipment required solely for coke would be installed as needed. The gasifiers designed for coke will have oil firing capability. This will permit an orderly commissioning of these units as well as minimizing interruptions to hydrogen production in the case of Fluid Coker upsets.

In Oil Gasification, hydrocarbon feedstock is partially oxidized (combusted) in relatively small, high capacity reactors. As high purity hydrogen is required, oxygen must be used instead of air. Oxygen, steam, and hydrocarbon react at high temperatures in a refractory-lined combustor to produce carbon monoxide and hydrogen. After cooling and carbon clean-up, the raw gas enters a conventional work-up system, first to remove the hydrogen sulphide, then to convert or shift the carbon monoxide to carbon dioxide with the release of additional hydrogen and finally to remove the large quantities of carbon dioxide.

The methanization step combines hydrogen with trace carbon oxides which have passed through the system, the latter being a poison to the downstream hydrotreating catalysts.

A small amount of carbon is produced in the POX reactor on oil feedstock. The reaction section of the unit includes carbon recovery and recycle facilities for the oil operation.

For the proposed future coke operation, new or add-on facilities required include a coke grinding and feeding system. In addition, the reactors and waste heat exchangers will be replaced or modified on conversion to coke, and slag/ash removal and handling facilities added.

The plants have a high environmental acceptability, with no new sulphur emission source associated with the Coke Gasification. The hydrogen sulphide is routed to the Sulphur Plant and net water from the carbon scrubber section is small in volume and amenable to treatment. In the Coke Gasification design, there are no tars or phenols produced and the ash in the feedstock is removed as an inert slag.

Oxygen Plant

The oxygen required by the Partial Oxidation Unit is produced by the low temperature separation of oxygen from air. The principal operations in this process are air compression, removal of water and carbon dioxide, expansion refrigeration, separation and rectification via distillation, removal of trace carbon dioxide and hydrocarbon, and product oxygen compression. The design and size of the plant is similar to packaged oxygen units in operation at major steel mills.

Total oxygen demand of 1,780 t/sd (1,750 LT/SD) in the initial gas oil phase of operation, and 2,240 t/sd (2,205 LT/SD) in the ultimate coke case, is split between two parallel trains.

Hydrotreaters

Hydrotreatment of various crude oil components is required to reduce sulphur, nitrogen, olefin, and aromatic concentrations to acceptable levels to produce a product suitable for downstream refining operations.

The Naphtha Hydrotreating Process combines hydrogen and naphtha in vapor phase over fixed bed catalysts with staging to obtain the optimum conditions for the specific chemical reactions in each reactor. Butanes/butylenes, coker naphtha, and cascade naphtha from the Gas Oil Distillate Hydrotreaters make up the feed to two parallel Naphtha Hydrotreaters, each producing 3,340 m^3/d (21,000 b/D) of hydrotreated butane and heavier liquid in the ultimate configuration. A Diluent Preparation Unit, common to both Naphtha Hydrotreaters, separates a narrow boiling range naphtha which is recycled to the extraction plant as diluent make-up. The remaining hydrotreated butane plus product is blended to synthetic crude. The naphtha is suitable for feed to conventional refinery Catalytic Reforming Units.

The Distillate/Gas Oil Hydrotreating processes combine hydrogen with distillates or heavy oils predominantly in the liquid phase over fixed bed catalysts, normally at higher pressures and residence times than for lighter feed.

While the economics of alternative upgrading configurations have been evaluated on the basis of Naphtha Hydrotreating and combined Distillate/Gas Oil Hydrotreating, it is possible that a separate Hydrotreater facility, treating a light distillate cut, could contribute to subsequent optimization, in the interests of better synthetic crude quality, more efficient hydrogen management, and internal optimization of the Hydrotreating design package.

Total volume of distillate and gas oil hydrotreated product would range from 17,500 to 19,100 m^3/d (110,000 to 120,000 b/d). Characteristic of hydrotreating heavy oil fractions is the generation of a cascade naphtha which is high in aromatics and low in other qualities. These streams are sent to the Gas Recovery Unit and subsequently to the Naphtha Hydrotreaters for retreatment. This will ensure that the final naphtha product meets product quality sulphur and nitrogen criteria. There would also be a small propane and lighter gas production which is routed to the Gas Recovery Unit.

Recycle gas is scrubbed with amine solution to remove hydrogen sulphide and the rich amine is returned to a common regeneration system for sulphur recovery.

Gas Recovery/Treaters

Feeds to the Gas Recovery/Treaters sections include the following streams:

- Gas and naphtha from the Fluid Coker
- Gas and cascade naphtha, from the Distillate/Gas Oil Hydrotreaters
- Gas from the Naphtha Hydrotreaters
- Amine treating solution cascaded from the Tail Gas SCOT Absorber

The Gas Recovery fractionation system is designed to produce the following products:

- Ethane and lighter gas which pass to the Fuel Gas Treater
- Liquid propane/propylene (LPG) which is routed to storage after amine treating, caustic treating, and extractive Merox treating
- Liquid butanes, butylenes, and naphtha which pass to the Naphtha Hydrotreater

The Gas Recovery fractionation heat requirements are met in large part by a circulating stream of Fluid Coker fractionator gas oil.

Sulphur Recovery

Acid gases from the Gas Recovery/Treaters and the Hydrogen Plant work-up section are routed to the Sulphur Plant. The Sulphur Recovery system consists of two parallel trains comprising three-stage Claus units followed by SCOT (Shell Claus Offgas Treating) units. Each unit will normally operate on 50 percent of the plant throughput. The design Sulphur Recovery for the Claus unit is 96 percent and, when combined with the SCOT unit, the overall Sulphur Recovery efficiency will be increased to 99.9 percent.

Each Sulphur Plant will be designed for 60 percent of total production for the ultimate case when coke is being gasified. As a result, during the initial phase, each plant will have a capacity of 75 percent of total production; therefore, the greatest capacity margin will be available during the commissioning period.

The conventional Claus process is widely used. The SCOT process is a more recent (1972) development, designed to recover sulphur compounds which have passed through the Claus plant. These compounds are reduced over a cobalt/molybdenum catalyst to hydrogen sulphide which is then recovered in an amine absorber. Hydrogen sulphide stripped from the amine is then recycled to Claus Plant feed. The tail gas which remains after hydrogen sulphide absorption contains 200 to 300 ppmv hydrogen sulphide, and is routed to the CO boiler for incineration of the hydrogen sulphide to sulphur dioxide.

Sour Water Handling

Process sour water containing ammonia and hydrogen sulphide is collected from the Diluent Recovery, Fluid Coker, Naphtha Hydrotreater, Gas Oil Hydrotreater, and Gas Recovery Units and sent to the main sour water stripper. In addition, the Partial Oxidation Unit (POX) has its own dedicated sour water stripper.

The stripped gas is sent to the Sulphur Plant and the stripped water is routed to the recycle water storage basin at the rate of 7360 m^3/d (1,350 USGPM) from the main stripper and 680 m3/d (125 USGPM) from the POX stripper.

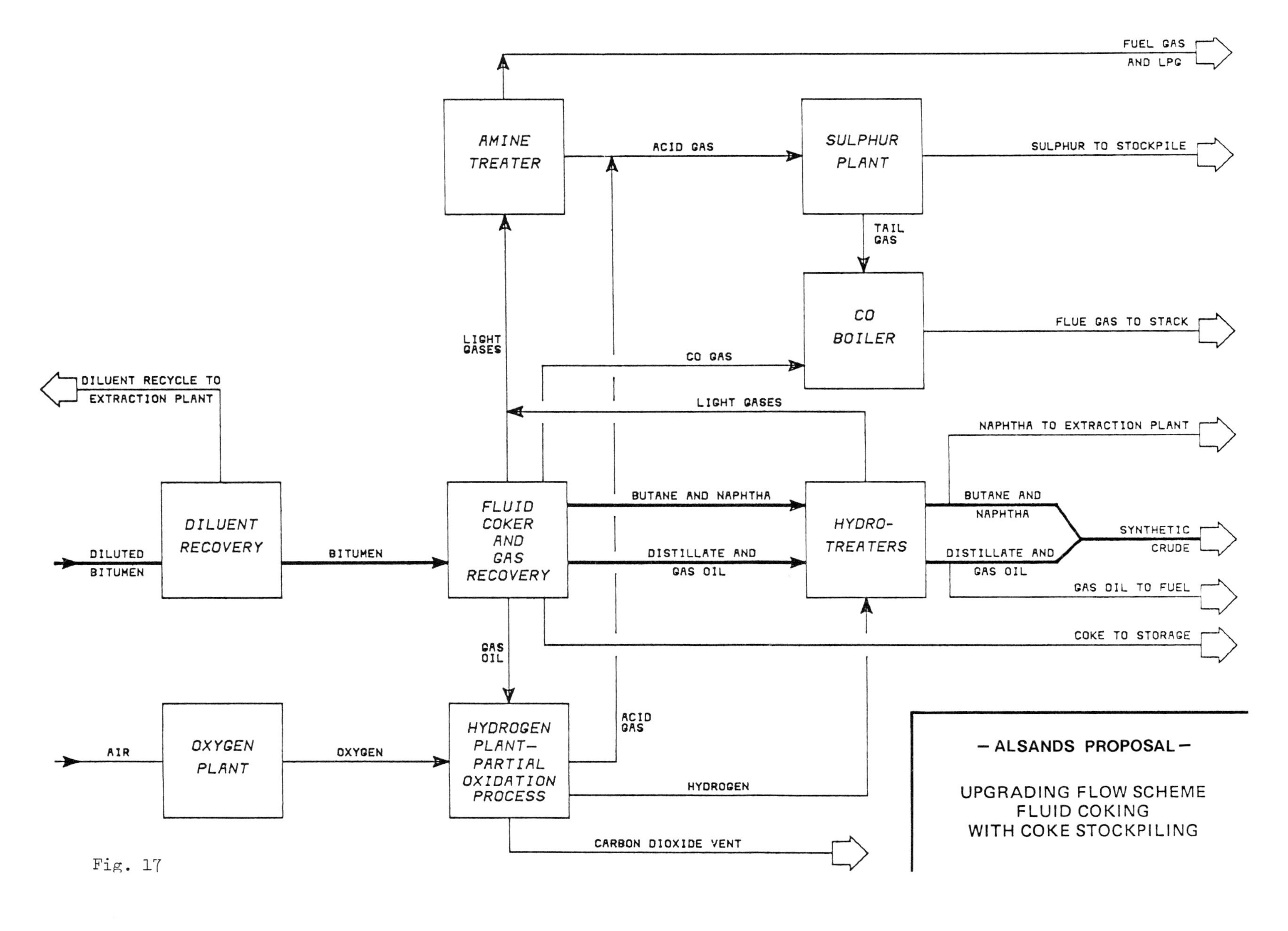

FUEL GAS
AND LPG
AMINE TREATER
ACID GAS
SULPHUR PLANT
SULPHUR TO STOCKPILE
TAIL GAS
CO BOILER
FLUE GAS TO STACK
LIGHT GASES
CO GAS
DILUENT RECYCLE TO
EXTRACTION PLANT
LIGHT GASES
NAPHTHA TO EXTRACTION PLANT
DILUTED
BITUMEN
DILUENT RECOVERY
BITUMEN
FLUID COKER AND GAS RECOVERY
BUTANE AND NAPHTHA
DISTILLATE AND
GAS OIL
HYDRO-TREATERS
BUTANE AND
NAPHTHA
DISTILLATE AND
GAS OIL
SYNTHETIC
CRUDE
GAS OIL TO FUEL
COKE TO STORAGE
GAS OIL
AIR
OXYGEN PLANT
OXYGEN
HYDROGEN PLANT– PARTIAL OXIDATION PROCESS
ACID GAS
HYDROGEN
CARBON DIOXIDE VENT
– ALSANDS PROPOSAL –
UPGRADING FLOW SCHEME
FLUID COKING
WITH COKE STOCKPILING

Fig. 17

7. CURRENT EXTRACTION PROBLEMS

The isolation of existing problems within this new industry to those only concerned with the bitumen extraction component is a logical one. It is one area, as the reasons will later unfold, where the greatest impact of new technology could be felt.

The Clark Hot Water Process, despite its very real success, possess distinct limitations of efficiency — a bitumen recovery capability of little more than 90% — as well as imposing a severe and costly need for the containment of contaminated waste product in the so-called tailings.

In the Syncrude facilities, tailings containment is provided in a man-made lake occupying an area of 30 sq. km (11.7 sq. mi.). The Alsands proposal considers, for the 25-year life of the plant, a tailings pond area of 20.8 sq. km (8.1 sq. mi.). Dyke heights for the Alsands case will range from 35 m (115 ft.) to 55 m (180 ft.) and, where a minimum of 8 years of mining operation is required to provide sufficient tailings sand to reach the required dyke height. The 8-year period is consistent with the span of time necessary before any part of the mined area could be additionally utilized for this purpose.

Initial construction, the maintenance, and the extension of these dykes, constitute a costly and continuous exercise in order to cope with both plant and environmental requirements.

From another point of view, large areas of mineable deposits are rendered to be non-recoverable by this necessity for tailings ponds.

Again, in the Syncrude configuration, some 7 miles (11.2 km) of 24-inch pipeline are necessary to transport the tailings to the pond. Five separate lines are used in this service and a significant maintenance involvement is incurred where line renewals are contemplated every 15 months due to the abrasive nature of the transported sludge. Within this period the lines are turned so as to equalize the wear on the bore of the pipe.

The Alberta Energy Resources Conservation Board, in its response to the Alsands proposal, made it clear that the Clark process raised serious environmental issues and that it also possessed a high water requirement.

In terms of the environmental issue, the process would produce 55,000 USGPM of tailings consisting of water slurry with sand and sludge (at 45-50 percent solids content) containing losses of bitumen and diluent as well as being further contaminated with caustic.

Tailings ponds constitute a considerable hazard to man, beast, and bird and provide a sizeable burden of maintenance responsibilities that a 20th century industry can well do without.

While it is recognised that the need for improvement spans all aspects of this industry, as with any other, there is no doubt that an extraction process — capable of a much higher bitumen recovery rate; possessing a lower diluent loss; producing a dry tailings or waste product, dispatched directly into the hole in the ground from whence it came — will be of important benefit to the overall productivity, efficiency, and acceptability of the integrated tar sand process plant scene.

8. AN OVERVIEW OF ACTIVE RESEARCH AND DEVELOPMENT ACTIVITIES

This presentation has confined itself to a discussion of tar sand methods and technology in the surface mined sphere of activity.

Equally, the overview will relate to the same confines despite the very great activity currently underway in the in situ recovery field of technology. However, in the absence of direct involvement in in situ work it may be appropriate to say that the subject is another story, to be related at another symposium, at another time and, more importantly by another person.

Major activities of note, therefore, pertinent to this overview are described below.

8.1 Bitumen Extraction by Spherical Aglomeration

The author's company, MHG International Ltd., has for a number of years been associated with research and development activities on this process, which is the brainchild of the National Research Council of Canada, Chemistry Division.

In an 18-month project completed during 1978, a laboratory and small pilot plant project (1.0 ton / day) was used to successfully evaluate the technical feasibility of the process. Since November, 1979, the company has been involved in further research and development work to examine the commercial viability and potential of the process under the auspices of a federal government contract.

The principle of spherical agglomeration is not new, being widely used commercially, in the fertilizer manufacturing industry in particular.

In the bitumen extraction environment the process is solvent and is capable of bitumen recovery in excess of 96%. The tailings are dry and contaminated only by unrecovered bitumen. Water requirements are minimal. The process would be capable of tolerating less rich ores, thereby reducing the quantity of overburden that it would be necessary to remove from above the ore bed.

The commercial application is being pursued to satisfy three requirements:

(1) as a supplementary process to further refine and recover bitumen from the tailings streams of the Clark Hot Water Process;

(2) as a direct replacement system for the hot water process; and

(3) as a means of recovering bitumen from existing tailings ponds.

We are convinced that the ultimate requirement should be for its application as a direct replacement for the hot water process and supplementarily as a process to clean up and eliminate the hazards of the tailings ponds.

We believe also that the agglomeration component of the process will not be unlike the conditioning vessel of the hot water process and may not require to be as large to accommodate the same throughput of material.

Due to the use of solvent, however, containment will be an important part of the configuration to ensure minimal solvent losses.

In our deliberations on the application of the process, we believe its potential could be almost revolutionary in comparison with the established extraction process, and indeed, could have significant impact on both the mining and the material handling concepts as currently practiced.

It is conceivable, in the ultimate, that the extraction phase could be located at the mine itself, in units served directly by the prime excavating equipment. Conveying systems could be drastically reduced in length and the dry waste returned directly to the mine.

A flexibility of pipework, bringing solvent and water to the units and conveying product to storage (and further treatment) prior to upgrading, would replace the more expensive and extensive conveyor systems. Of course, tailings slurry pipework would be eliminated as would be the need for the dykes and ponds.

At this point in time, the latter is a conceptual challenge but it is endorsed by the trends evident from our ongoing research and development.

Considerable work remains to be done in order to produce an economic appraisal by the end of the third quarter of this year.

8.2 Dry Retorting Process for Extraction

The Alberta Oil Sands Research Authority (AOSTRA), a provincial government agency dedicated to the motivation of research in the industry, is pursuing research into the retorting of tar sands involving a proprietary rotating heated kiln.

Umatac Industrial Processes Limited is the specialist company involved who were responsible for the construction of a pilot plant, again in Calgary, during 1978. Test work on various hydrocarbon feedstocks are continuing at this plant.

AOSTRA are hopeful that the process may also be useful for upgrading heavy oil by using hot circulating sand as the heat carrier.

8.3 Dry Distillation Process

Again under the direction of AOSTRA, a study has been commissioned to conduct a preliminary evaluation of the comparative economics of the Lurgi-Ruhrgas direct dry distillation process and the most advanced hot water process.

Further to this study, the agency has also negotiated an agreement with Lurgi to carry out a test run on a pilot unit located in Essen, West Germany.

This latter test is expected to be completed in June, 1980, and it is intended that the results be used to determine if construction of a demonstration unit is warranted to test the commercial viability of the process for the extraction and primary conversion of bitumen from the Athabasca tar sands.

The research and development net is being cast far and wide in an endeavor by provincial and federal government agencies in the development of new and improved processes.

Equally, considerable resources are being applied by commercial corporations in the same quest and extensive work is being undertaken by their research arms as well as by sponsored programs in Canadian and other universities.

Without question, foreign investors, in particular from Japan, are showing significant interest in Canada's unconventional resource and to its ultimate availability as an alternative source of energy.

9. COSTS

Project labor and cost magnitudes and the phenominal size of these tar sand projects are undoubtedly realised from the statements of bulk materials required to be mined or handled for processing: In both the Suncor and Syncrude plants a requirement, in round numbers, of 2 tons of excavated material is necessary to produce one single barrel of synthetic crude oil.

In terms of project costs and the input of both engineering and construction man-hours the figures are equally staggering.

9.1 Syncrude Canada Ltd.

This plant, completed in 1978, reflected the following statistics:

o Total erected cost, plant (ie., Mine, Extraction, Upgrading)	... $ 2.5 Billion (Can)
o Infrastructure (i.e., New Town, Roads, Bridges, etc.)	... $ 1.0 Billion (Can)
o Engineering/Design Man-hours	... 5.5 Million
o Construction Man-hours	... 43.0 Million
o Project Duration	... 5 Years

9.2 Alsands Project

For a plant of almost identical capacity and configuration to the Syncrude facilities, the following statistics are envisioned:

o Total project cost (including infrastructure)	... $ 5.5 Billion (Can)
o Engineering/Design Man-hours	... 6.0 Million
o Construction Man-hours	... 35.0 Million
o Project Duration	5+Years

9.3 Esso Resources Canada Ltd (Cold Lake)

The magnitude of this first commercial in situ recovery project is estimated as follows:

- o Total project cost ... $ 6.6 Billion (Can)
- o Engineering/Design Man-hours ... 8.0 Million
- o Construction Man-hours 45.0 Million
- o Project duration 5 + Years

All these projects are horrendous in both work content and cost. Where the two mined tar sand projects are compared, it is of significance that a lag of some six years will have doubled project costs.

While it is now almost inevitable that both the Alsands and the Esso Colk Lake projects will be proceeding simultaneously, considerable demands will be imposed upon both Canadian and U.S. resources to provide engineering and construction manpower equal to the challenge.

ACKNOWLEDGEMENTS

The author is indebted to his many colleagues within MHG International Ltd. for their interest, specialized contributions, and support in the preparation of this work.

SYNTHETIC FUELS FROM COAL

J.R. BOWDEN AND E. GORIN

Conoco Coal Development Company

ABSTRACT

This paper presents an overview of the technical and engineering status of the various coal-based synthetic fuel processes, in particular:

- o Gasification
- o Direct Liquefaction
- o Indirect Liquefaction

Emphasis in the paper is given to an assessment of current developments. The impact of the nature of the coal feedstock on the performance of processes described is discussed where appropriate information is available.

Comparative information on process yields and thermal efficiencies are given for the major processes. These results are based on projected commercial performance wherever adequately supported by available pilot plant performance.

The paper does not treat the relative economics of various competing processes, because these cannot be generally assessed due to the differing product yields in each process and the dependence of results for each process on the specific plant sites and the different types of available coal feedstocks.

ISBN 0-12-467101-2

1. NATURE OF COAL

Coal has been formed from plant materials in an environment and under conditions where it has been preserved from complete decay by biochemical processes.

Peat is generally considered to be the first stage in coalification. The coalification process proceeds through metamorphosis of the original peat to coal of successively higher rank from lignite through subbituminous to bituminous and anthracite. The metamorphosis proceeds through the action of geological forces of time, heat,and pressure.

Coal rank is usually defined in terms of its elemental composition. The general trend is for coals to lose oxygen and become richer in carbon as the coals move up in rank. Parallel with this is a reduction in inherent moisture with increasing rank. Inherent moisture may be defined as the moisture remaining after drying in ambient air.

The hydrogen content tends to peak out slightly in the bituminous range and then decreases again as one moves up to anthracite.

The above generalities are illustrated for some typical U.S. coals by analyses given in Table 1 (Whitehurst, 1977).

A number of formalized schemes for relating coal rank to chemical composition have been developed by Seyler and others. These have been reviewed in Lowry (Parks, 1962).

It is clear from the considerations on elemental analysis that coal as an organic material relative to petroleum is very rich in heteroatoms,such as,oxygen and nitrogen,and very deficient in hydrogen. The removal of oxygen, nitrogen,and sulfur consumes hydrogen while even more hydrogen is required to saturate the carbon structures to bring them up to the 12-14 wt % hydrogen required for most specification grade transportation fuels.

It is also clear that the lower rank coals will generally be more expensive to process to synthetic fuels than the bituminous coals. A larger thermal requirement is obviously required to dry the coal while more hydrogen is required to remove the heteroatoms and particularly the oxygen.

The chemical constitution of coal has been a subject of many research studies and considerable debates. The hydrogen deficient character alone strongly shows that it possesses a highly condensed cyclic structure. There is, of course, no such thing as a true structure for coal. Due to its highly diverse origin and history it consists of a mixture of literally thousands of compounds. The chemical constitution of coal can, therefore, only be represented statistically in terms of atomic types and functional groups. It is beyond the scope of this paper to review this voluminous information. Suffice to say that a large fraction of the carbon is present in aromatic rings (Vanderhart, 1976) while about 40-50% of the oxygen is phenolic (Roberto, R. G.,1977). Nitrogen and sulfur are present in heterocyclic ring structures.

Coal is a solid organic substance with a molecular weight in excess of 1500. The lower rank coals decompose thermally without undergoing fusion, at least at

TABLE 1

ULTIMATE AND PROXIMATE ANALYSES
U.S. COALS OF VARIOUS RANKS

State Seam	North Dakota	Wyoming	Illinois #6	West Virginia Pittsburgh #8
Rank	Lignite	Subbituminous	Bituminous	Bituminous
Mine	Velva	Wyodak	Burning Star	Loveridge
Ultimate Analysis - MAF Basis				
% C	71.40	75.39	80.10	85.51
% H	4.78	5.46	5.78	5.41
% O	22.58	17.97	10.65	5.97
% N	1.01	0.93	1.43	1.44
% S (organic)	0.23	0.24	2.04	1.67
Proximate Analysis - MF Basis				
% Moisture (as Rec'd.)	35.66	22.03	7.33	1.63
% Ash	5.43	3.63	9.77	10.29
% Vol. Matter	43.30	47.44	38.82	35.27
% Fixed Carbon	48.26	47.90	50.64	54.27

atmospheric pressure. Bituminous coals, however, melt to form a high viscosity liquid when heated to thermal decomposition temperatures of the order of 673°K. This phenomenon is of importance in many conversion processes.

Coals have also been classified by petrographic composition. This is a method of classification by microscopic examination analogous to the petrographic examination of inorganic rocks. One method (Parks, 1962) classifies the banded coal structure by reflectance measurements according to litho-types. The litho-types are analogous to rocks and consist of maceral groups which in turn are analogous to minerals in inorganic rocks. The three most important maceral groups (Parks, 1962) are vitrinites, exinites, and inertinite. The vitrinites and exinites are the reactive part of coal and comprise the major fraction of most coals. The inertinite contains such macerals as fusinite and semifusinite which are refractory and difficult to convert, particularly in liquefaction.

The mineral components in coal are also very complex but of importance in their subsequent processing. Some of the more important minerals are clay minerals, pyrite, quartz, gypsum, etc. (Ode, 1962). The above are all present as discrete particles distributed in various particle sizes in the coal matrix. The pyrite minerals tend to be more abundant in the bituminous coals and are important in that they possess catalytic properties for liquefaction.

There are also organically bound minerals, such as, alkaline earth and alkali metal salts of organic acids. These are especially important in low rank coals, such as, subbituminous coals and lignites. They can be troublesome in forming reactor deposits in coal conversion systems and in fouling of catalysts which are employed in some systems.

Also present are organically bound metals in the form of chelates of as yet undetermined structure. Examples are chelates of titania, alumina, and silica. These materials can also cause difficulties by deposition within catalyst pores in some coal conversion systems.

Chlorine is also present in most coals but in greatly varying amounts. The bulk of the chlorine is present as sodium chloride.

II. GASIFICATION

A. Definition and History

Gasification may be defined as a process for the conversion of coal into a combustible gas. Two sequential unit processes are involved which may be conducted separately or together depending upon the type of process or equipment employed. The first is in reality a pyrolysis process wherein the volatile matter is converted to a mixture of light fuel gases, oil and tar vapors, and a solid coke residue. The coke is then converted to fuel gases by reaction with steam by the water-gas reaction:

$$C + H_2O = CO + H_2$$

This reaction is endothermic and in practice heat must be supplied from either an internal or external source. The most usual, but by no means the only method, is to provide the heat by in situ combustion of part of the carbon with air or oxygen. If air is used, the product is usually referred to as producer gas, and if oxygen is used, the product is usually referred to as synthesis gas.

Producer gas has a relatively low calorific content in the range of 975-1600 $kcal/m^3$, and sometimes is referred to as low BTU gas. Its main application in the past has been as an industrial fuel gas. Strictly speaking, low BTU gases may not be regarded as a synthetic fuel since they cannot be readily stored or transported. The discussion that follows will accordingly de-emphasize processes of this type.

When oxygen or other means of supplying the endothermic heat of gasification are employed, the resulting gases are generally of a higher BTU content, in the range of 2250-4000 $kcal/m^3$. The higher figure can be achieved particularly in high pressure processes by virtue of in situ formation of hydrocarbon gases. Mechanisms by which hydrocarbon gases may be produced are discussed below.

Coal gasification, next to coal pyrolysis and coal combustion, is one of the oldest unit processes for conversion and utilization of coal. A very sizeable industry for the gasification of coal had been in operation in all of the advanced industrial countries prior to the advent of relatively low cost natural gas and petroleum fuels. The principal uses for the gases were as industrial fuel gases, town gas, and as feedstocks for production of NH_3-based fertilizer and chemicals. The gasification industry has now been abandoned in many countries but a sizeable revival is expected in the future. Even now a sizeable industry exists or is presently being constructed in a number of countries (Union of South Africa and India) which are rich in coal but lack oil or natural gas resources.

The development of gasification technology up to the year 1945 is reviewed in Lowry (van der Hoeven, 1945), and for the period 1945-1960 in the supplemental volume (von Fredersdorff, 1963). The discussion below will focus on developments of the last 25 years.

B. Fundamentals

An excellent review of gasification fundamentals is given in a recent book which has compiled the work of the late J. L. Johnson (Johnson, 1979). Reference should be made to this book for those interested in more details.

1. Thermodynamics of Carbon Gasification

The thermodynamics of gasification provides an inherent restraint on the design of a gasification process. The thermodynamics apply only to the second stage of the process referred to in section II.A., namely, the gasification of fixed carbon.

The first stage of gasification, or more broadly speaking, the devolatilization of the coal, involves complex and ill-defined chemical processes which cannot be treated thermodynamically.

The gasification of fixed carbon is treated thermodynamically as equivalent to graphite. This is conservative since, in practice, as has been observed (Johnson, 1979), the equilibria are somewhat more favorable for gasification of fixed carbon than graphite. In other words, the fixed carbon in coke is in a metastable condition relative to graphite. The simplified thermodynamic treatment also neglects the complications due to the presence of hydrogen, nitrogen, and sulfur in the fixed carbon.

The complete equilibrium in the gasification of carbon from the thermodynamic point of view may be defined by the simultaneous equilibria in the following four reactions:

(a) $C + O_2 = CO_2$

(b) $C + H_2O = CO + H_2$

(c) $CO + H_2O = CO_2 + H_2$

(d) $C + 2H_2 = CH_4$

The heats of reaction and equilibrium constants in the above reactions at two temperatures, 1100 and 1200°K, are given in Table 2.

It is immediately obvious from Table 2 that the principal limitations are the high endothermic reaction heat of the water-gas reaction (b) and the unfavorable equilibria in the formation of methane via reaction (d). The very favorable equilibria in carbon combustion and the high exothermic heat makes this the easiest method of supplying heat for fixed carbon gasification. Accordingly, air or oxygen injection is the most common method used in practical gasification systems. This method, however, has the disadvantage of reducing the methane and calorific content of the gas and in increasing the CO_2 content.

The ideal gasification process, at least where methane is the desired end product, is described by reaction (e) which in effect is a combination of reactions (b), (c), and (d).

(e) $2C + 2H_2O = CO_2 + CH_4$

It is noted in Table 2 that the above reaction is still slightly endothermic and the equilibrium is not highly favorable.

The equilibrium constant in reaction (e) increases very slightly with temperature. Low temperatures are thermodynamically more favorable for production of a methane rich gas, however, since the production of the other major components

TABLE 2

EQUILIBRIUM CONSTANTS K AND HEATS OF REACTION IN CARBON GASIFICATIONS

Temp., °K	1100		1200	
	ΔH (1)	K (2)	ΔH (1)	K (2)
$C + O_2 = CO_2$	-94.64	7.21×10^8	-94.66	5.52×10^7
$CO + H_2O = CO_2 + H_2$	-8.35	0.944	-8.06	0.697
$C + H_2O = CO + H_2$	+32.47	11.58	+32.45	39.77
$C + 2 H_2 = CH_4$	-21.65	0.0368	-21.79	0.0161
$2 C + 2 H_2O = CO_2 + CH_4$	+2.47	0.402	+2.60	0.446

(1) Enthalpy units in kilocalories/mol.

(2) Equilibrium constants pressure units in atmospheres.

CO and H_2, is enhanced by increasing temperature. Likewise, the equilibrium in reaction (e) is independent of pressure. In practice, however, for similar reasons the methane concentration at total equilibrium is increased by an increase in pressure.

Reaction (e) can be approached in practice by removing CO_2 and separating CH_4 from the effluent gases and recirculating the CO and H_2 until they are consumed.

The use of coal rather than coke as feed to such a system can become slightly exothermic. This results from the added hydrogen contained in the coal feed and the exothermic in situ reaction of hydrogen gas with the coal volatile matter to produce methane. The above is the basis for the Exxon catalytic gasification process to be discussed in section II.F.

Reaction (e) can also be approached in part, and the overall process made exothermic, by removal of the CO_2 in situ with a lime-bearing acceptor. This is also discussed in section II.E. under the CO_2 Acceptor Process.

2. Kinetics of Carbon Gasification

The kinetic limitations in fixed carbon gasification are such that equilibria in reaction (b), the water-gas reaction, cannot be approached except at very high temperatures or very long and impractical residence times. It is thus the controlling reaction in gasification systems. Water-gas shift reaction, (b), is relatively rapid and equilibrium generally is approached in most gasification systems.

The kinetics of gasification of carbon with steam and carbon dioxide has been studied by numerous investigators. Both pure graphite and fixed carbon associated with actual chars and cokes have been studied. The voluminous literature in this field has been exhaustively reviewed (Johnson, 1979, and von Fredersdorff, 1963). Only the salient conclusions are given:

> The water-gas reaction is highly inhibited by the reaction products, i.e., CO and H_2 and is, of course, rapidly accelerated by increase in temperature. The reaction is accelerated somewhat by an increase in pressure but the rate is largely determined by the partial pressures of the inhibitors relative to steam.
>
> The reaction rate is also a function of the coal rank, since reactivity increases with decreasing rank. The above increase in reactivity is due in large part to an increase in organically bound calcium and sodium in low rank coals. Calcium, in particular, has a strong catalytic effect on the water-gas reaction.

Practical gasification rates can be achieved at temperatures as low as 1100°K with lignites and subbituminous coals. Bituminous coal chars require operating temperatures in excess of 1200°K to achieve practical rates.

Coal chars generally have a large pore volume and internal surface area which is developed initially by devolatilization and enhanced by subsequent gasification with steam. The result is that the internal surface is made readily available and the rate is not strongly dependent upon particle size. The above is true at least for particles below about 1.5 mm in diameter and for temperatures below about 1225°K. The temperature at which diffusion control takes over is a very complex function of the operating conditions, nature of the feed coal, etc., and has never been adequately studied.

The gasification rate tends to fall off with carbon burnoff. This makes complete gasification difficult in a single stage fluidized bed system.

The inherent reactivity of the fixed carbon from a specific coal may be enhanced by addition of a catalyst. The literature in this field has also been reviewed (Johnson, 1979). The most effective common catalysts employed are calcium hydroxide (Chauhan, 1977) and potassium carbonate. The latter, in particular, is employed in the Exxon process to be discussed in section II.F.

The formation of methane by the interaction of hydrogen with coal and char was first studied by Dent (Dent, 1944) and more extensively by subsequent investigators (Johnson, 1979). Methane is rapidly formed in the devolatilization stage by interaction of hydrogen with the volatile matter. The yield of methane increases, of course, with the partial pressure of hydrogen.

The formation of methane by direct reaction of hydrogen with fixed carbon is, however, relatively slow and the rate is even less than that of the water-gas reaction.

It has been observed that the presence of steam enhances the rate of formation of methane in the C-H_2O-H_2 system (Zielke, 1957). A study of the kinetics of lignite char gasification indicated that methane is largely produced by simultaneous interaction of steam and hydrogen with carbon (Curran, 1968). That is,

(f) $C + H_2O + H_2 = CO + CH_4$

The reaction of oxygen with fixed carbon under ordinary gasification conditions is extremely rapid and the rate is predominantly mass transfer controlled. There is considerable evidence that the primary reaction is formation of CO. That is,

(g) $C + 1/2\ O_2 = CO$

and that the subsequent combustion of CO to CO_2 under certain conditions may be relatively slow (Bridger, 1948).

It is not at all unlikely in slagging type gasifiers that a large fraction of the CO is formed by reaction (g).

3. Influence of Coal Properties

There are a whole host of other coal properties that are of importance in designing and choosing a gasification system.

The fusion point of the ash has a strong influence on operability in fluidized bed oxygen-blown gasifiers and in the choice of dry ash or slagging type fixed bed gasifiers. It also dictates in certain cases the requirement for addition of lime as flux.

The plastic properties of the coal are of importance in choosing and designing a gasification system. Most bituminous coals exhibit plasticity when heated to devolatilization temperatures of 670°K and higher. The level of fluidity and its duration both increase with increasing total pressure and partial pressure of hydrogen. The plastic properties can cause operating difficulties due to particle agglomeration and coking.

The result of the above plastic phenomena is that special techniques, which may include mechanical devices, are required for processing bituminous coals in pressurized gasifiers. This is especially true for multistage or countercurrent type gasifiers.

The reactivity properties of coals as a function of rank as discussed previously, are also of importance in selecting a gasification system. Highly reactive coals are required for those processes that operate at low temperatures and are generally beneficial for non-slagging type gasifiers. Reactivity considerations are generally of relatively minor importance for high exit-temperature systems and for slagging systems in particular.

The aerodynamics of the coal particles is determined by their size, particle shape factor, and density. These have a bearing on the choice of contacting system as discussed below.

C. Generic Processes

There are three basic types of systems for contacting coal with the gasification medium and examples of all three will be discussed below in sections II.D., E., and F.

The three systems are cocurrent entrained phase, countercurrent fixed or moving bed, and fluidized bed. The fixed bed system can also be operated in a cocurrent manner but it is usually more advantageous to operate in the countercurrent mode.

All three systems may be operated at elevated pressures up to 30 or more atmospheres. The three systems differ largely in exit temperature and in particle size of the feed. The latter is dictated largely by aerodynamic considerations.

The entrained phase operates with coal dust carried in suspension in the gasification medium. The low concentration and cocurrent flow inherent in these systems requires a high effluent temperature of the order of 1650°K, or higher, to obtain efficient steam and carbon conversions. The oxygen requirements are generally high and the steam usage may be low. The latter follows since the principal reaction may be represented as follows:

$$CH_xO_y + \frac{(1-y)}{2} O_2 = CO + \frac{x}{2} H_2$$

A small particle size is required in the coal feed; that is, the coal must have a top size of less than about 200 microns.

The fluidized bed system will usually have outlet temperatures between about 1085 and 1300°K depending on the specific process and reactivity of the feed. The fluidized bed dynamics usually require a coal feed size between about 2.5 mm to 100 microns. Fine particles below about 100 microns are difficult to maintain in the bed due to excessive entrainment. It is generally difficult to obtain very high carbon conversions in single stage fluidized bed systems due to attrition and entrainment of carbon bearing fines. Ash-agglomerating fluidized bed systems may be an exception, as discussed below.

Moving bed gasifiers require a larger diameter feed in the range between 5 and 40 mm. The lower size limit may be relaxed somewhat when feeding caking coals due to agglomeration of the fines.

Two types of moving bed gasifiers may be distinguished on the basis of the operating temperature of the ash withdrawal system at the bottom of the gasifier. A dry ash system will operate at a temperature well below the ash fusion point or usually at less than about 1470°K. The slagging gasifier, of course, must operate at higher temperatures and above the ash fusion point. It is necessary to add flux in many cases to increase the fluidity of the slag.

It is clear that the dry bottom system will be more demanding of a reactive feedstock than the slagging system.

The most efficient system, especially where a high calorific gas is desired, is one wherein the methane content of the raw product gas is maximized. This makes it desirable to operate with a reduced exit temperature in an actual or simulated countercurrent manner. Methane yield is maximized by interaction of hydrogen with the volatile matter. The moving bed systems accomplish this feature "automatically". The other systems require two or more stages in series to accomplish this same result.

A drawback of the above "countercurrent" mode, however, is that the effluent gas contains considerable quantities of tars and phenols. This complicates the downstream recovery section. Some systems, such as the CO_2 Acceptor system, can still achieve high methane yields without tar formation. This is achieved by regulating the time and temperature in the gasification zone to destroy all of the tar.

Systems also differ in the manner in which the endothermic gasification heat is supplied. The most common practice is, of course, to use direct air or oxygen injection. Another system is to supply the heat via a circulating solid heat carrier. The CO_2 Acceptor system is a particular version of the heat carrier system where heat is supplied in addition by the chemical heat of the acceptor reaction:

$$CaO + CO_2 = CaCO_3$$

The Exxon catalytic system is unique in that an autothermal reaction is sustained by two expedients. First, the reaction temperature is lowered by the addition of a catalyst and second, the methane is separated cryogenically and the CO plus H_2 is recirculated to produce additional methane.

D. Commercial Systems

1. Lurgi Dry-Bottom Gasifier

The Lurgi gasifier was developed in 1927-1935 by Lurgi Gesellschaft für Warmetechnik GmbH of Frankfurt (Main), Germany. It was initially commercialized in 1936. Since then, over 100 Lurgi gasifiers have been constructed in commercial plants throughout the world (Rudolph, 1974). Many of these plants have been closed down, however, due to age or the ready availability of crude oil and natural gas. The Lurgi gasifier has been proposed for use by a number of U.S. gas transmission companies for large commercial plants to produce synthetic methane from coal. It is also the gasifier used to produce synthesis gas at the world's only commercial oil-from-coal plant, the SASOL complex near Johannesburg, South Africa.

Figure 1 is a schematic diagram of the Lurgi gasifier. It is a high-pressure (24-31 atm), fixed bed, non-slagging, steam-oxygen or steam-air gasifier. The coal and gases flow countercurrent to each other. The coal enters the top of the gasifier and steam and oxygen are introduced at the bottom. Crude synthesis gas leaves the top of the gasifier, and ash is removed from the bottom. Temperature at the top of the reactor is 590-870°K and at the bottom 1255-1475°K, depending upon the coal feed.

The coal feed to the gasifier is normally crushed and screened to a size range of 0.6 x 3.2 cm and maximum recommended fines content is 7 percent, preferably 3 percent. Smaller lumps down to 0.3 cm and larger lumps up to 6.4 cm have been successfully gasified in commercial installations, but to obtain good bed permeability for the upward flowing gases, size range should be limited to about sixfold.

Since the gasifier operates at 24-31 atm gauge pressure, pressurized lockhoppers are used to feed coal to the reactor and remove ash from it. These lockhoppers operate on a pressurizing/depressurizing cycle. Fully automated lockhoppers have been in operation at the Kellerman Power Plant in Luenen, Germany since 1971. A high pressure gasifier, up to 100 atm, has been under development at a 300 ton/day pilot plant at Dorsten, in the Ruhr area.

As coal enters the gasifier, a mechanical device distributes it uniformly across the reactor. The coal gradually moves downward as it is gasified. Steam and oxygen are introduced at the bottom of the gasifier to effect the gasification reactions. A revolving grate supports the coal bed, cleans out the ash, and distributes the steam-oxygen mixture. Excess steam is added to keep the ash from slagging.

The gasifier is contained in a water jacket which generates high pressure steam. The water jacket and attendant steam production minimize heat losses from the reactor.

Temperatures in the combustion zone at the bottom of the reactor are held at 1255-1475°K by the introduction of excess steam over that required for the steam-carbon reactions. The temperature must be controlled at this point to prevent the ash from melting and forming a slag. Particulate ash is necessary for proper operation of the revolving grate. The temperature limit and steam requirements for a specific coal are dictated by its ash fusion temperature--the higher the ash fusion temperature, the higher the permissible combustion temperature and the lower the steam requirements.

A revolving grate distributes the ash particles to its outside edges from which the ash falls into the ash lock. Ash is transferred to the ash receiver on a cyclic basis and is periodically removed from the system to disposal via a hydraulic sluicing system.

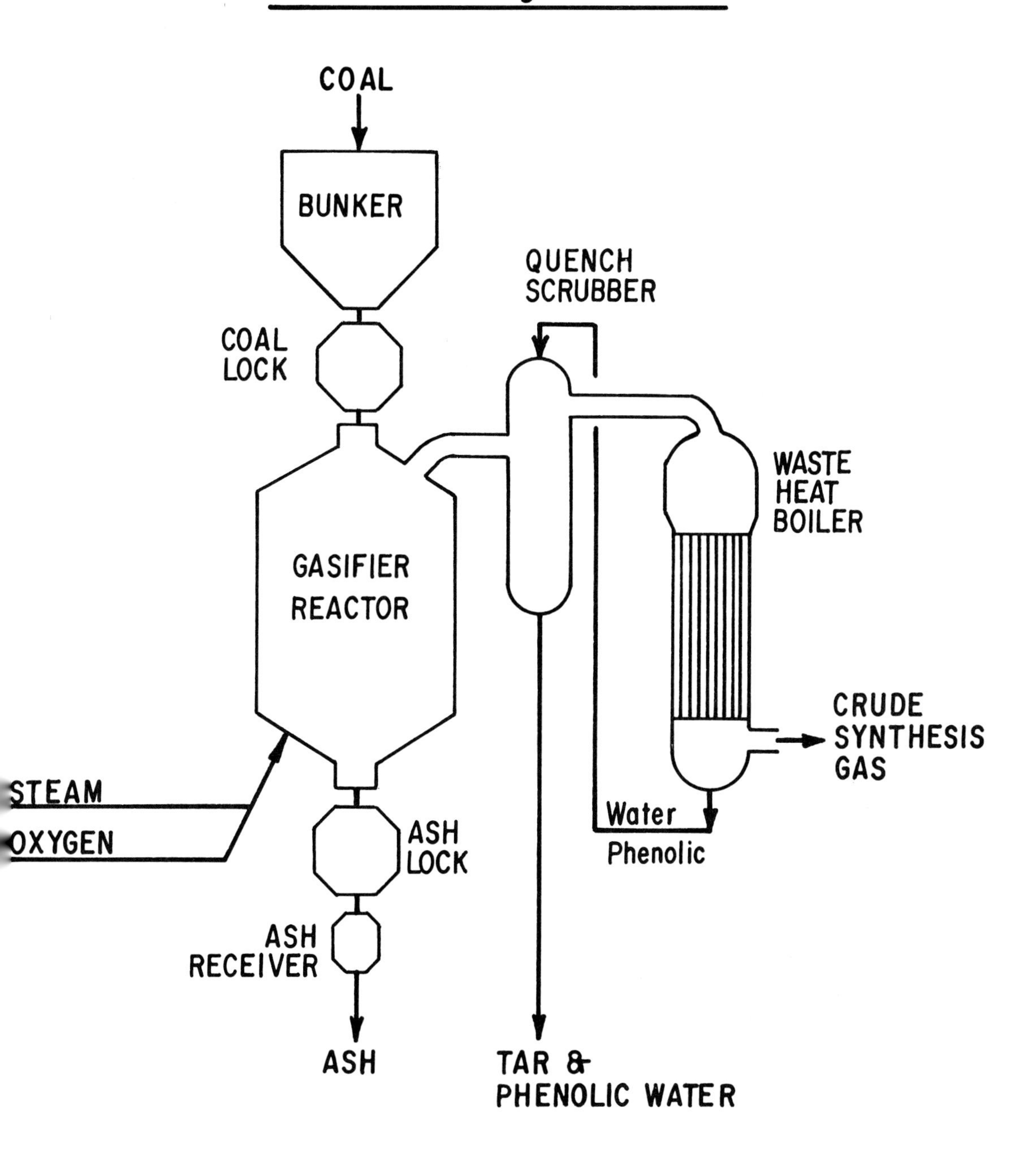
Sketch of Lurgi Gasifier
COAL
BUNKER
COAL
LOCK
QUENCH
SCRUBBER
GASIFIER
REACTOR
WASTE
HEAT
BOILER
CRUDE
SYNTHESIS
GAS
STEAM
OXYGEN
Water
Phenolic
ASH
LOCK
ASH
RECEIVER
ASH
TAR &
PHENOLIC WATER

Figure 1

The hot crude synthesis gas exits at the top of the gasification reactor at a temperature of 590-870°K and immediately enters a quench scrubber where it is contacted with a large quantity of a circulating phenolic water stream. The crude synthesis gas entering the quench vessel contains tar, oil, naphtha, ammonia, phenols, and coal dust. The tars are condensed and the particulate matter occludes to the tar. The water dissolves some of the phenols and ammonia. Tar, particulate matter, and phenolic water are separated from the gas and are withdrawn from the bottom of the quench vessel. Part of the tar and particulates are recycled to the top of the gasifier.

The gas exits the quench scrubber at about 475°K or less and passes through a waste heat boiler which further cools the gas. Low pressure steam is generated in the waste heat boiler. Additional phenolic water, ammonia, and light tar are condensed and removed from the gas in the waste heat boiler.

The Lurgi gasifier is suitable for producing both fuel gas and synthesis gas. Of the several SNG-from-coal plants that have been proposed in the U.S., the Great Plains Gasification Project in North Dakota is the only venture to receive approval by the Federal Energy Regulatory Commission (Foder, R. J., 1979). At this facility, 14 Lurgi gasifiers will generate 125 million cubic feet of pipeline-quality gas per day from local lignite. Total plant cost will be about $1.5 billion. Barring delays, construction will be complete in 1984.

Lurgi gasifiers have been operating since 1955 at the SASOL plant (Ricci, 1979) in South Africa, gasifying low-rank coals to synthesis gas for subsequent conversion to transportation fuels via the Fischer-Tropsch process. The present complex will be expanded in two stages to 85 operating gasifiers and a total product capacity of about 90,000 bbl/day by 1982. Synthesis gas from Lurgi gasifiers can also be used to manufacture chemicals, such as ammonia and methanol. Fuel gas from the gasifier may be used for industrial purposes or to provide fuel for combined cycle (gas turbine and steam turbine) power plants.

If equipped with a stirrer, the Lurgi gasifier can potentially handle (Rudolph, 1974) caking coals, but there is little commercial experience with processing moderately to highly caking coals, and there is probably some reduction in capacity for processing such coals. Also, steam and oxygen requirements for processing low reactivity, low ash fusion-point coals, such as Illinois No. 5 and Pittsburgh No. 8 coals, are relatively high.

Other disadvantages of the Lurgi gasifier are: (1) its size and capacity limitations require numerous units for commercial plants; (2) only a limited amount of fines can be tolerated at economically feasible gasification rates; (3) the demand for excess steam is a major source of heat loss; (4) the gasifier generates significant quantities of by-products which must be either used internally as fuel or sold; and (5) the synthesis gas from Lurgi gasifiers contains 8-15 volume percent methane which must be removed or reformed before the gas can be used for chemicals manufacture.

In addition to a wealth of commercial experience, Lurgi gasifiers benefit from construction of low alloy or ordinary steels, a relatively high turn-down ratio (as low as 30-40 percent of maximum capacity), a high methane content in the raw gas if fuel gas or SNG is desired, and the ability to handle high-ash and high-moisture content coals.

2. Koppers-Totzek Process

The Koppers-Totzek gasifier was developed by Heinrich Koppers GmbH of Essen, Germany under the direction of Dr. Friedrich Totzek. The first commercial plant was built in 1950 in Finland for the purpose of producing synthesis gas for ammonia manufacture. Currently, over 40 gasifiers have been built in industrial plants around the world. In 1978, large, four-headed gasifiers were commissioned in India to provide synthesis gas for two identical 900 ton/day ammonia plants. Similar gasifiers will provide feedstock for a 600 ton/day ammonia plant in Brazil, by 1983 (Wintrell, 1974) (Mitsak, 1975).

Figure 2 is a schematic diagram of the Koppers-Totzek coal gasification process. The gasifier is an atmospheric-pressure, entrained-bed, slagging, steam-oxygen gasifier. The gasifier operates at over 1750°K and yields a crude synthesis gas which is free of tars and other liquid hydrocarbons. Finely ground coal, steam, and oxygen are introduced near the bottom of the gasifier. The coal dust is entrained in the gasification medium and is swept upwards. Molten ash falls from the bottom of the gasifier into a water quench sluice. The gasifier is lined with 2-inch thick heat resistant ceramic material.

Coal feed to the K-T must be pulverized to a powdery state so that over 75 percent is finer than 75 microns. Any coal, regardless of its caking properties or ash fusion point, may be used providing its ash content is less than 40 percent. The coal must be dried to a moisture content of between 1 and 10 percent, depending upon the type of coal. After drying and pulverizing, the coal must be maintained under an inert atmosphere such as nitrogen to prevent spontaneous com bustion. The pulverized coal is transported in a fluidized state with the inert gas.

The dry pulverized coal is delivered to a storage bin and then is distributed to the service bins which are maintained under a slight nitrogen pressure. The coal falls by gravity to each feeder bin from which a screw conveyor conveys the coal into a stream of oxygen which feeds the coal into the gasifier. The proportions of oxygen and coal must be maintained at a constant ratio in order to produce a synthesis gas of uniform composition. The coal immediately ignites upon entering the gasifier. Temperatures in the combustion zone reach 2200°K or higher. Low pressure steam is admitted to the reactor in a manner which allows it to envelope the combustion zone. Steam addition, however, may not be necessary if the coal contains 8-10 percent moisture.

Each gasifier has 2 or 4 feed entry points. The feed entry points are termed "burners". The coal is entrained in the gases in the gasifier and is carried upward. Since the temperature in the gasifier above the combustion zone is about 1750-1920°K, gasification is very rapid. Over 95 percent of the carbon content of low rank coals is usually gasified.

About half of the ash in the coal is slagged and falls from the bottom of the gasifier into a water seal where it is quenched. The water-ash slurry is pumped to settling tanks.

The remainder of the ash is quenched with steam and passes overhead in particulate form with the crude synthesis gas. Sensible heat in the gas leaving the reactor is recovered in a waste heat boiler. Sufficient steam is generated to supply gasifier needs and to generate power for the oxygen plant.

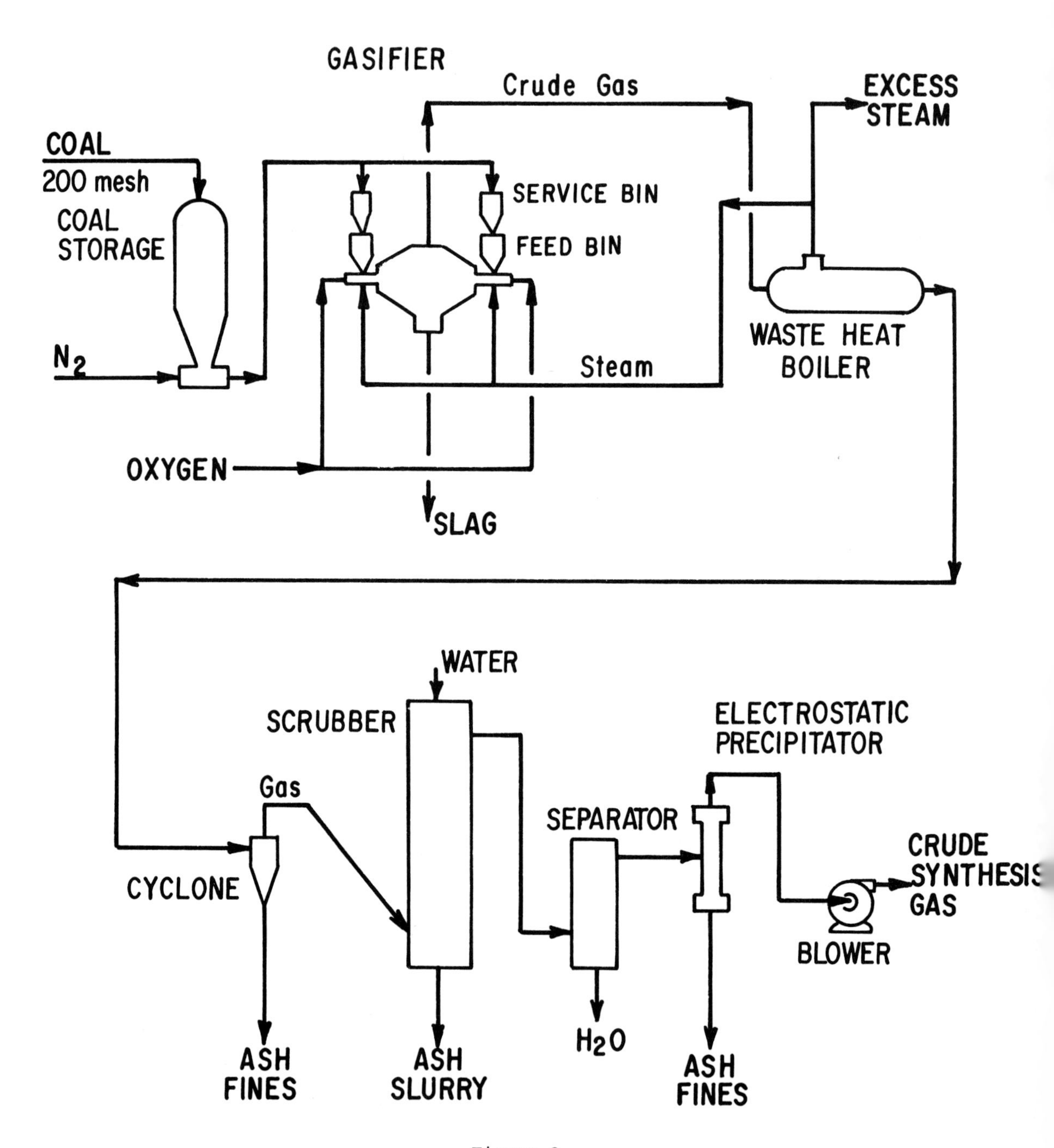
Koppers-Totzek Coal Gasification Process
GASIFIER
Crude Gas
EXCESS STEAM
COAL
200 mesh
COAL STORAGE
SERVICE BIN
FEED BIN
N_2
WASTE HEAT BOILER
Steam
OXYGEN
SLAG
WATER
SCRUBBER
ELECTROSTATIC PRECIPITATOR
Gas
SEPARATOR
CYCLONE
CRUDE SYNTHESIS GAS
BLOWER
ASH FINES
ASH SLURRY
H_2O
ASH FINES

Figure 2

The gas from the waste heat boiler is cleaned of particulate matter by passing through a cyclone, water scrubber, and electrostatic precipitator. A blower is included at the end of the gas clean-up train to raise the gas pressure from about 0.75 atm absolute to slightly above atmospheric pressure. The gas must be compressed for further processing.

The crude synthesis gas consists essentially of hydrogen, carbon monoxide, and carbon dioxide. Minor quantities of methane, nitrogen, and hydrogen sulfide are present.

Large, four-headed gasifiers with outputs up to 45 million SCFD of raw synthesis gas have been built. Coal feed rates to these units is 600-850 tons/day (Staege, 1979).

The ash from low rank coals usually contains 5 percent or less carbon. If its carbon content is high, the ash can be recycled to the gasifier or burned for fuel.

The K-T process requires about 30 percent more oxygen than the Lurgi process, but much less steam.

The drawbacks of the K-T coal gasification process include the following: (1) coal must be pulverized and dried; (2) loss of carbon to the ash can be significant, particularly with high rank coals; (3) high exit gas temperature reduces overall thermal efficiency; (4) entrained dust can cause plugging problems in the water scrubber; and (5) it is difficult to remove all particulate matter from the synthesis gas.

The K-T coal gasification process accounts for about 4000 ton/day of coal-based ammonia capacity, or almost 90% of the world's total. It also has potential for providing a synthesis gas for manufacturing methanol and other chemicals, but it cannot compete economically with the Lurgi gasifier for manufacturing high BTU pipeline quality gas (SNG). It may find application for gasifying the coal fines which cannot be processed in a Lurgi gasifier, and it has potential for producing a fuel gas for industrial and steam-power plant uses.

3. Winkler Gasifier

The Winkler gasifier was invented by Dr. Fritz Winkler in Germany in the early 1920s. It was developed to a commercial status by I. G. Farbenindustrie. The first commercial plant consisting of five gasifiers was constructed in 1926 in Leuna, Germany. Since then, 31 additional gasifiers have been built in 14 plants throughout the world (Anwer, 1979).

Figure 3 is a schematic diagram of the Winkler coal gasification process (Flesch, 1962). The gasifier is an atmospheric-pressure, fluidized bed, non-slagging, steam-oxygen or steam-air gasifier. Finely ground coal, steam and oxygen (or air) are introduced near the bottom of the gasifier. Crude synthesis (or producer) gas is withdrawn from the top and particulate ash from the bottom.

The coal feed to the Winkler gasifier must be crushed to less than 10 mm particle size. A size range of 0.2-4 mm is recommended. The coal feed should contain a maximum of 8 percent moisture although coals containing up to 14 percent moisture have been gasified commercially, and higher moisture contents may be tolerable. The particulate coal must be sufficiently dry to permit it to flow

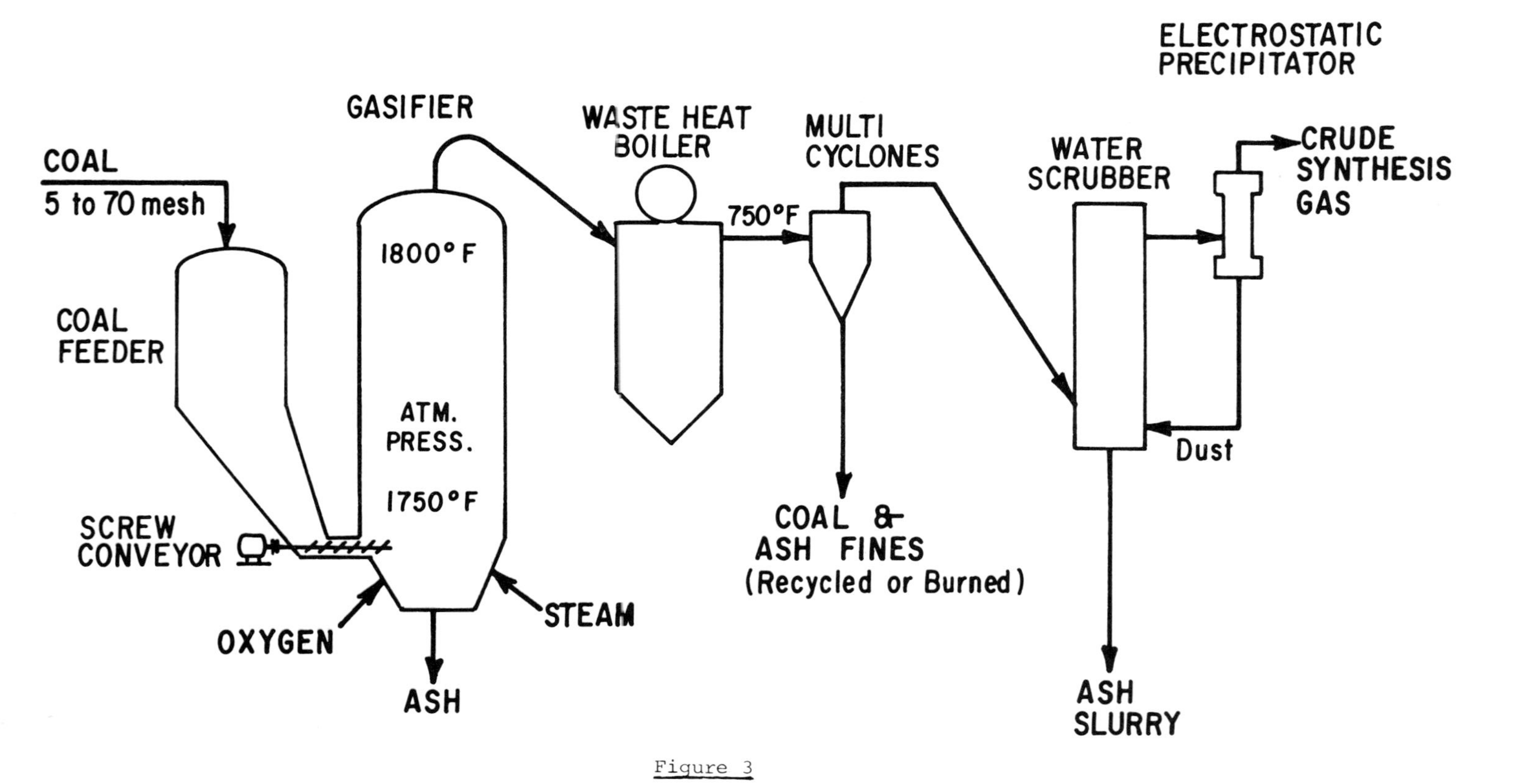
The Winkler Gasifier
COAL
5 to 70 mesh
COAL FEEDER
SCREW CONVEYOR
GASIFIER
1800°F
ATM. PRESS.
1750°F
OXYGEN
STEAM
ASH
WASTE HEAT BOILER
750°F
MULTI CYCLONES
COAL & ASH FINES
(Recycled or Burned)
WATER SCRUBBER
ELECTROSTATIC PRECIPITATOR
CRUDE SYNTHESIS GAS
Dust
ASH SLURRY

Figure 3

freely down the feed bunker, and there is an upper limit on the amount of moisture that can be vaporized in the gasification reactor. Thus, some coals would have to be partially dried before being fed to a Winkler gasifier.

The coal is conveyed to a feed bunker which is at atmospheric pressure and is blanketed with an inert gas to prevent spontaneous combustion in the bunker. The coal is transported to the gasifier via a variable speed screw conveyor that permits a somewhat wide variation in coal feed rate.

The gasifier operates at or marginally above atmospheric pressure. It is lined with ceramic bricks to protect the shell from heat and erosion. Reactors measuring about 5.5 m I.D. by about 18 m high have been constructed, but smaller diameter reactors have also been operated commercially.

Steam and oxygen (or air) are introduced separately through high-velocity injection nozzles which are located just below the coal feed entry point. The steam and oxygen intimately mix with and fluidize the entering coal. Gasification reactions commence immediately. A large excess of steam over that required for gasification is necessary to keep the fluidized bed temperature below the melting point of the coal ash.

Gasification temperature varies between 1090°K and 1280°K and is dependent on the reactivity and ash fusion properties of the coal being processed. Coals having high reactivity can be gasified at lower temperatures, and the higher the ash melting point, the higher the allowable gasification temperature.

Additional gasifying medium is introduced into the upper part of the gasifier about 2 m above the fluidized bed. The additional gasifying medium gasifies entrained, unreacted coal particles and carbon-laden ash particles present in the product gases leaving the fluidized bed. It also decomposes any tar or other hydrocarbons present in the crude gas. About 10 percent of the total gasifying medium is introduced above the bed. The additional gasification of entrained particles increases the temperature of the crude gas about 80°K; so the maximum temperature in the gasifier occurs in the space above the fluidized bed.

About 30 percent of the ash falls to the bottom of the gasifier and is continuously discharged into an ash receiver by means of a variable speed screw conveyor. Only the heavier, larger ash particles are discharged in this manner. The lighter ones, representing about 70 percent of the ash, are entrained in the crude gas and leave the gasifier with the gas.

The crude synthesis gas containing fine ash and char particles exits from the top of the gasifier at 1030-1145°K, and passes through a waste heat boiler, multi-cyclones, a water scrubber, and an electrostatic precipitator to cool the gas and remove the remaining particulate matter. The ash stream contains some carbon--the amount depending upon the efficiency of gasification operations.

The synthesis gas leaves the precipitator at atmospheric pressure and must be compressed before further processing. A blower may be sufficient if the gas is to be used on-site as a fuel.

The Winkler process was originally limited to highly reactive non-caking coals such as lignite which are relatively easy to gasify. A new design feature consisting of a grate and agitator arm installed at the bottom of the gasifier will reportedly permit processing caking coals. The addition of a radiant steam boiler in the top of the gasification reactor to cool the exiting crude gas reportedly permits higher gasification temperatures so that less reactive coals can be processed. Higher gasification temperatures also reduce the carbon content of the entrained ash.

The final product gas leaving the precipitator consists in the main of hydrogen, carbon monoxide, and carbon dioxide. Nitrogen content depends upon whether oxygen or air is used for gasification. The gas also contains minor amounts of methane and hydrogen sulfide. Very little, if any, liquid hydrocarbons are present in the crude gas leaving the gasification reactor.

There have been a number of drawbacks in the Winkler gasification process: (1) sintered ash deposits on the reactor walls, particularly at the coal feed entrance and at the gas exit points; (2) carbon losses to the ash can be significant; (3) high exit gas temperature and excess steam requirement reduce overall gasification efficiency; (4) potential exists for oxygen breakthrough into the product gas; (5) the presence of calcium oxide in the ash causes plugging problems in the scrubber and water recirculation system; and (6) it is difficult to remove all particulate matter from the synthesis gas.

While the Winkler gasifier will yield a synthesis gas suitable for manufacturing methane, ammonia, methanol, and other chemicals, it will most likely find application for generating on-site fuel gas for industrial and steam-power plant uses.

E. Advanced Systems

1. British Gas/Lurgi Slagging Gasifier

The British Gas/Lurgi slagging gasifier is a high pressure (25 to 30 atm) fixed bed process. It is similar to the Lurgi gasifier except that the bottom half has been modified to allow operation at slagging conditions. The ratio of steam to oxygen entering the bottom of the gasifier is reduced to increase the temperature above the melting point of the char ash. This high temperature in the gasification region enables the process to use low reactivity feedstocks such as bituminous coals without penalty.

The British Gas Corporation began developing the slagging gasifier in 1955 (Hebden, 1964). In 1974, a group of companies based in the United States sponsored further development of the gasifier at the Westfield Development Centre in Scotland. In 1977, Conoco Coal Development Company entered a contract with the United States Department of Energy to further test the process and to design, construct, and operate a demonstration plant based on this technology in Noble County, Ohio. The additional experimental work proved the operability of the British Gas/Lurgi slagging gasifier while feeding highly caking bituminous coal (Pittsburgh 8) and provided data necessary for the design of the demonstration plant (CCDC, June 1978). Design of a 940 ton MAF coal/day demonstration plant is currently underway.

A diagram of the British Gas/Lurgi slagging gasifier is shown in Figure 4. The coal feedstock is sized at 8 x 0.6 mm. However, when caking coal is fed, the gasifier has been shown to be tolerant of 25% or more material less than 0.6 mm. For some coals, flux must be added to the coal feed to insure a free flowing slag.

BG - Lurgi Slagging Gasifier

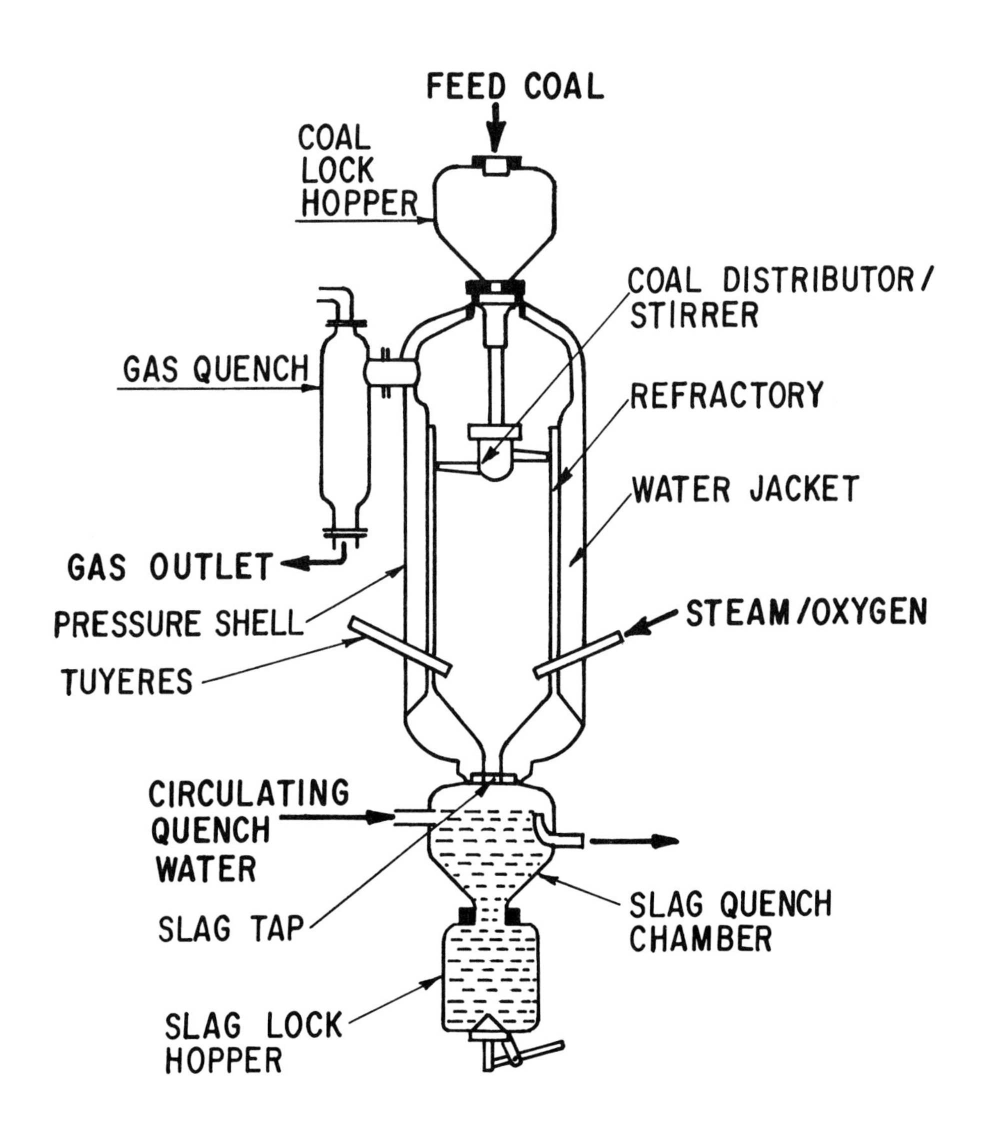

Figure 4

Coal enters the gasifier through a lockhopper and passes downward against the rising flow of gas. The coal is dried and devolatilized in the upper portion of the gasifier. A stirrer is employed to ensure free flowing char if caking coal is fed. Also, recycled tar is added to control dust evolution.

Steam and oxygen, introduced through tuyeres at the bottom of the gasifier, react with the descending char in a violently agitated region called the raceway, in front of the tuyeres. The steam conversion is very high at the temperatures required for slagging. It is in this region that the ash becomes molten and falls to the hearth of the gasifier. The liquid slag is maintained in a pool which is tapped periodically. The tapped slag is quenched in water and discharged as a frit through a lockhopper.

The raw gas exits the top of the gasifier and is quenched in the same manner as gas from a dry bottom Lurgi gasifier.

Tar separated from the raw gas can be totally consumed by introducing it into the gasifier through the tuyeres.

The advantages of the British Gas/Lurgi slagging gasifier are as follows:

(a) High coal throughput (4300 $Kg/hr/m^2$) of cross section with low reactivity Pittsburgh 8 coal.

(b) Tolerance of coal fines when caking coal is fed.

(c) Low steam requirement.

(d) High steam conversion and thus low liquor yield.

(e) Reduced oxygen consumption over a dry bottom Lurgi gasifier when low reactivity coal is used.

(f) High thermal efficiency.

(g) The gasifier can consume its own tar by-product.

These advantages are partially offset by the fact that the product gas is CO-rich, having a H_2/CO ratio of about 0.5. This coupled with the low level of unreacted steam in the product gas means that additional steam must be added downstream in many process applications.

The gas composition, product yields, oxygen, and steam requirements are given in Table 3. The overall plant efficiency is given in Table 4. The figures are based on a commercial design from Westfield pilot plant data (CCDC, October 1978).

The oxygen and steam consumptions are given in terms of mols/mol CO + mol H_2 + 4 mols CH_4. The above emphasizes the value of methane since 4 mols of CO + H_2 are required per mol of methane.

TABLE 3

COMPARISON GAS COMPOSITION, YIELDS, STEAM, AND OXYGEN REQUIREMENTS - VARIOUS GASIFICATION PROCESSES

Process	Dry Bottom Lurgi	Lurgi-BGC Slagger	Shell-Koppers	Texaco	CO_2 Acceptor
Coal	N. Dakota Lignite	←Illinois #6	Illinois #6	Illinois #6→	N. Dakota Lignite
Oxygen Requirement					
m^3/m^3 $(CO + H_2 + 4\ CH_4)$	0.12	0.20	0.28	0.34	Nil
Steam Requirement					
Mols/Mol $(CO + H_2 + 4\ CH_4)$	1.09	0.26	0.045	--	Nil
Liquid Yields					
Wt % Coal (MAF)					
Tar	1.89	(1)	--	--	Nil
Oil	3.23	2.08	--	--	Nil
Naphtha	0.96	1.43	--	--	Nil
Typical Raw Gas - Dry Basis Comparison - Vol %					
C_2^+	0.92	0.50	0.0	0.0	Nil
CH_4	11.01	6.14	0.0	0.1	11.58
CO	16.03	59.01	65.0	51.8	15.92
CO_2	32.46	6.49	0.8	10.6	6.20
H_2	39.18	25.91	32.1	35.2	66.25
$H_2S + COS$	0.40	1.95	1.4	1.3	0.04
$N_2 + A$	--	--	0.7	1.0	--
Reference:	(CCDC, Apr., 1978)	(CCDC, Oct., 1978)	(Voght, 1979)	(Chandra, 1978)	(CCDC, Apr., 1978)

(1) Tar recycled to gasifier.

2. Texaco and Shell-Koppers

Texaco and Shell-Koppers both are developing pressurized entrained phase coal gasification units.

The Texaco process is a development based on the extension to coal of the commercial technology developed for gasification of petroleum residual oils.

The Shell-Koppers process is a joint development of Shell International Petroleum Company and Krupp-Koppers. It is designed to take full advantage of the commercial technology of Shell in pressurized oil gasification and of Koppers in the atmospheric pressure Koppers-Totzek coal gasification process.

The Texaco and Shell-Koppers processes are both oxygen blown coal gasification units which operate at pressures of the order of 20 atmospheres or higher. 150 T/D coal gasification pilot plants are currently in operation for both processes. The Texaco pilot plant has been in operation at the Ruhrchemie Chemical Plant Complex in Oberhausen-Holten, West Germany since January, 1978. The Shell-Koppers plant has been in operation since November, 1978 at the Hamburg refinery of Deutsche Shell.

The Texaco process has the backup support of two 15-20 T/D pilot gasifiers in Montebello, California which have been in periodic operation for a large number of years. The Shell-Koppers process is supported by a 6 T/D pilot gasifier which has been in operation at Amsterdam since December 1976.

The principal distinction between the two processes is in the feed system and the detailed design of the gasifier. Texaco feeds the coal as a concentrated slurry in water. This is done at some sacrifice in oxygen requirement and thermal efficiency to assure a steady uniform feed. Texaco considers this to be essential for both process reliability and safety.

Shell Koppers, on the other hand, feeds the coal predried through a lockhopper system. The pressurized pulverized coal in the pilot plant is fed into the gasifier entrained in nitrogen. It is proposed to use recycle gas for this purpose in the commercial unit.

Shell-Koppers envisions commercial operation at pressures of 30 atmospheres although their Hamburg plant has so far not been operated above 20 atmospheres.

Texaco, because of their feed system, apparently has more flexibility in choice of operating pressure. Their pilot gasifiers in Montebello, California have operated with a range of coals at pressures up to 80 atmospheres.

Both processes operate in the slagging region with exit gases from the gasifier in the neighborhood of 1400°C. The Texaco unit is a single refractory lined vertical vessel. The coal slurry is downfired with oxygen through a nozzle at the top of the gasifier. The slag is collected in water at the bottom of the vessel and lockhoppered out as a water slurry.

The Shell-Koppers unit uses two horizontally fired diametrically opposed burners into a vertical pressurized shell. The shell is protected with a tube wall in which saturated steam at 50 atmospheres is generated. The tube wall is covered with a thin layer of refractory. Molten slag is removed from the bottom of the gasifier as in the Texaco process, but the gases leave through the top of the gasifier in the Shell process.

Schematic flow sheets of both the Texaco and Shell-Koppers gasifiers taken from recent publications are shown in Figures 5 and 6, respectively. The downstream facilities are similar but not identical for both systems. Both processes when operated in the fuel or syngas mode generate steam from the gasifier effluent via a waste heat boiler. Entrained particulates and unburned carbon are removed via a water quench in both cases. Both processes envision recycle of the unburned carbon to the gasifier although this has not been done in pilot plant operation as yet.

The steam and oxygen requirements for Illinois coal and dry raw gas composition are given in Table 3. Those are for projected commercial operation as given in references (Chandra, 1978) and (Voght, 1979) for Texaco and Shell-Koppers, respectively.

It is noted that both processes, in contrast to the moving bed systems, produce a tar- and essentially methane-free gas. The lack of tar production is economically favorable but the low methane production is not necessarily beneficial. Low methane formation is undesirable particularly when the ultimate desired product is SNG. It reflects itself, for example, in a higher oxygen consumption as compared with the slagging fixed bed system. The oxygen consumption is given in mols O_2/(mol CO + mol H_2 + 4 mols CH_4). Such a comparison emphasizes the value of methane since 4 mols of CO + H_2 are required per mol of methane.

The oxygen consumption for the dry Shell process is as might be expected, lower than for the Texaco slurry process by about 20%. The commercially projected oxygen consumption in the Texaco process is based on feeding a slurry containing at least 69.5 wt.% coal (Cornils, 1979). The use of such a concentrated slurry has not been achieved in the Oberhausen-Holten pilot plant operations and the lowest oxygen consumption so far achieved is about 25% higher than the figure given in Table 3.

The gas composition also shows significant differences between the two processes. The Shell-Koppers unit produces a significantly higher CO/H_2 ratio in the raw gas and a significantly lower CO_2 concentration. It is actually, except for its lower methane content, similar in composition to the gas from the slagging fixed bed gasifier.

The plant thermal efficiency to raw gas has been given by Shell (Voght, 1979) and is 78% as shown in Table 4. No comparable figure has been given by Texaco, but in view of the higher oxygen requirement it is undoubtedly lower.

It should be noted that the above efficiency excludes energy losses in the acid gas cleanup systems and other downstream processing. The efficiency, for example, would be substantially reduced in a plant producing hydrogen, methanol, or SNG. For one thing, the waste heat boiler which provides steam to drive the oxygen plant would be eliminated by a water quench to provide steam for water gas shift. Secondly, there is a significant loss of heating value in converting synthesis gas to methanol or SNG.

The Shell-Koppers process requires drying of the feed coal and thus thermal efficiency decreases as the drying loading increases. The lower the coal rank, the more loss in efficiency occurs for this reason. Shell, for example,

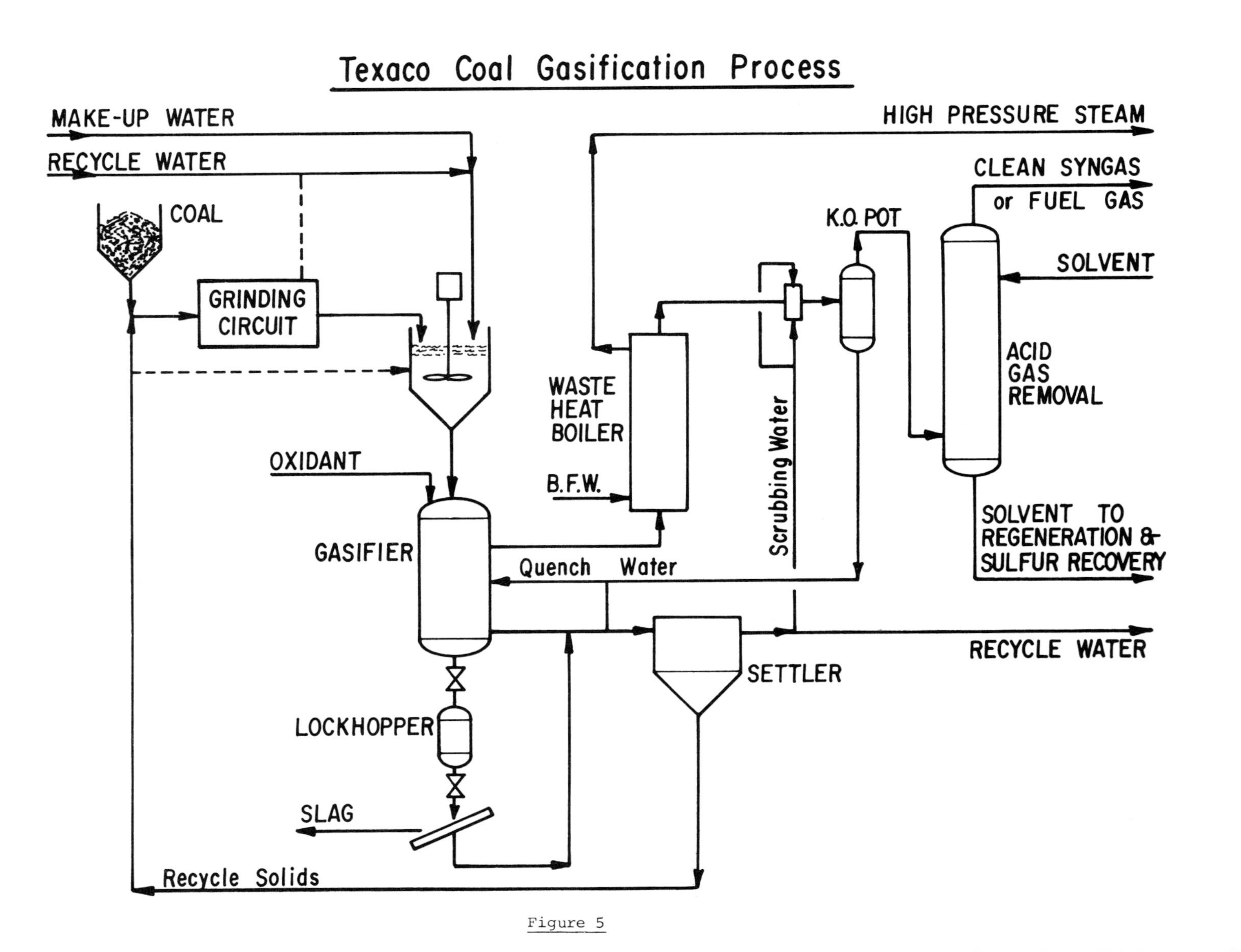
Texaco Coal Gasification Process
MAKE-UP WATER
RECYCLE WATER
COAL
GRINDING CIRCUIT
OXIDANT
GASIFIER
LOCKHOPPER
SLAG
Recycle Solids
WASTE HEAT BOILER
B.F.W.
Quench Water
Scrubbing Water
K.O. POT
SETTLER
HIGH PRESSURE STEAM
CLEAN SYNGAS or FUEL GAS
SOLVENT
ACID GAS REMOVAL
SOLVENT TO REGENERATION & SULFUR RECOVERY
RECYCLE WATER

Figure 5

Schematic Diagram Shell-Koppers Coal Gasification

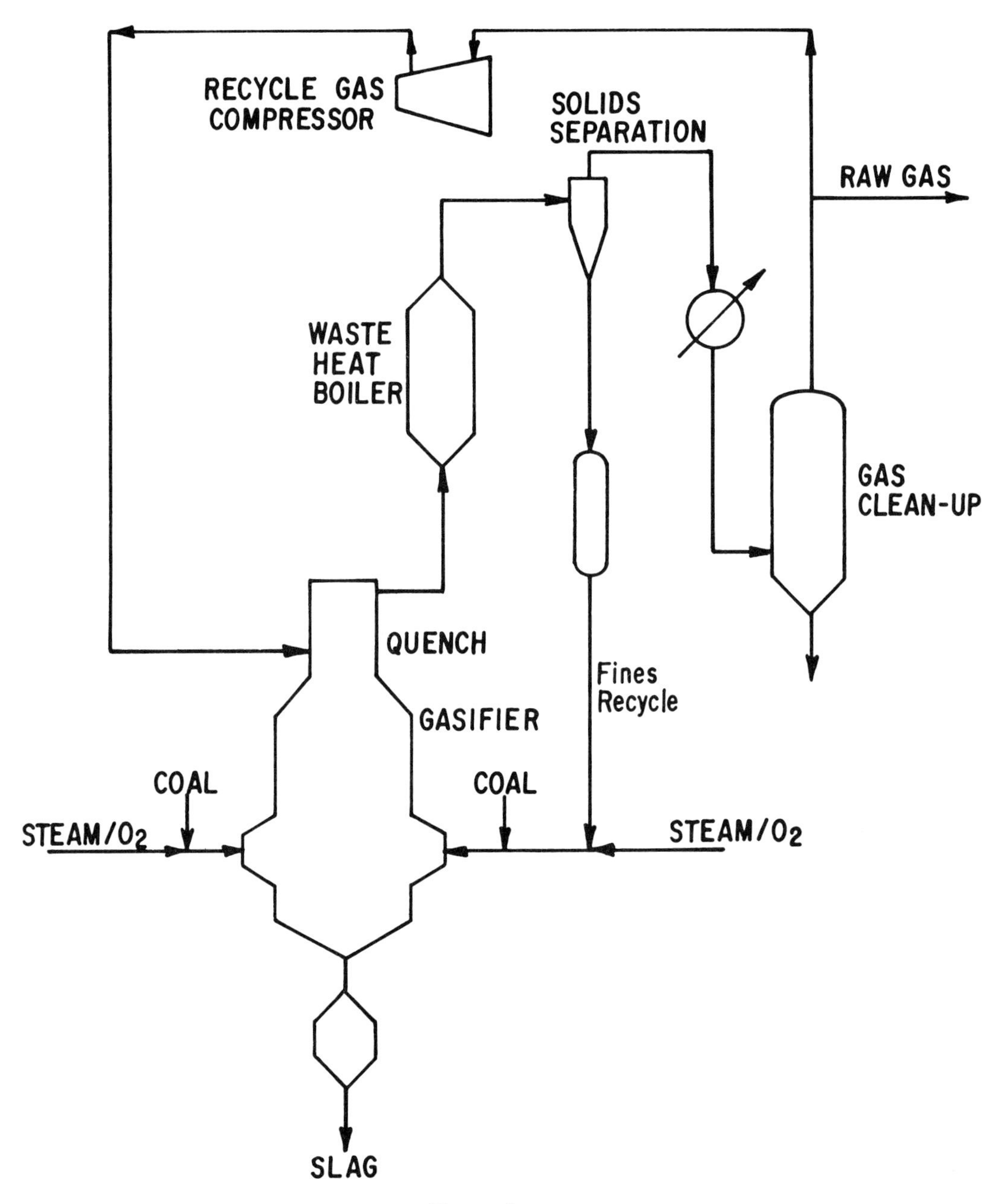

Figure 6

TABLE 4

COMPARISON OVERALL THERMAL EFFICIENCY
VARIOUS GASIFICATION PROCESSES

Process	Lurgi Dry Bottom	Lurgi-BGC Slagger	Shell-Koppers	CO_2 Acceptor
Coal	N. Dakota Lignite	←Illinois #6	Illinois #6→	N. Dakota Lignite
Thermal Efficiency % to Products				
Raw Gas	68.5	68.8	78	79.3
Hi-BTU Gas	58.3	53.3	(1)	67.5
By Products	6.8	5.8	--	0.8
Reference:	(CCDC, Apr., 1978)	(CCDC, Oct., 1978)	(Chandra, 1978)	(CCDC, Apr., 1978)

(1) No data available.

estimates an efficiency loss from 78% to 72% in going from bituminous to German brown coal.

Texaco would likewise lose efficiency with decreasing rank. No figures have been given, however, by Texaco. The problem is that the slurry feed will become less fuel rich/lb of water as the inherent moisture and oxygen content of the coal feed increases, as it does with decreasing rank.

Excellent performance has been obtained in the Texaco 12 T/D pilot unit in the partial oxidation of vacuum bottoms from several liquefaction processes, i.e., from H-Coal (Robin, 1976), SRC-I (Robin, 1978), and SRC-II (Schlinger, 1979) pilot plants.

A higher thermal efficiency and lower oxygen requirement is projected with this feedstock than with coal. The reason is that the feedstock is fed dry and preheated in molten form rather than as a water slurry.

The above tests have been relatively short-term tests of up to 30 hours duration and are thus not wholly conclusive. The process, however, as will be noted below, is widely counted on as a source of hydrogen for various liquefaction processes.

The Texaco plant at Oberhausen-Holten has had over 2000 hours of successful plant operation through July 1979. Not all problems have been fully resolved and particularly some refractory and boiler tube fouling problems remain. Another two years of test work is visualized in the above plant.

Another large pilot plant, 170 T/D, is currently also being installed in the U.S. at TVA. The purpose is to demonstrate the process as a source of hydrogen for synthetic ammonia.

A larger demonstration plant is now in the planning stage. This is a 1000 T/D combined cycle plant to be located in Barstow, California. The project is being supported by EPRI and others.

Shell's operating experience is not yet as extensive as Texaco's. Their longest run in the Hamburg pilot plant is 108 hours. High carbon conversions in the range of 96-98% have been achieved in both the Shell and Texaco pilot units. Ultimate carbon conversions of over 99% are projected via recycle of carbon fines.

The Shell experience has been sufficiently encouraging such that 1000 T/D prototype units are being designed for commissioning in 1983/1984.

The Shell system appears to be more efficient and applicable to a wider range of coals than the Texaco process. The Texaco system, however, because of its slurry feed system, likely would give more dependable operation and may be regarded as being somewhat more advanced in its state of development.

3. CO_2 Acceptor Process

The CO_2 Acceptor process has been developed by Conoco Coal Development Company under the auspices of the United States Government and the American Gas Association. A 35 ton/day pilot plant with an output of 42,500 m^3/d synthesis gas was built and successfully operated at Rapid City, South Dakota. The pilot plant program demonstrated all of the key features of the process. The process is now ready for larger scale demonstration (CCDC, Nov. 1978).

The CO_2 Acceptor process is shown in Figure 7. There are two fluidized bed reactors, a gasifier and a regenerator, which operate in the pressure range of 9-11 atm absolute at the top of the gasifier bed. The gasifier operates in the temperature range of 1080-1120°K.

Raw, dry lignite or subbituminous coal (sized to nominally 0.15 x 2.4 mm mesh) is fed at the bottom of a fluidized bed of char in the gasifier. After the rapid hydrodevolatilization reactions occur, fixed carbon in the char is gasified by the endothermic reactions:

$$C + H_2O_{(g)} = CO + H_2,$$

$$2C + H_2O_{(g)} + H_2 = CH_4 + CO.$$

Carbon dioxide is formed by the water gas shift reaction,

$$CO + H_2O_{(g)} = CO_2 + H_2.$$

The overall heat of reaction is about 18,300 cal/mol (288°K) of coal carbon gasified. This endothermic heat of reaction is supplied by the CO_2 Acceptor reaction,

$$CaO + CO_2 = CaCO_3 \quad \Delta H = -42,200 \text{ cal/mol } (288°K)$$

The acceptor, which can be derived from either limestone or dolomite (sized nominally to 2 x 3.4 cm), enters above the fluidized char bed, showers through the bed, and collects in the gasifier boot. Steam flow to the bottom of the boot is adjusted to strip cleanly the char from the acceptor so that a sharp, stable interface exists between the acceptor and the char-acceptor mixture above it. The remaining steam needed for gasification of fixed carbon enters through a distributor above the boot.

The recarbonated acceptor flows through a standleg and then is conveyed pneumatically by air to the bottom of the regenerator. Residual char from the gasifier also flows through a standleg and is conveyed by a stream of regenerator recycle gas to the regenerator where combustion raises the temperature of the fluidized bed of acceptor to about 1280°K. At this temperature, the acceptor is calcined by reversal of the CO_2 Acceptor reaction. The calcined acceptor is returned by gravity through a standleg, thus completing the acceptor "loop".

Seals between the gasifier and regenerator are maintained by the three solids standlegs which are purged with recycle gas. Solids flow rates are controlled by butterfly valves.

After combustion, the char ash is finely divided (80% <0.15 mm) and is elutriated from the acceptor bed and is removed by external cyclones.

As the acceptor circulates between the process vessels, it loses activity toward the CO_2 Acceptor reaction. Inert CaO forms by crystallite size growth. To maintain activity, some acceptor is purposefully withdrawn from the regenerator and fresh stone makeup is added to the acceptor lift line. Makeup requirements are 2 mols of CaO per 100 mols of CaO flowing through the gravity return line.

The CO_2 Acceptor process consumes substantially all (99%+) of the carbon bed to the gasifier, and is totally self-sufficient with respect to gasifier steam

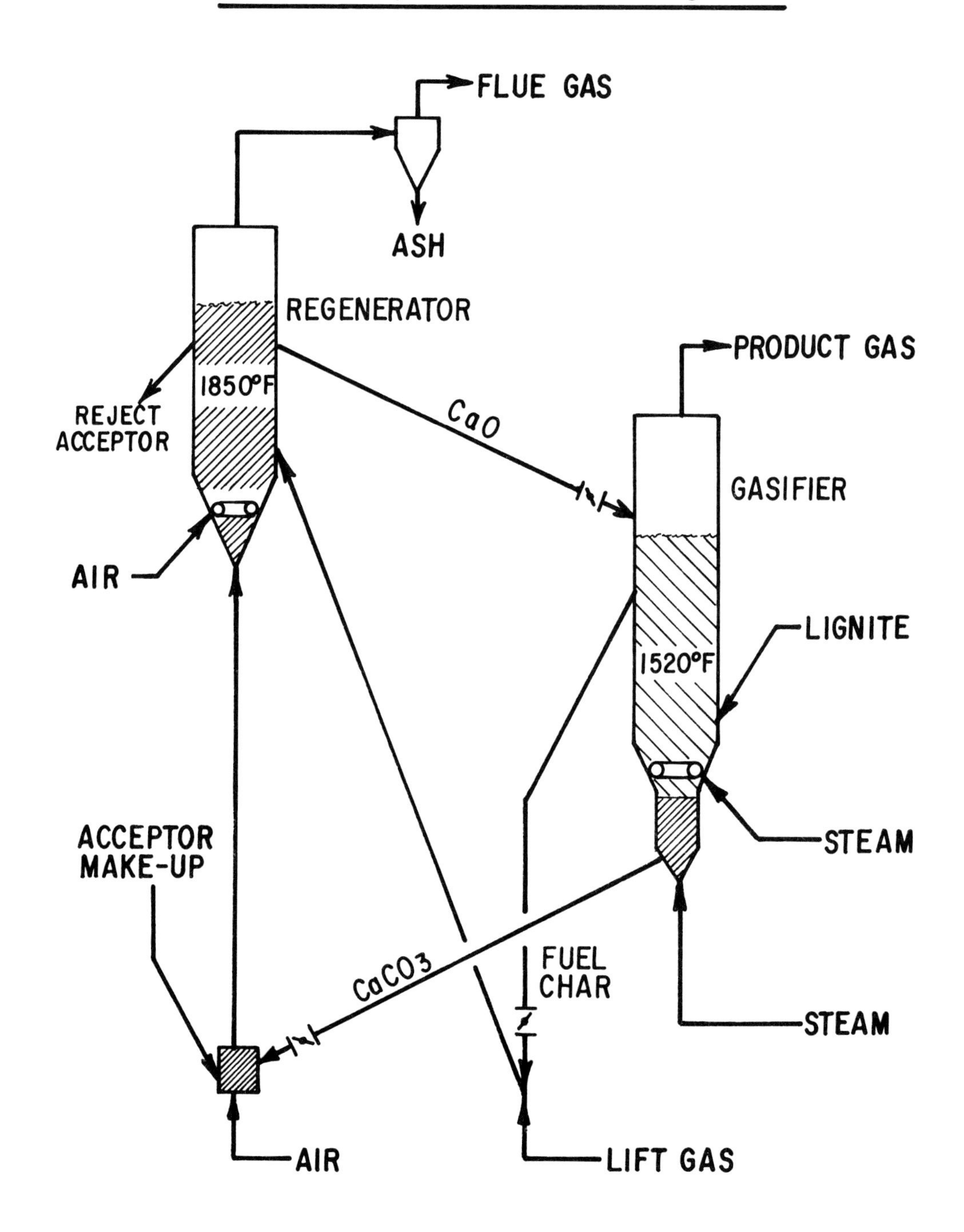

CO_2 Acceptor Process Diagram
FLUE GAS
ASH
REGENERATOR
PRODUCT GAS
1850°F
CaO
REJECT ACCEPTOR
GASIFIER
AIR
LIGNITE
1520°F
ACCEPTOR MAKE-UP
STEAM
FUEL CHAR
CaCO3
STEAM
AIR
LIFT GAS

Figure 7

generation and regenerator air compression requirements. By heat exchange with the gasifier and regenerator offgases, the gasifier process steam is generated and superheated. The partially cooled regenerator gas is expanded through turbines which drive the air compressors. There is no fuel fired boiler.

The CO_2 Acceptor process offers several advantages for the gasification of reactive coals. These advantages are listed below:

(a) Ability to produce synthesis gas or high BTU pipeline gas without use of oxygen.

(b) The absence of hydrocarbons other than CH_4 in the product gas. Trace amounts of heavier hydrocarbons are removed by absorption on the char fines recovered in the quench system and will be ultimately used as fuel in the regenerator.

(c) Production of synthesis gas without production of unwanted tars and phenols.

(d) Substantially total carbon utilization as shown by the fact that the carbon contained in the ash removed from the regenerator is less than one percent of the carbon in the feed.

(e) A ratio of H_2 to CO in the product which exceeds 3:1, such that all of the CO and part of the CO_2 can be methanated with no water-gas shift required.

(f) Low concentrations of CO_2 and H_2S in the product gas reduce costly gas cleanup problems.

(g) It is a relatively low pressure process.

(h) Fines present in run of mine coal could be totally utilized.

Some of the above advantages are illustrated in the comparison of the commercial design for the CO_2 Acceptor process (CCDC, April 1978) with that of the Lurgi process (CCDC, April 1978) as given in Tables 3 and 4. Particularly noteworthy is the higher thermal efficiency as compared with Lurgi.

The major disadvantage of the CO_2 process is that its application is restricted to high reactivity coals.

No plans have yet been formulated for construction of a demonstration plant.

4. Illinois Coal Gasification Group

The above is a consortium (McCray, 1978) formed to develop the FMC COGAS process in Illinois. It currently is developing an engineering study for DOE of a proposed demonstration plant to produce up to 1,700,000 m^3 of SNG per day to be located in Illinois.

The COGAS process has been developed by FMC (Eby, 1978) on a 2 ton/hr pilot scale. It is an outgrowth of the FMC-COED pilot plant development which was funded by the OCR (FMC, 1975) from 1971 to 1975. The COED process is essentially a multi-stage fluidized coal carbonization process designed to produce coal tar and a reactive char. The tar is upgraded to synthetic crude quality by subsequent hydrotreating.

The COGAS process was developed to gasify the char from the COED process. A schematic flow sheet of the overall process is shown in Figure 8. The upper part of the flow diagram represents the COED or pyrolysis part of the process.

The multi-stage process includes char recycle between stages which is not shown. The above feature permits the process to function without preoxidation with at least mildly caking coals such as those from Illinois.

The approximate temperatures in the various process units are indicated in the flow sheet. The process is a relatively low pressure operation, i.e., approximately 1.4-3.4 atm gauge.

The gasification system is a fluidized bed unit and may be classified as a moving bed process in that heat is supplied by recirculating hot char. Thus, an oxygen plant is not required which is one of the virtues of the process.

Various methods of reheating the hot char have been studied. A favored method is indicated on the flow sheet. The recirculating fines are reheated by direct contact with hot combustion gases. The latter are generated in a slagging unit by combustion with air of the fines generated in the char gasifier.

The overall process accomplishes a similar result in a very complex series of operations to what is accomplished in a single unit in a moving bed gasifier.

5. HYGAS Process

Development of this process was begun on a pilot plant scale by the Institute of Gas Technology (IGT) in 1971 and has continued since then. The pilot plant program is scheduled to be completed in mid-1980. The work has been sponsored by the Department of Energy and its predecessors, and by the American Gas Association.

In its present embodiment, the process gasifier contains four sections through which the coal and char move downward countercurrent to the uprising gas stream. The feed coal is slurried with a light oil which is produced in the process and is pumped to the upper section where the oil is evaporated. The dry coal then flows to the first gasification stage where the more reactive part of the coal is converted to give a high yield of methane. Heat is supplied by the hot gas from the second gasification stage where the partially converted coal (char) is contacted with a hydrogen-rich gas to form more methane and carbon oxides. Heat is supplied by the hot gas from the third stage which is a steam/oxygen gasifier. Temperature in the bottom stage is in the range of 1175-1275°K.

The process concept is to maximize the direct yield of methane to increase the thermal efficiency and to decrease the cost of conversion of the cleaned product gas to SNG.

Pilot plant feedstocks have been lignite, subbituminois coal, and Illinois Basin coals. The process requires that the latter coals be pretreated by preoxidation with air at about 673°K.

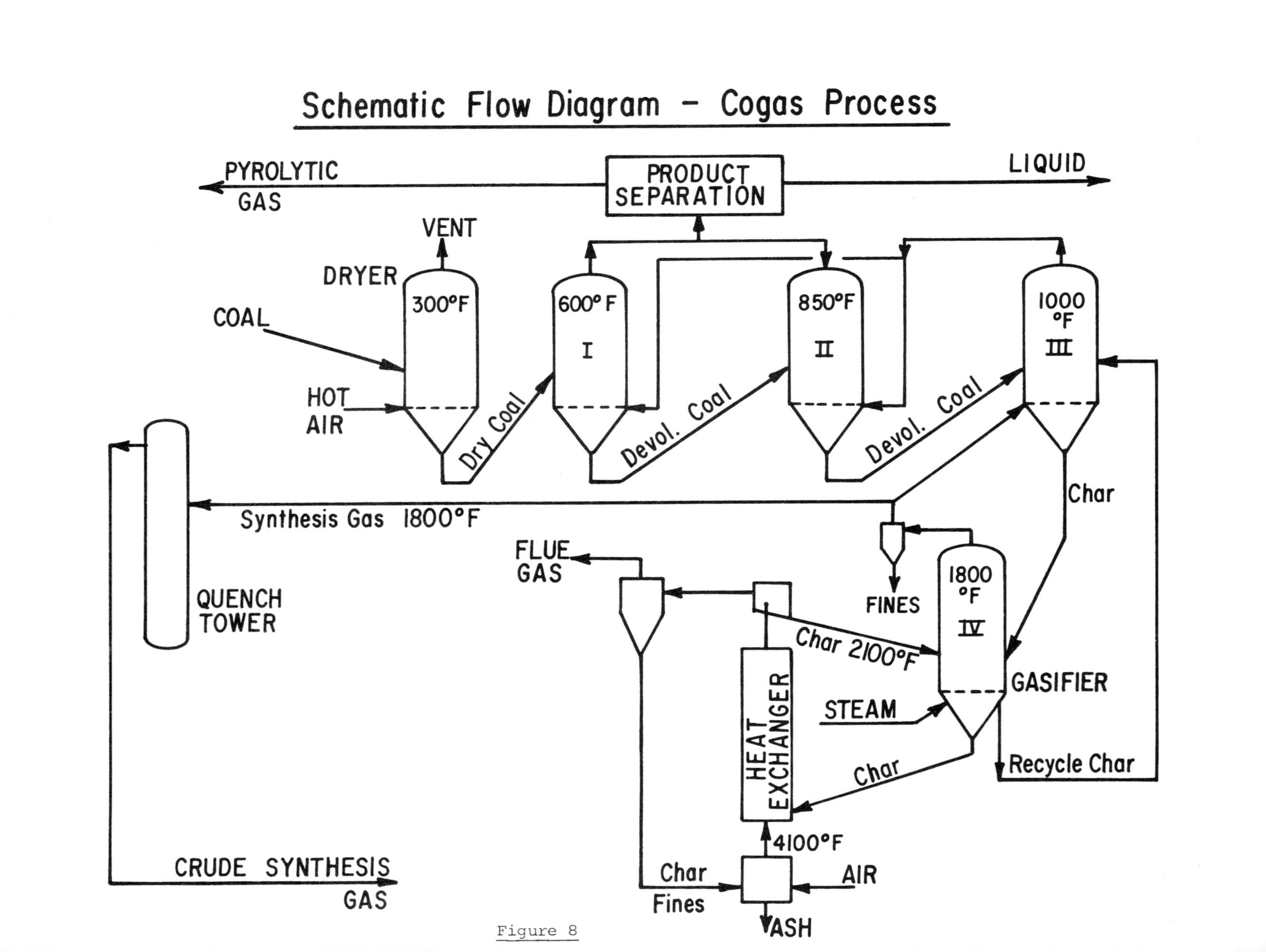
Schematic Flow Diagram - Cogas Process
PYROLYTIC GAS
PRODUCT SEPARATION
LIQUID
VENT
DRYER
300°F
COAL
HOT AIR
Dry Coal
600° F
I
Devol. Coal
850°F
II
Devol. Coal
1000 °F
III
Char
Synthesis Gas 1800°F
QUENCH TOWER
FLUE GAS
FINES
1800 °F
IV
Char 2100°F
GASIFIER
STEAM
HEAT EXCHANGER
Char
Recycle Char
4100°F
Char Fines
AIR
ASH
CRUDE SYNTHESIS GAS

Figure 8

The pilot plant has operated in the pressure range of 35-70 atm and at dry coal feed rates of up to about 2500 kg/hr.

The major process problem in the pilot plant has been the formation of ash clinkers in the steam/oxygen gasifier and has been especially severe with the Illinois Basin coals which require higher temperature operation because of the lower reactivity of the char. Over the past three years considerable progress has been made to eliminate the problem.

Under sponsorship of DOE, a detailed design of a demonstration plant having a capacity of about 20 x 10^9 kcal/day of SNG is underway.

Much has been published about the HYGAS process. Good reviews have appeared in the AGA Symposia on Synthetic Pipeline Gas (Lee, 1977, Lee, 1976, Bair, 1977, and Bair, 1978). More detailed information is available from IGT reports (IGT, 1975-1979).

F. Developing Systems

1. Exxon Catalytic Coal Gasification Process

Exxon Research and Engineering Company is developing an innovative process for converting a bituminous coal to substitute natural gas without the use of O_2 and without separate shift and methanation reactors (Gallagher, 1979).

The key feature of the process is operation of the gasifier at the low temperature of about 975°K and at the moderately high pressure of 34 atm. At these conditions, a high conversion of coal to CH_4 is obtained (the dry, raw gas contains about 30 mol% CH_4) and the overall heat of reaction is approximately thermoneutral. The reactions can be expressed as follows.

$2C + 2H_2O = 2H_2 + 2CO$	Highly Endothermic
$CO + H_2O = CO_2 + H_2$	Mildly Exothermic
$3H_2 + CO = CH_4 + H_2O$	Highly Exothermic
Sum $2H_2O + 2C = CH_4 + CO_2$	Slightly Endothermic

The sum is reaction (e) discussed in section II.B. In the pure graphite system, the process is slightly endothermic. As was discussed above in section II. B., the process becomes thermoneutral or slightly exothermic in the coal-steam system.

To obtain commercially attractive gasification rates at the low temperature of about 975°K, a catalyst is required. In this process, a relatively massive amount of potassium is added to the coal in a ratio of about 0.12 pound KOH per pound of dry coal. Most of the catalyst is recovered and then is recycled. Some of the potassium reacts, in the gasifier, with coal ash components. Recovery of KOH is enhanced by adding $Ca(OH)_2$ to an aqueous slurry of the spent char, ash, and catalyst.

Figure 9, adapted from (Gallagher, 1979) is a schematic flow diagram of the process. After heat recovery, condensation of unreacted steam, CO_2 removal, and gas cleanup, the gasifier product gas is cryogenically separated to provide a pure CH_4 stream and a CO-H_2 stream. The latter is recycled to the gasifier after being mixed with steam.

Experimental and process design work currently is being funded by the Department of Energy and the Gas Research Institute. The process development unit contains all the components shown in Figure 9 and operates in a completely integrated manner. Coal throughput is about 45 kg/hr.

The process is attractive because scale-up problems should be minimal, since only a "simple" fluidized bed gasifier, uncomplicated by ash slagging problems, is involved. Catalyst recovery appears at this time to be relatively straightforward and is carried out at mild conditions. Heat recovery, CO_2 removal, gas cleanup, and cryogenic separation are well-developed technologies. The long retention time of the gas in the gasifier which is characteristic of this process and the catalytic action of the K_2CO_3 tend to eliminate production of tars and oils, even at the low temperature which is used. The cost of downstream cleanup is reduced accordingly. Operations of the process development unit must be monitored closely to ensure that this situation will indeed occur.

There are two possible disadvantages:

(a) The catalyst probably will not decake sufficiently bituminous coals of higher rank than the Illinois No. 6 coal used so far in the development program.

(b) The thermal efficiency based on total coal fed to the process probably is lower than any of the gasification processes which are considered here. There are three reasons for this:

- Carbon conversion is limited in practice to not more than about 90% because the gasifier is a single stage fluidized reactor in which the char solids are highly backmixed.
- Fuel gas is needed to preheat the incoming feed gas to the gasifier to provide the sensible heat requirement for preheating the coal feed to reaction temperature.
- To achieve the required gasification reaction kinetics, even with catalysis, a high steam/coal inlet ratio and a low steam conversion are needed. Much of the steam is generated offsite by combustion of coal.

The unfavorable impact of the low thermal efficiency possibly can be overcome because of the relative simplicity of the process flow scheme, which in turn could lead to a low capital investment. Since all gasification processes are highly capital-intensive, the selling price of SNG may very well be lower than for other processes. For this reason, the development of the Exxon process deserves careful attention by the energy community.

2. Westinghouse Fluidized Bed Gasification Process

Since 1974 the Westinghouse gasification process development unit has been operated at Waltz Mill, Pennsylvania. Most of the work has been directed toward production of clean, low-BTU fuel gas for power generation, using an air-blown, two-vessel reactor system. Under Department of Energy sponsorship beginning in 1978, a single fluidized bed reactor, using O_2 and steam, has been operated to produce synthesis gas. A summary of the work through 1978 has been given (Archer 1978). Further information on 1979 operations has also been obtained (Salvador, 1980)

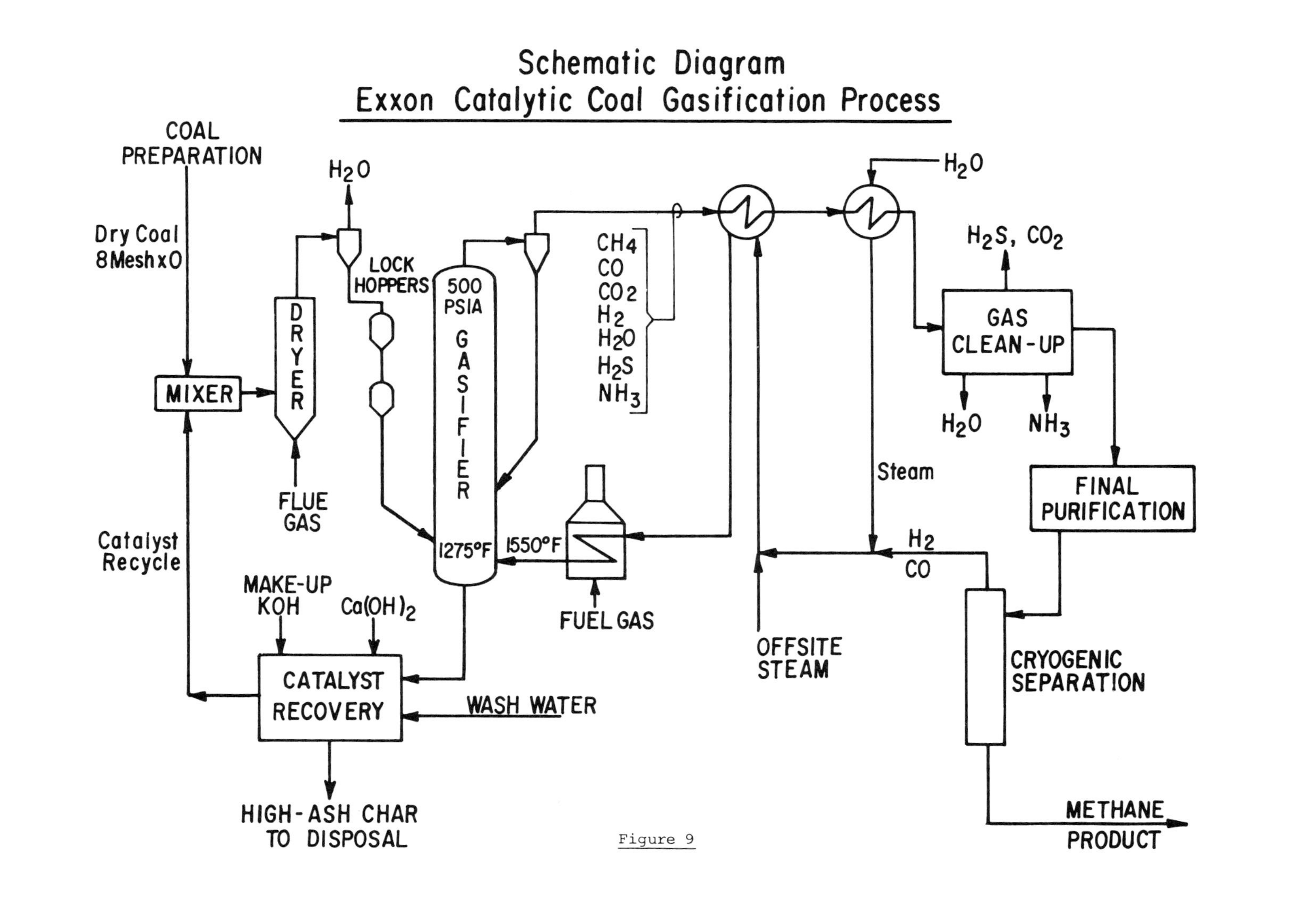

Schematic Diagram
Exxon Catalytic Coal Gasification Process
COAL PREPARATION
Dry Coal 8Mesh x 0
MIXER
Catalyst Recycle
DRYER
FLUE GAS
H_2O
LOCK HOPPERS
500 PSIA
GASIFIER
1275°F
1550°F
CH_4
CO
CO_2
H_2
H_2O
H_2S
NH_3
FUEL GAS
MAKE-UP KOH
$Ca(OH)_2$
CATALYST RECOVERY
WASH WATER
HIGH-ASH CHAR TO DISPOSAL
H_2O
H_2S, CO_2
GAS CLEAN-UP
H_2O
NH_3
Steam
FINAL PURIFICATION
H_2
CO
OFFSITE STEAM
CRYOGENIC SEPARATION
METHANE PRODUCT

Figure 9

This discussion is concerned solely with the single-vessel synthesis gas system. The key feature of the gasifier is the ability to convert about 95% of the coal carbon in a single, relatively small, fluidized bed reactor by using the high temperature generated by the incoming oxygen to agglomerate the coal ash in a controlled manner such that it readily separates from the gasifier char bed. Figure 10 is a simplified diagram of the gasifier.

At a nominal coal feed rate of 450 kg/hr, in runs lasting up to 200 hours each (voluntary shutdown), the following raw coals have been gasified successfully without the formation of clinkers or other deposits. (Gasifier conditions were 10 atm pressure at about 1000°C.)

- Montana subbituminous
- Indiana No. 7 bituminous
- Ohio No. 9 bituminous
- Pittsburgh No. 8 bituminous

Long term stability of ash agglomeration is accomplished by separate control of:

- O_2/H_2O ratio in the oxygen jet pipe;
- Stripping gas flow rate;
- Position of the coal feed pipe.

In view of the variability of coal properties which include the amount and nature of the ash, which literally can change minute-by-minute, successful operation with the above variety of raw coals must be regarded as a technical triumph.

A possible process problem is that of scale-up of the agglomerating section of the gasifier. Information gained from operation of a cold model was valuable in perfecting the 51 cm diameter (76 cm diameter gasifier) agglomerating section of the process development unit. A large (3 m diameter) cold model now is under construction. Information from this model should aid the scale-up problem.

3. U-GAS

The U-GAS system is being developed by IGT and is another example of an ash agglomerating, fluidized bed system. It is similar to the Westinghouse process when operated in the single stage mode. It differs in the detailed design of the ash agglomerating section of the gasifier.

The system was originally conceived as a scheme to provide a stream of hot synthesis gas, including unconverted steam, to the HYGAS process from the char produced in that process. The objective has been broadened to generate an industrial fuel gas.

The process has been tested (Patel, 1979) on a 450 kg/hr pilot unit at IGT. The gasifier is 1.22 m in diameter, but operating pressure has been limited to a maximum of 3 atm gauge pressure. Success has been achieved in processing moderately caking coals (W. Kentucky 9 and Illinois 6) in this unit. The discharged ash agglomerates were low in carbon. The throughput rates, however, were relatively small--440 $kg/m^2/hr$. The projected commercial design calls for operation both at much higher pressures, ca. 30 atm, and throughputs. The scale-up

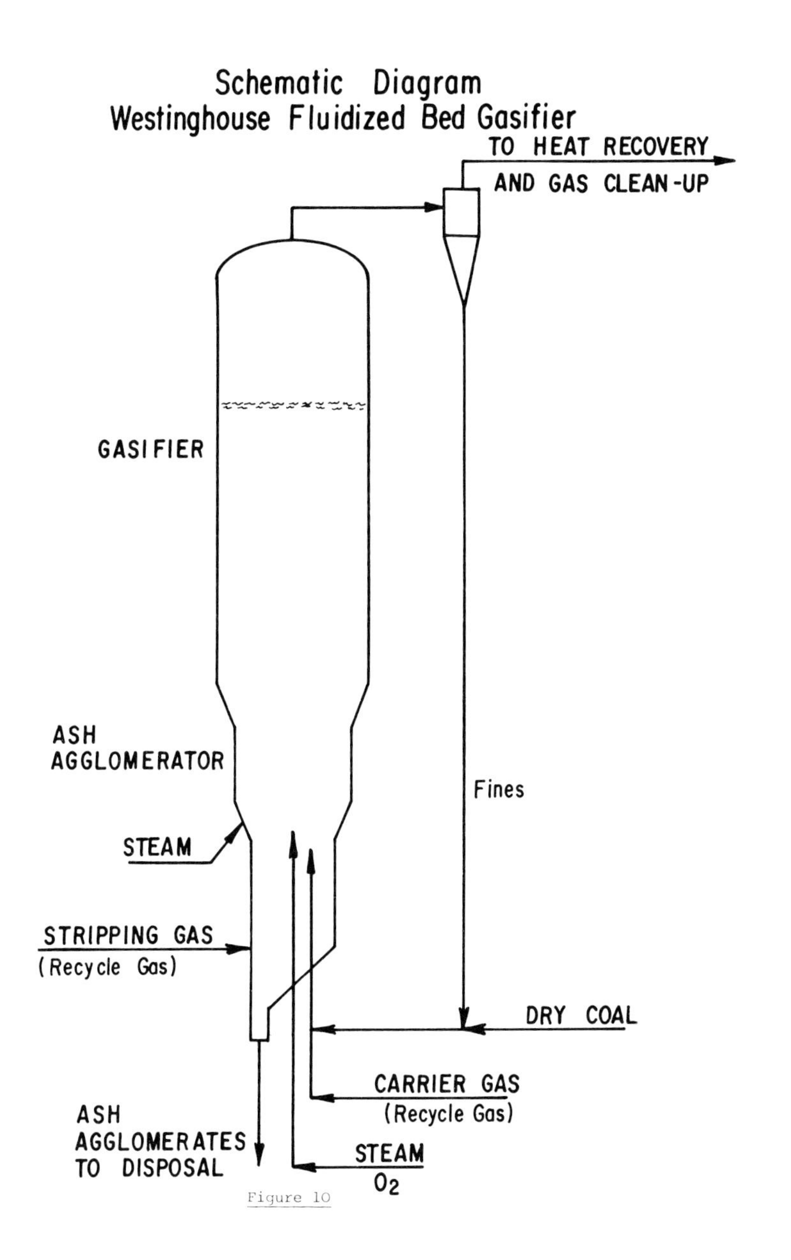
Schematic Diagram
Westinghouse Fluidized Bed Gasifier
TO HEAT RECOVERY
AND GAS CLEAN-UP
GASIFIER
ASH
AGGLOMERATOR
Fines
STEAM
STRIPPING GAS
(Recycle Gas)
DRY COAL
CARRIER GAS
(Recycle Gas)
ASH
AGGLOMERATES
TO DISPOSAL
STEAM
O_2

Figure 10

uncertainty with respect to size is thus compounded with uncertainty with respect to the effect of pressure on throughput rates.

A demonstration plant is currently being designed for the Memphis Light Gas and Water Division under contract with DOE (Loeding, 1979).

4. Bi-Gas and Others

A plethora of other gasification processes are operating or have operated recently in the U.S. and abroad.

One of the largest pilot plant programs currently in progress is the Bi-Gas process.

The Bi-Gas process is a high pressure (50 to 100 atm), two stage, entrained, slagging gasifier developed by Bituminous Coal Research, Inc. Coal is fed to the upper stage where it is devolatilized and partially gasified. Char entrained from the upper stage in the product gas is captured in a cyclone and returned to the lower stage where it reacts with steam and oxygen at slagging conditions. The hot gases from the lower stage provide the heat required in the upper stage. The Bi-Gas gasifier will operate on any coal and should offer a high methane yield. Operation of a 5 ton per hour pilot plant at Homer City, Pa., began in late 1976. Due to problems associated with feeding coal and recycling char, the program has not yet produced useful process data. Recently, however, progress has been made in overcoming the key process difficulties and useful results should be forthcoming.

Space limitations prevent an adequate discussion of other alternative systems. A summary of other important gasification process developments has been recently given (Loeding, 1979, Clark, 1979).

G. Gas Processing

1. Gas Purification

a. Composition of Raw Gas

The major components from coal gasification are nitrogen, carbon monoxide, carbon dioxide, hydrogen, methane, and unreacted steam. The nitrogen content is low when oxygen-steam rather than air-steam gasification is used.

The following components are present in all gasification processes. These are mostly considered as undesirable impurities and must be removed in downstream processing. They are: particulate matter, i.e., ungasified carbon and ash H_2S, COS, CS_2, SO_2, S_2, S_6, S_8, NH_3, HCN, NO, HCl, and argon.

The composition of the coal gas depends in part upon the composition of the coal and the type of gasification unit employed. The above list comprises substantially all components present in the effluent from an entrained gasifier. Ash will, of course, contain all of the metals, usually as oxides and sulfides, which are present in the original coal. Certain toxic elements have been of some environmental concern such as Hg and As. These are present in trace quantities in most coals. They do tend to be concentrated in the effluent gas, however, particularly in slagging gasifiers because of their volatility.

Some of the components listed above are present in very small quantities, from 5-20 ppm. This includes sulfur vapor, SO_2 and NO.

A single stage high temperature fluidized bed gasifier operated at temperatures above about 1200°K will also have substantially no other impurities present than those listed above.

Moving bed gasifiers and multistage fluid bed gasifiers operate in a countercurrent or simulated countercurrent manner with lower gas outlet temperatures. This introduces a whole host of other impurities, as noted below: thiophenes, mercaptans, and other organic sulfur components, C_2H_4 and C_2H_6, C_3 plus hydrocarbons, light oils, tars, monohydric phenols, dihydric phenols, etc.

The quantities, both relative and absolute, of the various components vary of course with the coal feed and the specific process.

The NH_3 and CH_4 concentration will also tend to be higher as compared with entrained type gasifiers while the HCN concentration will tend to be lower.

b. Removal of Particulates and Tars

The usual practice is to cool the exit gas either via a waste heat boiler or by direct injection of water to generate steam in situ. The use of a waste heat boiler prior to tar condensation is not practical with a tar containing gas because of fouling of the heat exchange surfaces.

The particulates are normally removed by water scrubbing after gas cooling to about 180°C. Venturi scrubbers in one or two stages are normally employed for this purpose.

Fluidized bed processes are designed to remove solids as efficiently as possible with dry cyclones before water scrubbing for final purification. The Winkler process places the cyclone after the waste heat boiler. Most of the developing fluid bed processes have used hot cyclones in the pilot plant operations.

A more complex purification system is required when the gases contain tar and light oil vapors. The recovery system in this case is illustrated by that which is used in the Lurgi process. The system was briefly described in section II. E. and in more detail in a recent paper (Becker, 1979).

c. Acid Gas Removal System

These are used to remove carbon dioxide and hydrogen sulfide from the raw gas. It is essential to remove the hydrogen sulfide and other sulfur compounds before the gas is finally available for its end use. The degree of sulfur removal required varies with the application. The sulfur content of the treated gas must be reduced to a level well below one ppm when the gas is used for either methanol or SNG synthesis. When the gas is to be used for hydrogen manufacture, the requirement is not quite as severe.

A number of different classes of acid gas removal systems have been employed in the natural gas industry and for purification of synthesis gases produced from oil feedstocks. These are summarized below as follows (Fleming, 1979):

(1) Aqueous Amine Solutions

There are a number of proprietary processes developed by various vendors using mostly various alkanol amine solutions. None of these processes is currently in use in actual commercial coal gasification units.

A variant of the above type of process is in use in a Winkler process for generation of ammonia synthesis gas in India (Baily, 1979). The process employs the alkazid process. The reagent is the aqueous solution of a sodium salt of an amino acid. The advantage of this process is its high selectivity for removal of hydrogen sulfide relative to carbon dioxide.

(2) Hot Potassium Carbonate Process

The hot carbonate system was originally developed at USBM (Field, 1962) in the 1950s. Its advantage over the amine systems is that it operates at higher temperatures, i.e., at 100-125°C, such that quenched gasifier offgases can be directly processed. This system also shows a reduction in steam requirements for regeneration as compared with other processes.

The hot carbonate process is capable in principle of somewhat selective removal of H_2S by a two stage operation. This variant has, however, not been practiced commercially.

A number of hot carbonate processes (Gas Processing, 1975) are offered by various developers. These differ by the type of promotor used to accelerate acid gas absorption and the detailed engineering of the process. The most common processes are the Benfield, the promoted Benfield, Catacarb, and Giammarco-Vetrocoke.

The Benfield process has been used commercially in oil gasification plants. The H_2S concentration has been reduced to ten ppm and the CO_2 concentration to about 0.8 vol% in these operations. The Benfield process has also been used commercially for purification of town gas produced in the Lurgi plant at Westfield, Scotland. The Westfield plant is no longer in operation.

The Giammarco-Vetrocoke process is used in the Winkler plant in India (Baily, 1979). It is used to remove CO_2 after shift conversion. The H_2S had previously been removed by the alkazid process followed by treatment with iron oxide.

(3) Physical Solvents

A number of physical solvent processes have been developed. The Lurgi Rectisol process is based on the use of refrigerated methanol at ca. -45°C. A number of other organic solvent based processes are offered by various vendors which mostly operate at or near ambient temperatures. The Selexol process of Allied Chemical uses the dimethylether of polyethylene glycol. The Lurgi-Purisol process uses N-methyl pyrrolidine while Fluor proposes to use propylene carbonate.

All of these solvents show the desirable property of selective removal of hydrogen sulfide. Process staging can be used such that on regeneration two separate streams are produced : one rich in hydrogen sulfide and the other rich in carbon dioxide. The removal of organic sulfur compounds is another desirable feature.

The Lurgi Rectisol process, however, is the only one that has been commercially used in coal gasification plants. The largest installation is in the SASOL plant in South Africa (Hochgesand, 1970) where it is used to clean up the gas from Lurgi gasification prior to its use in the Fischer-Tropsch reactors.

The process removes organic sulfur, H_2S and COS down to a very low level, i.e., below 0.1 ppm, and the CO_2 concentration to 0.1 vol% or less (Gas Processing, 1975). Sulfur poisoning of the Fischer-Tropsch catalyst has not been a problem in SASOL operations. The Rectisol process also removes naphtha, HCN, and NH_3 from the treated gas.

The Rectisol system is, however, high in capital and operating costs and lower cost alternatives are being sought.

(4) Mixed Solvents

A number of processes have been developed which combine the functions of physical solvents and chemical reagents, such as amines. The Shell-Sulfinol is an example of such a process. None of these have been used yet with coal generated gases. An installation is planned using the sulfinol process to desulfurize the ammonia synthesis gas in the TVA ammonia from coal project.

(5) Selective Removal of H_2S by Soda Solutions

A number of processes have been developed for selective removal of H_2S from CO_2 bearing gases. These are based on the use of sodium carbonate solutions which are ineffective for removal of CO_2. The absorbed sulfur is released as elemental sulfur by oxidation.

The Stretford process is a prime example of such a process and contains sodium vanadate and anthraquinone disulfonic acid. The purpose of the additives is to promote the oxidation of the absorbed H_2S to elemental sulfur.

The Stretford process and an improvement, the Holmes-Stretford process (Srini, 1979), has been used for desulfurization of coke oven gases. The Stretford process is also being used to remove H_2S from the tail gases of Claus plants and from the CO_2 streams released in the regeneration of hot carbonate systems in coal gasification pilot plants.

Most acid gas systems require a Claus plant to convert the H_2S to sulfur. The Stretford process may be used to remove residual hydrogen sulfide from such systems.

The commercial design of the CO_2 Acceptor process (CCDC, April 1978) is somewhat unique. The CO_2 in the offgas is required for the methanation system since the gas is rich in H_2. The H_2S concentration is rather low also, i.e., 405 ppm. A Stretford process is accordingly used to treat the gas and reduce its concentration to 5 ppm.

(6) Removal of Organic Sulfur Compounds and Naphthas

The Rectisol and perhaps other physical solvents are effective in removing both naphtha and organic sulfur from the gas.

When chemical acid gas removal systems are employed, separate systems must be used to remove these materials prior to acid gas removal. The naphtha may be removed by oil scrubbing. Organic sulfur removal requires treatment of the gas over a cobalt molybdate on alumina catalyst at ca. 370°C. Since the above is also a water gas shift catalyst, removal of organic sulfur compounds may be combined with shift conversion.

(7) Final Purification

Many of the processes other than Rectisol will still leave residual traces of sulfur compounds, principally H_2S and smaller quantities of COS. A final cleanup system is required with the use of a chemical guard chamber. Various materials have been used in the past, such as activated carbon and specially treated iron oxide. The technique which is most usually employed now, however, is ZnO at about 530°K. This has become commercial practice in the gas industry. The technique has proven successful in pilot plant tests of the CO_2 Acceptor process and at Westfield when applied to coal gases. In particular, it has been shown that COS may be removed as well as H_2S as long as steam is present in the gas.

2. Water-Gas Shift and Methanation

a. Water-Gas Shift Reaction

The water-gas shift reaction:

$$CO + H_2O = CO_2 + H_2$$

$$\Delta H = -9830 \text{ cal/mol}$$

has been in use for more than 100 years to produce hydrogen and to adjust the hydrogen/carbon oxides ratio in coal gasification systems. The reaction is especially important for SNG production. It is used in conjunction with carbon dioxide removal to give the proper amounts of CO and CO_2 for reaction with H_2 in the methanation section (Figure 11). The shift reaction is important in all partial oxidation and hydrocarbon reforming processes used for production of hydrogen, ammonia, methanol, and other chemicals.

Satisfactory reaction rates, catalyst life, and a close approach to equilibrium have been achieved with the following catalyst types:

- FeO/Cr_2O_3
- CoO/MoO_3 on alumina
- $Cu/Zn/Cr_2O_3$

The iron based catalyst was the standard shift catalyst until about 1960 when the cobalt/molybdate catalyst came into use. The more recently developed copper-zinc-chrome catalyst is for low temperature (478°K) operation.

The shift reaction section at the Westfield Lurgi coal gasification complex used the cobalt/moly catalyst over a temperature range of about 670-755°K, and at a pressure of 24 atm gauge (Ricketts, 1963). The Westfield system contained

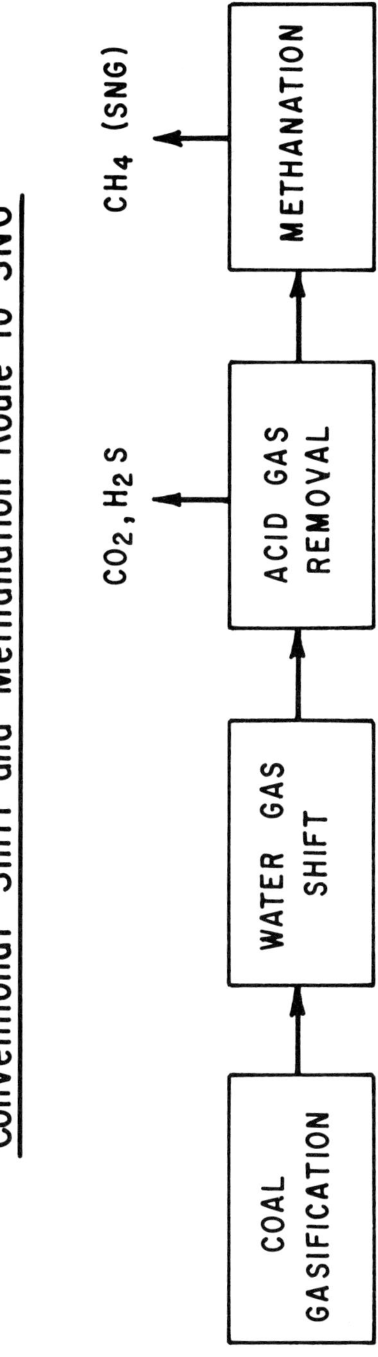

Conventional Shift and Methanation Route to SNG
COAL GASIFICATION
WATER GAS SHIFT
ACID GAS REMOVAL
METHANATION
CO_2, H_2S
CH_4 (SNG)

Figure 11

two adiabatic fixed bed reactors in series with intermediate cooling. This arrangement is typical of shift conversion systems.

b. Methanation

(1) Chemistry and Catalysis

The methanation reactions of CO and CO_2 are given below:

$$CO + 3H_2 = CH_4 + H_2O \qquad \Delta H = -49,300 \text{ cal/mol}$$

$$CO_2 + 4H_2 = CH_4 + 2H_2O \qquad \Delta H = -39,400 \text{ cal/mol}$$

These reactions were first discovered over 80 years ago and had been studied extensively. As shown above, both are highly exothermic which means that the equilibrium content of the methane product varies inversely with temperature. Methane production is also favored by high pressure. Nickel, iron, cobalt and ruthenium all catalyze the methanation reactions, but supported nickel catalysts are the most active and give the best selectivity. All of these catalysts also promote the water gas shift reaction. However, from an equilibrium viewpoint only two of the three reactions (i.e., the two methanation reactions and the shift reaction) are independent, since subtracting the CO_2 methanation reaction from the CO methanation reaction yields the water gas shift reaction.

The methanation reactions are rapid and make a close approach to equilibrium with an active catalyst. There are many problems in attaining satisfactory catalyst performance in SNG production:

- The feed gas must be free of poisons, especially sulfur. Catalyst life will be unsatisfactory unless the sulfur content is below 0.2 ppm.
- High temperature can deactivate catalyst by sintering and can cause carbon formation if other conditions are not satisfactory. High H_2O contributes to catalyst sintering.
- Carbon formation must be prevented by the proper control of carbon oxides/hydrogen ratio, temperature, and feed-gas water content.

c. Coal Gas Methanation Demonstration

A Conoco led consortium demonstrated the commercial scale conversion of coal derived gas to SNG in 1973-1974 (Landers, 1974). This work was done at the Westfield, Scotland coal gasification plant operated by British Gas Corporation. British Gas has considerable expertise in methanation, having done some of the pioneering work in the late 1940's (Dent, 1948). At the Westfield demonstration plant shifted gas from the existing unit was purified in a Lurgi Rectisol system (low temperature methanol absorption) and then converted to high BTU gas in a fixed bed adiabatic reactor system which used gas recycle for temperature control. A commercially available high nickel content catalyst was employed. Prior to this demonstration, methanation reactions had been conducted to remove low levels of carbon oxides from synthesis gas used in ammonia plants. However, this type of methanation is considerably less demanding on the catalyst and reactor systems than coal gas methanation to produce pipeline quality gas.

In August and September of 1974, several thousand Scottish families used 71,000 m^3 per day of SNG produced in the Westfield methanation demonstration unit. This work was a milestone in coal gasification since it demonstrated the last steps necessary for applying commercial coal gasification technology to produce high BTU, pipeline quality gas.

The demonstration plant and parallel pilot plant work showed that coal gas could be purified and converted to SNG very efficiently. Scale up factors were developed, good catalyst life demonstrated, and there were no temperature control or carbon formation problems. The process is available for licensing in commercial plants.

d. Other Methanation Systems

Lurgi developed and tested a methanation system on a slip stream at the SASOL coal gasification complex in South Africa (Moeller, 1974). The purified feed gas contained a large excess of CO_2, which would be removed after methanation in order to produce SNG. Although this work was done on a much smaller scale than the Westfield methanation project, it did indicate satisfactory catalyst life and system performance.

Many companies have done pilot plant studies of methanation (Watson, 1977). The reactor types include fixed bed adiabatic reactors with gas recycle, once-through fixed bed reactors, once-through fluidized bed reactors, and catalyst-in-tube reactors with gas recycle. A slurry system in which catalyst is recycled in an inert heat-transfer liquid has also been investigated.

3. Combined Shift/Methanation Process

a. Comparison with Conventional Methanation

The reactions which occur in a so-called "conventional" methanation system, such as described in the previous section and in the combined shift/methanation system, discussed here are the same. In either case, the CO methanation reaction and the water gas shift reaction are sufficient for describing system behavior since both systems give essentially the same approach to equilibrium when an active catalyst is used. The product gases differ considerably because the feed gases are quite different, as shown in Table 5. The relatively high H_2/CO ratio in the conventional methanation system feed means that carbon oxides are consumed and water is produced as a co-product with the methane. The combined shift/methanation system consumes CO and H_2O while producing a net quantity of CO_2.

In the conventional Conoco methanation system described in the last section, there is essentially stoichiometric hydrogen available for reacting with both CO and CO_2, so the CO_2 content in the final product gas is quite low. In the SUPER-METH® process, the outlet gas from the bulk methanators contains a large amount of CO_2 which must be removed before a final methanation step is performed to produce SNG. Figures 12 and 13 show the conventional and SUPER-METH processes.

The temperature control, carbon formation and catalyst deactivation problems which are present in the conventional methanation system also occur in the combined shift/methanation case, but to a different degree. Similar catalysts

TABLE 5

TYPICAL COMPOSITION OF FEED GASES TO "CONVENTIONAL" METHANATION AND SUPER-METH UNITS

	Mol %	
	"Conventional" Methanation	SUPER-METH Process
Hydrogen	67.1	28.8
Carbon Monoxide	19.8	60.9
Carbon Dioxide	1.8	2.4
Methane	6.8	7.0
$C_n H_m$	0.3	0.3
Nitrogen	0.6	0.6
Water	3.6	--
Totals	100.0	100.0

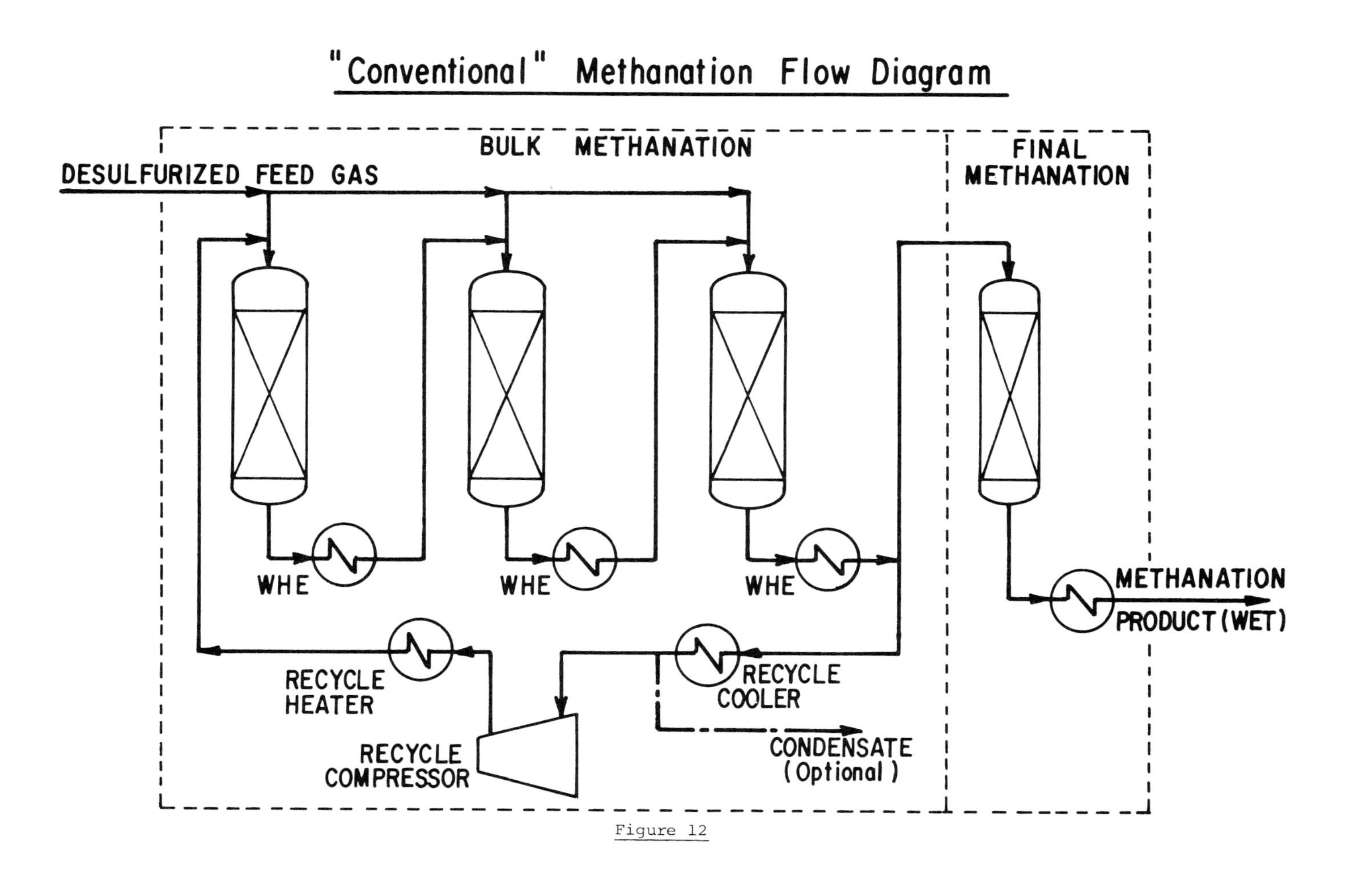
"Conventional" Methanation Flow Diagram
BULK METHANATION
FINAL METHANATION
DESULFURIZED FEED GAS
WHE
WHE
WHE
METHANATION PRODUCT (WET)
RECYCLE HEATER
RECYCLE COOLER
RECYCLE COMPRESSOR
CONDENSATE (Optional)

Figure 12

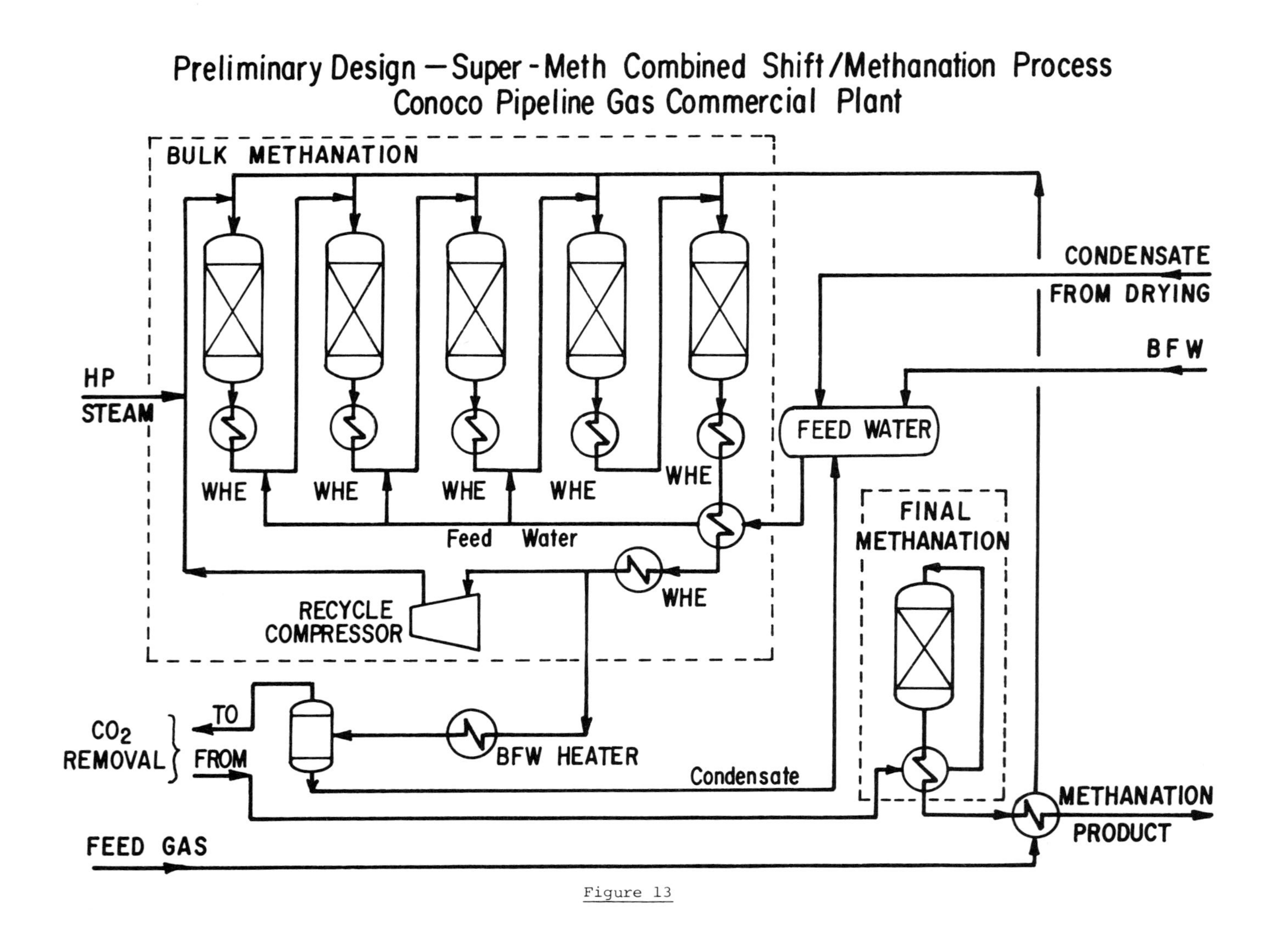
Preliminary Design — Super-Meth Combined Shift/Methanation Process
Conoco Pipeline Gas Commercial Plant
BULK METHANATION
HP
STEAM
WHE
WHE
WHE
WHE
WHE
Feed Water
WHE
RECYCLE
COMPRESSOR
CONDENSATE
FROM DRYING
BFW
FEED WATER
FINAL
METHANATION
CO_2
REMOVAL
TO
FROM
BFW HEATER
Condensate
METHANATION
PRODUCT
FEED GAS

Figure 13

are used in both cases, preferably high nickel supported catalysts. The much greater carbon monoxide content in the SUPER-METH feed means that much more steam must be present than in the conventional methanation case. The high steam content increases the possibility of catalyst deactivation via sintering. The high CO content also means that the minimum inlet temperature for SUPER-METH must be greater than for conventional methanation because the potential for nickel carbonyl formation is greater with the higher inlet CO in SUPER-METH. Temperature can be effectively controlled via gas recycle in SUPER-METH as in conventional methanation and the degree of gas purification for removal of sulfur and other poisons is essentially the same for both processes.

b. SUPER-METH Experimental Results

Koch, et al, (1979) have reported pilot plant results for eight commercially available catalysts which were tested under SUPER-METH conditions. Several of these catalysts are expected to give 1-2 years life in commercial operation. Three catalysts were rated excellent including one that had been tested for more than 900 hours. Most of the catalysts were highly active and gave the expected approach to equilibrium. Process variable studies were done to define the best operating conditions for SUPER-METH when feeding British Gas/Lurgi slagging gasifier composition gas. These tests confirmed similarities between SUPER-METH and conventional methanation, but also showed there were many important differences between the behavior of the two systems.

c. Lower Slagging Gasifier Product Cost with SUPER-METH

Preliminary design and cost estimates for conventional and combined shift/methanation routes of producing SNG via the British Gas/Lurgi slagging gasifier were done as part of Conoco's contract with DOE to design a demonstration plant for Noble County, Ohio. SUPER-METH produced a savings of about 15% ($1/MM BTU) in SNG cost. About 90% of the savings is in plant investment, which is reduced 20% in the SUPER-METH case as compared with conventional methanation. The reduced plant investment results from eliminating the shift reactors, reducing steam and power requirements, reducing the size of the CO_2 removal units and decreasing the system pressure drop.

In the base case study, which used the conventional methanation system, crude gas from the British Gas/Lurgi gasifier was mixed with steam and passed through a shift reactor. CO_2 and H_2S were removed in a dual Rectisol unit and the purified gas methanated in the conventional system. A large excess of steam must be added to the shift reactor inlet in order to produce a satisfactory approach to equilibrium. This excess steam must then be condensed, which results in an increase in the gas liquor stream which must be treated. In the SUPER-METH case, the crude slagger gas is purified in a one stream Rectisol system and then fed to the SUPER-METH unit. CO_2 removal is performed in a relatively low cost hot potassium carbonate system. The water necessary for the shift reaction and for preventing carbon formation is added via condensate injection and steam in the SUPER-METH unit; it is not necessary to use surplus steam as in the conventional unit. This more selective use of steam in the SUPER-METH process is one of the main advantages of combining shift and methanation.

The net thermal efficiency in the SUPER-METH case was 3.4% higher than in the base case with separate shift and methanation units. The thermal efficiency of the SUPER-METH unit was 99%.

Combining shift and methanation in a single unit is most efficient for a synthesis gas which is high in carbon monoxide and low in water, such as slagging gasifier product. The advantage of combining the units decreases as the CO decreases and the H_2O increases. In the standard Lurgi gasifier where ash is removed as a solid, large amounts of steam are added to the gasifier to prevent the ash from melting. In this case, the excess steam required for promoting a good approach to shift equilibrium is already present and would have to be condensed and treated whether the shift and methanation units were separated or not. There would be much less advantage to combining shift and methanation, but a detailed economic comparison has not been made for this case.

d. Other Shift/Methanation Systems

Others have also recognized the benefit of combining shift and methanation for high CO gases. British Gas Corporation has a process with generic similarities to SUPER-METH, but there are important differences in the method of adding H_2O to the system, reactor configuration and operating conditions (Roberts, 1979). Once through, high temperature systems with several reactors in series have been developed by Imperial Chemical Industries (Woodward, 1977) and Ralph M. Parsons Company (White, 1974).

III. DIRECT LIQUEFACTION

A. Definition and History

The direct liquefaction of coal may be defined in the broadest sense as the conversion of coal to a liquid product via reaction with molecular hydrogen. There are many variations in the manner by which the interaction of coal with hydrogen may be conducted and these will be described below. Likewise, the product from coal liquefaction can vary all the way from a heavy residuum "liquid" to a highly refined light distillate such as high octane gasoline. The residuum "liquid" is in reality a relatively low melting solid which is most often referred to as SRC or solvent refined coal.

The commercial interest in coal liquefaction developed in Germany with the invention of the Bergius process. The Bergius process involves the interaction of coal with molecular hydrogen at high pressures, normally greater than 150 atmospheres, high temperatures of the order of 723°K or higher, and in the presence of an iron catalyst.

Development of the Bergius process proceeded in Germany in the years 1927-1943 to a full commercial status largely under the direction of M. Pier of I. G. Farben. By 1943 the total installed hydrogenation capacity, including coal and coal tar feedstocks, amounted to 4,250,000 tons of annual capacity in Germany and the territories she occupied during WWII.

An excellent review of the German hydrogenation art for the period up to 1945 has been given (Kronig,1950). Two more recent reviews cover the state of the art on a broader international basis up to the year 1965 (Donath,1963), (Wu,1968).

The hydrogenation of coal and coal tar as a commercial industry ceased to exist after 1945, primarily since it was not competitive with natural petroleum.

A very extensive development effort is now underway in many countries, but mostly in the USA, to provide a basis for a new and improved commercial coal liquefaction industry. The basic technology involved is still based on the original German developments. Specifics of this development effort will be given in detail below.

B. Fundamentals

The liquefaction of coal is a very complex chemical process and is accordingly not wholly understood. The origin of the coal, its rank, its petrographic composition and the nature of its mineral matter all profoundly affect its performance in a liquefaction system.

Most coals of bituminous or lower rank and with a volatile matter on MAF basis greater than about 30 wt.% respond reasonably well to liquefaction.

A characteristic feature of these coals is that there are a number of weak points of attack within their chemical structure. Fragmentation of the coal into lower molecular weight pieces at the above weak points can occur either by purely thermal pyrolysis or by direct attack via specific chemical reagents.

Thermal pyrolysis of the coal in the absence of a stabilizing reagent leads to the formation of unstable fragments which undergo subsequent degradation to produce coke. Liquefaction processes are all conducted under conditions where stabilization of the fragments of the primary conversion of coal is readily effected. Such stabilization is usually effected by conducting the liquefaction in the presence of a hydrogenated solvent with or without the simultaneous presence of free hydrogen under pressure. Stabilization is effected by capping the unstable fragments with active hydrogen donated from the solvent. It is generally considered that under normal circumstances a free radical mechanism is involved in this process.

Molecular hydrogen is much less reactive than a hydrogen donor solvent as a stabilizing or "capping" reagent. It is in fact likely that the main role of hydrogen is to replenish the donor solvent by in situ hydrogenation of the hydrogen depleted solvent. A catalyst is usually required to carry out this process. In many cases the mineral matter of the coal, particularly if it is rich in iron-bearing minerals such as pyrite, provides sufficient catalysis for the purpose.

Liquefaction of coal via molecular hydrogen has been conducted in the absence of a solvent provided it is either intimately mixed with an added catalyst or if it contains a finely dispersed natural catalyst in the form of iron minerals (Mukherjee, 1976), (Okutani, 1979). The mechanism involved here may involve intermediate production of hydrogen donor materials via catalytic hydrogenation of the coal or fragments derived from it. Some processes have been proposed which avoid the use of a solvent but for the most part they suffer from the disadvantage that relatively large amounts of coke are produced (Schroeder, 1976).

Active donor materials have been identified as partially hydrogenated polynuclear aromatics or heterocyclic compounds. Typical examples of donor compounds are such materials as tetrahydronaphthalene, di- and tetrahydrophenanthrene, tetrahydroquinoline, etc. The partially hydrogenated aromatics are also excellent solvents for the primary products of coal liquefaction. Solvents of this general type are fortunately produced as natural products of coal liquefaction. Completely hydrogenated ring systems such as decalin and perhydrophenanthrene are unacceptable as donor solvents.

One of the objects in development of practical coal liquefaction processes is proper management of the solvent. It is necessary to maintain both solvent balance and solvent quality. The latter is dictated primarily by the hydrogen donor activity of the solvent and secondarily by its solvency for coal-derived products.

Other active hydrogen donors were considered prior to the development of active hydrogenated polynuclear systems. The earliest was the use of hydrogen iodide for liquefaction of coal (Bertholet, 1870). Fischer (1921) also showed that active hydrogen from sodium formate or generated from the interaction of CO with H_2O is effective for liquefaction of German brown coal. The latter materials in all likelihood operate through an ionic rather than a thermal or free radical mechanism as will be noted later.

The fundamentals of the primary liquefaction of coal via the non-catalytic donor solvent mechanism has been studied more extensively than any other aspect of liquefaction. The original work was conducted with the use of tetralin and tetralin cresol as model solvents (Pott, 1934).

A study of the reaction kinetics of donor extraction using tetralin as a model compound was carried out by investigators at Consolidation Coal Company in connectio with the development of the CSF process for liquefaction (Curran, 1967, OCR, 1968, Gorin, 1971). More recently an extensive study of the chemistry and kinetics of the donor extraction process was carried out by investigators at Mobil Oil (Whitehurst, 1977). A number of other investigators have also studied donor extraction processes usually in the presence of impressed hydrogen gas (Hariri, 1964, Cronauer, 1978).

Although all investigators are not in complete agreement, the following general picture emerges about the nature of primary liquefaction:

> The process is initially very rapid and the major primary product is a heavy non-distillable cresol-soluble residuum usually referred to as SRC. Relatively smaller quantities of hydrocarbon gases and distillate oils are simultaneously produced.
>
> High volatile bituminous coals are most readily converted. The initial conversion rate is very high at temperatures above 673°K if an active donor solvent is maintained. Conversions of MAF coal of 80% or better to cresol soluble and lighter products can be achieved, for example, at residence times as low as two minutes at 700°K.
>
> The ease of conversion decreases generally as the coal rank is decreased. This is largely a consequence of the high oxygen content of the lower rank coals which has the effect of requiring a larger hydrogen input to achieve a given level of conversion. A rough rule of thumb, however, is that the rate of hydrogen donation is independent of the coal rank from high volatile bituminous to lignite.
>
> Coal conversion also decreases when the feed coal is rich in the maceral group termed inertinite and particularly when large quantities of the macerals fusinite and semifusinite are present.
>
> The initial rate of elimination of oxygen as H_2O, CO_2, and CO and of sulfur as H_2S is also relatively rapid. The rate of elimination of sulfur and oxygen slows down very markedly after about 50% of the sulfur and oxygen are removed. Thus, highly refined products cannot be achieved in the absence of selective catalysis.
>
> In contrast to elimination of sulfur and oxygen, the elimination of nitrogen occurs to only a very minor extent. The SRC product is actually richer in nitrogen than the parent coal.
>
> High conversions to cresol solubles are more readily achieved with higher molecular weight donors such as partially hydrogenated phenanthrenes as compared with tetralin. The presence of heterocyclic compounds as well as hydrogen gas under pressure also tend to improve conversions. The use of hydrogen gas appears to be essential to achieve conversions above 90% in practical systems.
>
> The conversion of the SRC to distillate oils proceeds very much more slowly than the original conversion to SRC. The first fraction of the SRC is readily converted to distillates by a purely thermal hydrogen donor cracking process. The process

slows down rapidly as the more reactive portions of the SRC are converted. High conversions to distillate therefore cannot be achieved without use of hydrogen pressure and use of a catalyst. Many of the processes to be discussed below are content to achieve conversions to distillate plus gas of the order of 70% or less. For highly reactive coals this apparently can be achieved by the natural catalytic action of the coal ash. No externally added catalyst is required.

The fundamentals of the conversion of the SRC to distillate oils has been studied but not reported in detail. Kinetic models for design purposes have been developed for some of the processes, such as H-coal for example. These are, however, considered to be proprietary and cannot be referenced. Consolidation Coal investigators (Gorin, 1971) have studied the hydrocracking of SRC in an ebullated bed catalyst system. The hydrocracking reaction can be represented empirically as a second order reaction. Such an expression, in essence, takes cognizance of the decreasing ease of hydrocracking as the conversion proceeds.

The nature of the mineral matter in coal is obviously of importance in determining the catalytic effect it may have in the coal conversion process. Coal also contains certain organo-metallic compounds such as chelates of titanium and organic salts of calcium, magnesium, and sodium. These have an adverse effect on catalyst life when a contact catalyst (Gorin, 1980) is used with the process.

A number of other considerations are of importance in designing a commercial liquefaction system such as vapor-liquid equilibrium, mass transfer of hydrogen into the liquid, rheological properties of coal slurries, etc. Space does not permit consideration of these topics.

C. Products and Characterization

Direct liquefaction (i.e., hydrogenation) produces coal liquids which, depending on the severity of the hydrogenation, would be expected to reflect in part the composition of the original coal. Because of the resultant variety of chemical components, most methods of analyses are devised to separate coal liquids into a limited number of component categories.

The most commonly applied method of separation is on the basis of volatility. Typically, direct coal liquefaction products consist of gases (C_1 x C_3), naphtha (C_4 x 200°C), heavy distillates (200 x 500°C) and resid(500°C+). The net yields of products in these boiling ranges are highly dependent on the nature of the process, and the coal processed, as discussed below. However, certain generalizations can be made concerning the material in the different fractions.

Naphtha

Coal-derived naphthas differ from straight run petroleum naphthas in their higher aromatic character and greater ratio of naphthenes to paraffins. Because of its aromatic nature, coal is potentially better suited to the production of high octane gasoline, than to higher molecular weight transportation and boiler fuels. The chemical composition of a naphtha can be well specified by gas chromatography (GC) analysis (particularly by capillary column GC) and/or mass spectrometry (MS). Various tests, standardized by the ASTM, are used to measure the suitability of the naptha for use as a fuel or as a feedstock for further upgrading. Such tests include insolubles, gum formation, acid number, and copper corrosion, in addition to measurement of physical properties, combustion characteristics, and elemental analyses for potential pollutants.

Middle Distillates

The middle distillates from coal liquefaction are important both as net products (fuel oil or synthetic crude) and for recycle in the liquefaction process as coal slurry vehicles. In their latter function, the recycle distillation must serve as physical media for hydrogenation and have chemical properties conducive to the liquefaction reaction. The recognition that the chemical properties of recycle distillates, such as their ability to donate hydrogen in the liquefaction reaction, can have an important effect on process performance has created interest in analytical methods for assessing these properties. ^{1}H- and ^{13}C nuclear magnetic resonance (NMR) spectroscopy and gas chromatography/mass spectrometry are useful in characterizing these materials.

Heavy coal distillates consist largely of condensed aromatic molecules, and their alkylated homologs. Phenanthrene and pyrene are usually major components, and 50% to 75% of the carbon is found in aromatic ring structures. Oxygen containing compounds include phenolic and furan functionalities, with sulfur occurring as aromatic thiophenes. Nitrogen is found in aromatic and non-aromatic rings. Coal hydrogenation distillates also contain a perhaps surprisingly large percentage of paraffins and naphthenes, between 5% and 15% depending on the processing conditions, with normal paraffins ranging from about C_{14} to C_{30}.

It is possible, by high resolution low voltage (Lumpkin, 1977) or field ionization mass spectrometry (St. John, 1977, and Anbar, 1977) to obtain an extremely detailed compositional analysis of heavy coal liquefaction distillate products. If information on individual isomers of the same exact mass is desirable, the high resolution mass spectrometric techniques can be combined with gas chromatography and/or other methods of preseparation, to increase the specificity of the data. A common method of preparative separation is by liquid chromatography, either on an absorption or reverse phase column, to obtain fractions of the liquid differing on the basis of chemical functionality.

Assessing the heavy distillate as a chemical feed to the liquefaction reactor (i.e., recycle slurry oil) requires some measurement of its ability to donate hydrogen and to act as a solvent for the coal liquids produced. The importance of the hydrogen donor function depends on the process. Exxon has reported a relationship between process performance and solvent quality defined by a proprietary "solvent quality index", presumably derived from mass spectrometric data on the composition of the hydrogenated recycle distillate in the EDS process (Furlong, 1974). Conoco Coal Development Company (Kleinpeter, 1979) has studied recycle slurry oil properties in the SRC process. In the CCDC method a standard coal is extracted at(673°K) with the recycle solvent in question and conversion of the coal to THF solubles is determined as a measure of donor hydrogen concentration and kinetic activity. The extraction data are found to correlate with the ratios of condensed-to-uncondensed aromatic protons and cyclic-to-alkyl aliphatic protons as determined by ^{1}H-NMR analysis of the recycle solvent. ^{1}H-NMR analysis is a valuable technique for rapidly assessing solvent quality of recycle slurry oils.

Resid

All direct hydrogenation process yield a heavy resid (500°C$^+$) which can be separated from the unreacted coal by dissolving it in a solvent such as pyridine, tetrahydrofuran (THF), or cresol. A common analysis of the soluble resid is to separate it quantitatively into hexane solubles (oils), hexane insoluble-benzene solubles (asphaltenes), and benzene insolubles (preasphaltenes, asphaltols) by Soxhlet extraction. These fractions are operationally defined, and various techniques and solvents have been used in analogous separations. Because of the range in experimental methods and liquefaction processes it is only possible to give general ranges, but soluble resids usually contain 20 to 80% oils, 20 to 50% asphaltenes, and 10 to 50% benzene insolubles.

Separation of resid on an absorption chromatographic column (such as alumina or silica) can give information on its chemical constituents. A method of this type, called SESC, has been developed by Mobil Research and Development Corporation (Farcasiu, 1977). In the SESC analysis a sequence of nine solvents is used to separate the liquefaction resid into nine fractions of varying chemical functionality, using a silica gel column.

The molecular size distribution of the resid can be measured by gel permeation chromatography (GPC) (Coleman, 1977) although the lack of suitable standards makes it difficult to obtain accurate molecular weight data. Low voltage (Lumpkin, 1977) and field ionization mass spectrometry (St. John, 1977) have also been used with some success to measure molecular weight profiles, and in some cases, determine empirical formulae of resids. Two factors can compromise the quantitative validity of this approach. First, it is, in general, necessary to assume that each component gives the same response. If they do not, the quantitation will be in error accordingly. Second, only that portion of the sample which can be volatilized will be observed. The heavier liquefaction resids are unlikely to fully meet this criterion. Average molecular weights can also be determined by vapor phase osmometry (VPO). The accuracy of this method will depend upon the degree to which the resid is truly soluble (not associated or colloidal) in the solvent used for the measurement. VPO data are likely to give high estimates of the average molecular weight of a liquefaction resid. Data obtained on samples from various processes and by different methods would suggest that coal liquefaction resids have average molecular weights between 500 and 1500 amu.

D. Generic Processes

Specific processes now under development for the liquefaction of coal will be discussed in a later section. Generic type processes will be discussed in this section. Only limited examples of specific processes will be given.

A practical coal liquefaction process will normally, but not always, require multistage processing. The number of processing stages is a function of the type of process and the desired end product.

1. Non-Catalytic-Hydrogen Donor Systems

This type of process involves the thermal solvation of coal in a solvent which possesses hydrogen donor activity but in the absence of impressed hydrogen gas. The process is non-catalytic and proceeds by virtue of the thermal decomposition of the coal in the presence of the donor solvent. The solvent used normally is one that is indigenous to the overall process. The active donors consist of partially hydrogenated polynuclear compounds. Hydrogen accordingly must be added to the solvent by a catalytic process but this is external to the coal conversion process.

The coal conversion is conducted under relatively mild conditions, i.e., at 10-30 atmospheres pressure, 673-723°K and 5-30 minutes residence time. The major product is a heavy residuum or coal extract. Conversions, because of the mild conditions, are usually limited to values below 80-85%. The CSF process is an example of such a process (OCR 1968, Gorin, 1971).

2. Other Donor Systems

Work in recent years has been conducted with other donor systems as a follow up to the original work with sodium formate as a donor and CO plus H_2O as a source of "nascent" hydrogen (Fischer, 1921). It has been shown, for example, that secondary alcohols (Curran, 1967) also may function as donors. The process has been shown to be accelerated and extended to primary alcohols via base catalysis (Ross, 1979). These systems likely involve an ionic mechanism such as by hydride ion transfer. No attempt to date has been made to develop a practical liquefaction system based on the above principles.

Work is underway also on the so-called CO-steam (DelBel, 1975, Willson, 1979) process. It is likely that a formate intermediate is generated in situ in this type of process.

3. Solvation of Coal under Hydrogen Pressure - Hydroextraction

A process of this type involves the solvent extraction of coal under an applied pressure of hydrogen. Operating conditions are more severe than in the pure donor mode of extraction. Higher pressures, in particular in the range of 100-170 atmospheres, and somewhat higher temperatures are usually employed.

There are two types of processes. The first generates the donor solvent via in situ hydrogenation of the recycle solvent. The so-called SRC-I process is an example of such a process (Lewis, 1978). The mineral matter of the coal provides the catalysis required.

The second type provides for more control of the donor content by external catalytic rehydrogenation of the recycle solvent. Undoubtedly some in situ rehydrogenation takes place. Catalysis via the mineral matter is again depended upon to provide this "secondary" function. The Exxon-donor solvent process is the prime example of this type of operation(Exxon,1979,Epperly, 1979). One of its virtues is that it reduces dependence upon the somewhat uncertain catalysis via the coal mineral matter.

Like the pure donor systems, improved performance is obtained with high volatile bituminous coals as compared with lower rank coals.

4. Contact Catalyst Systems

The above systems are generally operated in a similar manner to the hydroextraction processes discussed above. A major difference is that a catalyst is added to intensify the catalytic action. The catalyst performs two functions, i.e., rehydrogenation of the solvent and hydrocracking of the SRC residuum. The latter function is particularly important when the major product desired is a distillate. More severe operating conditions, i.e., lower coal space velocities and higher temperatures and pressures are required than when the major product is a residuum.

The original Bergius process (Kronig, 1950, Donath, 1963) as further developed by I. G. Farben is an example of such a process. The catalyst used in this case is either red mud or ferrous sulfate impregnated on the coal. In neither case is recovery of the catalyst sought, i.e., relatively small "throwaway" amounts of the order of 1-2 wt.% of the coal are used.

Other processes have been carried out, at least on a pilot scale with other catalytic ingredients impregnated directly on the coal. Metal halides (Wu, 1968), such as, $SnCl_2$, $NiCl_2$ and $ZnCl_2$, have been employed in this manner. Molybdenum catalysts have also been used via impregnation of the coal with $(NH_4)_2MoO_4$.

Other catalytic processes have been tested on a laboratory scale in which the catalyst is impregnated on the coal and then the coal is processed in the dry phase entrained in hydrogen (Schroeder, 1976, Wood, 1976). Relatively large amounts of coke are produced in such processes.

Finally, several processes are being developed wherein the coal slurry in recycle oils is contacted with hydrotreating catalysts. The catalyst is of the type used in petroleum hydrofining operations and is usually a mixture of the sulfides of cobalt-nickel and molybdenum supported on anhydrous alumina base. The catalyst is normally used in the form of an ebullated bed. A prime example of such an operation is the H-coal process which will be discussed in detail below.

5. Homogeneous Catalysis

A number of homogeneous type catalysts are known to be effective for hydrogenation of pure compounds in organic chemistry. Homogeneous catalysis in coal conversion has been mostly limited to the use of molten halides of the Lewis acid type. Finally divided coal can dissolve in such materials by way of forming complexes between the Lewis acid and the heteroatoms in the coal.

Strong Lewis acids are very effective in promoting the hydrogenolysis of coal to light distillate products. The process occurs at much lower temperatures than with other catalytic systems. Significant rates of hydrogenolysis are achieved at temperatures as low as 600°K. Various metal halides were tested both by Conoco Coal Development Company (Zielke, 1966 and 1976, Struck, 1969) and Shell Development Company (Wald, 1973, Kiovsky, 1973). Zinc chloride was selected as the most practical of these catalyst systems for further development. It has the interesting property of converting coal directly (Zielke, 1980) to high quality gasoline in a single hydrocracking step. This interesting process will be discussed in more detail below.

It has also been shown that in strongly acid systems high conversions of coal can be achieved at even lower temperatures, as low as 473-523°K. Examples of such systems are "super" acids such as $HBr\text{-}AlBr_3$ (Ross, 1976) and methanolic solutions of $ZnCl_2$ (Shinn, 1979). The major product under these mild conditions is, however, a residuum rather than a distillate oil.

6. Hydropyrolysis of Coal

This is a non-catalytic process in which coal is treated in the entrained state in a stream of hydrogen under pressure. Emphasis is now on flash hydropyrolysis where the process is conducted at high temperatures, i.e.,1673°K at very short contact times of the order of seconds or less (Oberg, 1978).

The process produces interesting quantities of light aromatic hydrocarbons but the principal products are hydrocarbon gases and char. It is therefore strictly speaking, not a liquefaction process and will not be discussed further.

E. Advanced Systems

1. SRC-I

The Solvent Refined Coal-I (SRC-I) process produces a clean solid fuel from coal as a power plant fuel for the generation of electricity. The schematic flow diagram of a conceptual commercial plant is as shown in Figure 14. Product distributions and operating conditions for processing an Illinois #6 coal are compared with other coal liquefaction processes in Table 6. Overall thermal efficiency based on Kentucky 9 and 14 coal is compared in Table 7.

a. Process Description

Raw coal is pulverized and dried in Coal Preparation as shown in Figure 14. The dry pulverized coal is then weighed and continuously fed to Slurry Preparation where it is mixed with sufficient recycle process solvent to produce a coal-solvent slurry of about 40 wt.%. The coal slurry is pumped to Hydroextraction where it is reacted with hydrogen at reaction temperature and pressure to produce primarily extract or solvent refined coal (SRC). Some gas, distillate and water are also formed. The remainder is undissolved coal and ash.

Reactor hot effluent vapors are separated from the reacted slurry containing solvent extract and residue solids. Distillate and water are condensed

SRC-I Process Schematic Block Flow Diagram

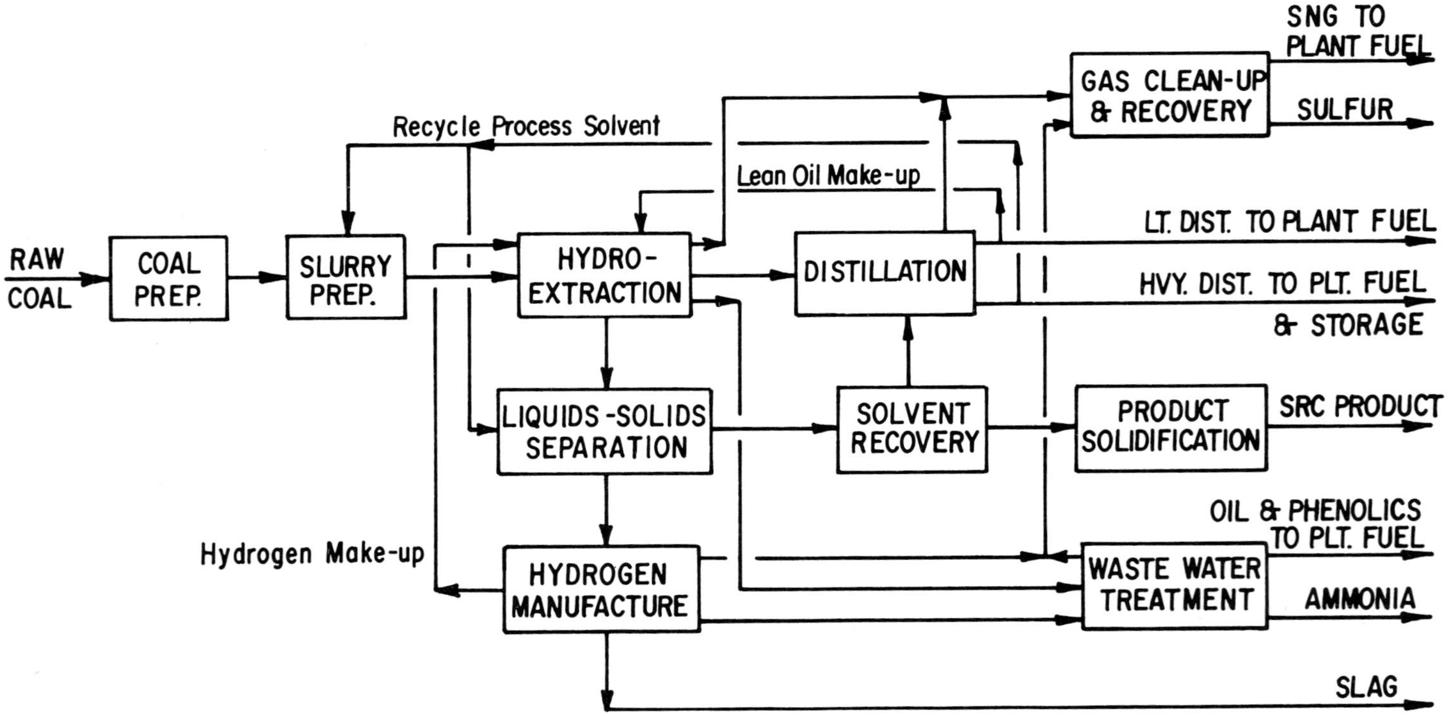

Figure 14

TABLE 6

PRODUCT DISTRIBUTION - OPERATING CONDITIONS
VARIOUS LIQUEFACTION PROCESSES

Process	SRC-I	SRC-II	Exxon[1] -EDS	H-Coal Syncrude	H-Coal Fuel Oil	$ZnCl_2$
Coal	← Illinois #6 →					Colstrip Subbitu-minous
Pressure, atm gauge	129	129	102	204	204	190
Temperature, °K	716	727	722	727	727	685
Coal Rate, $kg/hr/m^3$	625	440	560[2]	497	1188	850
Product Yields, wt % MAF Coal						
C_1 x C_3	2.9	12.5	7.3	13	7	7.1
C_4 x 200°C	3.7	8.6	23.8	21	16	68.5
200°C x 538°C	12.2	34.7	15.0	33	28	--
+538°C	68.7	33.3	41.8	22	33	4.1
H_2 Consumption, wt % MAF Coal	2.3	4.0	4.3	5.3	4.5	8.4

(1) Once-through data from 100 lb/day RCLU unit, excluding flexicoking yield data.

(2) Estimated from residence time data.

(3) Pure H_2 feed — commercial pressure would be higher.

TABLE 7

OVERALL THERMAL EFFICIENCY
VARIOUS DIRECT LIQUEFACTION PROCESSES

Process	SRC-I	SRC-II	H-Coal Fuel Oil	Exxon EDS		$ZnCl_2$ Hydrocracking
Coal	Kentucky 9-14	←	Illinois #6		→	Colstrip Mont Rosebud Seam
Thermal Efficiency, % to Products				(a)	(b)	
Gas	--	9.6	--	--	NA[1]	8
C_4 x 200°C	--	10.8	14.0	34	"	55
200 x 538°C	9	49.9	59.8 (200 x 538°C + SRC)	21.6	"	--
SRC	64	--		--	--	--
Total	73	70.3	73.8	55.6	63.6	63
Reference:	(Schmid, 1977)	(Schmid, 1976)	(HRI, 1979)	(EDS, 1979)		(Pell, 1979 Greene, 1979)

(1) NA = No figures available.

(a) Gas used for H_2 manufacture.
(b) Vacuum bottoms used for H_2 manufacture.

from the flashed vapors. Distillate is sent to distillation and water is sent to waste water treatment. The now hydrogen-rich gas is scrubbed in an oil absorber to remove most of the acid gas and heavier hydrocarbon gases. The hydrogen-rich gas is then compressed and recycled to minimize hydrogen solubility losses and to maintain hydrogen partial pressure in the reaction. Fresh hydrogen from gasification is mixed with the recycle gas to make up chemical, solubility, and mechanical losses. The absorbed acid gases are flashed from the absorber-rich oil at low pressure and sent to gas clean-up. The clean gas is used as plant fuel.

The hot letdown slurry after being cooled to 590°K is sent to solids separation where the residue solids, undissolved carbon and ash, are removed. The raw SRC product containing extract and a large amount of solvent is sent to solvent recovery. The separated residue solids are gasified to produce fresh hydrogen for the process in hydrogen manufacture.

The raw product is heated and vacuum flashed to recover the solvent and to produce a low-ash, low-sulfur, liquid SRC product for solidification. Liquid SRC is pumped to solidification where it is solidified and stored for shipment as a clean power plant fuel.

Recovered solvent and distillate streams are processed in distillation to produce a process recycle solvent (200 x 538°C boiling range) for slurry preparation. Solvent losses to SRC product and residual solids are made up from process produced heavy distillate. Excess heavy distillate and light distillate (C4 x 200°C boiling range) are sent to storage for fuel and sales.

b. Summary of Status

The process development has successfully progressed from the laboratory through the pilot plant stage. The process has been demonstrated on a pilot plant scale at the Southern Services Co./EPRI/DOE 6 TPD pilot plant at Wilsonville, Alabama and at the DOE 50 TPD pilot plant at Fort Lewis, Washington (EPRI, 1975, 1977, 1978, and 1979).

The Wilsonville pilot plant has been in operation since January, 1974, and has successfully processed nine different coals. The Fort Lewis pilot plant has been in operation since October, 1974, and has processed two different coals in the SRC-I mode.

c. Future Development

Plans for future operation of the Wilsonville pilot plant include continued evaluation of improved solid-liquid separation and product solidification processes. Improvements in the SRC process, such as SRC hydrogenation to produce a clean fuel oil, also will be evaluated.

The construction and operation of a 6,000 TPD commercial-type demonstration plant is in the design and planning stage. The possibility of funding such a demonstration plant is currently under discussion by government and industry.

d. Commercial Plans

Commercial plans for production of a clean, low-sulfur boiler fuel for electrical generation depend on the successful demonstration of the 6,000 TPD plant.

e. Advantages Over Prior Art

Compared to the German Bergius process the SRC process operates under much milder conditions with no added catalyst. The SRC process is an improved version of its forerunner, the Pott-Broche process.

f. Major Process Uncertainties

The major process uncertainty is the design of the feed-slurry fired heater. Direct scale-up of the pilot plant heaters is not possible. Careful design based on current operating experience will result in a workable heater but design confirmation and optimization can be achieved only by operation of the first commercial-type heater. Commercially reliable and economic methods of solid separation and product solidification also remain to be demonstrated. There are size limitations for the slurry charge pumps and metallurgy problems for some critical service slurry pumps in solids separation. Pumps could be a major operation and maintenance problem in a commercial-size plant (Air Products, 1977).

Some difficult problems remain to be solved, but with continued intense research and developmental work do not appear to be insurmountable.

2. SRC-II

The Solvent Refined Coal-II (SRC-II) Process produces primarily a clean distillate fuel oil in contrast to the solid fuel produced in the SRC-I process. The process also produces significantly more gas and light distillate products. To achieve the increased conversion of dissolved coal (extract) to liquid and gaseous products, a portion of the product slurry is recycled as a solvent for the coal instead of a solids-free heavy distillate as is done in the SRC-I process. This simplifies the process scheme and makes possible the omission of the solids separation and product solidification steps. The schematic flow diagram of a conceptual commercial plant shown in Figure 15 differs from the SRC-I process in this respect. Product distributions and operating conditions for Illinois #6 coal are compared with other liquefaction processes in Table 6. Overall thermal efficiency is compared in Table 7.

a. Process Description

Raw coal is pulverized and dried in coal preparation as shown in Figure 15. Pulverized coal is then mixed in slurry preparation with hot recycle slurry and enough distilled solvent to maintain a pumpable slurry consistency. The coal-recycle slurry/solvent mixture is pumped to hydroextraction where it reacts with hydrogen at reaction temperature and pressure to produce primarily a clean liquid fuel product. The hydrogenation and hydrocracking reactions are enhanced by the recycle of dissolved coal to the reaction zone, the increase in the per-pass residence time, and the resulting increase in overall effective residence time.

Hot vapors are separated from the reactor effluent slurry product. Light distillate condensed from the hot vapors is sent to distillation and the condensed sour water is sent to waste water treating. The noncondensed gas is sent to gas purification where, first, acid gas is removed and, second, pipeline gas and LPG are recovered in the cryogenic unit for sales. The remaining hydrogen-rich gas is mixed with fresh hydrogen make-up from hydrogen manufacture and recycled to hydroextraction.

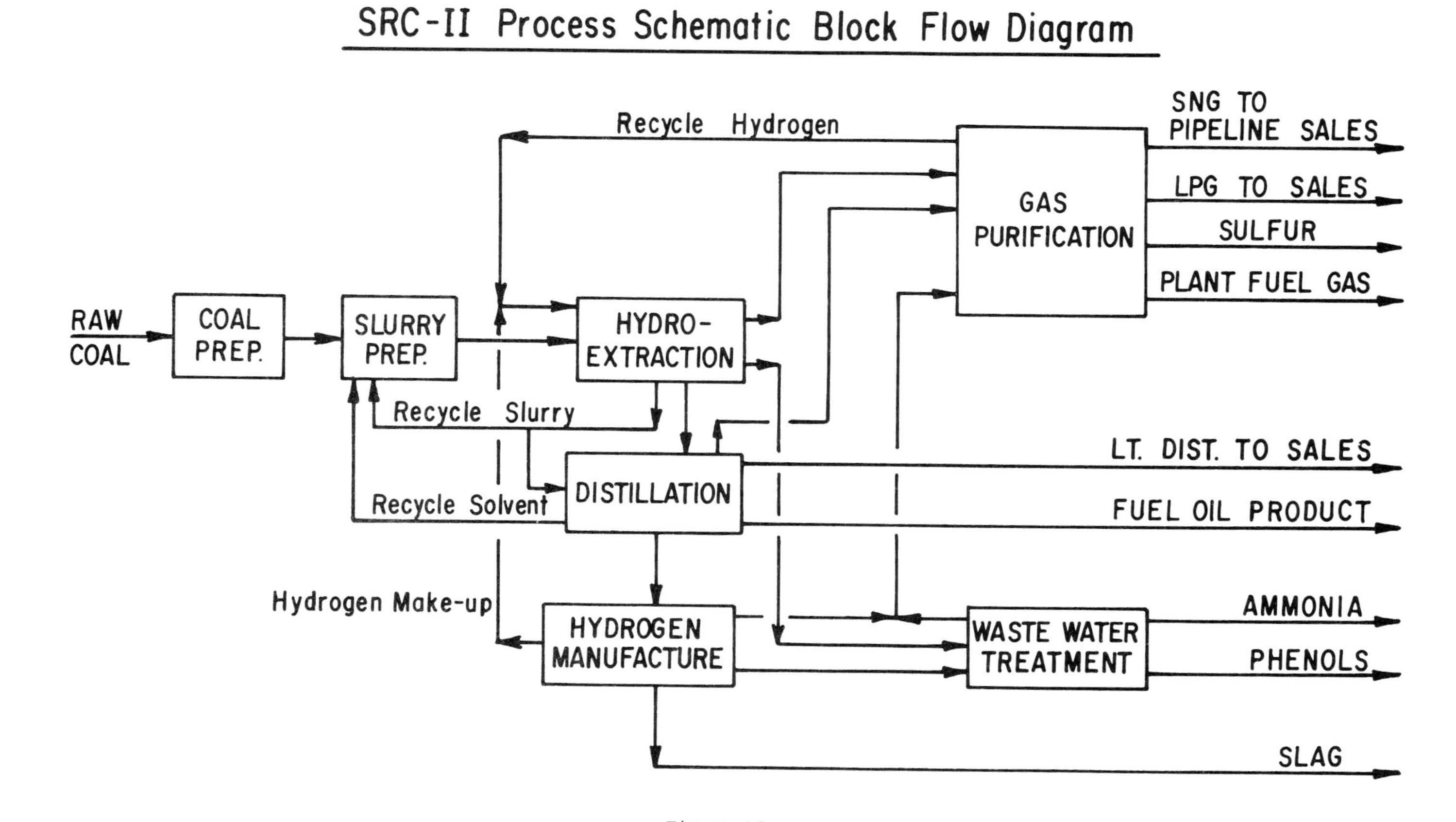
SRC-II Process Schematic Block Flow Diagram
RAW COAL
COAL PREP.
SLURRY PREP.
HYDRO-EXTRACTION
DISTILLATION
HYDROGEN MANUFACTURE
GAS PURIFICATION
WASTE WATER TREATMENT
Recycle Hydrogen
Recycle Slurry
Recycle Solvent
Hydrogen Make-up
SNG TO PIPELINE SALES
LPG TO SALES
SULFUR
PLANT FUEL GAS
LT. DIST. TO SALES
FUEL OIL PRODUCT
AMMONIA
PHENOLS
SLAG

Figure 15

The slurry product, after pressure letdown, is split into two major streams. One slurry stream is used as recycle "solvent" for the process, while the other stream is sent to distillation for separation into the major products. Heavy distillate from vacuum flashing together with middle distillate makes up the clean fuel oil product. Light distillate is also produced as a product. A portion of the fuel oil product can be recycled as solvent to maintain pumpability of the feed coal slurry.

The bottoms from the vacuum flash, containing a small amount of distillate, all of the SRC and residue solids, are sent to hydrogen manufacture to produce the fresh hydrogen for process make-up. The Texaco process is one of choice for hydrogen manufacture.

b. Summary of Status

The process development has successfully progressed from the laboratory through the pilot plant stage. The process has been demonstrated on a pilot plant scale in the DOE 50 TPD pilot plant at Fort Lewis, Washington. The pilot plant has been in operation since October, 1974.

Operation was switched from the SRC-I mode to the SRC-II mode on May 11, 1977. Three different coals have been successfully processed in the SRC-II mode. (Pittsburg and Midway, 1979), (Schmid,1976, 1977)

c. Future Development

Plans for future operation of the Fort Lewis pilot plant include installation of a new slurry fired heater. A Lummus deashing system has been installed and is now operating.

The construction and operation of a commercial-type SRC-II demonstration plant is also planned. The possibility of funding the demonstration plant is currently under discussion by government and industry. West Germany and Japan have each pledged to fund 25 percent of the demonstration plant cost.

d. Commercial Plans

Plans to build a multi-train commercial plant depends on the successful demonstration of the proposed 6,000 TPD plant.

e. Advantages and Disadvantages Over Prior Art

Like the SRC-I process,the SRC-II process operates under much milder conditions than the German Bergius process. Use of recycle slurry "solvent" in this process results in an equilibrium activity of the mineral residue "catalyst" sufficient to reduce the SRC and residue solids to the exact amount necessary for hydrogen production. This results in elimination of the difficult liquid-solids separation and product solidification steps and a process simplification.

The disadvantages are high hydrogen consumption and feedstock restrictions. Only a few coals (high-iron, Eastern bituminous coals) are reactive enough to give the high conversions required. If conversion to vacuum distillate is not adequate and too much SRC vacuum bottoms are produced for the amount of hydrogen make-up required, methanation of the excess syngas would be required.

f. Major Process Uncertainties

Some of the same uncertainties that apply to the SRC-I process also apply to the SRC-II process. The major process uncertainty is again the design of the feed-slurry fired heater. There is also the same charge pump size limitation problem. There possibly will be more severe pump operation and maintenance problems due to recycle of slurry for process solvent.

Although the problem of liquid-solids separation and product solidification have been eliminated, difficult problems remain to be solved. These problems do not appear to be insurmountable.

3. H-Coal

Start-up is underway on the H-Coal pilot plant -- the largest coal liquefaction plant in the United States. The process was developed by HRI as a modification of its H-Oil process for heavy oils (HRI, 1976). Work began on the process in 1964 and has progressed through bench-scale (11 kg of coal/day) and the process development unit (3 tons of coal/day) stages (Johanson,1979). The pilot plant -- located at Catlettsburg, Kentucky -- has a capacity of 600 tons of coal/day. The Catlettsburg plant was constructed by Badger Plants, Inc., and will be operated for DOE by Ashland Synthetic Fuels, Inc. It is adjacent to an Ashland refinery and will use hydrogen from the refinery instead of generating its own.

Major financial support for the Catlettsburg pilot plant comes from DOE. Industrial participants are Electric Power Research Institute, Ashland Synthetic Fuels, Inc., Conoco Coal Development Company, Mobil Oil Corporation, and Standard Oil Company (Indiana). The Commonwealth of Kentucky also is a financial participant. The purpose of the pilot plant is to develop the process to the point where full-scale commercial applications can be undertaken.

A simplified schematic flow diagram is shown in Figure 16. The unique feature of the H-Coal process is the use of a catalyst in an ebullated bed reactor. The catalyst is a 1.6 mm dia. extrudate of a commercial type hydrofining catalyst. Either $NiS\text{-}MoS_2\text{-}Al_2O_3$ or $CoS\text{-}MoS_2\text{-}Al_2O_3$ type catalysts may be employed. The catalyst is maintained in a fluidized state by means of an ebullating pump which recycles catalyst-free slurry from the top of the reactor. The finely ground coal accompanies the slurry oil and hydrogen into the reactor, recycles through the ebullating pump, and passes to a two-stage let-down system where the residual hydrogen and the liquid, gaseous, and solid products are roughly separated. The rough separation is followed by more refined separations as indicated on the flowsheet. The gas is separated into recycle hydrogen, fuel gas, ammonia, and hydrogen sulfide which is converted to sulfur. The liquid is separated into light and heavy fractions; and the solids are concentrated by means of a hydroclone and vacuum distillation. An alternative anti-solvent deashing process by the Lummus Company also will be tested at the Catlettsburg pilot plant.

The product distribution can be adjusted from the light end to the heavy by adjusting the space velocity as shown in Table 6. A syncrude is produced at the lower space velocity and a less hydrogenated boiler fuel at the higher. The gas and the bottoms slurry may be used to manufacture hydrogen for the process; and the thermal efficiency of the process then is about 74%, as shown in Table 4.

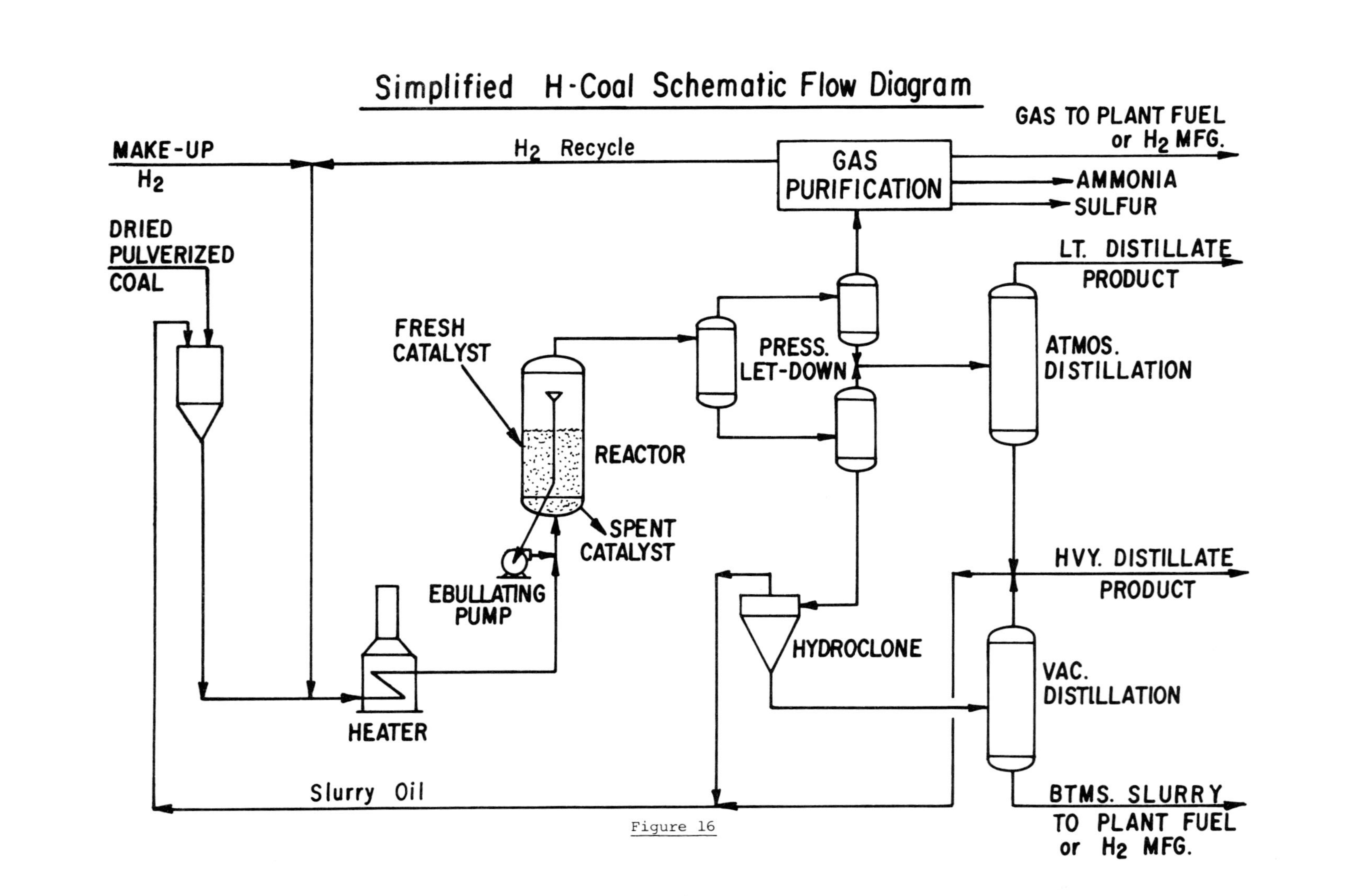
Simplified H-Coal Schematic Flow Diagram
MAKE-UP
H_2
H_2 Recycle
GAS PURIFICATION
GAS TO PLANT FUEL
or H_2 MFG.
AMMONIA
SULFUR
DRIED
PULVERIZED
COAL
LT. DISTILLATE
PRODUCT
FRESH
CATALYST
PRESS.
LET-DOWN
ATMOS.
DISTILLATION
REACTOR
SPENT
CATALYST
EBULLATING
PUMP
HVY. DISTILLATE
PRODUCT
HYDROCLONE
VAC.
DISTILLATION
HEATER
Slurry Oil
BTMS. SLURRY
TO PLANT FUEL
or H_2 MFG.

Figure 16

The major advantage of the H-Coal process is the flexibility provided by the use of a catalyst. The major uncertainties are the long-term reliable operation of the ebullated bed reactor and the economics of the process in comparison with other liquefaction processes. Uncertainty remains also with regard to catalyst make-up requirements under commercial conditions.

4. Exxon EDS

The Exxon EDS project evolved from coal liquefaction research conducted at Exxon Research and Engineering since 1966 (Exxon,1979). The prior Exxon work formed the technical basis for definition of the process to be described below. A jointly funded effort (Epperly,1979) involving, in addition to Exxon, the U.S. Department of Energy, the Electric Power Research Institute, Japan Coal Liquefaction Development Company, and Atlantic Richfield Company, was organized in 1976 to expedite development of the process. The schedule called for construction and operation of a 250 T/D pilot plant along with continuing supporting efforts on laboratory, bench-scale and smaller pilot scale units at a cost of $240,000,000 over a six year period. The pilot plant was scheduled to begin operation late in 1979 but is currently about three months behind schedule.

A pioneer commercial plant for final demonstration of the process is being planned for possible operation in the mid-1980s . Such a plant would process 10,000 T/D of coal and produce 25,000 bbl/day of synthetic coal liquids. A full commercial plant would produce 50,000 bbl/day. Assuming successful results in pilot and pioneer plant operations, Exxon projects the construction and operation of a multiplant synthetic oil industry in the 1990s.

The process is illustrated in the schematic flowsheet as given in Figure 17. It may be categorized as a hydrogen donor extraction system operated under hydrogen pressure. The recycle solvent is rehydrogenated in a conventional fixed bed hydrotreating system using commercially available catalysts such as Co-Mo or Ni-Mo sulfides on an alumina support. This system operates at the same pressure as the coal extraction system, i.e., 100-150 atmospheres.

The process is designed to eliminate contact and poisoning of active catalyst sites with coal bearing minerals. The primary conversion of the coal is described by Exxon as non-catalytic. As was pointed out above, this is likely not wholly true in reality. Some rehydrogenation of the solvent undoubtedly takes place under the catalytic influence of the coal minerals.

The detailed configuration and operating conditions in the Exxon process are considered proprietary. The extraction step is operated in a manner to simulate plug flow.

The operating conditions in extraction given in Table 6 represent the best estimate from published data.

Another important feature of the Exxon process is the management of solvent quality. This is done by control of the boiling range of the recycle solvent and catalytic hydrogenation conditions. Quantitative control is effected by means of a solvent quality index (SQI), but this has never been publicly defined.

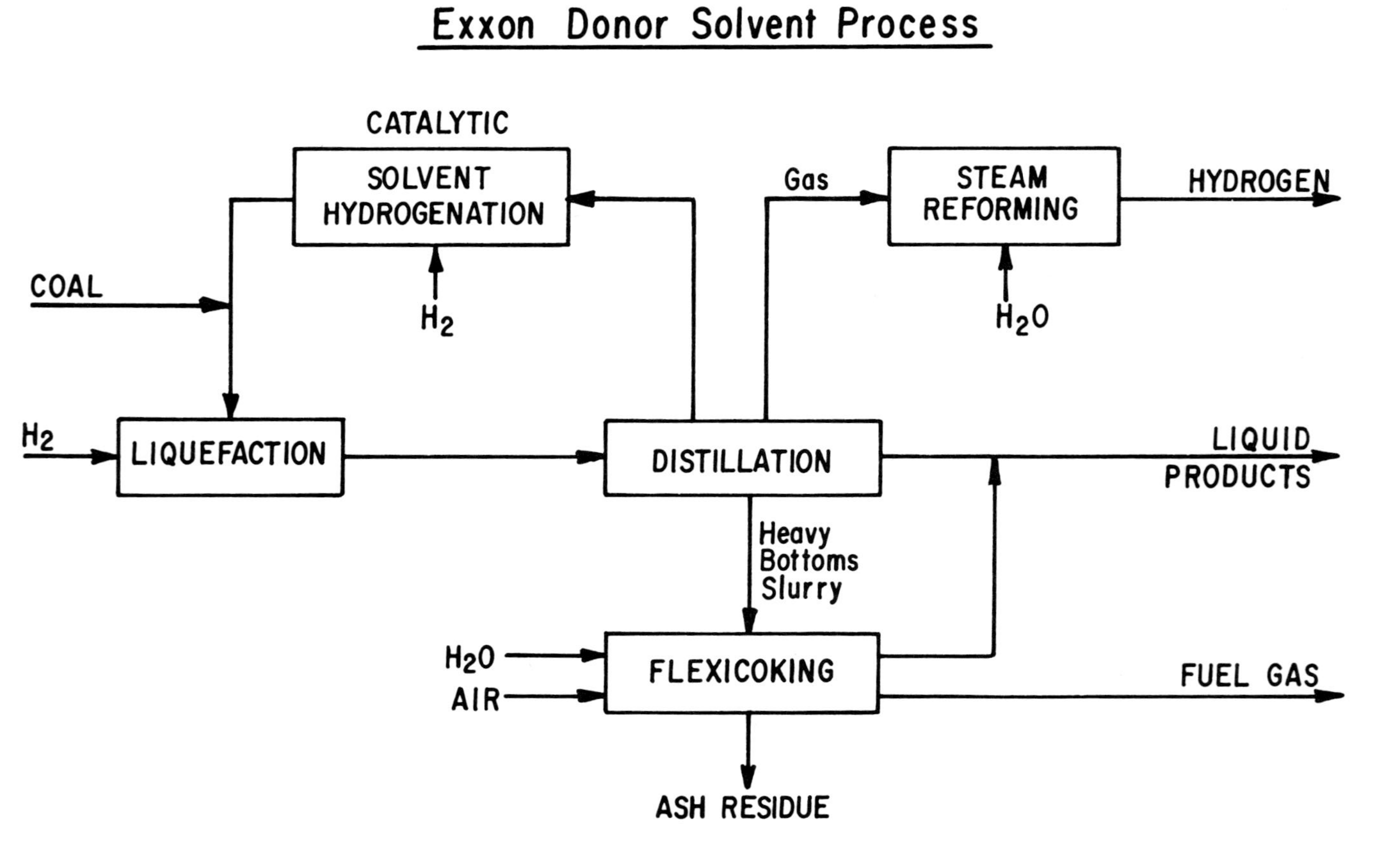
Exxon Donor Solvent Process
CATALYTIC
SOLVENT
HYDROGENATION
COAL
H2
H2
LIQUEFACTION
DISTILLATION
Gas
STEAM
REFORMING
H2O
HYDROGEN
LIQUID
PRODUCTS
Heavy
Bottoms
Slurry
H2O
AIR
FLEXICOKING
FUEL GAS
ASH RESIDUE

Figure 17

The recycle solvent is fixed to have a final boiling point below 454°C. This is likely done, however, to protect the solvent hydrogenation catalyst from deactivation by carbon laydown and deposition of trace minerals rather than to achieve a better donor quality.

The net products from the Exxon process are naptha and either a #4 or #6 fuel oil depending upon whether a 454°C or 538°C end point on the distillate product is employed.

One feature of the Exxon process is the production of additional distillate oils by coking of the vacuum bottoms obtained by distillation of the extraction effluent. It differs in this respect from most other processes which omit this unit operation. A notable exception is the CSF process which also has a coking step and will be described below (Consol, 1971).

The flexicoking process used by Exxon is a combined coking and gasification process which is a commercial process in the petroleum industry. The products are distillate oils and a fuel gas.

The flexicoking process is considered attractive by Exxon since it can increase the liquid yield by as much as 30-35% over that achieved without it.

The flexicoking as applied to the coal derived vacuum bottoms is a much more difficult operation than with petroleum feedstocks. This is a result of their much higher viscosity, ash, and solids content. Thus far, only a mini-plant of 380 kg/D capacity has been operated. Slagging problems have been encountered in operation of the gasifier as might be expected. Demonstration of the applicability of flexicoking will require construction of a pilot plant at the 250 T/SD liquefaction plant site.

The yield and operating conditions given in Table 6 represent the optimum liquid yield obtained in once-through operation in the 45 kg/day continuous unit with Illinois #6 coal from the Monterey Mine. It does not include incremental liquid yields from coking of the bottoms. Nor does it include improved yields that can be achieved, according to Exxon, by recycle of bottoms to liquefaction. The bottoms recycle would, of course, make the Exxon process similar in this respect to SRC-II.

The base commercial case (Exxon, 1979) in the Exxon process calls for the production of hydrogen via steam reforming of the C_1-C_3 gases. The fuel gas generated in the flexicoker is used conceptually as fuel for all fired heaters and for the steam reforming tubes. The overall thermal efficiency is rather low as shown in Table 7, i.e., 55.6%.

A higher thermal efficiency is achieved by limiting the flexicoking operation to that required for process fuel and taking the C_1-C_2 gases and C_3-LPG as part of the product slate. The hydrogen is then generated by Texaco partial oxidation of the remainder of the vacuum bottoms. As indicated in Table 7, this increases the overall thermal efficiency to 63.6%.

The net thermal efficiency will probably be somewhat further reduced since in practice further refining will be required to provide marketable products. The naptha, in particular, will require hydrofining and catalytic reforming before it can be marketed as gasoline. The 200°C x 538°C fuel oil is high in nitrogen and may have undesirable carcinogenic properties. For one or both of the above reasons further hydrotreating may be required.

The commercial case is based on Illinois #6 coal. A number of different coals have been tested in Exxon's small continuous experimental units (Exxon, 1979). All have been shown to show adequate but in most cases inferior performance to the base coal.

Subbituminous coals require higher residence times and give substantially lower net liquid yields for the same hydrogen input. They also tend to give problems due to deposition of calcium-rich solids in the reactor. This problem is not unique to the Exxon process, however, and procedures for adequate handling of it are being worked out.

The Exxon process, in summary, does not show an exceptional yield structure or performance with respect to competing processes. Its principal merits are a somewhat lower operating pressure and a careful engineering design to minimize potential operating problems. One principle uncertainty is whether the flexicoking unit will provide satisfactory performance with the difficult vacuum bottoms feedstock.

5. Comparison of Advanced Processes

The four processes described above are all variations or improvements on the World War II German liquefaction technology. Tables 6 and 7 are meant to provide some basis of comparison of the different processes. This is difficult to do because of differences in operating pressure. Only one thing is clear and that is it is beneficial to recycle the reactor slurry as in SRC-II if it is desired to maximize liquid yield.

The use of a catalyst as in the H-Coal process does not have a large beneficial effect on product yields.

The thermal efficiency, generally speaking, improves if the gases are included in the product rather than for hydrogen manufacture.

Likewise, the thermal efficiency tends to decrease, irrespective of the process, as one moves in the direction of producing lower boiling products.

F. Developing Systems

1. Molten Zinc Chloride Process

From the early German work to the present, the direct hydrogenation of coal has used either no catalyst, or a solid catalyst. The solid catalysts become contaminated with ash and their cracking activity is immediately poisoned by nitrogen from the coal. The various processes now being developed differ largely in how they handle these problems, and none of them yield motor gasoline in one step. The zinc chloride process to be described below is unique in that it is the only process that is distinctly removed from the original Bergius and Pott-Broche processes.

The use of molten zinc chloride to catalyze coal liquefaction provides a completely different type of process. Zinc chloride is a homogeneous catalyst which reacts with nitrogen and sulfur impurities in the coal, and yet provides "Lewis-acid" cracking activity for liquefying coal when used in a "bath" type reactor. This concept was developed in the 1960s at Conoco Coal Development Company (Zielke, 1966). It was immediately recognized as a breakthrough toward an entirely different and attractive route to gasoline from coal. Since then, work at Conoco has shown the process applicable to all kinds of coals, and continuous equipment up to one ton of coal per day has been tested.

The basic steps in the process are shown in Figure 18. Coal, hydrogen, and molten zinc chloride are fed separately to a series of liquid pool reactors at 660 to 765°K under a pressure of 190 atm. The coal reacts rapidly, evolving hydrocarbon vapors overhead. Hydrogen gas is fed to the "finishing" reactor before being sent through the others in series to aid in final removal of product oils from the catalyst. Hydrogen gas leaving is separated from the overhead vapors and recycled, while the hydrocarbons are cooled, depressured, and fractionated. The gases are free from hydrogen sulfide and ammonia, since these react with zinc chloride and are retained in the catalyst pool as zinc sulfide and zinc chloride ammoniate. This results in hydrocarbon products with unusually low levels of sulfur and nitrogen impurities. These hydrocarbons are fractionated into synthetic natural gas (SNG or methane), "liquified petroleum gas" (LPG) and/or other gases for use as chemical feedstock, gasoline, and heavy distillates. The heavy distillates are recycled to the reactors as a means for feeding the coal via pumps.

The unconverted coal, heavy oil, ash, and "used" catalyst flow from the finishing reactor as a stable suspension. The catalyst is regenerated by feeding the spent melt to a fluidized sand bed at 1200-1255°K where the zinc chloride is vaporized and air burns the unconverted organics to carbon dioxide and water. Sulfur and nitrogen are vaporized as sulfur dioxide and nitrogen gas. Hydrogen choride is recycled and fed with the entering air to drive zinc compounds back to zinc chloride. Ash comes overhead and is removed by a cyclone before zinc chloride is condensed. The gases after the condenser are treated to recover hydrogen chloride and then sulfur dioxide before discharge.

One of the unique properties of zinc chloride is that it readily hydrocracks polynuclear compounds present in coal, but does not crack single-ring compounds in gasoline. Also, the butanes, pentanes, and hexanes produced have a high ratio of isoparaffins to normal paraffins. These features produce in one step a very high yield of motor gasoline having Research Octane Numbers in the 90-92 range. This process then by-passes the steps of distillate hydrocracking, hydrotreating, and reforming required by other direct liquefaction processes.

Liquefaction by the molten zinc chloride process was developed further by Conoco and Shell under a U. S. Department of Energy contract from 1975 to 1980. Commercial yields from a Western U.S.A. subbituminous coal were projected from results on a continuous unit feeding 1 kg/h of coal. These are compared with those from other processes in Table 6. Noteworthy, are a high yield of gasoline (C_4 x 200°C) and the modest consumption of hydrogen considering that this coal has a high oxygen content and that the product is nearly all gasoline, containing over 13% hydrogen. The coal used is one which is much more difficult to liquefy than Kentucky or Illinois bituminous coals. An estimate of gasoline costs from a commercial plant based on this coal indicated that they would be lower than from other liquefaction processes, if all technical uncertainties are resolved favorably (Biasca, 1979).

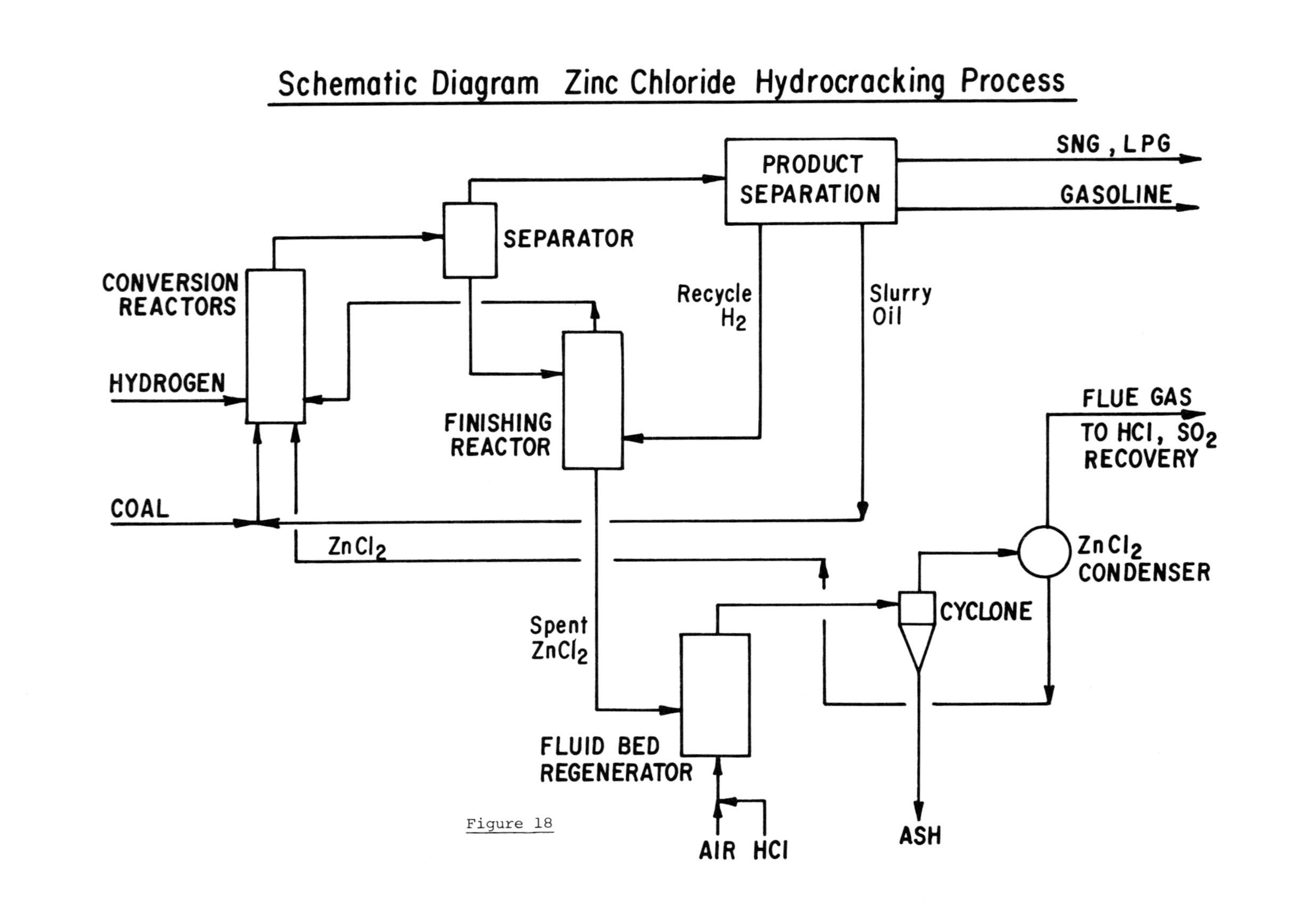
Schematic Diagram Zinc Chloride Hydrocracking Process
PRODUCT SEPARATION
SNG, LPG
GASOLINE
SEPARATOR
CONVERSION REACTORS
HYDROGEN
Recycle H2
Slurry Oil
FINISHING REACTOR
COAL
ZnCl2
FLUE GAS
TO HCl, SO2 RECOVERY
ZnCl2 CONDENSER
CYCLONE
Spent ZnCl2
FLUID BED REGENERATOR
AIR
HCl
ASH

Figure 18

Following the above encouraging results, a 45 kg/h process development unit (PDU) was built. The liquefaction and continuous fractionation sections were operated over 100 hours feeding solvent-refined coal, confirming the high gasoline yield and quality expected (Pell, 1979). Mechanical difficulties were encountered with leaks at the reactor stirrer seal, valves and pumps in molten service, and with hydrogen compressors. The catalyst regeneration section was separately operated briefly. Good removal of impurities was achieved, but optimum conditions were not defined.

The route to commercialization requires a year or two of small scale study to obtain additional design data, demonstrate HCl recovery processes and to define steady-state levels of ash metals in the catalyst (Greene, 1979). Following this the PDU must be operated to:

(a) Demonstrate high conversion rates in a non-stirred reactor (or a successful stirrer seal).

(b) Demonstrate steady-state operation with recycle of catalyst and heavy distillate. Define products and yields.

(c) Demonstrate two-stage regeneration to confirm $>99\%$ recovery of zinc. Define the need for an electrostatic precipitator in the recovery train.

(d) Demonstrate materials of construction and design of the zinc chloride condenser.

(e) Demonstrate the processes for recovery of HCl from offgases in both liquefaction and regeneration.

Following this, it is expected that a 200-500 TPD pilot plant would be operated before a commercial plant could confidently be designed.

In summary, the zinc chloride liquefaction process shows a high rate of catalytic conversion with a high yield of motor gasoline and high removal of nitrogen and sulfur. It works well with all coals and shows promise of being a relatively low-cost process. It is in a relatively early stage of development compared to the processes described earlier, and considerable time and money will be required on the route to a commercial plant.

2. CSF Process - Other Donor Systems

The CSF process as noted above (Consol, 1968), (Gorin, 1971),(Consol, 1971) is the principal example of a process which conducts the primary conversion at relatively mild conditions and in the absence of added hydrogen gas. It is most suited for operation with the high volatile bituminous coals of the Eastern United States.

A simplified block diagram of the CSF process is illustrated in Figure 19. The CSF process is actually the first synthetic fuels process to be extensively studied in the post WWII period. It evolved from extensive laboratory and continuous bench-scale unit studies conducted at the laboratories of Consolidation Coal Company in the period from 1949 to 1963. A 20 T/D pilot plant funded by the Office of Coal Research was designed, constructed and operated at Cresap, W. Va., subsequent to 1963. The pilot plant was shut down in early 1970 (Consol, 1971).

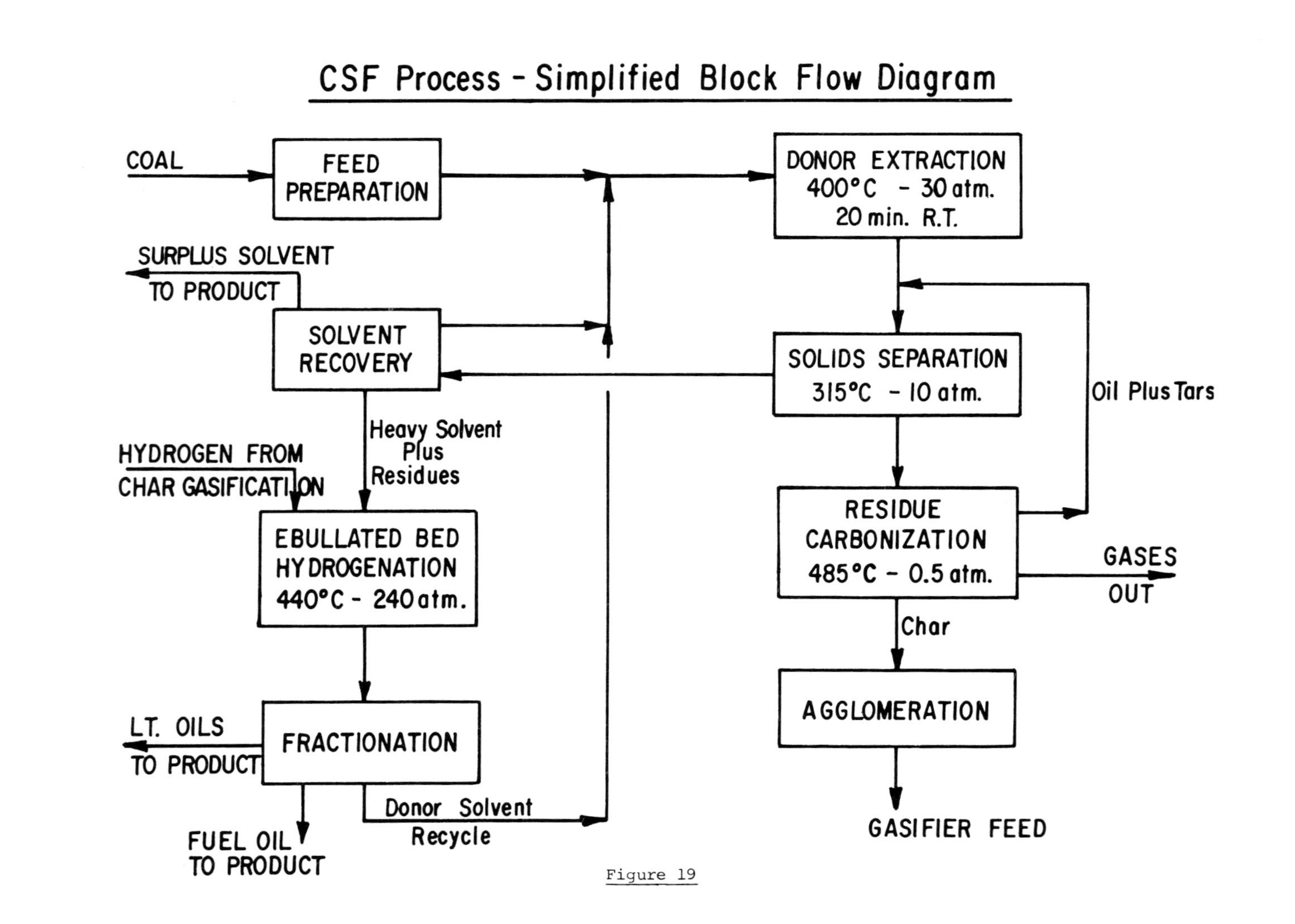
CSF Process - Simplified Block Flow Diagram
COAL
FEED PREPARATION
DONOR EXTRACTION
400°C - 30 atm.
20 min. R.T.
SURPLUS SOLVENT
TO PRODUCT
SOLVENT RECOVERY
SOLIDS SEPARATION
315°C - 10 atm.
Oil Plus Tars
Heavy Solvent
Plus
Residues
HYDROGEN FROM
CHAR GASIFICATION
RESIDUE
CARBONIZATION
485°C - 0.5 atm.
GASES
OUT
EBULLATED BED
HYDROGENATION
440°C - 240 atm.
Char
AGGLOMERATION
LT. OILS
TO PRODUCT
FRACTIONATION
Donor Solvent
Recycle
FUEL OIL
TO PRODUCT
GASIFIER FEED

Figure 19

The block diagram indicates the principal unit operations and their operating conditions. Successful demonstration of all indicated unit operations were achieved in continuous 4.5 kg/hr mini-plant operations prior to construction of the pilot plant. The Cresap pilot plant also achieved successful demonstration of the major unit operations indicated on the block flow diagram. Continuous demonstration of the whole process with all unit operations operating together in integrated fashion was not achieved, however, principally due to mechanical difficulties.

The Cresap pilot plant was refurbished under a contract signed by ERDA with Fluor in 1974. The purpose of renewed operations was not, however, demonstration of the CSF process but to reestablish the Cresap site as a test facility. The revised pilot plant has now also been shut down.

Subsequent evaluation of the process for DOE by Fluor concluded that the CSF process was one of the more promising synthetic fuels processes (Fluor, 1978) for commercial development. Funding to complete the development of the process has not been provided, however.

Another donor extraction system similar to the CSF process (Thurlow, 1979) but differing in detail is being developed by the National Coal Board in England. The whole extraction effluent after solids separation is subjected to catalytic hydrogenation. The distillate product boiling below 260°C is taken as net product. It is planned to construct a 25 T/D pilot plant.

3. Two Stage Liquefaction Processes

The CSF process is an example of a two stage liquefaction system. A two stage system is broadly defined as one where the first stage is conducted under relatively mild conditions to produce primarily a residuum product. The latter, after solids separation, is subsequently treated in a second hydrogenation stage to produce primarily distillate fuels. The primary advantage of the two stage systems is in the improved flexibility it provides to introduce efficient catalytic systems particularly in the second hydrogenation stage. The result is a more efficient use of hydrogen in the overall process.

A significant upsurge of activity sponsored by DOE and EPRI is now underway in the U.S.A. to demonstrate other two stage systems. These involve, for the most part, extraction operations conducted at lower residence times, referred to as SRT processes, but usually higher pressures than in the CSF process. Normally, but not necessarily, it is anticipated that hydrogen pressure will be applied.

The ability to carry out hydroextraction operation at low residence times has been demonstrated at the Wilsonville pilot plant. Bench-scale studies confirming this behavior have been conducted for (Styles,1968) EPRI by CCDC investigators. This system suffers from the disadvantage that neither solvent quality nor solvent balance can be sustained without addition of hydrogenated solvent from an external source.

Another alternative that has been studied for EPRI by CCDC is the incorporation of light SRC in the recycle solvent. The light SRC is recovered from the product SRC by the Kerr-McGee critical deashing process. The light SRC enhances extraction performance (Kleinpeter, 1978, Kleinpeter, 1979). It generates very active hydrogen donors by in situ hydrogenation.

Current plans are to demonstrate the SRT-extraction process at the Wilsonville pilot plant. An ebullated bed hydro unit is planned at the Wilsonville site to demonstrate hydrogenation of the SRC. Solids separation and SRC fractionation will be conducted by the Kerr-McGee critical solvent deashing system. The latter system is currently being operated at Wilsonville.

A number of other very similar two stage hydrogenation systems are being conducted under the sponsorship of DOE. It is probable, for example, that an ebullated bed hydrogenation system will be installed at the SRC-I demonstration plant to handle upgrading of the SRC-I product.

The Lummus Company also has entered into a contract with DOE to conduct a study of an integrated SRT-SRC hydrogenation operation on a mini-pilot plant scale. Solids separation will be conducted by the Lummus solvent deashing process. SRC hydrotreating will be carried out in an LC-fining process. The latter is a proprietary ebullated bed process.

4. Disposable Catalyst Systems

There is a revival of interest in slurry phase systems using disposable, usually iron based, catalysts. These essentially amount to a revival and modernization of the old I.G. Farben (Specks,1979) system. A large pilot plant, for example, is currently being constructed in West Germany.

The Dow Chemical process (Moll, 1979) is a somewhat unique and interesting example of such a process. The disposable catalyst is dispersed in the slurry oil by emulsification with a water-soluble molybdenum salt. It is present in the hydro reactor as a colloidal suspension of MoS_2. The reactor effluent passes to a hydroclone system for solids removal. The bulk of the coal solids are rejected as hydroclone bottoms. The overflow, however, contains the major part of the catalyst and is recycled to slurry preparation. The hydroclone underflow is passed to a countercurrent solvent deashing step which acts as a deasphalting unit. The solids, benzene insolubles, and some of the asphaltenes are rejected as bottoms product and the deasphalted oil is removed as overhead product. Part of the deasphalted oil is taken as net product and part is recycled.

Operating conditions in the liquefaction step are 460°C, 136 atmospheres total pressure and a coal space velocity of 560 $kg/h/m^3$. Conversions to hexane soluble oils and lighter products of the order of 80 wt.% of the MAF coal (Pittsburgh #8 bituminous) with a hydrogen consumption of 5.1 wt.% of the MAF coal are reported. The process has so far been tested only on a 90 kg/d continuous mini-plant scale unit.

G. Processing of Primary Liquids

1. Solids Separation

Solids separation is an essential unit operation in all coal liquefaction systems. A number of systems generate the bulk of the distillate products in the primary liquefaction step. The solid separation step in these processes is simply vacuum distillation. The solids along with residuum oils are rejected as the vacuum bottoms. Examples of such operations are the SRC-II, Exxon-EDS, and H-Coal processes. This certainly is the simplest of all solid separation systems but has several drawbacks and problems associated with it.

First of all, potential liquid yield is lost because of the rejection of the residuum. Some of the liquid may be recovered by coking as in the Exxon process, but most consider the bottoms as Texaco gasifier feed. In both cases, it is necessary to maintain a sufficiently low viscosity of the bottoms for satisfactory operation of the vacuum tower and for subsequent processing in either gasification or coking. This objective has as yet not been achieved in pilot operations without simultaneously retaining significant quantities of the distillate oils in the vacuum tower bottoms.

All other processes, and particularly SRC-I and two-stage liquefaction systems, require systems capable of separating the solids from the residuum oils.

Filtration is the most obvious method of separation. The high viscosity of the coal extraction effluents requires operation at relatively high temperatures of the order of 520°K or higher to obtain acceptable filtration rates. The above means that equipment must be provided for operation under pressures up to about 10 atmospheres. Filter blinding must also be provided for such that it is necessary to live with the complications of filter precoating. Filtration rates are low and precoat costs are high. Successful operation has been achieved with batch cyclic filtration equipment at both the original Pott-Broche pilot plant and in the Wilsonville SRC pilot plant. The latter (Lewis, 1978) uses an automated Funda filter that operated quite well. However, projected commercial costs are far from satisfactory.

The Gulf SRC-I plant has successfully operated continuous rotary precoat filters. The filtration rates were low and it is not clear that the operating costs are superior to that achieved with the batch filters.

The CSF process (Consol, 1971) used hydroclones which are in some respects the simplest solids separation equipment. The hydroclones performed satisfactorily but the efficiency of solids removal was relatively low, i.e., approximately 85-90%.

CCDC subsequently developed a solvent deashing technique. This uses a paraffinic anti-solvent to precipitate some of the SRC. The precipitated SRC acts as a binder to agglomerate the ash and other solids. The result is that satisfactory sedimentation rates can be achieved to make the use of gravity settlers feasible. Operating conditions are about 285-315°C and 10-15 atmospheres. The process was successfully demonstrated with both CSF (Gorin, 1977) and SRC-I effluents in a continuous mini-plant scale and also at the Cresap pilot plant (Kleinpeter, 1978).

A very similar solvent deashing process has been developed by Lummus (Peluso, 1978). It differs from the CCDC process principally in the mechanical design of the settler. In particular, a multi-compartment system with a large number of settlers in parallel is employed within a single tower. This has the desirable effect of increasing the throughput per unit area of containment vessel. The process has been demonstrated in continuous mini-plants and is now being tested at the 50 T/D Ft. Lewis SRC pilot plant. Initial pilot plant runs have given encouraging results. A Lummus deashing unit has also been installed at the H-Coal pilot plant.

A somewhat different solvent deashing system has been developed by Kerr-McGee (Adams, 1979). This is now undergoing successful trials at the 6 T/D Wilsonville SRC plant. The feedstock is the vacuum bottoms rather than the "topped" extraction effluent. The solids separation is conducted by adding a pure solvent, usually toluene. The ash separation is conducted near but slightly below the critical temperature of the solvent. This results in a much higher operating pressure but also a much higher separation rate than in the CCDC and Lummus solvent deashing process. The deashed SRC is carried overhead with the solvent. Separation between the SRC and the solvent is conducted at supercritical conditions. This feature makes it possible to fractionate the SRC, as was mentioned previously, into a light and heavy SRC fraction.

The Dow process uses a combination of a preliminary separation with hydroclones followed by a countercurrent solvent deashing of the hydroclone underflow.

The performance and the preferred deashing system likely depends on the nature of the liquefaction effluent. The Kerr-McGee process, for example, does not function nearly as well with SRT extraction products as with those that have been converted under more severe conditions. The deashing system accordingly must be custom tailored to fit each specific liquefaction system.

2. Hydrotreating, Hydrocracking, and Catalytic Reforming

The products from almost all coal liquefaction processes discussed above do not meet current specifications for petroleum distillate fuels. An exception to this rule is the $ZnCl_2$ hydrocracking process.

Current petroleum technology for hydrotreating, hydrocracking and catalytic reforming can be applied to coal derived liquids as long as they are distillates.

The most natural petroleum substitute product that can be made from coal liquids is high octane gasoline. This follows from the aromatic and cyclic nature of the coal liquids.

The conversion of coal derived distillate oils to high octane (Consol, 1968) gasoline was first demonstrated at Universal Oil Products using CSF liquids. The sequence of hydrofining, hydrocracking, and via the proprietary Isomax process gave a 121 volume percent yield of C_4 x 200°C gasoline. Reforming performance was not tested but excellent results were predicted by UOP from available correlations.

A number of other processing sequences are possible including hydrotreating plus catalytic hydrocracking. Although the processes are fully developed, optimum conditions must be defined for each particular feedstock.

Other transportation and distillate fuels than gasoline may be produced from coal liquids such as diesel oil, #2 fuel oil, and jet fuel via hydrogenation techniques. Usually, however, the major product of the overall liquefaction complex will still be gasoline.

UOP (de Rossett, 1979) and Chevron (Sullivan, 1979) have continued to test the applicability for DOE of various hydrotreating sequences for coal liquids. Technical feasibility has again been established, even though the hydrogen consumption and costs are relatively high.

IV. INDIRECT LIQUEFACTION

A. Definition and History

Indirect liquefaction may be defined as the catalytic synthesis of liquid fuels from carbon monoxide and hydrogen. The carbon monoxide and hydrogen are, of course, produced from coal or coke via gasification as discussed in the first section.

The original interest was in the production of hydrocarbon liquids via the Fischer-Tropsch process. The process was first discovered in 1923. The catalyst was alkalized iron turnings (Fischer, 1923, 1924). High pressure operation at 100-150 atmospheres and 400-450°C yielded largely oxygenated compounds. Milder operating conditions and particularly lower pressures of the order of seven atmospheres yielded largely hydrocarbons.

The Fischer-Tropsch process was developed in Germany and became an important source of synthetic hydrocarbon fuels in that country during World War II. The Germans used primarily a cobalt-kieselguhr catalyst because it operated at milder conditions and in contrast to iron catalysts did not yield any oxygenated compounds.

A large amount of research and development work was also carried out in the U.S. from about 1940-1950, both by government and industrial organizations.

The technology and development of the Fischer-Tropsch synthesis in the years prior to 1950 is thoroughly reviewed in a book by Storch and co-workers (Storch, 1951).

Commercial development since World War II has taken place in South Africa. This is reviewed in a separate section below.

Recently interest has also developed in the synthesis of alcohol fuels, principally methanol and in a two-stage synthesis of hydrocarbons via methanol as an intermediate. These are also discussed in separate sections below.

B. F-T Synthesis - SASOL

The post World War II commercial development of F-T synthesis has been concentrated largely in South Africa.

The original SASOL plant came on stream in 1955. Lurgi gasifiers are used as the source of synthesis gas. Synthesis is carried out with iron catalysts. The major problem in Fischer-Tropsch synthesis is temperature control due to the high exothermic heat of reaction.

The major reactions are the synthesis of paraffins and olefins by the respective reactions.

$$(2n + 1)\ H_2 + n\ CO = C_n\ H_{2n+2} + n\ H_2O$$

$$2n\ H_2 + n\ CO = C_n\ H_{2n} + n\ H_2O$$

The heat release in the above reactions, per mol of synthesis gas converted, is only slightly less than that of the methanation reaction discussed in Section III-G.

Two reactor systems are used: a Lurgi fixed bed system and a Pullman-Kellog circulating fluid bed system (Hoogendoorn, 1977). Cooling is by steam generation with tubes immersed in the fixed bed system. Oil cooling is practiced in the fluid bed system with tubes immersed in the circulating fluosolids system but external to the F-T synthesis.

Both reactor systems, as a result of SASOL experience and development, can be considered as commercially proven.

The two systems at SASOL give different products (Peacock, 1975). The fixed bed product has a high molecular weight with a predominance of waxes and waxy oils. The fluid bed product consists mostly of gasoline and lighter products but also contains a significant proportion of alcohols and ketones. The fluid bed process also makes a more olefinic product.

A decision was made in 1974 to expand the SASOL plant to produce a total of 1,500,000 metric tons/yr of gasoline plus diesel oil and jet fuel. Another 640,000 t/y of ethylene plus ammonia, tar, and chemical products will be produced. The expanded plant, SASOL II, will use the same Lurgi technology for synthesis gas production (Kronseder, 1976). The F-T synthesis section will use exclusively the fluid bed reactor system both for economic reasons and because it makes a more marketable product slate.

Fluor was accorded the contract in 1975 to design and construct SASOL II. The government of South Africa in 1979 designed to expand the production again. The final plant, SASOL III, is scheduled for completion in 1982 and will have double the capacity originally scheduled for SASOL II. The total coal production for the SASOL complex will simultaneously be expanded to 27 million tons.

The SASOL technology is the only one currently commercially available for the production of liquid hydrocarbons from coal.

C. Methanol Synthesis Reactions and Equilibria

The catalytic synthesis of methanol from carbon oxides and hydrogen was patented in Germany in 1913 by Mitasch and Schneider. The reactions are:

$$CO + 2\ H_2 = CH_3OH$$

$$\Delta H = -21{,}700 \text{ cal/mol}$$

$$CO_2 + 3\ H_2 = CH_3OH + H_2O$$

$$\Delta H = -11{,}800 \text{ cal/mol}$$

Since the reactions are exothermic, methanol formation is favored by low temperature. Applying LeChatlier's principle, methanol production is favored by high pressure. Figure 20 shows the equilibrium methanol concentration in the product gas for a feed gas containing 5 moles H_2 and 1 mole CO. Complete conversion of the CO corresponds to 25% methanol in the product gas for this binary case. The commercial case is more complicated because of the presence of CO_2 and other gases in the feed. The water gas shift reaction ($CO + H_2O = CO_2 + H_2$) occurs and affects the equilibrium composition of the product. In commercial reactors the approach to equilibrium is 60-70% and the reactor affluent contains 5-7% methanol.

Methanol Equilibrium for a $5H_2 + 1CO$ Mix

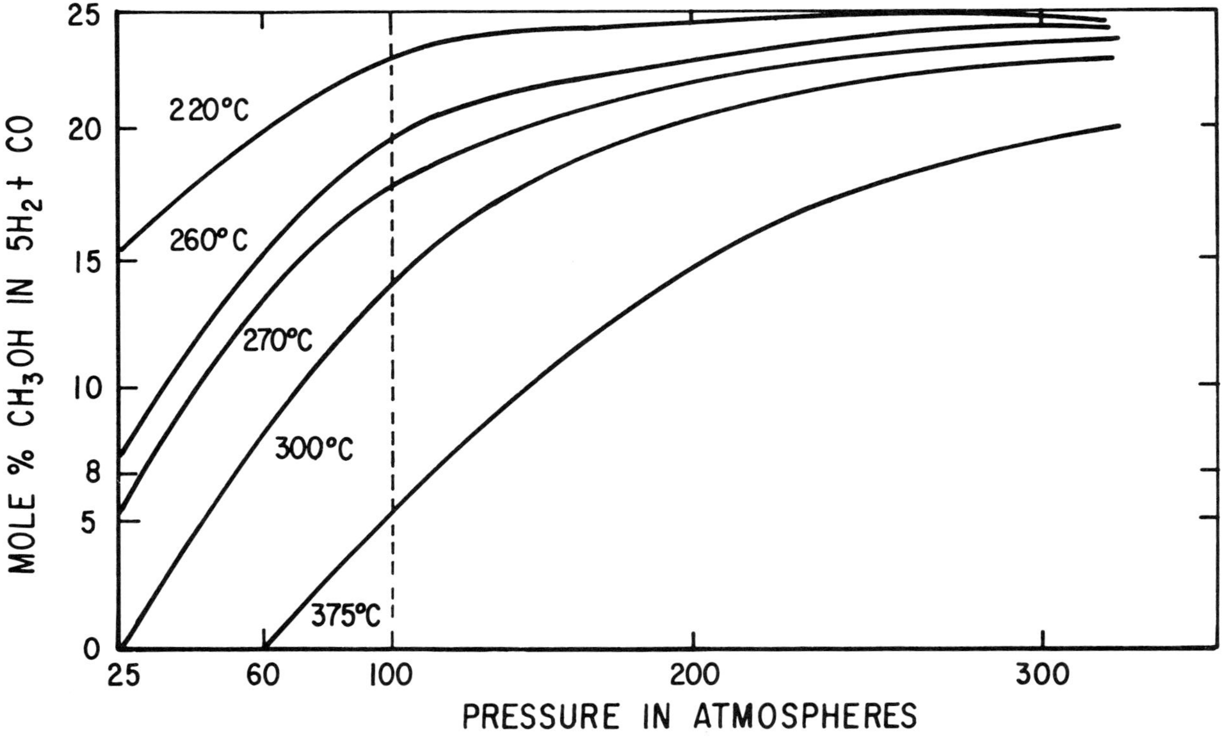

SOURCE: Stiles (1977)

Figure 20

The temperature must be high enough for satisfactory reaction rates but low enough to give a reasonable approach to equilibrium. The production of methanol is limited by the system kinetics but detailed studies of the reaction kinetics have been complicated by such factors as side reactions, heat and mass transfer phenomena, and the degree of catalyst activity, aging, sintering, and deposits. Kinetic expressions and literature references of kinetic studies on methanol synthesis are given by Styles (1977).

Side reactions are more prevalent in high pressure (300-350 atm) processes than in low pressure systems (50-100 atm). Methane, higher alcohols, and other oxygenated compounds can form in parallel to methanol formation. Iron pentacarbonyl can form from carbon monoxide reacting with iron in piping in vessels. The carbonyl decomposes in the higher temperature zones of the methanol synthesis reactor, causing iron deposits which can deactivate the catalyst, increase pressure drop, and catalyze the formation of methane and higher hydrocarbons via Fischer-Tropsch reactions between CO and H_2.

D. High Pressure Process Used for 50 Years

The first synthetic methanol plant was built by BASF in Germany in 1923 based on a patent by Mitasch and Schneider. Prior to this, all methanol was produced by the destructive distillation of wood. For the next 50 years most of the world's methanol was produced from this basic process (Supp, 1973). The process used a sulfur tolerant zinc/chromium oxide catalyst at 300-350 atm and 320-380°C. The partial pressure of carbon monoxide was 40-50 atm and the iron carbonyl problem discussed above was prevalent. As a result of the high pressure and iron carbonyl problem, there were considerable difficulties with producing higher alcohols and methane. The methane production sometimes led to runaway temperatures since the methane formation reaction is highly exothermic.

The earlier methanol plants were in the 50-300 T/D range. Reciprocating compressors were used because the plants were too small for centrifugals. Centrifugal compressors came into use with the design of 1,000 T/D methanol plants.

The iron carbonyl problem was minimized by using carbon monoxide resistant steels and copper linings; however, this did not completely avoid the problem. The Union Rheinische Braunkohlen-Kraftstoff AG (UKW) developed a high pressure methanol process which uses a CO partial pressure below 20 atmospheres. This prevented the formation of iron carbonyl and methane, while largely suppressing the formation of higher alcohols.

Synthesis gas (CO, CO_2, H_2) from many different sources have been used to produce methanol. Most of the methanol produced today is made from methane and liquid hydrocarbons. About 10-15% of the world methanol is produced from bunker C oil and vacuum residue which are converted to synthesis gas by partial oxidation. Many of the earlier plants used synthesis gas produced from coal or coke. Figure 21 is a schematic of a plant which produces methanol. The details of each process unit would vary considerably depending on the feed gas source, the synthesis system, and the degree of purity required of the final product.

E. Low Pressure Processes

It was known from the early days in methanol synthesis that copper based catalysts gave better performance than zinc/chromium catalyst if the feed gas sulfur content could be reduced to 0.25 ppm or less. However, the sulfur removal technology

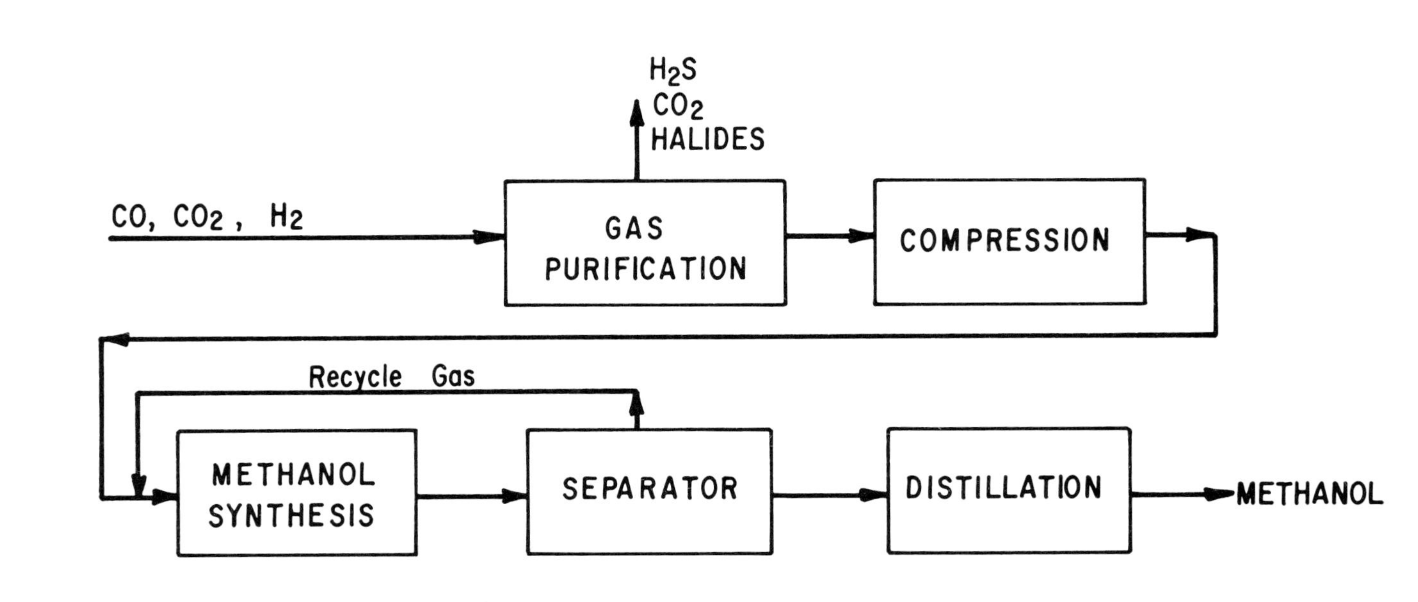
Schematic Diagram of Methanol Synthesis Plant
H_2S
CO_2
HALIDES
CO, CO_2, H_2
GAS PURIFICATION
COMPRESSION
Recycle Gas
METHANOL SYNTHESIS
SEPARATOR
DISTILLATION
METHANOL

Figure 21

did not reach this level until about 1960. This new sulfur removal ability led Imperial Chemical Industries (ICI) to develop a copper based catalyst that completely altered the technology of methanol synthesis. The first plant to use this catalyst was started up in 1966. It had a capacity of 300 T/D methanol production. Synthesis took place at 50 atmospheres with a temperature in the 250-300°C range. The catalyst used was a mix of copper, zinc, and chromium oxides with the copper constituting about 60% by weight.

In 1969 ICI started developing a 100 atmosphere process based on an improved copper, zinc, and alumina catalyst. The catalyst used in the 100 atm process has a life of about 2 years, compared with a catalyst life of about 3 years for the 50 atm catalyst (Rogerson, 1973).

Since 1967, more than 25 ICI low pressure methanol plants have been built throughout the world. Over 80% of the new methanol plants built in that time period have been ICI units. Some formerly high pressure plants have been converted to low pressure operation merely by changing the catalyst and reducing pressure and temperature. Plant sizes have ranged from 150 to 2,500 T/D of methanol. Even the smaller plants can effectively use centrifugal compressors because the actual gas volume is much greater than at high pressure operation. The standard size ICI methanol plant is 1,000 metric T/D. Carbon steel can be used for the system because the low pressure essentially eliminates the iron carbonyl problem present in high pressure plants.

F. Lurgi Low Pressure System

Lurgi Mineralenteknik GmbH of West Germany has a low pressure methanol synthesis system which operates at 40-55 atm and 230-265°C. This system also uses a copper based catalyst. Lurgi started low pressure synthesis development work in the late '50s after their Rectisol process was shown capable of producing very low (0.1 ppm or less) sulfur concentrations in synthesis gas. In 1970 Lurgi operated a 4,000 T/Y demonstration plant to confirm earlier pilot plant results (Hiller, et al., 1971). Since then, several commercial plants have been built, including two plants of nominal 1,000 T/D capacity in the United States.

Lurgi considers 50 atm pressure to be optimum since 99% CO conversion is achieved at that pressure and 260°C, and because the synthesis gas used per ton of methanol produced does not increase after 50 atm (Supp, 1973). As with the ICI process, the relatively low pressure in the Lurgi system eliminates the iron carbonyl problem and produces fewer side reactions to higher alcohols, methane, and oxygenated compounds. The high degree of purity of the feed gas when using the Rectisol low temperature methanol absorption purification scheme minimizes catalyst poisoning in the Lurgi system and gives good catalyst life.

G. Present and Future Uses of Methanol

In 1978 about 13 billion kg of methanol were produced worldwide. The United States produced about 3.5 billion kg. Methanol is used primarily as a chemical feedstock and as a solvent, with much of the methanol going to captive use in chemical plants. The major use of methanol is in the manufacture of formaldehyde, but it is also used to make methylhalides, methylamines, and acetic acid. The use of chemical grade methanol is expected to increase at about 5% per year for the next several years.

In the United States there is strong interest in producing fuel grade methanol from coal derived synthesis gas. This could be done using conventional technology for coal gasification, synthesis gas purification, and methanol manufacture. Methanol is a clean burning fuel with many other favorable properties. It is the liquid fuel from coal that could be produced at a relatively low cost.

Methanol is an excellent gas turbine fuel; it operates at a higher efficiency and produces fewer emissions, especially nitrogen oxides, than petroleum fuels now used to run gas turbines for peak production of electricity. The Southern California Edison Company has tested methanol in a 50 megawatt gas turbine. Results are forthcoming but should indicate excellent performance. The clean burning characteristics of methanol are especially well suited for minimizing pollution problems in the populated Southern California region.

Methanol is an excellent automobile fuel, although there are some problems to be solved before wide spread use of methanol in cars could occur. The use of pure methanol and blends of methanol in gasoline are being tested in the U.S., Germany, and Sweden. Methanol gives improved emissions control and better engine efficiency than gasoline.

Conoco has successfully tested methanol as a clean burning fuel and has identified the following positive features (Tillman, et al., 1975):

1. Significantly lower exhaust gas nitrogen oxides emissions with methanol than for gasoline when both cars were operated normally.

2. Methanol gave higher power output than for the same model car operated with gasoline and adjusted to give the same nitrogen oxides emissions as the methanol car.

3. Carbon monoxide emission at maximum power was lower for methanol than for gasoline.

4. Slightly better energy efficiency resulted with methanol.

There are also some negative features of using methanol for automobiles. A distribution system would have to be established. The use of methanol in captive fleets such as taxi cabs or buses would be the most practical first step. Methanol has about half the volumetric energy content of gasoline so a methanol car must have either a larger fuel tank or a shorter range than a similar gasoline car. Transporting large quantities of methanol from a remote coal processing center would also incur energy penalties as compared with transporting higher energy content materials. The low volatility of methanol is a safety advantage but a disadvantage in low temperature starting. Methanol cars will not start below 16°C unless butane or a similar material is added to increase volatility. Also, methanol attacks some materials currently used in automobile fuel systems, such as, elastomers and aluminum.

The above problems are for the use of pure methanol as a fuel in cars. Low amounts of water can cause a phase separation in methanol/gasoline mixes. This tendency is aggravated by the hydroscopic nature of methanol. Proper blending and care of fuel storage systems can overcome this problem.

Methanol can also be used to produce methyl tertiary butyl ether (MTBE) which has a high octane blending number and is an effective anti-knock agent for use in gasoline. MTBE is catalytically synthesized from 30% methanol and 70% isobutylene. It can be blended up to 7% with unleaded gasoline, where it upgrades unleaded gasoline octane numbers by 2-3. Two MTBE plants with a combined capacity of 5,500 bbl/d have recently come on stream in the U.S.

H. Commercial Plant Plans

Conoco is interested in building a coal processing facility that would produce about 32 x 10^9 kcal per day of methane plus methanol. The plant would process 15,000 T/D of raw lignite or 9,400 T/D subbituminous coal. The coal would be gasified by the Lurgi fixed bed process and the synthesis gas purified using conventiona technology including the Rectisol process for removing sulfur and other poisons to give good life for methanation and methanol synthesis catalysts. Substitute natural gas (SNG) would be produced by applying the Conoco methanation technology developed in the early 1970s . Total SNG production would be 2 million m^3 per day (17 x 10^9 kcal/day). Methanol production would be 3,000 T/D (15 x 10^9 kcal/day) using one of the existing low pressure synthesis processes.

Conoco has proposed a joint project evaluation study to the U.S. Department of Energy. The company is also exploring the possibility of a joint project with several other U.S. companies to make SNG and methanol from Western coal or lignite.

I. Methanol-to-Gasoline Process

The Mobil methanol-to-gasoline process is a straight forward conversion of crude methanol which contains 5-30% water and small amounts of higher alcohols and oxygenated compounds to hydrocarbons and water. The key to the process is the shape selec tive ZSM-5 aluminosilicate zeolite catalyst (Meisel, 1976). This catalyst is a member of the family of zeolite catalysts developed since the 1950s by Mobil and used in more than 95% of all United States petroleum cracking installations. The ZSM-5 catalyst has channels which are smaller in diameter than the 9-10 angstroms present in cracking catalysts. The methanol partially dehydrates to dimethylether over the catalyst. Then, methanol and dimethylether undergo dehydration to light olefins which in turn react to form heavier olefins. Finally, the olefins undergo hydrogen transfer reactions and rearrange to paraffins, cycloparaffins, and aromatics. The selective pore structure of the catalysts prevents any significant formation of hydrocarbons heavier than C_{10} (Chang, 1978).

Table 8 gives typical yields from a fluidized bed pilot plant which processed 4 BPD of crude methanol (Kuo, 1978). Operating conditions were:

Maximum Bed Temperature, °C	413
Pressure, atm gauge	1.7
Space Velocity (WHSV)	1.0

Methanol conversion was steady at 99.5%. The C_5^+ gasoline yield was 60 wt % bu the total gasoline yield would increase to 88.0% if the propylene, butene, and isobutane were alkylated. The thermal efficiency of the methanol-to-gasoline process, including the alkylation step, is estimated at 88.0% based on higher heating values.

The yields of hydrocarbons and gasoline from the fluid bed were higher than fro fixed bed runs (Wise, 1976). Fluid bed runs of 34 and 75 days were made, indicating good catalyst life.

TABLE 8

TYPICAL YIELDS FROM METHANOL
(4 BPD FLUID-BED)

Yields, wt % of Methanol Charge	
Methanol + Ether	0.2
Hydrocarbons	43.5
Water	56.0
CO, CO_2, Coke, Others	0.3
	100.0
Hydrocarbon Product, wt %	
Light Gas	5.6
Propane	5.9
Propylene	5.0
i-Butane	14.5
n-Butane	1.7
Butenes	7.3
C_5 + Gasoline	60.0
	100.0
Gasoline (Including alkylate)	88.0

Source: Kuo and Schreiner, 1978.

The 10 cm x 7.6 m fluid bed reactor used in this work represents a 380-fold scale up from preliminary bench-scale fluidized bed tests. The 10 cm reactor contained 23 kg catalyst vs. 60 grams in the bench-scale unit. A superficial gas velocity up to 0.55 m/sec was used in the 10 cm reactor. The 4 BPD fluidized bed gave good temperature control and no hot spots were observed. Six zone heaters were used to simulate adiabatic operation. The methanol-to-gasoline reactions generate about 390 cal/gm methanol. Fluid bed temperature was controlled by feeding mixed vapor and liquid. The vaporization heat required by the liquid cooled the system. In earlier fixed bed operations gas recycle was used to control temperature.

A small amount of carbon forms during the gasoline synthesis so it is necessary to remove the zeolite catalyst and regenerate it by burning off the carbon. In the 4 BPD pilot plant operations about 10% of the catalyst inventory was removed daily for regeneration with air.

The gasoline made in the 4 BPD pilot plant, when combined with alkylate produced from gaseous by-products, was a high quality fuel with 96.8 research octane number (RON). Properties of the gasoline are given in Table 9. The gasoline produc could be used directly or blended with conventional gasoline.

The process has to be controlled to keep the durene (1, 2, 4, 5 tetramethylbenzene) content below 4% or the durene will crystallize out of the gasoline and form deposits in the fuel system. This occurs because durene has a high freezing point, 175°F.

Agreements are expected soon between Mobil, the West German government, and the U.S. Department of Energy to build a 100 BPD fluidized bed pilot plant in Germany to provide the information needed to scale the methanol-to-gasoline process to full size operation. This program would take 3-4 years.

The New Zealand government has expressed an interest in converting offshore natural gas to methanol and then to make gasoline via the Mobil process. An economic study was performed which indicated a significant advantage of the Mobil route over reforming the natural gas and producing gasoline via Fischer-Tropsch synthesis

TABLE 9

TYPICAL PROPERTIES OF FINISHED GASOLINE
(4 BPD FLUID-BED)

Components, wt %		
Butanes		3.2
Alkylate		28.6
C_5^+ Gasoline		68.2
		100.0
Composition, wt %		
Paraffins		56
Olefins		7
Naphthenes		4
Aromatics		33
		100
Octane	Research	Motor
Clear	96.8	87.4
Leaded (3 cc Tel/US gal)	102.6	95.8
Reid Vapor Pressure, psi	9.0	
Specific Gravity	0.730	
Sulfur, wt %	Nil	
Nitrogen, wt %	Nil	
Durene, wt %	3.8	
Corrosion, Copper Strip	1A	
ASTM Distillation, °F		
10%	117	
30%	159	
50%	217	
90%	337	

Source: Kuo and Schreiner, 1978.

V. THERMAL EFFICIENCY COMPARISON

It is difficult to compare the thermal efficiencies of direct liquefaction processes discussed in Section II with the indirect processes discussed in Section III. The reason is that all the liquefaction processes for which thermal efficiency data are given in Table 7 used entrained type gasifiers and operated under conditions where the pipeline gas yield was minimized. Also, in most cases, bituminous rather than subbituminous coal feed to the process was used. Thermal efficiency comparisons are made in Table 10 of various indirect liquefaction systems with a single direct liquefaction system. All systems given in Table 10 use a low rank coal feed, usually subbituminous, and a common type gasifier, namely, Lurgi dry ash. The direct liquefaction system chosen for comparison is the only one where published data are available in which a Lurgi gasifier was used. The two-stage liquefaction process was developed by CCDC and the data given in Table 10 are those developed as a result of an engineering study by Catalytic, Inc. The process was termed Clean Fuels West and the study was conducted under the auspices of a consortium which included CCDC, EPRI, Mobil, and SCE.

Several conclusions may be drawn from the figures given in Table 10. The thermal efficiency is greatly improved by producing SNG as a coproduct. The improvement is much greater in indirect as compared with direct liquefaction. The thermal efficiency for direct liquefaction, everything else being equal, is somewhat greater than for indirect liquefaction. No really direct comparison is possible because of the differences in products and because the indirect systems are commercially proven while the direct systems are not. It is more efficient to produce MeOH rather than liquid hydrocarbons by indirect liquefaction. Gasoline production via the Mobil route is also a more efficient process than F-T via the SASOL process.

TABLE 10

THERMAL EFFICIENCY COMPARISON

Case	Two Stage Liquefaction	MeOH Only	SNG + MeOH	SNG Plus Mobil MeOH to Gasoline	SNG Plus F-T
Coal	Montana Subbituminous	Low Rank[1]		Wyoming-Subbituminous	
Thermal Yield - % HHV Feed Coal					
SNG + LPG	23.5	-	29.5	35.5	36.8
MeOH	-	56.4	27.2	-	-
Naphtha	16.7			-	-
Gasoline	-	-	-	27.6	15.0
Heavy Distillate	31.5	-	12.0[2]	-	3.5
Other By-Products				-	2.3
Total	71.7	56.4	68.7	63.1	57.6
Reference	(Clean Fuels) (West, 1976)	(Rudolph, 1975)		(Lee, 1978)	

(1) Coal not specified HHV = 9000 BTU/lb.

(2) Includes Pyrolysis Tar, Naphtha, and other by-products.

ABBREVIATIONS USED IN TEXT

AGA	American Gas Association
ASTM	American Society for Testing Materials
Consol	Consolidation Coal Company
CCDC	Conoco Coal Development Company
DOE	U. S. Department of Energy
ERDA	Energy Research & Development Administration
EPRI	Electric Power Research Institute
F-T	Fischer-Tropsch
GC	Gas Chromatography
HRI	Hydrocarbon Research, Inc.
IGT	Institute of Gas Technology
MAF	Moisture and Ash Free
MF	Moisture Free
MS	Mass Spectrometry
OCR	Office of Coal Research
NMR	Nuclear Magnetic Resonance
SASOL	South African Coal, Oil,and Gas Corp.
SCE	Southern California Edison Company
SNG	Synthetic Natural Gas
SRC	Solvent Refined Coal
SRT	Short Residence Time
USBM	United States Bureau of Mines
UOP	Universal Oil Products, Inc.

REFERENCES

Adams, R. H., Knekel, A. H., and Rhodes, D. E., Chem. Eng. Prog., 44, June 1979.

Air Products & Chemicals, "Assessment of Status of Technology Solvent Refining of Coal - Phase I," Prepared for Argonne National Laboratory, November 23, 1977.

Anbar, M., EPRI Report AF-390, February 1977.

Anwer, J. and Bogner, F., Ammonia From Coal Symposium, May 8-10, 1979, Mussel Shoals, Alabama.

Archer, D. H., et al., Proceedings of the Tenth Synthetic Pipeline Gas Symposium, Chicago, Illinois, October 30-November 1, 1978.

Bailey, E. E., et al., TVA Symposium on NH_3 from Coal, pg. 51, May 9-10, 1979.

Bair, W. G., Papers presented at Ninth and Tenth AGA Synthetic Pipeline Gas Symposia, October 31-November 2, 1977 and October 3-November 1, 1978.

Becker, P. D., TVA Symposium on NH_3 from Coal, pg. 48, May 9-10, 1979.

Berthelot, A., Annales d Chemie, 20, 526 (1870).

Biasca, F. E., Greene, C. R., Clark, W. E., and Struck, R. T., Paper to Sixth Annual Conference on Coal Gasification, Liquefaction and Conversion to Electricity, University of Pittsburgh, USA, August 2, 1979.

Bridger, G. W. and Appleton, H., Soc. Chem. Ind., 67, 445 (1948).

Chandra, K., McElmurry, B., Neken, E. W., and Pack, G. E., Fluor Corporation Report to EPRI, EPRI AF-642, Research Project 239, June 1978.

Chang, C. D., et al., Ind. Eng. Chem. Proc. Des. Dev., 17, No. 3, 255 (1978).

Chauhan, S. P., Feldmann, H. F., Stanbaugh, E. P., and Oxley, J. H., Preprints Div. Fuel. Chem; ACS 22, No. 1, 38 (1977).

Clark, E. L., TVA Symposium on Ammonia from Coal, pg. 12, May 9-10, 1979.

Clean Fuels West, A brochure issued by Consortium consisting of CCDC, EPRI, Mobil, and SCE, 1976.

Coleman, W. M., Wooton, D. L., Dorn, H. C., and Taylor, L. T., Anal. Chem., 49, 533 (1977).

CCDC - Stearns Roger, Inc., Plant Operations DOE Final Report - Executive Summary, Vol. 12, Contract EX 76-C-01-1734, November 1978.

CCDC - Stearns Roger, Inc., Executive Summary - Process Design and Economics, DOE Final Report, Vol. 13, Contract EX-76-C-01-1734, April 1978.

CCDC - Design and Evaluation of Commercial Plant, DOE Contract EF-77-C-01-2542, October 1978.

CCDC - Technical Support Program Report, DOE, Contract EF-77-C-01-2542, June 1978.

Consol OCR R&D Report No. 39, Vol. II, Part I (1968).

Consol OCR R&D Report No. 39, Vol. IV, Book 3, Interim Report No. 6, August 1971.

Cornils, B., Hikkel, J., Langhoff, J., and Seipenbusch, J., TVA Symposium NH_3 from Coal, pg. 184, May 8-10, 1979.

Cronauer, D. C., Ruberto, R. G., and Shah, Y. T., Proceedings EPRI Contractor's Conference on Coal Liquefaction, Paper No. 4 (1978).

Curran, G. P., Fink, C. E., and Gorin, E., ACS Div. Fuel Chem., Preprint 12 (3) 62-82 (1968).

Curran, G. P., Fink, C. E., and Gorin, E., Phase II Research on CSG Process, OCR Final Report, Contract No. 14-01-0001-415, R&D Report No. 16, Interim Report No. 3, Book 3, July 1964-March 1965.

Curran, G. P., Struck, R. T., and Gorin, E., Ind. Eng. Chem. Proc. Des. and Dev. 6, 166 (1967).

Del Bel, E., Friedman, S., Yavorsky, P. M., and Wender, I., Chem. Eng. Prog. Technical Manual 2, 104 (1975).

Dent, F. J., et al, "49th Report of the Joint Research Committee of the Gas Research Board and the University of Leeds," February, 1948.

Dent, F. J., Gas J, 244, 502 (1944).

de Rossett, A., Reprints of paper presented at DOE Conference on Refining of Syncrudes, May 7, 1979.

Donath, E. E., Lowry, H. H., Chemistry of Coal Utilization Supplementary Volume, Chapter 22, John Wiley and Sons (1963).

Eby, R., McClintock, N., and Bloom, R., Paper presented at 10th AGA Synthetic Gas Symposium, October 30, 31 and November 1, 1978.

Epperly, W. R. and Taunton, J. W., Paper No. 81C, AIChE National Meeting, San Francisco, California, November 27, 1979.

EPRI AF-585, Research Project 1234-1-2, Annual Report - January-December 1976, November 1977.

EPRI AF-867, Research Project 1234, Annual Report - January-December 1977, August 1978.

EPRI FE-2270-46, Annual Report - January-December 1978, October 1979.

EPRI Research Project 1234, Interim Report, May 1975.

EPRI AF-918, Vol. 1 and 2, Research Project 1234, Interim Report, November 1978.

Exxon - EDS Coal Liquefaction Process Development, Annual Technical Progress Report for Period July 1, 1978-June 30, 1979, U.S. Dept. of Energy - Agreement No. EF-77-A-01-2893.

Farcasiu, M., Fuel, 56, 9 (1977).

Field, J. H., et al., USBM Bulletin 597, 1952.

Fischer, F. and Schrader, H., Brennstoff - Chem. 2, 287 (1921).

Fischer, F. and Tropsch, H., Brennstoff - Chem. 4, 276 (1923), 5, 201 (1924), 5, 217 (1924).

Fleming, D. K., TVA Symposium on NH_3 from Coal, pg. 142, May 9-10, 1979.

Flesch, W. and Vellig, G., Erdoel u Kohle, 15, 710-713 (September 1962).

Fluor Corporation, Final Report to DOE Contract FE-2251-52 UC-904, June 1978.

Fodar, R. J., 1979, Lignite Symposium, May 30-31, 1979, Grand Forks, North Dakota.

FMC, Report to ERDA, Vol. I, Final Report FE-1212-7-9, September 1975.

Furlong, L. E., Effron, E., Vernon, L. W., and Wilson, E. L., Chem. Eng. Prog., 69, August (1976).

Gallagher, J. E. and Euker, C. A., Paper presented at 6th Annual Conference on Coal Gasification, Liquefaction & Conversion to Electricity, University of Pittsburgh, July 31-August 2, 1979.

Gas Processing Handbook, "Hydrocarbon Processing", 1975.

Gorin, E., Fundamentals of Coal Liquefaction, Second Supplement to Lowry, H. H. Chemistry of Coal Utilization, In Press, 1980.

Gorin, E., Lebowitz, H. E., Rice, C. H., and Struck, R. T., Proceedings - 8th World Pet. Congress, Preprints Session No. PD 10 (5) 4X (1971).

Gorin, E., Kulik, C. J., and Lebowitz, H. E., Ind. Eng. Chem. Proc. Des. & Dev. 16, 1, 95 (1977), Ibid 16, 1, 102 (1977).

Greene, C. R., Biasca, F. E., Pell, M., Struck, R. T. and Zielke, C. W., Paper 81e to the American Institute of Chemical Engineers, San Francisco, California, USA, November 25-29, 1979.

Hariri, H. and Hill, G. R., Tech. Report to OCR, Contract 14-01-0001-271, September 1964.

Hebden, D., Edge, R. F., and Foley, K. W., Research Communication GC 112, Gas Council of Great Britain, November 1964.

Hiller, H., et al., Chemical Economy and Engineering Review, September 1971.

Hochgesand, G., Ind. Eng. Chem., 62, 7, pg. 37 (1970).

Hoogendoorn, J. C., Paper presented at Ninth AGA Synthetic Pipeline Symposium, November 1, 1977.

HRI - 1976, OCR R&D Report No. 26, Contract No. 14-01-0001-477.

HRI - DOE Publication FE-2547-40, September 1979.

IGT Reports to DOE NTIS - FE-2434 Series, 1975 through 1979.

Johanson, E. S. and Comolli, A. G., HRI Report to DOE on PDU 5, Contract No. EX-77-C-01-2547 (1979).

Johnson, J. L., Kinetics of Coal Gasification, John Wiley and Sons, 1979.

Kiovsky, T. E., U.S. Patent 3,764,515, October 1973.

Kleinpeter, J. A. and Burke, F. P., Proc. EPRI Contractor's Conference on Coal Liquefaction, Contract RP-1134-1, May 1978.

Kleinpeter, J. A. and Burke, F. P., Proc. EPRI Contractor's Conference on Coal Liquefaction, Contract RP-1134-1, May 1979.

Kleinpeter, J. A., Jones, D. C., Dudt, P. J., and Burke, F. P., Paper at 85th Nat'l. AIChE Meeting, June 1978.

Kleinpeter, J. A., Burke, F. P., Jones, D. C., and Dudt, P. J., EPRI Report AF-1158, August 1979.

Koch, B. J., et al., "Application of Conoco's SUPER-METH® Combined Shift/Methanation Process to the BGC/Lurgi Slagging Gasifier," 2nd International Coal Utilization Exhibition, Houston, Texas, October 1979.

Kronig, W., Catalytic Hydrogenation, Springer Verlag (1950).

Kronseder, J. G., Hydrocarbon Processing, July 1976.

Kuo, J. C. W. and Schreiner, M., Presented at Fifth Annual International Conference on Coal Gasification, Liquefaction, and Conversion to Electricity, University of Pittsburgh, August 1-3, 1978.

Landers, J. E., "Review of Methanation Demonstration at Westfield, Scotland," Sixth Synthetic Pipeline Gas Symposium - AGA, October 1974.

Lee, B. S., Papers presented at Seventh and Eighth AGA Synthetic Pipeline Gas Symposia, October 27-29, 1975 and October 18-20, 1976.

Lewis, H. E., Weber, W. H., Usnick, G. B., Hollenack, W. R., and Hooks, H. W., EPRI AF-867, 1977 Annual Report, August 1978.

Loeding, J. W. and Stanfill, F. G., Reprint of paper presented at The Fourteenth World Gas Conference, May 27-June 1, 1979.

Lumpkin, H. E. and Aczel, T., (ACS Div. Fuel Chem. Preprints), 22, 135, (1977).

Lurgi - ANG Filing, 1975.

Matchele, W. N., Trachte, K. L., and Zaczepenski, Sam, Ind. Eng. Chem. Prod. Des. & Dev. 18, 311 (1979).

McCray, R. L., McClintock, Neil, and Bloom, Ralph, Jr., Paper at 85th National Meeting, AIChE, June 4-8, 1978.

Meisel, S. L., et al., Chemtech, February 1976.

Mitsak, D. M., Farnsworth, J. F., and Wintrell, R., Energy Communications, 1, No. 2, pg. 157-178, 1975.

Moeller, F. W., et al., Hydrocarbon Processing, Vol. 53, pg. 69, April 1974.

Moll, N. G. and Quarderer, G. J., Chem. Eng. Prog., 46, November 1979.

Mukherjee, D. K. and Chowddury, P. B., Fuel, 55, 4 (1976).

Oberg, C. L. and Falk, A. T., Paper 81B, AIChE National Meeting, San Francisco, California, November 28, 1978.

Ode, W. H., Chapter 5, Supplement Lowry, H. H., Chemistry of Coal Utilization, John Wiley and Sons, 1962.

Okutani, K., Yokoyama, S., Yoshida, R., and Ishli, T., Ind. Eng. Chem. Prod. Res. & Dev., 18, No. 4, 367 (1979).

Parks, B. C., Chapter 1, Supplement Lowry, H. H., Chemistry of Coal Utilization, John Wiley and Sons, 1962.

Patel, J. G. and Leppin, D., TVA Symposium Ammonia from Coal, pg. 63, May 8-10, 1979.

Peacock, C. G., Paper presented at Institution of Chemical Engineers, London, England, May 14, 1975.

Pell, M., Maskew, J. T., Pasek, B., Struck, R. T., and Greene, C. R., Paper to the Intersociety Energy Conversion Engineering Conference, Boston, Massachusetts, USA, August 5-10, 1979.

Peluso, M. and Ogren, D. E., Chem. Eng. Prog., 41, June 1979.

Pittsburg & Midway Coal Company, Quarterly Technical Progress Report to DOE for April 1, 1978 through June 30, 1978, FE-496-157, May 1979.

Pittsburg & Midway Coal Company, Quarterly Technical Progress Report to DOE for January 1, 1978 through March 31, 1978, FE-496-155, March 1979.

Pott, A., Broche, H., Nedelmann, H., Schmitz, H., and Scheer, W., Fuel, 13, 91, 125, 154 (1934).

Ricci, L. J. and Grover, R., Chemical Engineering, Vol. 86, No. 23, November 19, 1979.

Ricketts, T. S., Publication 633, Inst. Gas Engrs. (London), 1963.

Roben, A. M., Texaco Report to DOE under Contract EX-76-C-01-2247, February 1979.

Roben, A. M., EPRI AF-233, Project 714-1, December 1976 and EPRI AF-777, Project 714-3, June 1978.

Roberts, G. F. I., et al., 14th World Gas Conference, Toronto, Canada, May 1979.

Rogerson, P. L., Chemical Engineering, August 20, 1973.

Ross, D. S. and Blessing, J. E., Fuel, 58, 433 (1979), Ibid, 58, 438 (1979).

Ross, D. S. and Loro, J. Y., Quarterly Report, SRI to ERDA, Contract No. E(49-18)-2202, 1976.

Ruberto, R. G., Proceedings of EPRI Contractor's Conference, pg. 81, May 1977.

Rudolph, Paul, F. H., Sixth Synthetic Pipeline Gas Symposium, Chicago, Illinois, October 28-30, 1974.

Rudolph, Paul, F. H. and Herbert, Peter K., Symposium on Synthetic Fuels, London, England, May 14, 1975.

St. John, G. A., Buttrill, S. E., and Anbar, M., ACS Div. Fuel Chem. Preprints, 22, 141 (1977).

Salvador, L. A., Westinghouse Advanced Coal Conversion Department, Personal Communication (January 1980).

Schlinger, W. G., Paper presented at Pittsburgh Conference on Synthetic Fuels from Coal, August 1, 1979.

Schmid, B. K. and Jackson, D. M., Paper at 3rd Annual International Conference on Coal Gasification and Liquefaction, University of Pittsburgh, August 3-5, 1976.

Schmid, B. K. and Jackson, D. M., Paper at 4th Annual International Conference on Coal Gasification and Liquefaction and Conversion to Electricity, University of Pittsburgh, August 2-4, 1977.

Schroeder, W. C., Hydrocarbon Processing, 131-133, January 1976.

Shinn, J. H. and Vermuelen, T., Preprints Div. Fuel Chem. ACS 24, No. 2,74 (1979).

Specks, R. and Klusmann, A., Paper 99e, AIChE National Meeting, San Francisco, California, November 29, 1979.

Staege, Hermann, Chem. Engineering 86, No. 19, September 1979.

Stearns Rodgers, Commercial Design.

Storch, H. H., Golumbic, Norma, and Anderson, R. B., The Fischer-Tropsch and Related Syntheses, John Wiley and Sons, New York City, New York (1951).

Srini, V., TVA Symposium on NH_3 from Coal, pg. 157, May 9-10, 1979.

Struck, R. T., Clark, W. E., Dudt, P. J., Rosenhoover, W. A., Zielke, C. W. and Gorin, E., Ind. Eng. Chem. Proc. Des. & Dev., 8, 552 (1969).

Styles, A. B., AIChE J. 23, 3 (1977).

Styles, G. A., Weber, W. H., and Bases, H., Proc. EPRI Contractor's Conference - Coal Liquefaction, Contract RP-1234-1, May 1978.

Sullivan, R. F., Reprints of paper presented at DOE - Conference on Refining of Syncrudes, May 7, 1979.

Supp, E., Chemtech, July 1973.

Thurlow, G. G., paper 99d, AIChE, National Meeting, San Francisco, California, November 29, 1979.

Tillman, R. M., et al., Paper at Automotive Engineering Congress and Exhibition, Detroit, Michigan, February 24-28, 1975.

Vanderhart, D. L. and Petcofsky, H. L., SRI Coal Chemistry Workshop, Session III, Paper 15.

Van der Hoeven, B. J. C., Lowry, H. H., Chemistry of Coal Utilization, Chapter 36, Vol. II, John Wiley and Sons, 1945.

Voght, E. V. and Van der Burgt, M. J., Paper 99f, 72nd Annual Meeting, AIChE, San Francisco, California, November 29, 1979.

Von Fredersdorff, C. G. and Elliott, M. A., Lowry, H. H., Chemistry of Coal Utilization Supplementary Volume, Chapter 20, John Wiley and Sons, 1963.

Wald, M. M., British Patent 1,310,280, March 1973.

Watson, W. B., "The Manufacture of SNG from Coal," Gas Conditioning Conference, University of Oklahoma, March 1977.

White, G. A., et al., Preprints Division of Fuel Chemistry, 168th ACS Meeting, Vol. 19, No. 3, pg. 57, September 1974.

Whitehurst, D. D., Farcasiu, M., Mitchell, T. O., and Dickert, J. J., EPRI Annual Report, Project 410-1, July 1977, EPRI Annual Report, Project 410-1 in press.

Willson, W. G., Knudson, C. L., Baker, G. G., Owens, T. C., and Severson, D. E., Ind. Eng. Chem. Proc. Des. and Dev., 18, 297 (1979)

Wintrell, Reginald, Paper presented at National Meeting of AIChE, Salt Lake City, Utah, August 18-21, 1974.

Wise, J. J. and Silvestri, A. J., Presented at Third Annual International Conference on Coal Gasification and Liquefaction, University of Pittsburgh, August 3-5, 1976.

Wood, R. E. and Wiser, W. H., Ind. Eng. Chem. Proc. Des. and Dev., 15, 144 (1976).

Woodward, C., Hydrocarbon Processing, Vol. 56, pg. 136, January 1977.

Wu, W. R. K. and Storch, H. H., Bureau of Mines Bulletin 633 (1968).

Zielke, C. W., Struck, R. T., Evans, J. M., Costanza, C. P., and Gorin, E., Ind. Eng. Chem. Proc. Des. and Dev. 5, 151 (1966), Ibid, 5, 158 (1966).

Zielke, C. W., Rosenhoover, W. A., and Gorin, E., ACS Advances in Chem. Series, 151 153-166 (1976).

Zielke, C. W., Klunder, E. B., Maskew, J. T., and Struck, R. T., Ind. Eng. Chem. Proc Des. and Dev., 19, 85 (1980).

Zielke, C. W. and Gorin, E., Ind. Eng. Chem. 49, 396 (1957).

THE STATE OF THE ART OF PRODUCING SYNTHETIC FUELS FROM BIOMASS

DAVID C. JUNGE

Oregon State University

ABSTRACT

A variety of synthetic fuels can now be produced from biomass energy resources using existing technology. Many research workers are involved in projects to advance the level of technology, to further understand the processes involved, to increase productivity and efficiencies, and to reduce costs. This paper summarizes the state of the art and the principal areas of activity in the production of synthetic fuels from biomass resources in the United States of America. Emphasis is placed on the technologies of fermentation, anaerobic digestion, pyrolysis, gasification, and liquefaction. Brief comments are offered concerning the areas of hydrogasification, steam gasification, molten salt reactors, oil production from *Euphorbia* type plants, and biomass densification. Some of the restraints to increased use of biomass fuels are reviewed.

ISBN 0-12-467101-2

I. INTRODUCTION

"Biomass" is not a well defined term. It may be helpful, therefore, at the outset to place some limits on the interpretation of the term. In this paper it is used to refer to all energy resource bases which emanate from living matter. In that context, biomass includes those energy resource bases shown in Table 1.

TABLE 1
BIOMASS FUEL ENERGY RESOURCE BASES

FOREST BIOMASS	AGRICULTURAL BIOMASS	MISCELLANEOUS BIOMASS
Trees Grown for Fuel	Crops Grown Specifically As Energy Resources	Marine and Freshwater Biomass
Whole tree chips for fuel	Examples: Cattails, Jerusalem artichokes, cassava, sugar cane	Examples: Kelp, algae, water hyacinth
Cord wood for fuel	Unused Agricultural Crops	Animal Manure
Recoverable Forest Residues	Example: Surplus wheat	Municipal Sewage Sludge
Forest thinnings	Crop Residues Available For Energy Recovery	Some Refuse Derived Fuels
Dead timber removal	Examples: Grass straw, wheat straw, corn stover, corn cobs, whey	Examples: Wood fiber (cellulose based) residues and other organic wastes
Wastes generated by forestry practices, such as slash, stumps, etc.	Food Processing Plant Wastes	
Recoverable Forest Products Manufacturing Industry Wastes		
Bark, sawdust, slabs, and other fiber-based residue		
Spent sulfite liquor		
Kraft black liquor		
Other Plants, Trees, Shrubs		
Leaves, tree prunings, desert mesquite, etc.		

Some explanation of Table 1 is in order. Spent sulfite liquor and kraft black liquor are not typically considered to be biomass fuels. However, they are direct derivatives of biomass and are closely coupled with the forest products industry, a major user of biomass as a raw material.

Similarly, refuse derived fuels are not typically considered to be within the framework of biomass. However, a considerable percentage of refuse derived fuels in the United States are cellulose fiber based (paper, cardboard, etc.) and/or organic residue based (grass clippings, food wastes, etc.). These types of RDF can be converted to synthetic fuels and are biomass based.

There are several resource bases which have specifically been excluded from the list of biomass fuels but which are difficult to classify into other categories. Examples include used tires, peat, industrial waste treatment sludge (i.e., pulp mill clarifier sludge), and other industrial wastes not specifically included in Table 1. This paper will not consider synthetic fuels produced from biomass resources unless the general types of resources are indicated in Table 1.

There are two broad general classifications of conversion processes for making synthetic fuels from biomass: (1) Biochemical (microbial) conversion; and (2) Thermochemical conversion. Under the general category of biochemical conversion there are two subcategories of technologies: (a) Fermentation; and (b) Anaerobic digestion. Both have been in existence for many years but there are recent developments in each area which make them useful for synthetic fuels production.

The general category of thermochemical conversion technology includes such diverse areas as pyrolysis, gasification, liquefaction, solvent extraction, and others. These fields also include some recent developments and applications involving use of biomass fuels.

In summarizing the state of the art of these technologies, it is assumed that sufficient biomass exists as a feedstock. No attempt is made to assess the magnitude or potential availability of biomass resources. Neither is the technology related to increasing production of biomass resources summarized. Further, the technologies related to collection, preparation, storage, and handling of biomass are not considered. The emphasis is placed on the conversion technologies.

The synthetic fuels which can be produced from biomass using existing technologies are relatively limited in scope. Gaseous fuels of primary interest include hydrogen and carbon monoxide (synthesis gas), methane, and mixtures of various gases to produce low BTU gas (LBG), intermediate BTU gas (IBG) and high BTU gas (HBG). The liquid fuels include methanol, ethanol, heavy fuel oil, lighter oils, and gasoline.

By-products resulting from the production of synthetic fuels from biomass may be important to the economics of the production process. Brief attention is given to some of the more important by-products, however they are not stressed.

II. PRINCIPAL CONVERSION PROCESSES USED TO PRODUCE SYNTHETIC FUELS FROM BIOMASS

A. FERMENTATION

Fermentation is a biochemical process in which sugar is transformed to ethanol and carbon dioxide by the action of yeasts and other enzymes. The classic example is described by the equation:

$$C_6H_{12}O_6 \longrightarrow 2C_2H_5OH + 2CO_2$$

The process is influenced by the characteristics of the feed stock. Fermentable biomass can be divided into three categories of feedstocks according to their dominant chemical characteristics.

1. Sacchariferous materials which contain natural sugars. This category includes sugar beets, sugar cane, fruits, and similar materials.

2. Amylaceous materials which contain starch. This category includes wheat, corn, potatoes, other grains, and tubers.

3. Cellulosic materials,such as,wood, bark, corn stover, wheat straw, grass straw, beet pulp, etc.

It is important to note that fermentation occurs when sugars are converted to ethanol and carbon dioxide. As long as the biomass feedstocks contain natural sugars, the process can proceed directly. But if the feedstocks contain starch or cellulosic base materials, then it is necessary to precede the fermentation process by a process which will convert the starch to sugar or the cellulose to sugar.

Starch is a polysaccharide. Its basic molecules comprise several sugar units which are chemically bound. Cooking the starch molecule results in a breakdown of the cellular structure of the starchy biomass and the liberation of the polysaccharides. Enzymes are then used to rupture the starch molecules. The descriptive equation is:

$$(C_6H_{10}O_5)_n + nH_2O \longrightarrow nC_6H_{12}O_6$$

Enzymatic processes are used to prepare grains and other starchy materials for fermentation to beer and other alcoholic beverages.

If the biomass is cellulosic in nature, the cellulose must be chemically converted or hydrolyzed to a form suitable for fermentation. Cellulose can be hydrolyzed in a natural process that enables cattle and other ruminating animals to digest such materials as grass, grain, and paper. Enzymatic hydrolysis can also be carried out in an industrial process.

Acid hydrolysis is a common process used in the pulp and paper industry to convert wood into cellulose fibers for paper making. Acid hydrolysis involves the use of heat and pressure to aid in the separation of lignin and cellulose. Some of the cellulose molecules may be hydrolyzed to form additional wood sugars. The wood sugars can be recovered and converted into ethanol by conventional fermentation.

Figure 1 is a schematic illustration of a typical fermentation process for the production of ethanol.

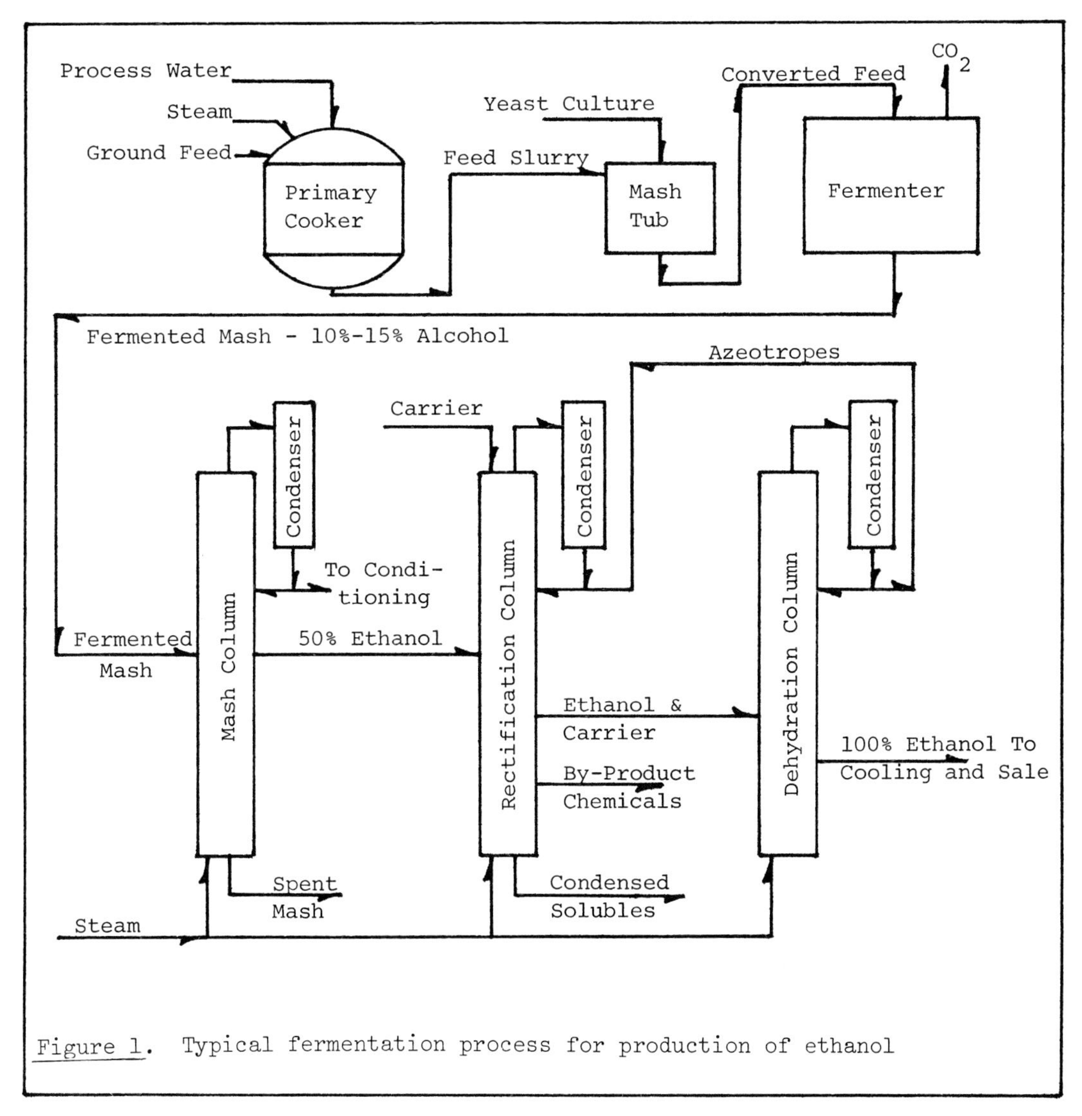

Figure 1. Typical fermentation process for production of ethanol

In the fermentation process, yeasts feed on the sugars and produce ethanol and carbon dioxide as waste products. In most fermenters, the process is carried out on a batch basis and is considered to be complete when the ethanol concentration of the liquid mixture reaches 10%-15% by volume. The fermented mash, still in a liquid slurry, is then transfered to the first of a series of distillation columns.

Alcohol is traditionally dehydrated by multistage distillation, a process that relies upon differences between the boiling temperatures of water and ethanol to achieve separation in the vapor phase. Commerical alcohol is distilled in a series of steam-heated columns or stills. In the first column (mash column) the fermented mash is heated to boiling with injected steam. Alcohol and water are liberated from the mash and the vapor product is simultaneously concentrated to about 50% alcohol by volume. The alcohol concentrate is sent to a second column (rectification column) in which additional boiling concentrates the alcohol to about 95% by volume.

Normal distillation cannot concentrate the alcohol beyond this level because 95% ethanol in water is an azeotropic or constant boiling mixture. Obtaining pure (100%) ethanol requires that the azeotrope be broken by adding a third component, a carrier, to the constant boiling ethanol/water mixture. The carrier combines with the alcohol and water to form two new liquid mixtures which have appreciably different boiling temperatures and which migrate to different locations in the rectification column. One mixture contains water, ethanol, and the carrier. The other mixture contains only ethanol and the carrier. This second mixture is withdrawn from the rectifier and is sent to a third column where the alcohol is separated from the carrier and is sent to market. The carrier is recovered and returned to the rectifier.

Pure ethanol is the principal product from this conversion process. But other by-product materials are also produced. The principal by-product is the solid component in the fermenter product. The solids are called "slops" in the fermentation industry. When they are dried and marketed, they are called spent grains, brewers grains, distillers dried grains and solubles, or fodder yeast. They are relatively low in sugar and starch but are high in protein and fiber. They can be an important livestock feed supplement and are being used for this purpose in the United States.

Fermentation processes also produce carbon dioxide and organic chemicals of the ketone, aldehyde, and fusel oil families. These valuable chemicals may be recovered from the rectification step and used in the manufacture of detergents, perfumes, and preservatives. (Sladek, 1978)[1]

Note that the process of fermentation is a fairly well standardized one in terms of the steps beginning with the fermenter tank and ending with the final distillation column. The process has been used in industrial operations for the production of a variety of grades of ethanol for many years. As long as the feed material is in the form of fermentable sugars or can be readily converted to fermentable sugars, the process works well.

And that brings the summary to the point of recent advances in the state of the art.

Interest in developing gasoline substitutes has led to research efforts to make possible the use of a broad range of feedstocks. By general category these include grains (i.e., wheat, corn, barley, etc.), high sugar content plants (i.e., sugar cane, sugar beets, sweet sorghum, and the sugar-bearing tuber manioc), cellulosic materials (i.e., corn stover, corn cobs, and wood-based materials), and aquatic crops (i.e., kelp).

The interest in grains stems from their high starch content and the ease with which they can be converted to fermentable sugars. The same rationale applies to the high sugar content plants. But both grains and high sugar content plants have moderately high prices in the market place and both are in demand as food for human and/or animal consumption. Cellulose base materials, on the other hand, are more plentiful on a worldwide basis, are not competing as food, and are available at relatively low cost. The great difficulty with cellulose base materials lies in converting them to fermentable sugar forms.

A research effort at the Massachusetts Institute of Technology's Dept. of Nutrition and Food Science is working on direct microbiological conversion of cellulosic biomass to ethanol. (Wang, 1979). The thrust of their efforts is summarized below:

> Different techniques are being considered to achieve the conversion of cellulosic fuels to ethanol. These include acid or enzymatic hydrolysis of the biomass followed by fermentation using soluble sugars to produce ethanol. This approach generally uses the yeast, *Saccharomyces cerevisiae*, due to its fermentative ability for the production of ethanol. Unfortunately, this yeast is unable to metabolize five carbon sugars such as xylose. It has been established that the major production cost of ethanol through biological conversion processes is the cost of the feedstock. The inability to utilize the hemicellulosic fraction in biomass detracts to a certain extent from such techniques for biomass conversion.
>
> In searching for novel bioconversion processes for ethanol production, the M.I.T. group has been exploring a different concept from those stated above. It was rationalized that a single step conversion of cellulosic biomass to ethanol could offer potential economical advantages over those using multiple processes. Within this single step conversion scheme, it was also the intent to utilize both the cellulosic (6-carbon sugars) and hemicellulosic (5-carbon sugars) fractions in biomass to produce ethanol. To achieve these objectives, we have focused our attention on an anaerobic and thermophilic (Temp = 60 ^{o}C) bacterium, *Clostridium thermocellum*, which is able to hydrolyze cellulose and hemi-cellulose in biomass. This microorganism is also able to catabolize the six carbon sugars to produce ethanol and other organic acids. However, it is not able to metabolize five carbon sugars such as xylose. Therefore, a second anaerobic and thermophilic (Temp = 60 ^{o}C) bacterium, *Clostridium thermosaccharolyticum*, is being examined for the conversion of the hemicellulosic fraction to ethanol. The overall goal is to use a mixed culture fermentation for the direct conversion of biomass to ethanol.

The research effort at M.I.T. is still in the relatively early stages — a long way from commercial application. The efforts report by Wang, et. al., indicate the principle concern at this time is to develop a suitable strain of bacterium which is resistant to the poisoning effects of alcohol and which can be used effectively with a variety of cellulosic materials.

Another research effort aimed at cellulose fermentation is underway at the University of Pennsylvania. It is summarized by Pye (1979).

> A novel, elevated temperature process for the total coversion of cellulosic biomass to liquid fuels has been devised and shown to be technically feasible by our research group. It has several unique features which allow reduced process energy costs in the simultaneous production of two liquid fuels - ethanol for use as a gasoline extender.... and a butanol/lignin slurry for use as a pumpable furnace or diesel fuel. The process relies on a hot aqueous butanol pretreatment of the biomass to yield an enzyme-degradable cellulose fraction, a high quality polymer grade lignin fraction, a partially degraded hemicellulose fraction for fermentation to butanol, and a butanol/lignin slurry for use as a fuel. The cellulose is fermented to ethanol either by an elevated temperature (- 60 °C) simultaneous saccharification/vacuum fermentation step using cellulase from *Thermoactinomyces*, and the anaerobic bacterium *C. thermocellum*; or by hydrolyzing it to a high glucose syrup (greater than 20%) with *Thermoactinomyces* cellulase for use in standard yeast fermentation. The process is currently being tested and optimized with wood chips from fast-growing poplar trees, but it also appears attractive for conversion of municipal solid waste and agricultural residues. Production costs for ethanol appear to be about 70¢ - 80¢ / gallon, depending on biomass costs, yields, and by-product credits.

This research effort is a joint project involving General Electric Co. and the University of Pennsylvania. It is in the early laboratory stages and has not even been proven in a continuous integrated bench scale facility. Like the work at M.I.T., it has a long way to go to commercialization but it does hold substantial promise for future use of cellulose materials.

Research efforts at Purdue University have gained considerable national attention. Dr. Tsao (1979) provides the following description of his efforts.

> Cellulosic materials such as cornstalks used in conversion processes are usually made of three major components: cellulose, hemicellulose, and lignin. In utilization of such crude mixtures, fractionation of the raw materials into pure components usually will result in upgrading the value. Cellulose, in fact, is difficult to hydrolyze due to two main obstacles, one of which is the presence and the interference of lignin that cements cellulose fibers together. The other factor that makes cellulose resistant to hydrolysis is the highly ordered crystalline structure of cellulose. In order to overcome the two obstacles, Purdue researchers have employed solvents to dissolve cellulose. Once it is dissolved into a liquid, there is no longer crystalline structure nor the interference by lignin. Consequently, the treated cellulose becomes easily hydrolyzable by either acids and enzymes to glucose in

high yields at a fast reaction rate.

Another important feature of the Purdue Program is its heavy emphasis on the utilization of hemicellulose hydrolysate. In cornstalks, there is actually more hemicellulose than cellulose. Unless we can get a reasonable by-product credit from hemicellulose, cellulose alone cannot yield alcohol at a low cost. We recognized this fact very early in our program, when everyone else was paying attention only to cellulose hydrolysis and fermentation of glucose. The Purdue Program actually was started as a study of fermentation of pentoses from hemicellulose. We are now investigating the production of butanediol, diacetyl, methyl ethyl ketone, ethanol, lactic acid, and pyruvic acid from hemicellulose hydrolysate.

No doubt, the Purdue efforts with corn stover have made significant headway in utilizing the 5 carbon (pentose) sugars found in hemicellulose as well as the other more readily fermentable sugars. But as with the other research efforts summarized (M.I.T. and U. of Pennsylvania) the work is still to be considered in the early developmental stages.

In fact, there are many related ongoing research projects in the United States dealing with fermentation technology which are still in early developmental stages but which show long-term promise. Some of the more visible projects are summarized in Table 2. (Bente, 1979)

TABLE 2
PARTIAL SUMMARY OF ONGOING FERMENTATION RESEARCH

Project Title	Personnel	Location
Enzymatic Decomposition of Lignin and Cellobiose in Relation to Hydrolysis of Cellulose	Wilke, C.R. Rosenberg, S. Riaz, M.	Univ. of Calif. Chem. Engr. Berkeley, CA
Enzymatic Degradation of Cellulose Using NOVO Cellulase	Posorske, L.H.	NOVO Labs, Inc. Wilton, CT
Design, Construction, and Operation of a Cellulose Fermentation Facility	O'Neil, D.J. Colcord, A.R. Bery, M.K.	Georgia Tech. Eng. Exp. Stat. Atlanta, GA
Conversion of Lignocellulose by Thermophilic Actinomycete Microorganisms	Crawford, D.L.	Univ. of Idaho Bact. & BioChem. Moscow, ID
Conversion of Agricultural By-Products to Sugars	Reilly, P.J. Fratzke, A.R. Frederick, M.M. Shei, J.C. Fournier, R. Oguntimein, G.B.	Iowa State Univ. Chem. Engr. Ames, IA

Table 2 (Cont.) PARTIAL SUMMARY OF ONGOING FERMENTATION RESEARCH

Project Title	Personnel	Location
Enzymatic Hydrolysis of Cellulose to Fermentable Sugars	Spano, L.A. Mandels, M. Reese, E. Sternberg, D. Allen, A. Tassinari, T. Blodgett, C. Macy, C.	U.S. Army Natick R & D. Command Natick, MA
Assessment of Bacteria for Lignocellulose Transformation	Crawford, R.L.	Univ. of Minn. F. Biol. Inst. Navarre, MN
Selection of Hyper-Cellulase Producing Microbes	Eveleigh, D. Montenecourt, B.S. Cuskey, S.	Rutgers Univ. Cook College New Brunswick, New Jersey
Enzymatic Transformations of Lignin	Hall, P.L. Drew, S.W. Glasser, W.G.	Virginia Poly. Institute Chemistry Blacksburg, VA
Thermophilic Fungi-Derived Cellulase for Solubilization of Cellulosic Wastes	Skinner, W.G. Takenishi, S.	SRI Int'l Life Sci. Div. Menlo Park, CA
Chemicals from Cellulose	Emert, G.H.	Gulf Oil Chem. Co Biochem. Tech. Shawneee Mission, Kansas
Microbial Production of Ethanol from Solvated Cellulosic Wastes	Zabriskie, D.W. Qutubuddin, S. Downing, K.	SUNY at Buffalo Engr. & Sci. Chem. Emgr. Amherst, NY

The summary of ongoing research efforts shown in Table 2 points out the emphasis placed on fermentation of cellulose base feedstocks. This reflects both the availability of such feedstocks and their cost.

A report by Bungay (1979) presents an overview of the status of fermentation technology and the degree to which it is ready for commercialization.

> "Perhaps the best way of providing an overview ..... is to show what is ready for commercialization, what is being tested, and what research

should be initiated. A factory built now to be competitive with petroleum-based enthanol should not use lignocellulosic biomass because there are no reliable markets for hemicellulose and lignin in the quantities that would be produced. The factory could use corn, grain, molasses, or any easily fermentable carbohydrate, but the fuel alcohol will cost over $1.50 per gallon. Looking through the U. S. Dept. of Energy contractor reports, there are many ideas that are ready right now to lower this cost. Groups at Battelle and Purdue have pointed out ways to get cheap sugars from sweet sorghum or from other biomass by taking the easily extractable material and finding uses for the insoluble residues. Since the raw materials make up at least 2/3 of the cost of manufacturing ethanol, these suggestions could impact rapidly on the economics of fuel alcohol.

Turning attention to what is almost ready, ... the order of discussion will try to approximate the probable chronological sequence of commercialization. A very new .... observation at Purdue .... may cut distillation energy requirements to 1/10 of that of today's factories. This is based on using a short column to distill to an intermediate ethanol concentration and removing the rest of the water with drying agents. This will dispell the criticism that fuel alcohol processing takes about as much fossil fuel energy as the energy content of the product.

.....Another chance for early commercialization is fermentation of sugars from hemicellulose. Several contractors have developed processes for fermentation of pentose sugars to organic chemicals such as ethanol, acetone, butanol, and 2,3-butanediol. Common strains of yeast do not ferment pentoses, but other organisms utilize these sugars well. The acetone/butanol fermentation does not give high yields because butanol is toxic; recovery costs are too high because the solutions are dilute. Ethanol from pentoses is beginning to reach good yields, and a commercial process may be near.

Hydrolysis of cellulose to glucose for fermentation works quite well and needs scale-up research mainly on the pretreatment step. A decision on which pretreatment is best may be possible within a year. By-product credits are crucial to alcohol fuels from biomass, thus several things must fall into place before a factory should be built.

Really outstanding progress has been made in improving cellulase enzyme titers (concentrations of enzymes in solution). However, acid hydrolysis has also been improved so that acid is a clear choice for early commercialization. For the longer range, enzymatic hydrolysis seems quite probable.

Taking the longer range to be 5 years, the most exciting development is direct fermentation of cellulose. Already, ethanol concentration is approaching 2 percent in the experiments at M.I.T. This could render separate hydrolysis obsolete. Other projects are dealing with reactor design, new methods for recovery of products, and processes for other petrochemical substitutes".[2]

Investigators at SRI, International (1978) reviewed the state of the art of fermentation technologies in an attempt to determine the most promising paths for production of synthetic fuels from biomass. In their final selection of the most promising technologies, 4 paths using fermentation were selected. These are shown in Table 3.

TABLE 3
COMPARISON OF FERMENTATION TECHNOLOGIES BY SRI, INT'L

Process Description	Estimated Product Cost for Comparison ($/MMBTU)*
Acid hydrolysis and fermentation of corn stover to produce ethanol	$25.55
Fermentation of sugar cane to produce ethanol	$27.00
Enzymatic hydrolysis and fermentation of wheat straw to produce ethanol	$52.60
Acid hydrolysis and fermentation of algae to produce ethanol	$25.90

* The costs shown are "base case" costs and include the estimated costs for plant capital and operation, raw materials, energy, etc. They are shown for comparative purposes only.

It should be understood that the 4 technological paths for production of ethanol by fermentation are based on an analysis of the probable market penetration of the alternative technologies and feedstocks. SRI's analysis included various factors, such as, biomass availability, process cost reduction potential, process energy balance, product marketability, and process environmental impact. So while the selection of most probable technologies and feedstocks for maximum market penetration may be applicable in the United States, it is not necessarily true that these same technologies would be most suitable for market penetration in other countries.

The U.S. Dept. of Energy has made economic estimates of ethanol production costs using varied feedstocks and technologies. These are summarized in Table 4. Note that the range of costs (in 1978 dollars) is from $1.13 - $2.41 per gallon. Plant sizes are typically forcast as 25 million to 50 million gallons per year except for the feedstock cheese whey which is forcast as 2.8 million gallons per year. There is an important aspect of these plant size projections. The diversity of potential biomass feedstocks and the problems associated with transportation of these feedstocks makes it uneconomical to build conversion facilities larger than 50 million gallons per year. By contrast, a typical petroleum refinery may process 100,000 BBLs/day or 1,512 million gallons per year — 30 times the production capability of an economical biomass fermentation plant.

TABLE 4

ECONOMIC ESTIMATES OF ETHANOL COST PRODUCTION -- 1978 COST BASIS**

[100 percent Company Equity—20-Year Plant Life—7 percent Inflation Rate]

Status of technology	Description	Feedstock for ethanol production	Feedstock price	Production rate MMG/Y	Plant capital, $MM		Components of selling price, $/gal.				Estimated Alcohol selling price, $/gal.	
					Fixed investment	Working capital	Feedstock costs	Direct operating costs	Fixed costs	By-Product credits	15% DCF	20% DCF
Commercial	(1)	Milo	$2.20/bu	50	58.0	5.5	0.88	0.28	0.11	0.42	1.02	1.13
Commercial	(1)	Wheat	$3.15/bu	50	58.0	7.9	1.26	0.30	0.11	0.53	1.31	1.44
Under development.	(1)	Sweet sorghum [2] and corn.	$14.42/ton—$2.30/bu.	50	91.6	6.4	1.00	0.22	0.18	0.22	1.40	1.59
Commercial	(1)	Corn with corn stover as fuel.	$2.30/bu—$25.00/ton.	50	57.0	5.8	0.89	0.30	0.11	0.38	1.09	1.21
Under development.	Purdue hydls BF.	Corn stover [3]	$25.00/ton	25	58.8	3.76	0.51	0.80	0.31	0.61	1.32	1.63
Under development.	Enzymtc hydls BF.	Wheat straw [3]	$25.00/ton	25	66.7	5.03	1.12	0.76	0.26	0.36	2.13	2.41
Commercial	Cane milling BF.	Sugar cane [3]	$65.00/ton	25	58.4	4.12	1.20	0.34	0.18		2.07	2.30
Under development.	Acid hydls BF	Aquatic biomass.[3]	$75.00/DAF/ton.	25	46.3	4.28	1.12	0.50	0.18		2.05	2.23
Commercial	Concentration BF.	Cheese whey [3]	$3.00/ton	2.8	10.2	0.41	0.42	0.89	0.31	0.83	1.25	1.63
Commercial		Corn (base case).	$2.30/bu	50	58.0	5.7	0.89	0.27	0.11	0.39	1.05	1.16

[1] Batch fermentation + energy conserving distillation design.

[2] Half year sweet sorghum; half year corn.

[3] SRI Process information. Ethanol product is 190° proof. Ethanol product for other cases is 199° proof. BF=Batch Fermentation. (A) Economic analysis in table is for processes under development. However, commercial processes do exist. (1) Bituminous coal—Base case, 10 year depreciation, 20 year plant life, 10% investment tax credit rate. (2) Bituminous coal—20 year depreciation period. (3) Bituminous coal—30 percent investment tax credit rate. (4) Bituminous coal—Debt financing, 80/20 debt-to-equity ratio.

** See Footnote 3 in Bibliography for source reference of this table.

Incidentally, the information found in Table 4 can be used to find the yields of ethanol from various feedstocks. For example, the first case shown is using milo as the feedstock with a cost of $2.20/bushel. Since the feedstock costs for the ethanol are estimated to be $0.88/gallon, the yield must be $2.20/$0.88 = 2.50 gallons/bushel. By the same logic, wheat yields an estimated 2.5 gallons/bushel and corn stover yields 49 gallons/ton.

Table 4 also lists 6 presently used commercial feedstocks for the production of ethanol.

In brief summary of the foregoing section dealing with fermentation technology it is noted that fermentation is the principal process used for the production of ethanol. The fermentation process itself is well developed and has been used in commercial application for many years. Recent developments deal with processes which can be used to make low cost non-food feedstocks (principally cellulosic in nature) into fermentable sugars. Substantial improvements have been made in reducing the energy requirements for the distillation processes involved with fermentation.

B. ANAEROBIC DIGESTION

Anaerobic digestion (decomposition) of organic materials involves the breakdown of the feedstock to form methane and carbon dioxide. The process takes place in the absence of oxygen. Since methane is a very desirable product which can be generated from a wide variety of organic materials through the process of anaerobic digestion, considerable interest in this technology has arisen in recent years.

For purposes of organizing and presenting the material in this section, the subject has been divided into two distinct aspects: (1) Controlled processes for generation of methane; and (2) Recovery of methane from sanitary landfills.

Controlled Processes for Methane Generation

Anaerobic digestion has been successfully used to treat municipal, industrial, and agricultural wastes. The process is a microbiological one involving three specific steps. First, the wastes are hydrolyzed by enzymatic action. Then facultative and anaerobic organisms act to convert the wastes to simple organic acids — predominantly volatile fatty acids. The acids produced are acetic, propionic, and butyric for the most part. The acetic and propionic acids are thought to be responsible for about 85% of the methane produced. In the final step, the organic acids are converted to methane and carbon dioxide by a group of substrate-specific, anaerobic bacteria (sometimes referred to as "methane formers"). The waste materials undergoing this anaerobic digestion process are typically carried in a liquid waste stream, however it is not a requirement for the bacterial actions to take place. As will be discussed later, anaerobic decompostion can take place in a dry landfill.

Anaerobic processes can operate in one of two ways. The conventional process provides for continuous or intermittant feeding without solids separation. There is no buildup of anaerobic seed sludge in the system and the minimum retention period may range from 3 to 5 days. More typical retention times are 10 to 30 days. An alternative to the conventional process is the anaerobic contact process which provides for separation and recirculation of seed organisms. The anaerobic contact process reduces retention time to typical periods of 6 to 12 hours. (Eckenfelder, 1966) The two processes are shown for comparison purposes in Figure 2.

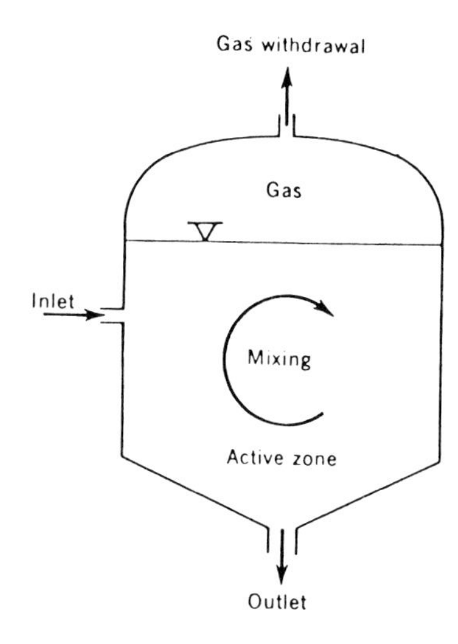

Conventional anaerobic digestion system.

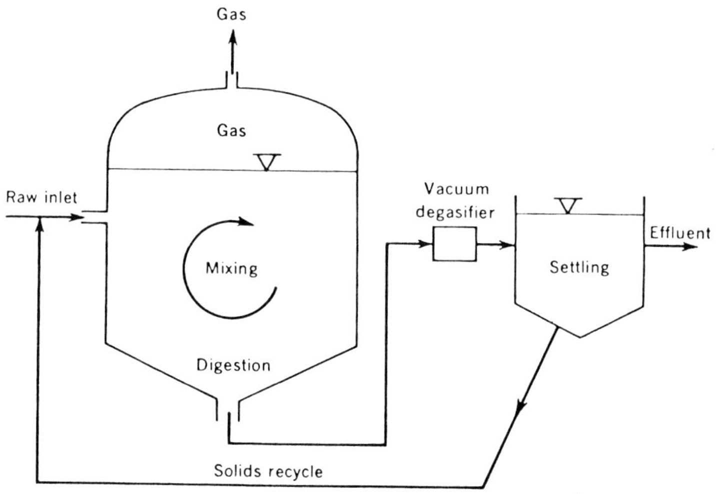

Anaerobic contact process.

Figure 2. Conventional and anaerobic contact disgestion systems (Eckenfelder, 1966)

The retention time of solids in anaerobic digestion systems is of concern due to its effect on the required size of digesters. Obviously if the solids retention time can be reduced by process control, then the digester size requirements diminish. The limiting factor in the biological system is the rate of growth of the methane-forming bacteria. These regenerate slowly and are very sensitive to changes in their environment. Table 5 lists the most significant parameters which must be controlled for successful anaerobic digestion. (Varani and Burford, 1977).

TABLE 5
PHYSICAL AND CHEMICAL FACTORS AFFECTING ANAEROBIC DIGESTION**

Physical Factors	Chemical Factors
Solids Retention Time	pH
Oxygen Free Environment	Alkalinity
Temperature	Volatile Acid Content
Solids Concentration	Nutrients
Degree of Mixing	Toxic Materials
Solids Loading Rate	

** This table taken from Varani and Burford (1977).

The gas which is produced by the anaerobic digestion process is typically in the range of 50% - 70% methane, 30% carbon dioxide, and contains trace amounts of hydrogen sulfide. The quality of the gas (i.e., its methane content) may vary with the operating conditions of the digester and with the characteristics of the feedstock. Methane contents of 62% - 65% by volume are considered to be typical of the off-gases from well controlled digester installations.

Anaerobic digestion processes have been used on a commercial basis since the end of the 19th century, principally for treatment of waste water from industrial operations and municipalities. More recently the technology has been used for production of methane from animal manures. Recent research and development activities have focused on methods to decrease the investment and operating costs for anaerobic digestion systems and to increase the gas yields from specific types of feedstocks. Sponsored research activities have analyzed reactor (digester) options, sought improvements in operation of specific types of reactors by use of better mixing methods, investigated improved process kinetics by operating at higher temperature levels, and promoted increased yields of methane by altering the physical and chemical characteristics of the feedstocks. (Jones, J.L., and Fong, W.S., 1978) Mathematical modeling of the kinetics and mass flow relations in digesters have interested many investigators.

A summary of the ongoing research and development efforts dealing with anaerobic digestion technology is provided by Bente (1979). Twenty-seven of the projects listed concern methane generation from manures. Fourteen are involved with sewage and refuse as feedstocks. Six deal with methane from plant matter, and sixteen are systems studies. Table 6 indicates the range of feedstocks being investigated.

TABLE 6
RANGE OF FEEDSTOCKS BEING INVESTIGATED FOR METHANE GENERATION VIA ANAEROBIC DIGESTION

Manures	Plant Matter	Refuse
Beef Cattle	Cellulosic Forest and Agricultural Residues	Municipal Sewage
Dairy Cattle	Sea Kelp	Industrial Wastes
Hogs	Bermuda Grass	Municipal Solid Waste
Chickens	Alligator Weeds	
Turkeys	Spoiled Hay	
Sheep	Cornstalks	

A review of recent published information on the yields of methane gas from anaerobic digester systems indicated that the yield could be expected to vary according to the type of feedstock used and the operating conditions. Little information was presented in the literature with the exception of yields for manure feedstocks. Gross yields of methane from manure-fed digester systems were in the range of 2.5 - 4.5 SCF of methane per pound of volatile solids of manure. Projected costs for a digester system capable of treating the manure from 10,000 cattle per day indicated that the methane would have to be sold for $4.90/MMBTU (Jones, J.L., and Fong, W.S., 1978). A comparable study by Jones and Fong (1978) indicated that methane generated from anaerobic digestion of prepared wheat straw would have to be sold for $22.10/MMBTU. It is interesting to note that the projected costs for methane generation by anaerobic digestion using varied feedstocks may be high in comparison to 1980 costs for methane (natural gas). Yet there are many production facilities currently being built and/or in operation using biomass feedstocks.

Closely related to this point is that while there is great interest and activity in the United States in building and operating anaerobic digestion systems for methane recovery from biomass, the anticipated maximum impact on the U.S. energy demand from this technology is comparatively small. Jones and Fong (1978) note that if all of the manure from all of the confined animals in the U. S. was collected on a frequent basis, it would amount to 70 million dry tons per year with a gross heating value of 1.05 Quads. However, in converting this material to methane rich gas, the energy efficiency is anticipated to be 26%. Therefore, all of the manure which might be collected would only yield 0.26 Quads if converted to methane rich gas. By comparison, the present use of wood residues for energy in the U.S. amount to 0.5 Quads/yr. Thus, the technology appears to be popular but the overall impact is likely to be small due to the low conversion efficiencies involved.

SRI, International (1978) has reviewed the system paths involving anaerobic digestion of biomass fuels and has sellected five paths which appear to show promise of the greatest market penetration in the coming years. The selected paths are indicated in Table 7.

TABLE 7
COMPARISON OF ANAEROBIC DIGESTION TECHNOLOGIES BY SRI, INT'L

Process Description	Estimated Product Cost for Comparison ($/MMBTU)
Cattle Manure to SNG** 100,000 Head Feedlot	$ 7.00
Cattle Manure to SNG** 10,000 Head Feedlot	$14.40
Cattle Manure to IBG*	$ 4.90
Wheat Straw to IBG*	$22.10
Kelp to SNG** (Speculative Design Case)	$21.20

* IBG - Intermediate BTU gas mixture. That would result from a typical mix of methane and carbon dioxide from an anaerobic digester with a heating value in the range of 200 — 600 BTU/SCF.

** Note: The description in this summary has centered on the production of methane from anaerobic digestion systems. However, as noted in the text, methane is generated in conjunction with carbon dioxide. The carbon dioxide can be removed from the gas mixture along with water and other gaseous components to leave a resultant high BTU gas which is essentially pure methane. This product is referred to as synthetic natural gas or SNG.

An noted in Table 7, carbon dioxide can be separated from the gas mixture leaving an anaerobic digester to enrich the gas quality. One of the more recently developed processes to accomplish this is described by Cheremisinoff and Morresi (1976).

"NRG NUFUEL of California has developed a process for using a molecular sieve to remove high concentrations of CO_2 from gas streams. The molecular sieves are hydrated metal aluminum silicates possessing unique structures with centrally located cavities whose only entrances are apertures of a specific size. The unique structure of the molecular sieve allows them to be used to separate gases or liquids selectively and with respect to: the diameter and configuration of the molecules, the degree of the unsaturation of the molecules, and the polarity of the molecules. Molecular sieves can effect these separation processes by means of their apertures or "pores" which can screen out large or irregularly shaped molecules, and by means of their preferential attraction for polar molecules.

The NRG gas purification process operates by circulating the contaminated gas through three vessels packed with adsorbent molecular sieves. The gas is pumped into the first vessel where impurities are adsorbed by the molecular sieve until the capacity of the bed is exhausted. The stream is then switched to a second vessel, while the first vessel is regenerated by depressurization and evacuation. When the second vessel is exhausted, the stream is switched back to the first vessel, while the second is regenerated. This "pressure swing cycle" is repeated continuously on a controlled basis.

Over a period of time, the molecular sieve slowly loses its capacity to adsorb all impurities, requiring each vessel to be removed periodically from the pressure swing cycle and heated to drive off residual contamination. This is called the "thermal swing cycle". A third vessel is employed to permit the process to continue without interruption. These cycles revolve continuously, with each vessel taking a turn through the thermal swing cycle while the other two alternate on pressure swing cycles. The net result of this procedure is a continuous flow of pipeline quality methane gas".

In brief summary of the foregoing discussion, the technology for using biomass in anaerobic digestion systems to produce methane and carbon dioxide is moderately well developed for some feedstocks. Municipal sewage and manure are especially well suited as feedstocks and are proven in full scale commercial operations. Other feedstocks have great potential in terms of their expected ability to produce methane but are not necessarily attractive economically nor well proven in commercial applications. Research and development efforts are underway to improve the reliability of systems, to reduce costs for construction and operation of the systems, and to improve the product yield. Significant research activities are under way to investigate pretreatment of potential feedstocks as a means of increasing yield of methane-rich gas. The technology is well developed and proven for upgrading digester off-gases by removing impurities, water, and carbon dioxide so that the resultant gas is essentially pure methane. Purified digester off-gases can be compressed and used as pipeline gas.

Recovery of Methane from Sanitary Landfills

In the mid-1970s recovery of methane from sanitary landfill became a commercially proven technology. The following summary is a brief review the technology involved.

Cheremisinoff and Morresi (1976) provide pertinent general information on landfills:

Sanitary landfilling is a method of controlled solid waste disposal in which four basic functions are performed: (1) The landfill site is prepared to accept the municipal solid wastes; (2) The solid wastes are deposited, spread out, and compacted into thin layers; (3) These wastes are regularly covered; and (4) The cover material is then compacted.

With the forehand knowledge that the organic materials buried in the landfill will undergo anaerobic digestion and that methane and carbon dioxide gases will be generated in the process, it is possible and desireable to construct sanitary landfills in a manner which will channel the gases to points where they can be conveniently and safely collected and processed. Gas movement can be controlled by proper construction, taking into account the available natural soils, the hydrologic conditions, and the geologic conditions of the site.

The anaerobic decomposition of municipal solid wastes in sanitary landfills is very similar to that which takes place in controlled anaerobic digesters. The primary differences lie in the rates at which methane is formed and in the resultant blend of methane and carbon dioxide. Figure 3 shows the time framework and relative concentrations of methane and carbon dioxide in the collected gases from a sanitary landfill. Note in Figure 3 that approximately 1.5 - 2 years are required to reach equilibrium conditions for methane and carbon dioxide generation. By comparison, methane generation in controlled anaerobic digesters may reach equilibrium conditions in 5 - 10 days. Note also in Figure 3 that typical off-gases from sanitary landfills are approximately 50% methane and 50% carbon dioxide, whereas, off-gases from controlled digesters are typically 60% - 65% methane with the balance being carbon dioxide.

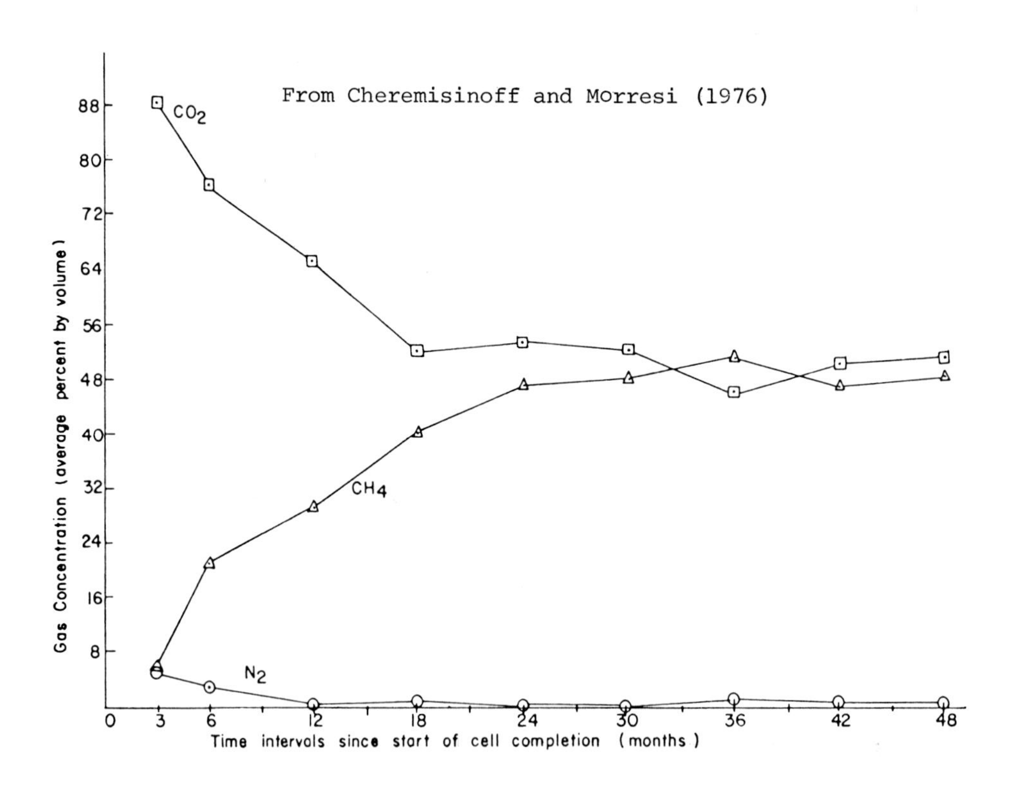

Figure 3. Gas composition from a sanitary landfill

The amount of methane gas which may be generated in the lifetime of a sanitary landfill is calculated by Cheremisinoff and Morresi (1976). According to their estimates, each pound of solid waste in a typical landfill may generate a maximum of 4.44 cubic feet of methane and 3.98 cubic feet of carbon dioxide over its lifetime. Using a half-life approach (wherein 23 years is the anticipated half-life), 1 pound of solid waste is estimated to generate 1.84×10^{-7} cubic feet/minute of methane. Typical sanitary landfills are compacted to a density of 29.63 pounds per cubic foot. Therefore, if it is assumed that the rate of methane generation is fairly uniform with time throughout the lifetime of the landfill, then the volume of a landfill required to continuously produce 1,000 cubic feet per minute of methane would be 1.83×10^{8} cubic feet. That is roughly equivalent to a landfill of 105 acres which is 40 feet deep. In metric equivalent, it would require a landfill with a volume of 5.2×10^{6} m^3 to generate a continuous flow of methane at a rate of 0.5 m^3/s.

> "From this it is concluded that the potential for sustained extraction of methane from a landfill for significantly large use is not possible in other than extremely large landfills with ideal conditions for biological decomposition and containment of gases." (Cheremisinoff and Morresi, 1976).

Commerical applications of methane recovery technology from municipal sanitary landfills have been underway since 1975. Calden (1976) reports that the world's first commercial landfill methane recovery facility was located in Southern California near Palos Verdes. The facility (as of 1976) was processing 2 million SCF/day of gas and upgrading it to 1000 BTU/SCF using molecular sieve technology. Bente (1979) reports on 6 projects involving the collection and use of methane from sanitary landfill sites. Five of these projects are located in California and one in New York. Selected comments taken from the report summaries may be of interest in that they demonstrate some of the specific variations of each site:

Location: Industry Hills, California

Project completion in March 1979. Goal is recovery of 400 CFM of 500 BTU/SCF landfill gas for use as a fuel source for heating on-site facilities (including a swimming pool).

Location: Palos Verdes, California

Two million cubic feet of refuse gas per day can be withdrawn when the plant is at full capacity. The purification process transforms this amount of gas into approximately one million cubic feet of pipeline standard methane which is enough to meet the natural gas energy needs of 3,500 homes.

Location: Monterey Park, California

The company's second facility is presently under construction at the Operating Industries landfill. This plant, which will produce four million cubic feet per day of pipeline quality gas is scheduled to go on stream in mid-1979.

Location: Sun Valley, California

By July 1979, a gas well collection system will be installed in Sun Valley to recover methane gas for delivery to the Valley Generating Station for use as boiler fuel. The gas recovery project will provide fuel to generate about 28 million kilowatt hours of electricity annually and provide gas migration control at the landfill. The gas well recovery system is composed of 14 wells drilled 80 to 100 feet deep.

Location: Wilmington, California

The installed methane gas recovery system is comprised of 27 gravel packed wells, approximately 50 feet in depth and a header system located on a 50 acre landfill. The goal is recovery of up to 1200 cfm of low BTU (500 BTU/SCF) sanitary landfill gas for sale to a nearby refinery. Gas will be supplied at 45°F and 80 PSIG.

Location: Staten Island, New York

Four 40-foot deep test wells and collection system were installed in a 40 acre section. Pumping resulted in a sustained flow rate of 150 CFM. Methane concentrations varied from 50% - 60%. Landfill gas will be directly fired in various industrial burners and appliances to determine combustion properties, including corrosive effects and flue gas emissions. Electricity generation will be demonstrated by fueling a 100 kW gas engine generator with landfill gas.

In summary, the technology for recovery of methane gas from sanitary landfill sites is new but acceptably well demonstrated. With the availability of proven systems for selectively separating the methane gas from the carbon dioxide, the landfill gas can be economically upgraded to pipeline quality SNG. There are some restraints on the pumping rate of landfill gas from existing landfill sites due to the relatively slow rate of bacterial decomposition of the municipal solid waste. Further, it is important to design and operate the landfill in a manner which promotes the generation and efficient collection of methane. The landfill gas may be used without cleanup for direct combustion purposes. Apparently, the economics of landfill gas recovery are attractive as evidenced by the recent developments in California.

This completes the discussion of the biochemical conversion processes for producing synthetic fuels from biomass. The reader should note that the fermentation process produces ethanol while the anaerobic digestion process results in the formation of methane as the synthetic fuel of interest.

The summary now turns to the thermochemical conversion processes. In contrast to the biochemical processes, the products of any one of the thermochemical processes may be quite diverse. For example, pyrolysis processes may result in the production of a mixture of gases, a blend of liquids, and solid carbon char. To produce specific synthetic fuels may require that a series of unit operations be used. In the discussion that follows, the principle conversion systems and their products are presented. Additional information is presented about unit operations used to manufacture specific synthetic fuels from the products of the primary conversion systems.

C. PYROLYSIS

Two very closely related thermochemical processes are: (1) Pyrolysis; and (2) Gasification. For purposes of this summary they are treated separately.

Pyrolysis is an irreversible chemical change brought about by the action of heat energy in an atmosphere devoid of oxygen. (Lewis and Ablow, 1976). It is often referred to as thermal decomposition, destructive distillation, and/or carbonization.

A wide range of biomass resources have been subjected to pyrolysis, and it has been shown that a variety of useful products can be obtained from this process. Such diverse feedstocks as manure, municipal solid waste, agricultural residues, and forest residues have been decomposed at moderate to high temperatures to form a range of gaseous, liquid, and solid products.

The steps which are thought to take place in typical pyrolytic reactors are shown in Figure 4. A principle rate-limiting factor in the process is the low thermal conductivity of most biomass materials.

> There are two kinds of pyrolysis: slow and fast. At slow heating rates or with large pieces of biomass, pyrolysis leads to a high proportion of charcoal..... At the most rapid heating rates, cellulose is largely converted to a gas containing a high proportion of olefins that are valuable as chemical feed materials; char production is minimal. (Milne, 1979)
>
> Although not yet proven quantitatively, it is commonly accepted that the pyrolysis of many complex forms of biomass can be understood as the sum of the breakdown of its three components; cellulose, hemicellulose, and lignin. This is borne out qualitatively by comparison of laboratory analysis of the pyrolysis of components with those of whole biomass. (Milne, 1979).

A great deal of laboratory research has been conducted in an attempt to fully understand the mechanisms that control and/or influence pyrolysis reactions. From evidence gathered to date it appears that the products of pyrolysis are sensitive to the physical size and state of the feedstock, inorganic impurities, heating rate, and the final temperature.

By strict definition, pyrolysis involves the decomposition of matter in the absence of oxygen. As such, it is often one of the initial steps in combustion and gasification processes. Both researchers and manufacturers of equipment tend to interchange the terms "pyrolysis" and "gasification" with the result that it is often difficult to review specific projects underway and make a determination as to whether they are in fact pyrolysis operations.

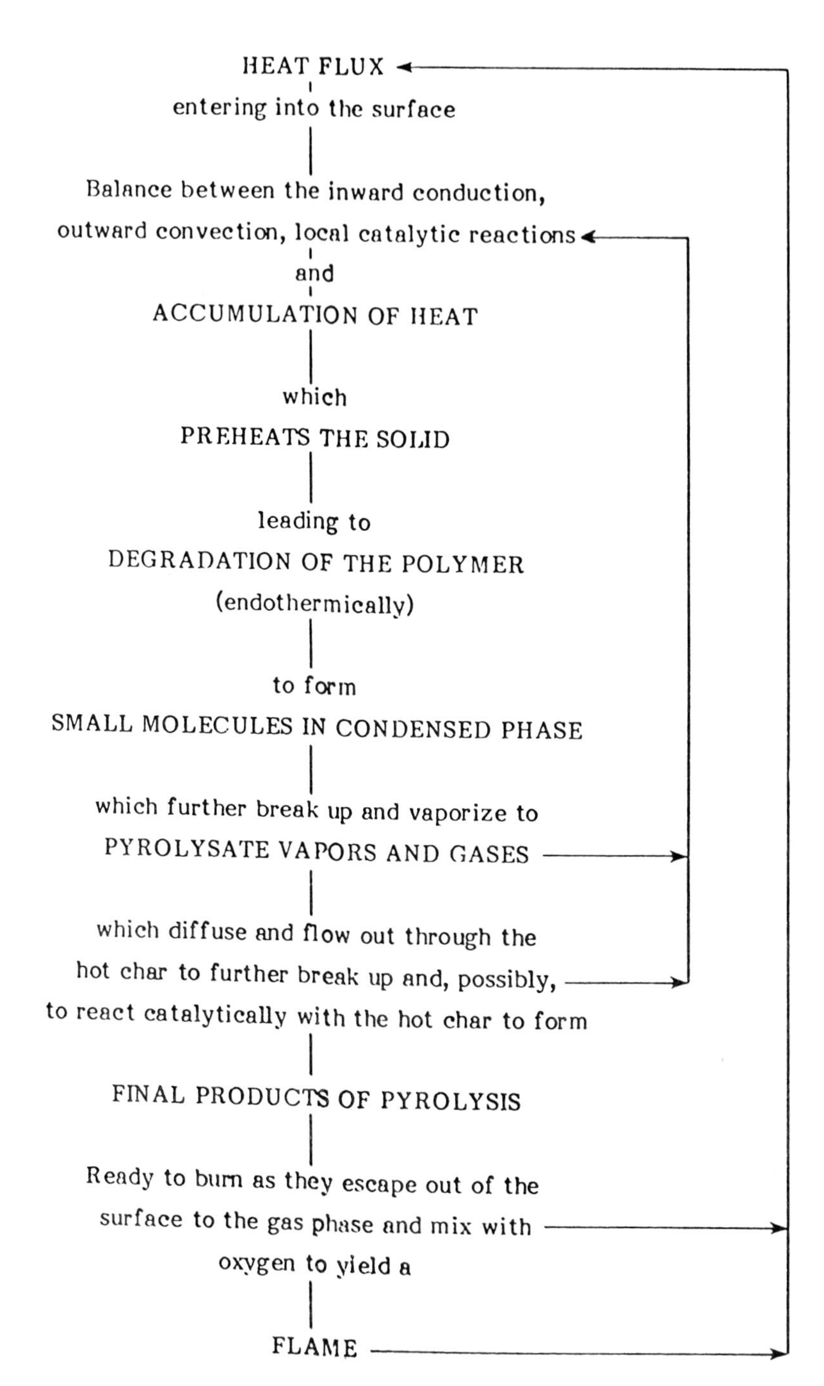

Figure 4. Process steps in slow gasification *

*From Milne (1979)

Taking that limitation into account, the following projects are listed by Bente (1979) as being related to pyrolysis.

Location: Ojai, California

Convert as received (moist and possibly cumbersome) biomass materials into a medium BTU gaseous product by means of a pyrolysis-water gas reaction. The experimental program has concentrated on cattle manure as its raw material. The process uses a modified 2-stage drying, multiple hearth furnace for the conversion. This process may also have application in the multiple hearth furnace processing of sewage sludge or municipal solid waste.

Location: Redwood City, California

Process municipal and automotive solid wastes by continuous pyrolysis to produce gas and a recyclable residue of carbon, ash, and metallics. The Pyro Sol, Inc. process utilizes waste as primary feed to a pyrolysis unit, which in the absence of oxygen and in the presence of heat will cause chemical decomposition of the waste into a gas and a char/ash residue. ...... the process is self-supporting. A portion of the produced gas is burned in eight radiant heat tubes to provide heat for the endothermic process. The produced gas exits the pyrolyzer at approximately 1100 ^{o}F.

Location: Atlanta, Georgia

Develop a vertical bed pyrolytic reactor for the conversion of agricultural and forestry residues and municipal wastes into char, oil, and gas. Research and development work was initiated in 1969. A pyrolysis process was developed which has been proven technically feasible on a continuous basis with a 40 ton/day field demonstration pyrolysis facility. The feed material is dried to less than 10% moisture on a wet basis for smooth, steady state operation. The energy recovered in the charcoal, oil, and gaseous products is greater than 95% of the input energy of the feed material on a dry basis. The technical feasibility and reliability of the process have been proven by the processing of pine bark-sawdust in a field demonstration pyrolysis system for several months, 24 hours a day, seven days a week, at better than 90% on-line time.

Location: Princeton, New Jersey

Determine the optimal conditions for the production of a hydrocarbon-rich synthesis gas from organic matter by steam pyrolysis/gasification. Kinetic models of gasification rates and product compositions are being formulated on experimental work. The study has application to all projects which rely on pyrolysis for the conversion step of organic solids to a liquid or gaseous fuel.

Research during the first two years of the project has shown biomass gasification to be a three step process:

1. Pyrolysis at low temperature (300 ^oC – 500 ^oC) producing gaseous volatile matter and char (10% – 40% by weight of the dry feedstock);
2. Steam cracking of the volatile matter at higher temperatures (600^oC) producing a hydrocarbon-rich synthesis gas; and
3. Char gasification at high temperatures (700 ^oC), producing synthesis gas.

Reaction rates for steps 1 & 2 have been measured for a variety of materials, including cellulose, hard and soft woods, and manures. Kinetic models have been developed based on experimental results.

Location: Poughkeepsie, New York

Reprocess region's municipal waste and sewage sludge into a medium BTU fuel gas using a Purox pyrolysis system. This proposed project will consist of two full-sized 300 tons per day pyrolytic converters.

Location: Lubbock, Texas

Produce ammonia synthesis gas, ethylene, and/or a low BTU gas (250-350 BTU/SCF), using thermochemical reactor technology developed at Texas Tech University. Feedstocks for the thermochemical process include manure, wood, and field waste. A countercurrent pyrolysis reactor for biomass has been invented which allows the volatile organic compounds to escape from the heating zone very rapidly. This results in a different product mix than has been observed in other pyrolysis research, containing unusually high concentrations of ethylene. The work includes studies in an existing 1/2 ton/day test reactor to determine the influence of temperature, residence time, pressure, and feedstock materials on the yield and quality of the products of the reaction.

These ongoing projects indicate the range of approaches to using pyrolysis as a process for converting biomass into more useable products. It is apparent that pyrolysis has been developed to the point where it is commercially demonstrated. Yet it is also apparent that there are many facets of the process and its control that are not yet fully understood.

One of the most widely publicized pyrolysis systems which is in commercial application is the PUROX System, handled by the Linde Division of Union Carbide Corp. It is shown schematically in Figure 5 and is described by Fisher, Kasbohm, and Rivero (1975) as follows:

Union Carbide's PUROX System employs a partial oxidation process using oxygen for converting solid waste to fuel gas and inert slag. The process has been successfully demonstrated on a commercial scale in a 200 Ton/day facility at Union Carbide's plant at South Charleston, West Virginia. One ton of refuse requires about 0.2 tons of oxygen and produces 0.7 tons of medium BTU fuel gas, 0.22 tons of sterile residue, and 0.28 tons of wastewater. Within the process 0.03 tons of oil are separated in the gas cleaning train

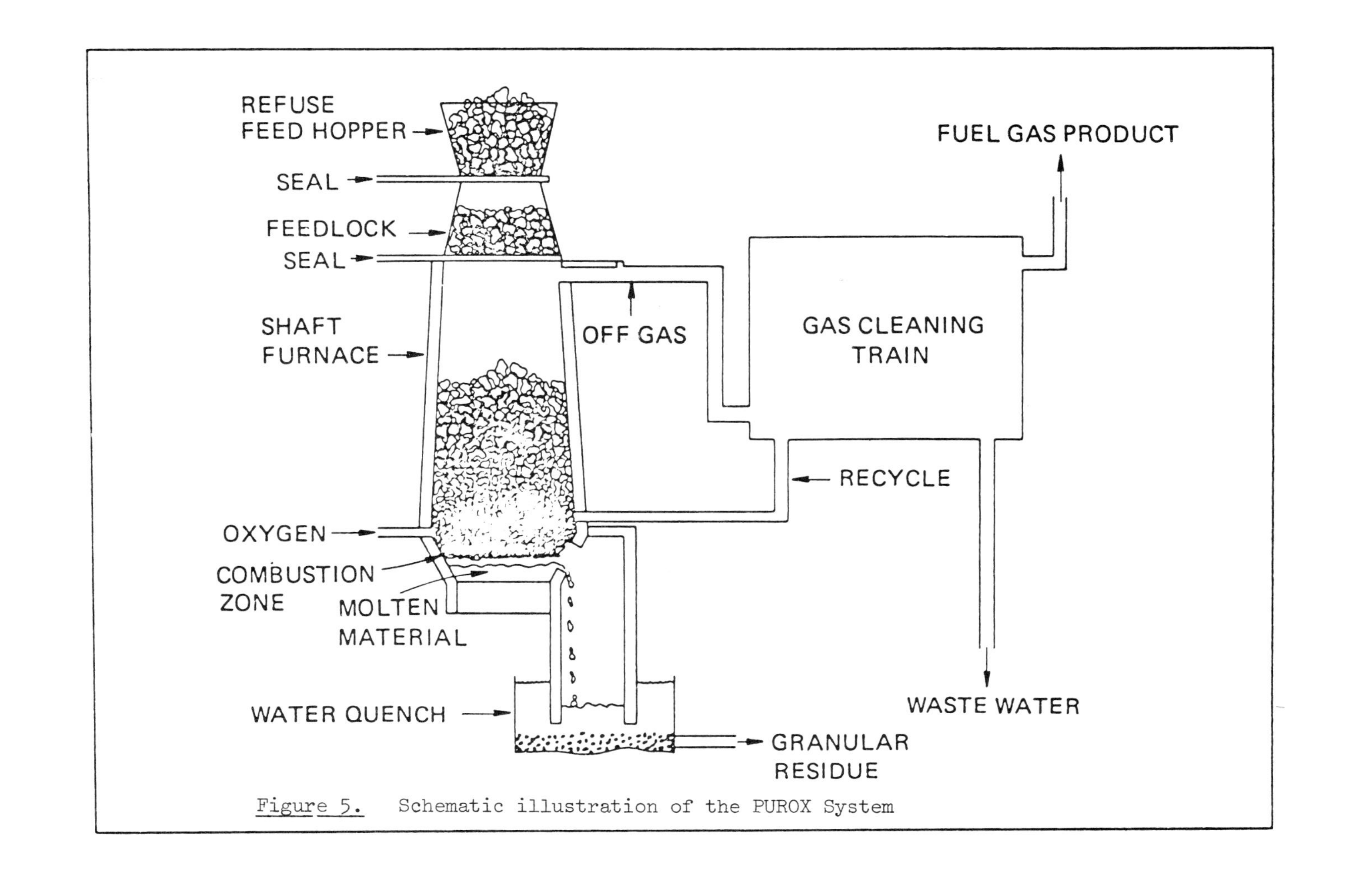

Figure 5. Schematic illustration of the PUROX System

and recycled to the furnace for cracking into additional gas. All of these numbers are approximate and are given in an effort to give a simplified view of the overall process.

The key element of the system is a vertical shaft furnace. Refuse, which has been preprocessed for recovery of materials such as iron, is fed to the top of the furnace through a gas seal to prevent escape of fuel gas. Oxygen is injected into the bottom hearth section to provide the partial oxidation that drives the reactor. The furnace is maintained essentially full of refuse which continually descends by gravity through the shaft. The oxygen in the hearth reacts with char formed from the particles of refuse. This reaction generates the high temperature in the hearth needed to melt the glass, metal, and other materials to give a molten residue. The molten residue drains continuously into a water quench tank where it forms a hard granular aggregate.

The hot gases from the hearth section rise through descending refuse, cooling the gas and pyrolyzing the refuse to yield a fuel gas. In the upper portion of the furnace, the gas is further cooled as it dries the fresh incoming refuse. This countercurrent heat exchange efficiently utilizes the energy of the gas and provides a top gas that is cleaned by the incoming refuse. The gases leaving the converter are further cleaned of their oil mist and excess water vapor by passing through a recirculating water scrubber system and an electrostatic precipitator. The liquid hydrocarbons and any entrained solids are separated from the scrubber water and recycled to the furnace for disposal. The condensate, which is the net water product discharge from the scrubber system, is cleaned of organics and sent to the sewer.

Typical gas analysis from the PUROX System are:

H_2	26% by Vol.
CO	40% " "
CO_2	23% " "
CH_4	5% " "
C_2+	5% " "
N_2 & A	1% " "

Gas Higher Heating Value = 370 BTU/SCF

A particularly important feature of pyrolysis systems is demonstrated by the off-gas analysis presented above. That is that the off-gases may contain high concentrations of hydrogen and carbon monoxide — the primary ingredients in synthesis gas.

The Solar Energy Research Institute (1979) has prepared a list of the pyrolysis systems for research, development, or manufacture. (See Table 8).

TABLE 8
SURVEY OF PYROLYSIS GASIFICATION SYSTEMS*

Organization	Gasifier Type		Fuel Products	Operating Units	Size
	Input	Contact Mode			Btu/h
Pyrolysis Gasification of SMW					
Monsanto, Landgard, Enviro-chem.	P, C	K	LEG, O, C	1 D	20 (375)
Envirotech, Concord, CA	P	MH	LEG	1 P	
Occidental Res. Corp El Cajon, CA	P	Fl	PO, C, MEG	1 C	
Garrett En. Res. & Eng. Hanford, CA	P	MH	MEG	1P	
Michiga Tech, Houghton, MI	P	ML	MEG		
U. of W. Va-Wheelebrator Morgantown, WV	P, G, C	Fl	MEG	1P	
Pyrox Japan	P, G, C	Fl	MEG	1C	
Nichols Engineering	P		MEG, C		
ERCO Cambridge, MA	P	Fl	MEG	1P	16
Rockwell International Canoga Park, CA	P	MS	MEG, C	1P	16
M. J. Antal Princeton, NS	P	O	MEG, C	2R	- -
Pyrolysis Gasification of Biomass					
Wright-Malta Ballston Spa, NY[a]	PG	O	MEG (C)	1R, 1P	4
Coors/U. of MO	P	Fl		1P	
U. of Arkansas	P	O	MEG (C)	IR	
A & G Coop Jonesboro, AR	P	O	MEG (C)	1C	
ERCO Cambridge, MA	P	Fl	PO, C	1P, (1C)	16, (20)
ENERCO Langham, PA	P		MEG, PO, C	1P, 1C	
Garrett Energy Research	MH		MEG	1P	
Tech Air Corporation Atlanta, GA 30341	P	U	MEG, PO, C	4P, 1C	33

* From the Solar Energy Research Institute (1979)--Table Notation See Next Page

Table 8 (Cont.)

TABLE NOTATION: (by columns)

Input: A = air gasifier; O = oxygen gasifier; P = pyrolysis process; PG = pyrolysis gasifier; S = steam; C = char combustion

Contact Mode: U = updraft; D = downdraft; O = other (sloping bed, moving grate); Fl = fluidized bed; S = suspended flow; MS = molten salt; MH = multiple hearth

Fuel Products: LEG = low energy gas (~150-200 Btu/SCF) produced in air gasification; MEG = medium energy gas produced in oxygen and pyrolysis gasification (350-500 Btu/SCF); PO = pyrolysis oil, typically 12,000 Btu/lb; C = char, typically 12,000 Btu/lb

Operating Units: R = research; P = pilot; C = commercial size; CI = commercial installation; D = demonstration

Size: Gasifiers are rated in a variety of units. Listed here are Btu/h derived from feedstock throughput on the basis of biomass containing 16 MBtu/ton or 8000 Btu/lb, SMW with 9 MBtu/ton. ( ) indicate planned or under construction.

Note in Table 8 that of the 25 pyrolysis units listed, only 6 are shown to be in commercial use. Five of the 6 commercial systems are producing medium energy gas.

SRI, International (1978) has reviewed the system paths involving pyrolysis of biomass fuels and has selected four paths which appear to show promise of the greatest market penetration in the coming years. The selected paths are indicated in Table 9.

TABLE 9
COMPARISON OF PYROLYSIS TECHNOLOGIES BY SRI, INT'L

Process Description	Estimated Product Cost for Comparison ($/MMBTU)
Wood to oil and char by-product with char valued at $1.25/MMBTU	$ 4.50
Wood to oil and char (total product basis)	$ 2.70
Wood to intermediate BTU gas at 280 PSIA	$ 4.00
Wood to intermediate BTU gas at 25 PSIA	$ 3.40

Note: All estimates based on plant size of 3,000 Tons/Day of feedstock.

It should be understood that the 4 system paths selected by SRI, Int'l are based on an analysis of the probable market penetration of the alternative technologies and feedstocks. The analyses included various factors such as the availability of biomass feedstocks, process cost reduction potential, process energy balance, product marketability, and process environmental impact. It is not necessarily correct to assume that the same system paths would be applicable to countries other than the United States.

In summary, pyrolysis processes deal with the thermal decomposition of matter (in this case biomass) in the absence of oxygen. The technology for pyrolysis has been used for many years but has seen substantial interest during recent years because it offers the opportunity to produce oils, gaseous fuels, and char from low-cost renewable biomass feedstocks. Many research workers are involved in attempting to more fully understand the processes which take place and are working on relatively new developments, such as, fast pyrolysis. The technology for conversion of biomass fuels by pyrolysis is commercially available and proven in large scale. It is predicted that in the United States pyrolysis conversion of wood feedstocks to medium BTU gas, char, and oil will have significant market penetrations in the near future.

D. GASIFICATION

Gasification, like pyrolysis, is an irreversible thermochemical process in which the feedstock is thermally decomposed and the end products are principally in gaseous form. For most gasification processes, the initial steps may involve pyrolysis of the feedstock (thermal decomposition of the feedstock in the absence of oxygen). However, the intermediate products from the pyrolysis steps (char and pyrolytic oils) are burned to form gaseous end products.

> The simplest form of gasification is air gasification, in which the excess char formed by pyrolysis is burned with a limited amount of air at an equivalence ratio of about 0.25, requiring 1.6 g of air per gram of biomass. (Reed, 1979)

If air gasification is used, the resultant product gas will have a relatively low BTU content, typically in the range of 80 - 200 BTU/SCF. The reason is that the product gas is diluted by the nitrogen from the gasifying air stream. An alternative to air gasification is oxygen gasification or gasification with oxygen enriched air. Oxygen gasification systems produce a medium BTU content gas (300 — 500 BTU/SCF) which may be used for synthesis gas for the manufacture of products such as methanol, ammonia, methane, or gasoline. Another alternative is to gasify with steam, a process that may generate intermediate BTU gaseous products under specific conditions.

From a technical standpoint, oxygen gasification is not difficult. However, it does require the facilities for manufacturing oxygen and these involve substantial investment. Steam gasification is less capital intensive than oxygen gasification.

When biomass feedstocks are gasified, a reactor is necessary to provide a proper environment for the reaction to take place. There are a wide variety of reactor designs available which have been used for experimental and commercial purposes. They are typically divided into five specific groups:

1. Fixed bed, downdraft
2. Fixed bed, updraft
3. Fluidized bed
4. Entrained flow
5. Molten bath*

*There is some question as to whether molten baths are gasifiers or pyrolyzers.

Each design has its own limitations and virtues. For example, the selection of gasifier design is limited by the size distribution characteristics of the feedstock. If the size distribution is small (i.e., less than 5 mm) then an entrained flow gasifier may be most suitable. Very small particle size of the feedstock will prevent uniform operation of fixed bed systems and may result in substantial elutriation of particles from fluid bed reactors. On the other hand, if the feedstocks are bulky (i.e., wood chips or corn cobs), then a fixed bed gasifier may be very suitable from a material flow and operational standpoint. For other than very fine particles, fluid bed reactors can tolerate a broad range of particle sizes in the feedstocks.

As a general rule, fixed bed updraft gasifiers produce gas which has a high oil and tar content. By comparison, fixed bed downdraft gasifiers have low oil and tar content.

Fixed bed updraft gasifiers have a relatively low temperature product gas; fixed bed downdraft gasifiers have moderately high temperature product gas. Fluid bed gasifiers and entrained flow gasifiers can have a wide range of exit temperatures depending on the equivalence ratio during operation.

Most of the gasifiers used today for research and/or commercial operation are operated at approximately atmospheric pressure. There are a few experimental systems designed to operate at pressures in the range of 10 - 100 atmospheres, yielding a gas that can be fed to a turbine or upgraded to pipeline quality gas. However, these units are substantially more costly than atmospheric systems.

One of the options available in gasifier design is the ability to mix, stir, or tumble the feedstock in the reactor. Herreshoff type multiple hearth furnaces, rotary kilns, and stirred bed reactors are typical of the variations which are possible and used.

Biomass fuels have a small inert, inorganic ash content which is typically in the range of 0.5% to 3% on a dry weight basis. When the fuel is fed to the gasifier system, this ash must be taken into account. In particular, it is important to note the temperature at which the ash will soften and fuse. Some gasifiers are designed to operate at temperatures below the ash fusion point. Others are designed to operate at temperatures well above the ash melting point. In the high temperature systems, the ash is collected as a molten liquid. Experience has shown that if the system is operated at temperatures between the ash fusion point and the ash melting point, the ash will agglomerate and generally plug up the gasifier, making it inoperable.

Air gasification systems are particularly popular at this time, probably due to the range of designs available, the range of feedstocks which can be gasified, the relatively low capital investment required, and the relative ease of operation. The Solar Energy Research Institute (1979) has listed the air-blown and oxygen-blown gasification systems in use for biomass feedstocks and municipal waste. These are shown in Table 10. Note that of the design options available, the most popular reactors are the updraft and downdraft fixed bed designs, with only 4 of the 26 companies using fluid bed reactors. None of the companies listed are using entrained flow reactors.

TABLE 10
SURVEY OF AIR AND OXYGEN-BLOWN GASIFICATION SYSTEMS*

Organization	Gasifier Type		Fuel Products	Operating Units	Size
	Input	Contact Mode			Btu/h
Air Gasification of Biomass					
Alberta Industrial Dev. Edmonton, Alb., Can.	A	Fl	LEG	1	30M
Applied Engineering Co., Orangeburge, SC 29115	A	U	LEG	1	5M
Battelle-Northwest Richland, WA 99352	A	U	LEG	1-D	—
Century Research, Inc. Gardena, CA 90247	A	U	LEG	1	80M
Davy Powergas, Inc. Houston, TX 77036	A	U	LEG-Syngas	20	—
Deere & Co. Moline, IL 61265	A	D	LEG	1	100kW
Eco-Research Ltd. Willodale, Ont. N2N 558	A	Fl	LEG	1	16M
Forest Fuels, Inc. Keene, NH 03431	A	U	LEG	4	1.5-30M
Foster Wheeler Energy Corp. Livingston, NH 07309	A	U	LEG	1	—
Fuel Conversion Project Yuba City, CA 95991	A	D	LEG	1	2M
Halcyon Assoc. Inc. East Andover, NY 03231	A	U	LEG	4	6-50M
Industrial Development & Procurement, Inc. Carle Place, NY 11514	A	D	LEG	Many	100-750kW
Pulp & Paper Research Inst.,[c] Pointe Claire, Quebec H9R 3J9	A	D	LEG	—	—
Agricultural Engr. Dept. Purdue University W. Lafayette, IN 47907	A	D	LEG	1	0.25M
Dept. of Chem. Engr. Texas Tech University Lubbock, TX 79409	A	Fl	LEG	1	0.4M
Dept. of Chem. Engr. Texas Tech University Lubbock, TX 79409	A	U	LEG	1	—
Vermont Wood Energy Corp. Stowe, VT 05672	A	D	LEG	1	0.08M

*From the Solar Energy Research Institute (1979)

Table 10 (Cont.) SURVEY OF AIR AND OXYGEN–BLOWN GASIFICATION SYSTEMS*

Organization	Gasifier Type: Input	Gasifier Type: Contact Mode	Fuel Products	Operating Units	Size Btu/h
Dept. of Ag. Engr. Univ. of Calif. Davis, CA 95616	A	D	LEG	1	64,000
Dept. of Ag. Engr. Univ. of Calif. Davis, CA 95616	A	D	LEG	1	6M
Westwood Polygas (Moore)	A	U	LEG	1	
Bio-Solar Research & Development Corp. Eugene, OR 97401	A	U	LEG	1	- -
Oxygen Gasification of Biomass					
Environmental En. Eng. Morgantown, WV	O	D	MEG	1P	0.5
IGT-Renugas	O,S	Fl	MEG		
Air Gasification Solid Municipal Waste (CSMW)					
Andco-Torrax Buffalo, NY	A	U	LEG	4C	100M
Battelle NW Richmond, VA 99352					
Oxygen Gasification of SMW					
Union Carbide (Linde) Tonowanda, NY	O	U	MEG	1	100M
Catorican Murray Hills, NS	O	U			9M

TABLE NOTATION: (by columns)

Input: A = air gasifier; O = oxygen gasifier; P = pyrolysis process; PG = pyrolysis gasifier; S = steam; C = char combustion

Contact Mode: U = updraft; D = downdraft; O = other (sloping bed, moving grate); Fl = fluidized bed; S = suspended flow; MS = molten salt; MH = multiple hearth

Fuel Products: LEG = low energy gas (~150-200 Btu/SCF) produced in air gasification; MEG = medium energy gas produced in oxygen and pyrolysis gasification (350-500 Btu/SCF); PO = pyrolysis oil, typically 12,000 Btu/lb; C = char, typically 12,000 Btu/lb

Operating Units: R = research; P = pilot; C = commercial size; CI = commercial installation; D = demonstration

Size: Gasifiers are rated in a variety of units. Listed here are Btu/h derived from feedstock throughput on the basis of biomass containing 16 MBtu/ton or 8000 Btu/lb, SMW with 9 MBtu/ton. ( ) indicate planned or under construction.

*From the Solar Energy Research Institute (1979)

Ongoing research and development efforts in the area of biomass gasification are summarized by Bente (1979). Selected projects are noted below in order to provide the reader with some insight as to the breadth of projects underway.

Location: California - Various demonstration sites

Demonstrate the technical and economic feasibility of converting wastes to low BTU gas to be used in industrial boilers, substituting fossil fuel. ... Gasification characteristics of approximately 12 agricultural and forest residues were investigated in a laboratory-size gas producer with output of 150,000 — 300,000 BTU/hr. A 531-hour test period was completed with a pilot plant gas producer producing up to 6.5 million BTU/hr. This gas was used to fire a steam boiler, heated air burner (2 million BTU/hr), and a dual-fueled 60 kW naturally aspirated diesel engine electric generator set.

Location: Yuba City, California

Convert biomass and lowgrade mineral fuels to low-BTU fuel gas for use either as replacement for natural gas or as engine fuel for in-plant and small community base load applications.

Location: Golden, Colorado

Carry out fundamental and applied research in the field of: (1) Thermal conversion of biomass to heat (gaseous, liquid, and solid fuels, and chemicals), and (2) electro-chemical conversion of biomass to electricity in fuel cells.

Riceboro, Georgia

Gasify wood wastes to produce charcoal and to fire combustible gas in a high pressure steam boiler to produce plant electricity.

Location: DeKalb, Illinois

Develop complete operating unit (hardware, controls, fuel preparation) to recover fuel values from corn cobs for use in drying operations. Manufacture of a clean producer synthetic gas from corn cobs is contemplated. Key issues are how to avoid slagging and whether to burn tars in-stream or separate them out prior to burning the gas.

Location: Johnston, Iowa

Corn cob gasifier. Convert corn cobs to produce gas for seed corn dryer fuel using two 9 million BTU/hr units. This project has been operating and producing a gas consisting of carbon monoxide, hydrogen, carbon dioxide, methane, and small percentages of other gases on a short run basis. Problems include vaporized tar in the gas, ash cake, and some slag on the grate.

Location: Keene, New Hampshire

Develop and market a wood-to-gas fuel generator and burner system to replace oil or gas burners for small to medium industrial boilers as well as some direct fired applications. Three units are on line with outputs of 1.5, 4.0, and 10-12 million BTU/hr output respectively. One unit has been continuously operating since May 1978.

Location: Ballston Spa, New York

Steam Gasification of Biomass. Develop a process and equipment for efficient and economical conversion of wood chips into medium BTU fuel gas. Accomplishments to date:

1. Process chemistry has been proven on bench scale (10 lbs/hr) in batch and continuous equipment.
2. The overall process has been shown to be sufficiently exothermic so that with recuperative feedback no external heat will be required.
3. Gasification is essentially complete; overall efficiency will be about 90%.
4. Heat transfer coefficients have been shown to be sufficiently high so that indirectly heated, rotary kiln reactors are feasible.
5. Design work has been completed on a 6 green ton/day process development unit.
6. Preliminary economic analysis indicates a 20-year gas selling price of $3.00 - $4.50 per million BTU.

Location: Columbus, Ohio

Convert forest residue into methane rich gas through a gasification reactor. Residue is catalytically treated before it is gasified. A proprietary catalytic pretreatment process was applied to forest residue products. In this process, shredded forest residues are reacted with a calcium oxide slurry at elevated temperature and pressure to produce a chemically incorporated and effective gasification catalyst. Study findings indicate that wood ash is as effective a catalyst as more expensive catalysts tried.

In the United States the coal energy resource base is significantly larger than the biomass energy resource base. For a variety of reasons the main thrust of gasification technology has centered in coal resource areas with the result that many new processes have been developed and are in process of being developed for coal gasification. So as not to overlook technological opportunities, one of the ongoing research efforts in the United States is to assess the coal gasification technologies to determine the potential for using these technologies for wood feedstocks. A portion of the comparative results from that study are indicated in Table 11 and show a comparison of air-blown and oxygen-blown producer gas generated from coal and from wood.

TABLE 11
COMPARISON OF AIR-BLOWN AND OXYGEN-BLOWN PRODUCER GAS GENERATED FROM COAL AND WOOD*

	Air-Blown % of Product Gas		Oxygen-Blown % of Product Gas	
Component	Coal	Wood	Coal	Wood
CO	30	26	38	45
H_2	15	14	38	35
CH_4	2	2	3.5	0.35
CO_2	3	9	18	19.5
N_2	50	49	2.5	0.15

*From Keene and Nyce (1979)

Note in Table 11 that the characteristics of producer gas from coal and wood are similar. Note also that the oxygen-blown producer gas has great potential for use as synthesis gas in the production of other needed synthetic fuels.

SRI, International (1978) has reviewed the system paths involving gasification of biomass fuels and has selected two paths which appear to show promise for market penetration in the production of synthetic fuels. Both of the paths (as shown in Table 12) involve oxygen-blown gasification systems coupled with additional unit processes to convert the synthesis gas into SNG and methanol.

TABLE 12
COMPARISON OF GASIFICATION TECHNOLOGIES BY SRI, INT'l (1978)

Process Description	Estimated Product Cost for Comparison ($/MMBTU)
Wood to Synthetic Natural Gas via Oxygen-Blown Gasification	$ 6.50
Wood to Methanol via Oxygen-Blown Gasification	$ 7.80

Note: All estimates based on plant size of 3,000 Tons/Day of feedstock.

It should be understood that the 2 system paths selected by SRI, Int'l are based on an analysis of the probable market penetration of the alternative technologies and feedstocks. The analyses included various factors such as the availability of biomass feedstocks, process cost reduction potential, process energy balance, product marketability, and process environmental impact. It is not necessarily correct to assume that the same system paths would be applicable to countries other than the United States.

In brief summary of the discussion on the state of the art of gasification technology for generation of synthetic fuels from biomass, it appears that technology exists to convert biomass fuels to gaseous products. There are many types of gasifiers and depending on their feedstock and mode of operation they can produce low to medium BTU products. The medium BTU gases can be used as feedstock in the production of synthetic natural gas.

By far the majority of existing gasification systems using biomass are air-blown systems. Most of the reactors used are fixed bed with less than 25% being of the fluid bed design. Much of the ongoing R & D efforts are directed toward demonstration of feasibility. Perhaps the most promising of the research projects deals with catalytic gasification of biomass.

Although not pointed out previously in the summary, one of the major difficulties in the field of gasification is the size limit of proven systems. At this time there is some risk associated with the purchase and installation of any biomass gasification system whose gross heat input exceeds 80 Million BTU/hr. Systems of that size and larger do not have long-term proven reliability on a variety of biomass feedstocks.

E. LIQUEFACTION

In the category of thermochemical conversion technologies there are three general approaches used to convert biomass feedstocks into synthetic fuels: (1) Pyrolysis; (2) Gasification; and (3) Liquefaction. From the foregoing discussion it is seen that pyrolysis and gasification are closely related. Pyrolysis operations produce gases, liquids (pyrolytic oils), and char. Gasification operations convert the biomass feedstock primarily to gaseous end products.

Liquefaction, the converting of solid biomass feedstocks to liquid fuels, can be accomplished in three general ways:

1. Use a reducing agent and catalyst under elevated temperature and pressure conditions to convert the biomass feedstock to liquid fuels;

2. Convert the biomass feedstocks to synthesis gases (hydrogen and carbon monoxide and/or olefins) in a gasification reaction and then use the synthesis gases to manufacture liquid fuels, such as methanol, ethanol, or gasoline;

3. Use pyrolysis techniques and adjust the operating variables to yield the maximum amount of pyrolytic oils.

Each of the techniques is reviewed.

Production of Synthetic Oils by Catalytic Liquefaction

In the 1930s laboratory experiments showed that cotton and sawdust could be converted to liquids if they were subjected to reducing agents under conditions of elevated pressure and temperature. Subsequent research has led to the conclusion that oxygen removal from the molecules of biomass solids will leave a product in a liquid state. In essence this amounts to changing the carbon, hydrogen, oxygen ratios.

Serious research efforts to convert biomass feedstock to liquid fuels using this general approach began in the late 1960s. Early bench scale experiments showed the potential for producing heavy synthetic oils from diverse feedstocks including wood, urban refuse, agricultural wastes (corn cobs and animal manure), plus other organic wastes. As a result of the initial research efforts, a process for the liquefaction of wood was conceived. In the process 30 parts of dry pulverized wood is blended with 70 parts of oil and 7.5 parts of 20% aqueous solution of sodium carbonate. The mixture is heated to about 700 ^{o}F under carbon monoxide (or synthesis gas) pressure to a final pressure of 2,500 ± 1,200 PSI. Typical residence times are 45-60 minutes. Complete conversion of the wood yields 17.4 parts of oil (58%) having a heating value of about 13,500 BTU/lb. This process has the potential to convert 1 ton of dry wood to 1.75 barrels of synthetic crude oil. (Ergun, 1979)

In the mid-1970s a pilot scale process development unit was constructed in Oregon for research and development work on liquefaction of wood wastes. Several research groups and organizations are carrying on investigations relative to this process. The process is described by Berry (1979):

> "The process can be represented by two basic equations which involve reacting carbon monoxide and hydrogen with cellulosic materials at temperatures up to 700 ^{o}F and at pressures up to 4,000 PSIG in the presence of sodium carbonate as a catalyst. Carbon monoxide reacts with sodium carbonate in the presence of water to form sodium formate which, in turn, reacts with cellulose to form oil and regenerate sodium carbonate. The following reactions are believed to be typical:
>
> 1. $Na_2CO_3 + H_2 + 2CO \longrightarrow 2HCOONa + CO_2$
>
> 2. $2C_6H_{10}O_5 + 2HCOONa \longrightarrow 2C_6H_{10}O_4 + H_2O + CO_2 + Na_2CO_3$
>
> The oil resulting from the process is low in sulfur, suitable for use in power plants, or as feedstock for conversion to gasoline and diesel fuels. Laboratory-produced samples show the following properties:
>
> | | | |
> |---|---|---|
> | Specific Gravity | | 1.1 |
> | Viscosity, CP (140 ^{o}F) | | 515 |
> | Heating Value, BTU/lb | | 15,000 |
> | Composition: | % Carbon | 76.62 |
> | | % Hydrogen | 7.05 |
> | | % Nitrogen | 0.13 |
> | | % Sulfur | 0.14 |
> | | % Oxygen | 20.05 |
>
> Potential applications as process feedstocks are possibly even more important economically than fuel uses.

Carrying out the apparently simple deoxygenation reaction in the process development unit (PDU), however, involves several unit operations functioning in somewhat difficult conditions because of constraints, such as, temperature, pressure, small size, and the requirement for commercial availability of all items of equipment. In its simplest terms the primary operating mode consists of:

- Drying the biomass (wood in the PDU at present) to less than 4% moisture
- Grinding to a -35 mesh flour
- Slurrying the wood flour with carrier vehicle oil
- Pressurizing and preheating
- Injection into the heated reactor at pressures up to 4000 PSIG and temperatures up to 700 ^{o}F
- Injection of catalyst solution into the slurry of wood and oil
- Injection of carbon monoxide and hydrogen into the slurry of wood and oil
- Pressure letdown
- Separation of solids (catalyst and unreacted wood) from the oil

In the most recent operations, both catalyst solution and reactant gases have been injected ahead of the preheater.

Process and Equipment Details

The biomass feedstock in present use is Douglas Fir wood chips purchased from a nearby paper mill. These chips are stored in an open shed. The chips are moved by a front-end loader to the feed conveyor for transfer to the wood storage bin. The wood chips are withdrawn from the storage bin by a controlled rate table-feeder and fed into a gas-fired rotary dryer where their moisture content is reduced from about 50% to a maximum of 4%, using hot combustion products for heating. The dried chips are then fed to a hammer mill for size reduction, and the resulting flour is screened to be sure it passes through a -35 mesh screen.

The dried wood, as well as the grinding and subsequent wood processing operations, is kept under a nitrogen atmosphere to minimize moisture pick-up and explosion hazard.

The wood flour is conveyed pneumatically to an elevated wood flour storage bin in a closed system which uses nitrogen as the conveying medium.

The catalyst, sodium carbonate, is prepared as a 10-20% aqueous solution for injection into the reaction system by means of a conventional high pressure metering pump.

Carbon monoxide and hydrogen are received as compressed gases in standard over-the-road tube trailers. A carbon monoxide-hydrogen mixture is compressed to reaction pressure in a non-lubricated, non-contaminating diaphragm-type compressor. The compressed mixture may be fed into the system continuously either at the preheater or at the reactor.

Wood flour is continuously fed from the wood flour storage bin through a weigh-belt feeder to the wood oil blender. Anthracene oil is used as a carrier vehicle for start-up in the process.

Anthracene is pumped to the wood-oil blender, where it is mixed with wood flour in a cone-shapped blender which is equipped with helical screws that sweep the side walls.

After start-up, product oil serves as the carrier vehicle.

The wood-oil slurry from the blender, at 10-20% wood content, is pumped into the reaction system through a feed slurry circulating pump and a high pressure slurry feed pump; thence through an electrically heated scraped-surface feed preheater to an electrically heated reactor, where temperature and pressure are maintained and reaction occurs.

Catalyst solution and the carbon monoxide-hydrogen mixture are injected ahead of the preheater. The high-pressure, high-temperature equipment is constructed of Type 316 stainless steel. The feed slurry circulating pump is a Tuthill positive-displacement lobe-type pump, while the high pressure slurry feed pump is a Bran and Lubbe variable stroke type equipped with ball check valves.

The reactor effluent is cooled in an air-cooled heat exhanger prior to pressure letdown. Water and light oils flash and are separated in a low pressure flash system. The vented gas is cooled for recovery of light oils and water. Vented gas and non-condensibles are incinerated.

As the PDU was originally designed, a centrifuge was installed to separate any sludge of catalyst and/or unreacted wood from the product oil since bench-scale batch experiments had shown this separation possible. To date (as of the time of the paper in June, 1979) no product has been successfully clarified by the centrifuge because of the high viscosity of the oil mixture.

The process as .. reviewed, feeds a slurry of 10 — 20% biomass in vehicle oil. This means that approximately 80% of the circulating load acts only as an inert carrier, and pumping, heating, and cooling this carrier results in the waste of considerable energy which could lead to a considable increase in plant size and cost for a given throughput of biomass. Recognizing this fact, two additional methods for solids injection were built into the PDU:

- A system for feeding pretreated wood; and
- A system for feeding dry wood flour directly to the reactor.

The first alternative system for pretreating the feed involves a batchwise cooking of the wood chips in water in an electrically heated, agitated autoclave. Original laboratory work indicated a requirement of exposing the wood to a temperature of 500 ^{o}F and a pressure of 680 PSIA for one hour.

The resulting slurry is cooled in an external pump-around circuit through an air-cooled heat exchanger. The cooked wood is separated from the aqueous waste by means of a small continuous vacuum filter. The wood solids are somewhat similar to charcoal in appearance and are more easily processed than the dried chips. The solids are dried and ground and then processed as previously discussed. The pretreated wood offers the advantage of making a slurry with improved rheological properties such that a 50% slurry in anthracene oil can be pumped. In addition, there may be enhancement of the chemical reactions.

Recent bench-scale experiments conducted........point to using these autoclaves for hydrolysis of the wood at conditions of 356 ^{O}F and 130 PSIG for 45 minutes, using sulfuric acid for pH control.

In the second alternative system, solids go directly into the Reactor at pressure and temperature.

For this part of the PDU, a mechanical feed arrangement incorporating a rotary feeder was designed. It basically consists of two pressure balanced lock hoppers which will be used alternately. To be able to use conventional rotary feeders, the lock hoppers were designed so that the feeders are installed inside and, therefore, are not exposed to high differential pressures. A rotary feeder capable of withstanding a 4000 PSIG differential would have been preferred for better access, improved maintenance, and a simpler lock-hopper design; but the equipment vendors expressed little interest in building such a feeder.

The lock hopper system functions in the following manner. A lock hopper is filled with wood flour and pressurized to 4000 PSIG, using a carbon monoxide-hydrogen mixture. The solids are then fed to the reactor by the rotary feeder. When the lock hopper is empty, the pressure is reduced, venting the pressurizing gas first to a surge tank, then passing the remaining gas to the incinerator. Once the lock hopper is vented, it is ready for the next cycle. The surge tank collects most of the carbon monoxide mixture from the lock hoppers, so that the gas can be recompressed and reused in the next cycle.

Heat is supplied by recycle of part of the product oil through the preheater as in the previous operational mode. We would expect high slurry concentration with this feed method.

Work now in progress ..... may eventually permit feeding a very high slurry concentration, thereby providing a more attractive alternative."

The process described is in what should be considered early development stages, far from being commercially available and well demonstrated. Berry (1979) reviewed some the problems recently experienced:

"Practically all the reaction product that was produced during the .... contract period was extremely viscous at ambient temperatures and virtually unpumpable even at elevated temperatures.

Early .... operations were hampered by inadequate heating of electrically traced piping. This caused premature shutdowns in some of our earlier operations but installation of proper high temperature tracing has eliminated this problem.

During a recent run, numerous leaks developed in high pressure welded fittings and screwed connections; these required shutdown for repair.

The wood processing equipment has generally functioned well.

..... pumps used for oil-wood slurry circulation were originally supplied with Buna stators which had poor resistance to the anthracene oil used as start-up vehicle. Replacement with Viton stators showed improvement. Replacement of the (original pumps) with .... lobe-type, all metal positive displacement pumps was eventually necessary, however, because of temperature limitations of the Viton.

Numerous failures of polymeric gaskets, rings, and seal parts were experienced; these required replacement by such materials as Teflon, graphite metallic parts, etc.

High pressure slurry feed pump check valves had short life until the stainless steel valve seats were replaced with Stellite. Packing life has been short, but is improving with changes in material.

Serious mechanical deficiencies with the scraped surface preheater have existed. The mechanical shaft seal was modified extensively to provide a double seal with intermediate pressure between the seals. The original scraper blades were mechanically inadequate and required frequent replacement. Spring-loaded scraper blades .... have been installed and are expected to provide better heat transfer and improved service life.

As supplied, the reactor head-to-shell closure was not adequate for the cyclic operations which are common in new process pilot plants. Replacement with a GrayLoc closure eliminated this problem.

One major problem that has existed in all runs to date concerns the reactor agitator shaft seal. Instead of the packless, magnetic drive specified, a conventional double seal with elastomeric parts was installed. This seal has been inadequate. The previous contractor removed the agitator, and when the reactor was used, its contents were not agitated. A magnetically driven, packless agitator is being installed and will be placed in service in the immediate future.

Reactor product pressure letdown valves have had short life. This is not surprising when the extremely severe one-step drop from 4000 PSIG to 7 PSIG is considered. The small valve required for low capacity throughput and high pressure differential compound the problem. Parallel valves were installed to permit replacement without process interruption. Various valve trim materials have been used and Stellite has been the most successful to date.

Several failures of the carbon monoxide compressor diaphragms occurred. Head modification by the manufacturer has partially solved this problem."

These problems, innumerated by Berry, are typical of the kinds of difficulties which would be expected in the startup of a high temperature, high pressure operation in the pilot plant stages. They serve only to illustrate that the process is far from commercial application. They should not be construed to indicate that the process is not technically feasible.

From the technical perspective, there are many important questions which remain to be answered concerning process parameters. For example, the following four concerns are still unresolved pertaining to the liquefaction process:

1. Selection of the best catalyst
2. Selection of the best carrier liquid
3. Techniques for reducing product oil viscosity
4. Techniques to improve the process yield and efficiency

If these process parameters can be optimized, then this particular liquefaction procedure may hold significant opportunities for future conversion of cellulosic wastes and other biomass fuels to useful liquid products.

Production of Synthetic Liquid Fuels Via Synthesis Gas

A mixture of gases containing carbon monoxide and hydrogen is often referred to as synthesis gas (or syngas) because it can be used as feedstocks to synthesize desired liquid hydrocarbon compounds. The particular product which is manufactured from synthesis gas depends in part on the ratio of carbon monoxide to hydrogen in the feedstock. For example, 1 mol of carbon monoxide and 2 mols of hydrogen can be converted to 1 mol of methanol.

$$CO + 2H_2 \rightarrow CH_3OH$$

Similarly, ethanol and methane can be produced from synthesis gas:

Ethanol: $2CO + 4H_2 \rightarrow C_2H_5OH + H_2O$

Methane: $CO + 3H_2 \rightarrow CH_4 + H_2O$

Table 13 summarizes synthetic fuels which can be commerically manufactured using synthesis gas as a feedstock. The table also lists the equilibrium temperatures for the conversion reaction. Conversions are normally made at temperatures below the equilibrium temperatures. Further, conversions are usually made at high pressures and in the presence of a catalyst. There is an energy loss associated with the conversion as shown in Table 13. This energy loss is often justified by the higher economic value of the product produced.

TABLE 13
SELECTED GAS CONVERSION SYNTHESES

Reaction	Approximate T °C At Which dF = 0	Percent of Heating Value of Syngas Lost
Methanol:		
$CO + 2H_2 \rightarrow CH_3OH$	140	15.2**
Ethanol:		
$2CO + 4H_2 \rightarrow C_2H_5OH + H_2O$	300	17.4**
Methane:		
$CO + 3H_2 \rightarrow CH_4 + H_2O$	690	18.2
Nonane:		
$9CO + 19H_2 \rightarrow C_9H_{20} + 9H_2O$	410	17.8
Decane:		
$10CO + 19H_2 \rightarrow C_{10}H_{22} + 10H_2O$	410	17.8
Ethylene:		
$2CO + 4H_2 \rightarrow C_2H_2 + 2H_2O$	380	12.4

Note: All species in standard gas states unless otherwise noted.
**Alcohol in liquid state.
Synthesis gas heating value is approximately 67.8 kcal/mol.

The information in this table is from Reed (1979).

Several catalysts have been used commercially. An example is CuO-ZnO which is used with natural gas feedstock at 300 °C and 1,500 PSIG to make methanol. Other methanol catalysts are available.

To produce most synthetic liquid fuels from biomass feedstocks via synthesis gas, requires that the biomass feedstock first be converted to a mixture of carbon monoxide and hydrogen. Other products of the combustion process, such as, water, reduced or oxydized sulfur compounds, nitrogen, oxygen, and/or particulate impurities, should be separated and removed from the synthesis gas. That serves to protect the catalysts from contamination and to insure the proper mix of carbon monoxide and hydrogen for the desired reaction. Note that the purification of the synthesis gases (starting from the products of biomass gasification) may require many physical and/or chemical unit operations with large capital investment requirements. Kohan and Barkhordar (1979) describe the synthesis gas purification steps which might be used as part of a biomass-fed methanol production plant:

Two thousand wet short tons per day of wood are gasified with oxygen and steam in a 500 PSIA fluid bed gasifier .... About 500 short tons per day of 98 percent oxygen are preheated to 610 ^{o}F .... and fed to the gasifier.

Hot product gases are cooled to 725 ^{o}F in a waste heat boiler, generating high pressure, superheated steam. About 70 percent of the water in the syngas is removed by cooling to 290 - 300 ^{o}F, and the syngas is reheated by waste heat boiler effluent to 700 ^{o}F before being sent to the high temperature shift converter. This condensation step removes sufficient water to permit the shift reaction

$$H_2O + CO \rightleftharpoons H_2 + CO_2$$

to convert a portion of the carbon monoxide in the gas to hydrogen. No steam is added because of the moisture content of the feed gases.

The mol ratio of H_2/CO in the high temperature shift effluent gases is about 1.8:1. This ratio is selected based on the desired 2:1 ratio of H_2 to CO for the .... synthesis step, plus an estimate of the losses of CO from the syngas in the cryogenic separation step. With the composition of the gases from the wood gasifier used in this work, the H_2:CO ratio of 1.8:1 cannot be attained at the shift effluent unless some water is dropped out. An alternative to this condensation step would be to remove more wood moisture in the wood drying step, entailing perhaps more severe thermal penalties for the design.

The shifted gases are cooled to 250 - 300 ^{o}F and sent to a hot potassium carbonate acid gas-removal system. Carbon dioxide is removed to 0.02 volume percent and 95 percent H_2S removal and 70 percent COS removal (hydrolysis) is assumed. The acid gases containing less than 0.2 mol percent H_2S are sent to a .... unit for sulfur recovery.

Purified syngas at 440 PSIA is compressed adiabatically to 735 PSIA and further desulfurized in a zinc oxide bed. The heat of compression is used to permit reasonable space velocities to be employed in the zinc oxide desulfurization step.

The sulfur-free....syngas is sent to a dew point depression step (dehydration or absorption chilling) to remove the bulk of its moisture content. This step is necessary to permit reasonable sizes for the cold box feed preparation step (molecular sieves) to be used.

The molecular sieves remove the last traces of water and CO_2 in the gas before cryogenic separation. The cryogenic separation step.... separates the syngas into three streams by using the pressure in the syngas to provide the necessary refrigeration. These streams are:

- A high pressure, hydrogen-rich stream containing about 95% hydrogen
- A low pressure, CO-rich stream containing about 93% CO and N_2

- A low pressure tails stream, containing most of the methane in the synthesis gas stream

From this process description, the reader can understand that the process of synthesis gas purification and separation can be complex if biomass feedstocks are used.

Methanol is the most important synthetic liquid currently produced in the United States. Natural gas is used as the primary feedstock. Annual production of methanol is approximately 1×10^9 gallons. Most of the methanol is used as feedstock in the plastics industry.

The present processes for producing methanol from natural gas is in the range of 50% - 70% efficient. It is anticipated that if biomass is used as a feedstock, the overall efficiency for conversion to methanol would be in the range of 30% - 50%.

> A number of studies have been made of the cost of methanol production from wood, refuse, gas, and coal in the past five years. The results of these studies, brought to a common basis for comparison, are presented in Table 14 . Here production costs from wood are projected to be $0.50 to $1.35/gallon based on feedstock costs from $20 to $48/dry ton. Methanol costs from refuse are projected to be $0.72 to $0.42/gallon based on a $6 to $14/ton credit for waste disposal.
>
> An interesting new concept in the manufacture of methanol is that of the hybrid biomass-methane plant. Syngas produced from biomass is hydrogen poor, and increasing the hydrogen content requires additional processing. Syngas from re-forming natural gas is hydrogen rich. Therefore, there would be considerable advantage in using a biomass-methane feedstock anywhere that isolated gas wells can be used. Depending on the gasification process, it is expected that the yield would be increased two to five times over that achieveable with the biomass alone, and processing costs would be reduced.
>
> Recently the Mobil Corporation has announced a new process for converting methanol to gasoline using molecular sieves. If the C_3 and C_4 olefins are alkylated with isobutane produced in the reaction, the process gives over 90% yields of high octane gasoline from methanol. Conversion is projected to cost $0.06/gallon of gasoline and requires 2.4 gallons of methanol per gallon of gasoline produced. Gasoline from methanol requires 23% more energy than is contained in the methanol feedstock. Since methanol can be burned in spark engines with 26% - 45% higher efficiency than gasoline, this is a severe energy penalty. The cost of producing gasoline from wood by the Mobil process has been estimated to range from $1.89 to $2.51/gallon. (Reed, 1979)

The United States Naval Weapons Center has also found a method for converting biomass to gasoline in a process described by Diebold and Smith (1979):

> This non-catalytic process involves the low pressure selective pyrolysis of organic wastes to gases containing relatively high amounts of

ethlyene and other olefins. After char, steam, and tars are removed at low pressure, the gases are compressed to 450 PSIA (3100 kPa) for purification. The concentrated olefins are then further compressed to 750 PSIA (5200 kPa) and fed to the polymerization reactor where they react with each other to form large molecules, 90% of which boil in the "gasoline" range. Using organic feedstock derived from trash, gasoline was produced on a bench scale system which had the same appearance and distillation characteristics as gasoline made from pure ethylene which had an unleaded motor octane of 90. Preliminary econcomic analyses indicate that the process is currently competitive with petroleum derived gasoline. Based on a plant size of 500 tons of municipal refuse per day as feedstock with a credit of $10/ton, the cost of producing the gasoline would be about $.50/gallon.

The work reported by Diebold and Smith is still in the early stages of development. Like the efforts of Mobil Corporation, it is promising for the future but technically complex, requiring significant developmental efforts.

In summary, several liquid synthetic fuels can be produced from biomass feedstocks including methanol, ethanol, and gasoline. Methane can also be produced using synthesis gas. The approach reviewed in the preceding discussion involves the gasification of the biomass to form carbon monoxide and hydrogen, separation of the carbon monoxide and hydrogen from the products of the gasification reaction, blending of the carbon monoxide and hydrogen to proper ratios in the synthesis gas, and finally a low temperature, high pressure catalyzed reaction to form the desired end products. Gasoline can be produced either by methanol conversion (Mobil Corp. process) or by controlling the initial gasification process to produce olefins which are then polymerized to form "gasoline".

The technology for synthetic fuels production starting from synthesis gas appears to be at the demonstrated commercial stage for methane, methanol, ethanol, and some of the higher molecular weight hydrocarbons. Synthetic gasoline production does not appear to be as well demonstrated, particularly at the commercial level.

When biomass feedstock is used to generate synthesis gases, the technology for gasification is demonstrated but only on a limited scale. The technology for cleanup and separation of the carbon monoxide and hydrogen portions of the gasification products also appears to be demonstrated only on a small scale. Substantial developmental efforts would be required to demonstrate this technology on a commercial scale.

Production of Synthetic Liquid Fuels Via Pyrolysis

From the technical summary of pyrolysis presented in section II.C. of this summary, it is apparent that biomass feedstocks can be converted to end products of char, pyrolytic oil, and gas using pyrolytic process equipment. Under the general discussion of liquefaction technology, pyrolysis is again presented with specific emphasis on the formation of pyrolytic oils.

The range of feedstocks which has been used in experimental test facilities to generate pyrolytic oils is relatively broad and includes materials summarized in Table 14.

TABLE 14

FEEDSTOCKS WHICH HAVE BEEN USED TO GENERATE PYROLYTIC OILS

Bark	Rice Hulls	Municipal Solid Waste
Wood Chips	Manure	Tires
Sawdust	Corn Cobs	Sewage Sludge
Newspaper	Sucrose	Waste Oil
Cellulose	Polyethylene	

The preparation of feedstocks prior to their introduction to the pyrolysis reactors depends on the following factors:

1. The initial physical characteristics of the feedstock:
   A. Moisture content range
   B. Size distribution
   C. Bulk density
2. The design of the pyrolysis reactor:
   A. Type of reactor (i.e., moving bed, entrained flow, or fluid bed)
   B. Size of reactor — reflecting the size limits of material which can be fed without plugging problems
   C. The desired mix of pyrolysis products (i.e., percent of char, oil, and gas)

As an example of these factors taken into consideration, assume that an entrained flow reactor is used to pyrolyze rice hulls or grass straw. In either case, it is desirable to use small particle size feedstock of low to medium bulk density. If a high yield of low moisture content gas is to be generated, then predrying of the feedstock may be beneficial.

In a second example, assume that manure is to be fed to a fluid bed for production of pyrolytic oils. The manure should be predried sufficiently to maintain a reasonable energy balance in the reactor with the use of auxiliary fuel. The size of the manure is not critical in the fluid bed reactor as long as it is small enough to feed without plugging the feed mechanisms. The moisture levels of the feedstock should not be too low because some moisture in the collected pyrolytic oils will aid in keeping the viscosity of the product oil in the pumpable range.

The yield of pyrolysis oils is controlled principally by the temperature of the pyrolysis reaction. This has been established by repeated testing under reasonably well controlled conditions. As a general rule, yield of pyrolysis oils is highest in a temperature range from 300 ^{o}C to 450 ^{o}C. Figure 6 shows the relationship of oil yield to temperature for wood sawdust pyrolyzed in a fluid bed reactor. For this particular feedstock, note that the maximum oil yield is approximately 36% of the initial mass of the feedstock (dry basis) and that maximum yield occurs in a temperature range of 450 °C - 500 °C. At 800 °C the oil yield drops to 5% of the initial feedstock mass.

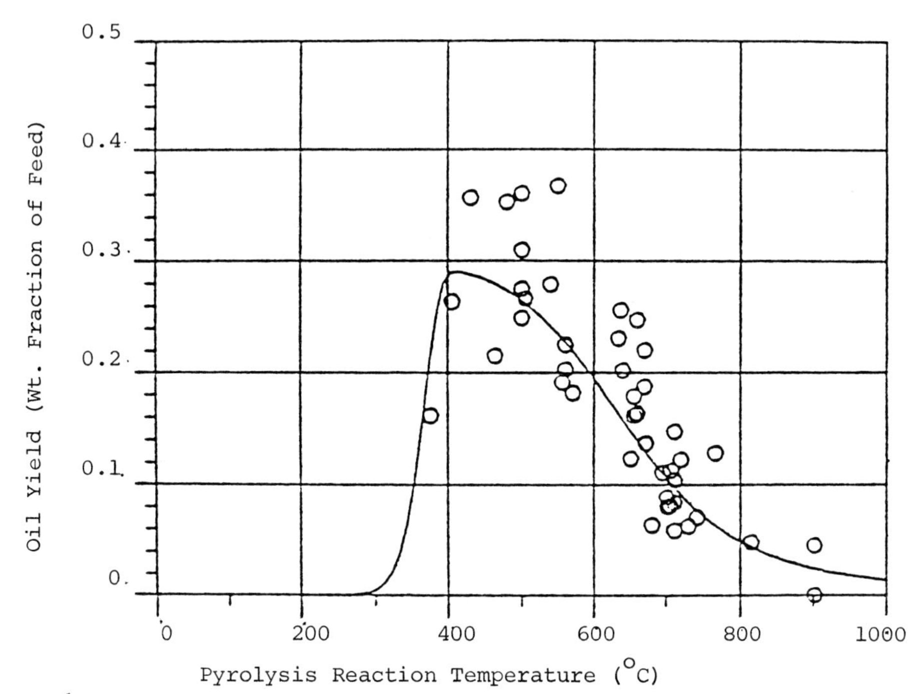

Figure 6. Pyrolysis oil yield data for sawdust (ERCO, 1977)

Figure 7 shows a similar graph for paper feedstock. Note that the maximum oil yield is about 50% of the initial mass of the feedstock compared to 36% for sawdust feedstock. The maximum oil yield for paper occurs at around 500 °C for paper. At 800 °C the yield drops to less than 5%.

The yield for pyrolysis oil from municipal solid waste is substantially less that that for sawdust or paper as shown in Figure 8. At 500 °C the maximum yield is 22% dropping to 2% at 800 °C.

Figures 6, 7, and 8 indicate that the yield of pyrolysis oil is indeed very dependent on the temperature of the pyrolysis reaction. But it also depends on the feedstock. Figure 9 shows a comparison of pyrolysis oil yields for six feedstocks as a temperature function. For the six feedstocks shown the maximum yield of pyrolysis oils (in a fluid bed reactor) is 63%. The feedstock which produced the maximum yield was waste oil and that would be expected to have a high yield relative to the other feedstocks. The average expected yield for a broad range of biomass feedstocks under well controlled conditions is in a range of 25% - 30%.

The information presented on yield of pyrolytic oils from biomass feedstocks in Figures 6, 7, 8, and 9 is measured as a percent of the initial feedstock mass.

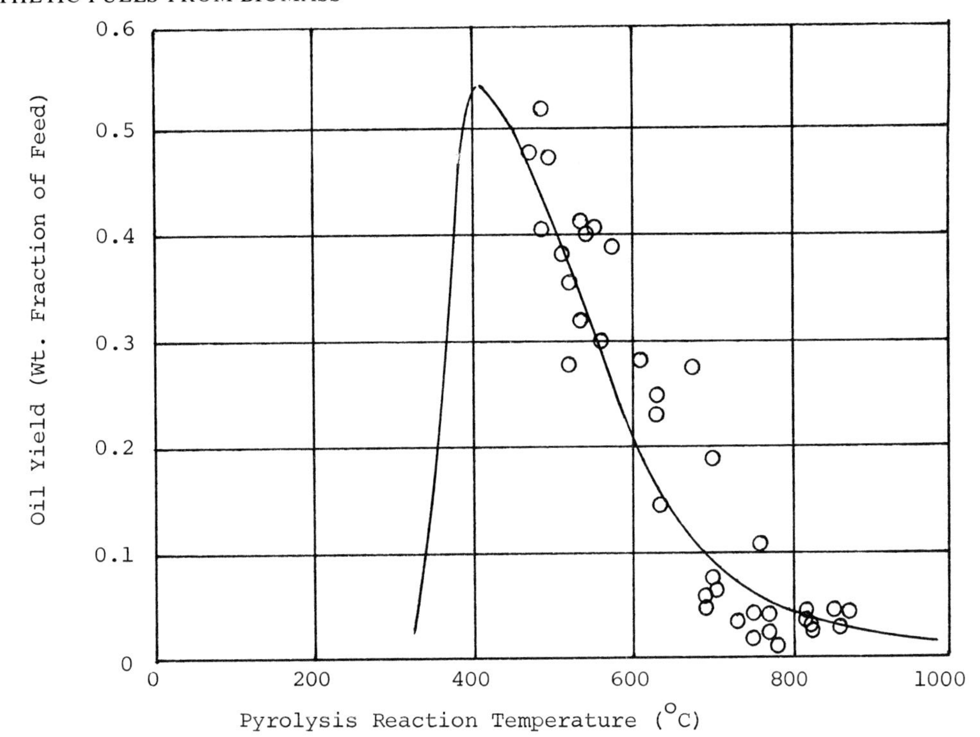

Figure 7. Pyrolysis oil yield data for paper (ERCO, 1997)

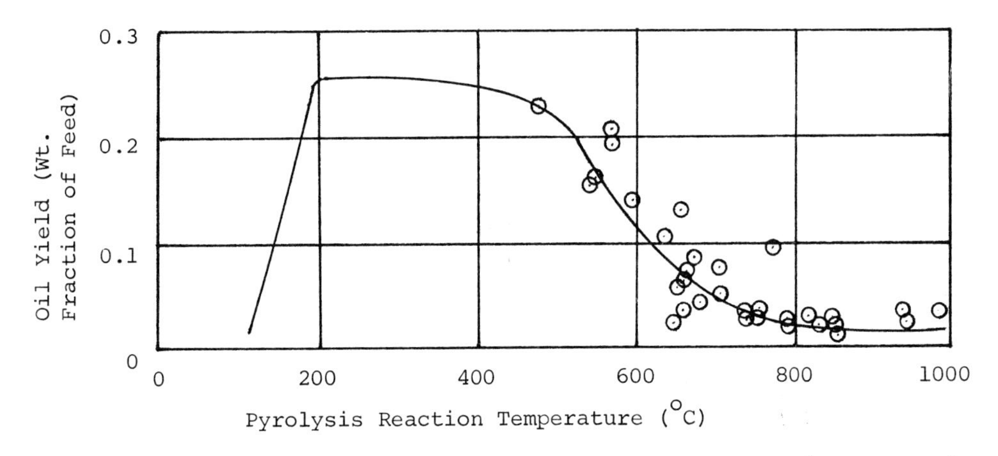

Figure 8. Pyrolysis oil yield data for municipal solid waste (ERCO, 1977)

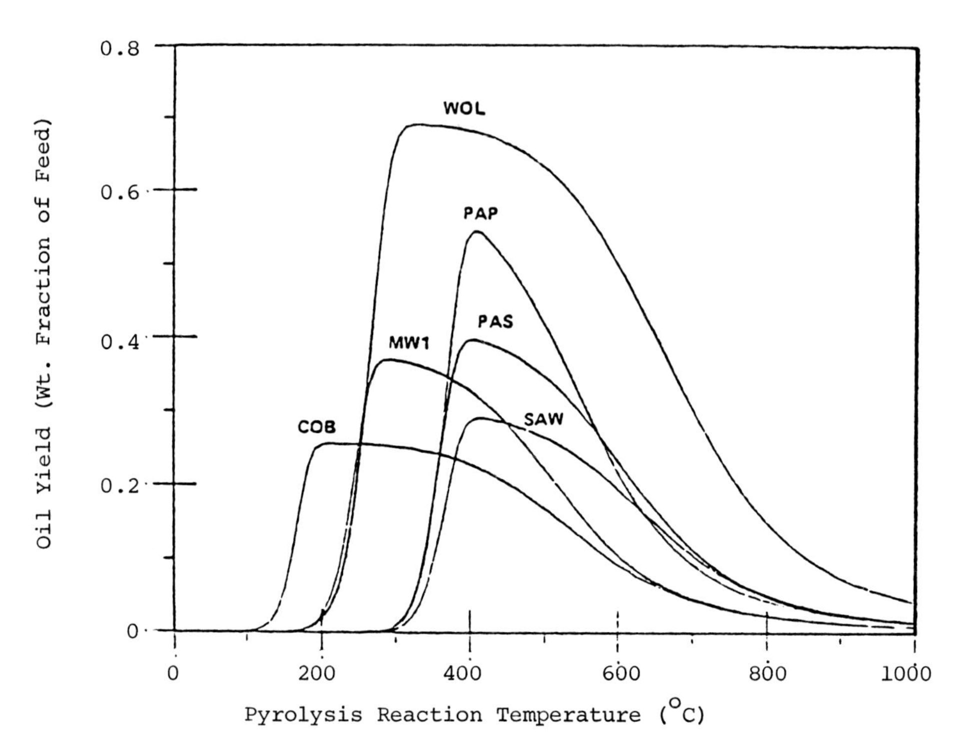

Figure 9. Comparison of pyrolysis oil yield from six feedstocks (ERCO, 1977)

Notation: WOL = Waste Oil
PAP = Paper
PAS = Paper and Sawdust
MW1 = Municipal Solid Waste
SAW = Sawdust
COB = Corn Cobs

The pyrolysis products (char, oil, and gas) have different heating values per unit mass. Therefore, the energy yield of the pyrolysis oils is not necessarily the same as the mass yield of the oils when considered as a fraction of the total energy of the feedstock. The heating value of pyrolysis oils is reported to range from 9,100 to 10,600 BTU/lb.

The data on yield of pyrolytic oils was (in most cases) collected under reasonably well controlled experimental conditions. Sampling trains were used to collect as much of the condensible vapors from the pyrolysis process as could be collected. The sampling trains typically included cold condensation surfaces and filter systems. However, experience has shown that collection of condensible organic materials from pyrolysis may be difficult on a continuous basis. Cold condensing surfaces may be quickly fouled by the heavy oils and lose their heat transfer (condensing) function. When this occurs, condensible oils are lost in the gaseous effluent and the yield levels will decrease. One means of alleviating this problem is to continuously wash the collector surface with a solvent material which will dissolve the heavy oils and carry them to a collection point. The solvent can be recovered by distillation or steam stripping and returned to the wash cycle.

The presence of moisture in the feedstock will result in water vapor as a constituent gas from the pyrolysis process. If cold condensing surfaces are used for pyrolytic oil recovery, water vapor may also condense with the oil to form a water-oil emulsion. This aids in reducing the viscosity of the heavy oils and thus helps to keep heat transfer surfaces clean. Therefore, if maximum yield of pyrolytic oils is desireable, then some moisture content in the feedstocks may be desireable (i.e., in the range of 30% - 45% rather than say less than 20%).

There are some equipment design considerations affecting the formation of pyrolytic oils. Figures 6, 7, 8, and 9 show the temperature dependence of pyrolytic oil formation. The low temperatures required for high yields can be achieved in fluid bed reactors, entrained flow reactors, and in moving bed updraft gasifiers. But they are difficult if not impossible to achieve in moving bed downdraft gasifiers.

The characteristics of pyrolytic oils are expected to vary with the biomass feedstock characteristics and with the conditions of temperature and pressure in the reactor. Thus, it is not possible to specify that pyrolytic oils from biomass feedstocks will have characteristics parameters lying within fixed ranges. The published data base is relatively small. Table 15 lists data on the characteristics of pyrolytic oils from two different biomass feedstocks and compares the synthetic oils with No. 6 fuel oil.

TABLE 15

COMPARISON OF PYROLYTIC OIL PROPERTIES WITH NO. 6 FUEL OIL

	Wood Oils - Georgia Tech.[1]		Occidental[2]	
Property	Condenser	Draft Fan	Flash Pyrolysis	No. 6 Oil[3]
Water Content (%)	14	10.4	14	2
BTU/lb	9,100	10,590	10,600	18,590
BTU/gal	86,700	97,850	114,900	148,900
Density, g/ml	1.142	1.108	1.30	0.96
Density, lb/gal	9.53	9.25	10.84	8.01
Pour Point, ^{o}F	80	80	90	65-85
Flash Point, ^{o}F	233	240	133	150
Viscosity, cP	225	233	-	2,262
Elemental Analysis				
Carbon, %	51.2	65.6	57.0	87.0
Hydrogen, %	7.6	7.8	7.7	11.7
Nitrogen, %	0.8	0.9	1.1	-
Sulfur, %	-0.01	-0.01	0.2	0.9-2.3

(1) From Knight, Hurst, and Elston (1977)
(2) From Preston (1976)
(3) From North American Combustion Handbook (1952)

As of January 1980 there are no facilities in operation in the United States whose purpose is to manufacture pyrolytic oils from biomass fuels on a commercial basis. However, there is a demonstration project under construction which is designed to convert agricultural residues and wood and bark residues into saleable pyrolytic oils and char. The system is a mobile pyrolysis waste conversion system and is described in a report by ERCO (1979):

"The unit is designed to accept waste/feedstock in the form and at the moisture contents at which they are generated. Some of the feedstocks are timber/lumbermill wastes, rice straw, wheat straw, cotton gin residues, and feed-lot wastes. The capacity of the system is 100 tons of dry feedstock per day. The dry feedstock has a maximum moisture of 10 percent by weight.

[A] front-end loader takes the waste material from the storage pile and empties it into [a] shredder conveyor hopper. The shredder conveyor hopper is mounted on a horizontal section of [a] drag conveyor. The hopper and the conveyor are set up on the ground near the shredder. The drag conveyor acts as a live bottom in the bin and conveys the feed out of the hopper and then upward into the top of the shredder.

Tramp iron is removed from the feed by a magnet mounted in the shredder inlet hopper and by a magnet mounted over the drag conveyor. The shredded feed and dust are removed from the shredder by [a] shredder blower.

After being removed from the shredder, the feed material is pneumatically conveyed into a dryer. The drying air is provided by exhaust air from [a] gas turbine. Maximum inlet gas temperature to the dryer is 500 ^{o}F. The dried feed leaves the dryer and is removed from the gas stream by a skimmer and cyclone combination. In the skimmer, the solid material is thrown against an outer wall by centrifugal action. In the vicinity of the outer wall, a small stream is collected and sent to the cyclones where the solid material drops out of the gas stream. Next, the gas discharged from the cylones is drawn through ... dust filters where any residual dust in the stream is removed. Induced draft fans on the filters create a negative pressure in the ducts and pull the gas through the filters.

The feed material drops out of the gas in the cyclones and passes through rotary airlocks. The rotary airlocks prevent air from being pulled through the bottom of the cyclones by the filter fans. After the airlocks, the feed falls onto a belt conveyor which carries the feed to a weigh belt conveyor. The weigh belt conveyor is set up ..... to isolate the weighing mechanism from the vibration of the shredder and to ensure weight measurement accuracy. The weigh belt discharges onto an enclosed screw [conveyor] which is used to carry the feed to the top of the pyrolyzer feed bin. The screw [conveyor] consists of a horizontal section which crosses the distance between the [two] trailers [on which the entire assembly is mounted] and a vertical section which lifts the material into the feed bin. The feed bin is sized for a 20-minute surge capacity at full design operating capacity. Bin vibrators are mounted on the feed bin to assure a smooth flow of feed material from the bin and through a rotary airlock. The rotary airlock conveys the feed material into the surge hopper while locking the pressure in the pyrolyzer. Then the feed falls through a surge hopper into the trough of a screw feeder which feeds material directly into the pyrolyzer bed. The surge hopper is a transition piece which prevents bridging between the rotary air lock and the screw feeder.

[A] gas turbine compressor provides air for the fluidized bed pyrolyzer. The bed can be brought up to temperature in approximately 2 hours by using a 1 million BTU/hr start-up burner that is fired on propane. A propane tank holds the propane needed for this and other uses.

All [pyrolysis] products are removed overhead from the fluidized bed. However, if the bed height is increased by retained material, an ash drain valve removes ash directly from the bed. This valve will operate only intermittently and will dump into a water-filled enclosed trough alongside the trailer [mounted assembly]. After the gas, oil, and char exit overhead, product char [and ash] is removed with a high temperature cyclone. The char is removed through the bottom of the cyclone through a rotary airlock designed to operate at high temperatures and still provide a pressure seal. The char is cooled in an enclosed char quench screw. Water is sprayed into the enclosed screw which churns the hot char/ash mixture and conveys it to the screw outlet. Sufficient water is provided from a water tank by a fractional horsepower char quench pump to cool the char to below its ignition temperature but to still allow vaporization of all the water used. Where local water is not available for quenching, a storage tank will be used. Water is introduced into the char quench by a portable gasoline pump. The vaporized water is vented from the screw to the atmosphere through the quench filter. An enclosed screw conveyor is used to convey the quenched char from the cooling screw to the [char] storage vehicle.

The gas and oil vapor products leaving the cyclone are sent to a venturi oil scrubber. The scrubbing system is used to cool and condense out the oil from the gas stream. The product oil itself is used as the cooling medium although a No. 2 fuel oil is used to start up the system. Start-up oil is stored in a tank.

Because of the high-temperature pour point of the product oil, all oil lines and valves are steam traced. Two different oil flows are brought into the vertical venturi. The high-rate flow enters the top of the venturi through a central nozzle and the low-rate flow enters through tangential nozzles. The sprays of cool oil from the nozzles and the hot gas stream are forced through the throat of the venturi. The sprays from the nozzles help provide momentum for thorough mixing of cooling oil and hot gas in the throat of the venturi. The vaporized oil condenses out of the cooled gas stream and then the oil droplets are separated from the gas by the oil scrubber cyclonic separator. The oil collects in the oil scrubber sump at the base of the separator.

In the three lines the oil is pumped from the sump to the air-cooled heat exchanger where it is cooled. Then the cooled oil is carried back to the scrubber venturi nozzles in two lines. [An] oil pump provides the high capacity flow which enters the central nozzle of the venturi. [Two] oil pumps combine to produce the low-rate flow which enters the tangential nozzles of the venturi. The low-rate flow is carried by two pumps to allow greater turn-down capability. The oil which flows into the oil cooler must be turbulent or the oil will not cool sufficiently. If a flow is turned down more than 20%, it will change from turbulent to laminar flow and will not be adequately cooled. Therefore, by splitting the low-rate flow into two equal flows, greater turndown capability is provided in the scrubber system. [One] oil pump has a greater discharge pressure than [the other] because there are more pressure dropping elements in the first pump line. An oil filter is placed between the oil pump and the oil cooler to prevent a buildup of solid material in the scrubbing oil circuit. The filter consists of two loaded cartridge containers in parallel which allows for filter cleaning and replacement during operation. After the oil filter, but before cooling, a slipstream of excess product oil is piped to a heated tank trailer for storage.

The gas leaving the oil scrubber is sent to a baffled impingement mist eliminator. The mist eliminator removes the last portion of the vaporized oil entrained in the gas.

After the mist eliminators, the gas is sent to the gas turbine system. The gas turbine system consists of two parallel turbine compressor trains; each train has two compressor stages and a turbine stage, all on the same shaft. Each gas turbine, therefore, powers a set of compressors. Air for controlling the temperature of the two combustors is brought in through the two compressor sets. Compressed air which is tapped from the exit lines of the compressors is used to fluidize the pyrolyzer. Compressed air for the air instruments is tapped from the exit lines of the compressors. The pyrolytic gas stream enters the gas turbine system and is compressed in the fuel turbocharger. Kerosene which is stored in a tank is used to fuel the turbines during startup. The kerosene is compressed in the turbocharger. The compressed fuel and air mix in the combustors and the fuel is burned and subsequently cooled to the turbine inlet temperature. Two streams are taken from the combustors to the turbine stages.

Each turbine shaft is joined through a speed-reducing gear. The generator produces the electricity for the entire mobile pyrolysis system. The exhaust gas from the turbines is recombined with the expanded combustion gas in [a] mixing nozzle and the gas mixture is used as the hot air for the [feedstock] dryer. If the gas temperature in the mixing nozzle is greater than the desired dryer temperature of 500 ^{o}F, 95 ^{o}F air is added to cool the gas. The 95 ^{o}F air is compressed in a compressor which is driven by an air motor. The air motor is powered by compressed air which is tapped off the compressor.

A small steam boiler could be placed downstream from the exhaust of one of the turbine stages; however, steam may be required while the gas turbine system is down. Therefore, a small low-pressure, propane-fired

B. STEAM GASIFICATION

With sponsorship from a U. S. Government research contract, steam gasification of biomass is being studied by Wright-Malta Corporation. The project is summarized by Coffman (1979):

> The project purpose is development of a process and equipment for conversion of as-harvested green wood chips into medium BTU fuel gas. The chemistry that has proven to be most satisfactory is wood ash-catalyzed steam gasification (1150 ^{o}F, 300 PSI, with 1 hour dwell time). The equipment that has been evolved is a slender, jacketed rotary kiln, heated regeneratively and by the wood decomposition exotherm.
>
> Process chemistry has been defined in two bench-scale (10 lb/hr) facilities. The "mini-kiln", a 1 ft by 3 ft rotating autoclave, is a batch reactor, and provides data as a function of temperature. The "biogaser", a 2 inch by 10 ft auger reactor, is continuous and provides integrated gas composition data. The "mini-kiln" has also given useful information on the exotherm and heat transfer.
>
> Calculations have shown that the product gas, if raised to 1500 ^{o}F and 200 PSI, shifts to a H_2/CO ratio of 2/1, with only a small residual methane content. Condensation of this syngas will yield methanol at an overall energy efficiency (green wood to methane) of 70 - 75%.
>
> The process development unit (6 green tons/day), presently under construction, will be highly instrumented for complete process definition. Its mechanical features include a cylinder/lock valve feed, lock valve discharge, small diameter stuffing-box seals, and provision for steam as well as regenerative gas heating.
>
> With operation of the process development unit in the second half of 1980, development of the fuel gas process will have been completed. Syngas shifting and methanol systhesis will be project tasks in 1981.

Steam gasification, like hydrogasification , is in the early developmental stages. It will likely require 10 years of efforts to bring this technology to the point where it can be considered as commercially viable and proven technology. But like other technologies currently being investigated, it holds promise for relatively high conversion efficiencies.

C. MOLTEN SALT REACTOR TECHNOLOGY

In the preceeding discussions of pyrolysis and gasification technology, fluid bed systems were reviewed. Fluidized beds have become popular during the past 35 years because they provide reasonably good mixing of the reactants, uniform temperature profiles, and an opportunity to carry out a reaction in the presence of a catalyst which can be regenerated on a continuous basis. A parallel technology revolves around molten salt reactors. Yosim and Barclay (1977) provide a summary of molten salt reactor technology:

The concept of molten salt gasification is shown schematically in Figure 11. Shredded combustible waste and air are continuously introduced beneath the surface of a sodium carbonate-containing melt at about 1000 ^{o}C. The waste is added in such a manner that any gas formed during combustion is forced to pass through the melt. Any acidic gases, such as HCl (produced from chlorinated organic compounds) and H_2S (from organic sulfur compounds), are neutralized and absorbed by the alkaline Na_2CO_3. The ash introduced with the combustible waste is also retained in the melt. Any char from the fixed carbon is completely consumed in the salt.

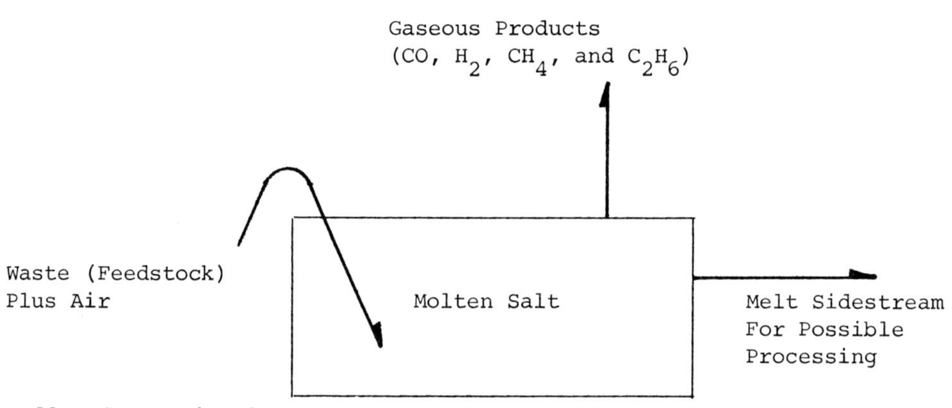

Figure 11. Schematic diagram of a molten salt reactor

Gasification of the waste is accomplished by using deficient air, i.e., less than the amount of air required to oxidize the waste completely to CO_2 and H_2O. Thus, the waste is partially oxidized and completely gasified in the molten salt furnace. The product gas flows to a conventional gas-fired boiler in which it is combusted with secondary air.

As a possible option, a sidestream of sodium carbonate melt can be withdrawn continuously from the molten salt furnace, quenched, and processed in an aqueous regeneration system which removes the ash and inorganic combustion products retained in the melt and returns the regenerated sodium carbonate to the molten salt furnace. The ash must be removed to preserve the fluidity of the melt at an ash concentration of about 20% by weight. The inorganic combustion products must be removed at some point to prevent complete conversion of the melt to the salts, with an eventual loss of the acid pollutant removal capability.

The advantages of molten salt gasification are as follows:

1. Intimate contact of the hot melt, air, and waste provides for complete and immediate destruction of the waste.

steam generator is employed. The steam generator provides steam for tracing the oil lines, purging the pyrolyzer, and purging the gas turbine system. Water for the steam generator is provided from the storage tank and pumped to the generator by the pump that is part of the complete package. Return lines from the steam tracing system are sent to the tank.

A flare is provided for burning the product gas during start-up. For system upsets, in which kerosene can be used to maintain the turbine, the flare is used for burning any product gas beyond the requirements of the turbine system."

This system described by ERCO (above) is currently under construction. The purpose of the system is to demonstrate that pyrolytic oils and char can be produced in the field near to the feedstock source on an economic basis. With the possibility of pyrolytic oil yields of 30% of the initial mass of the feedstock, the system should be capable of collecting 30 tons per day of this synthetic oil (maximum). If the oil has a density of 9.5 lbs/gal, and the overall system operates with an 80% load factor, the net production of pyrolytic oil from the mobile operation should be 120 barrels per day. Of course, it remains to be seen as to whether the concept will be successful. At this point, the individual components are commercially available and reasonably well proven, but the complete assembly is not demonstrated as being feasible.

Some of the physical properties of the pyrolysis oil are shown in Table 15. ERCO (1979) reports that the physical and chemical properties of the pyrolytic oil are such that the best methods for storing, transporting, and utilizing it is not clear without further analysis. The oil is viscous (essentially a solid at ambient temperature), oxygenated, and therefore acidic (corrosive), and is partially water-soluble. The properties of the oil are strongly dependent on the feedstocks and pyrolysis operating conditions.

ERCO (1979) suggests that two markets are potentially viable for the oil: (1) Boiler fuel; and (2) Asphalt. In each case the product might be utilized alone or it might be blended with conventional petroleum products. The most important properties of boiler fuel from the viewpoint of the utility or industrial boiler operator are its viscosity, heating value, corrosiveness, and requirements for special handling, if any. Based on test data available, it appears that pyrolysis oils which have been produced from wood feedstocks are suitable for use in existing boilers which typically use No. 6 fuel oil without changes to the burner configuration or other components of the system. It is important to note, however, that the heating value of "typical" pyrolysis oils is only 65% - 75% of the heating value of No. 6 fuel oil on a volumetric basis. This is due to the high oxygen content of the pyrolysis oils produced from cellulosic feedstocks.

ERCO (1979) comments on some of the specific characteristics of pyrolytic oils:

"Viscosity — Pyrolytic oil is more viscous than a typical residual oil. However, its fluidity increases more rapidly with temperature than does that of No. 6 fuel oil. Hence, although it must be stored and pumped at higher temperatures than are needed to handle heavy fuel oil, it can be atomized and burned quite well at 200 $^{\circ}$F. This is only about 20 $^{\circ}$F higher than the atomization temperature for electric utility fuel oils.

Corrosion — Because this fuel is more corrosive to metals than natural residual oils, particularly at high storage and handling temperatures, the use of corrosion-resistant materials in the storage and handling equipment will be required.

Temperature degradation — At temperatures above 200 °F the fuel begins to undergo chemical changes which adversely affect its viscosity. It will, therefore, be necessary to store the fuel at lower temperature and heat it when it is used.

Mixing — Pyrolytic oil can be blended with several different No. 6 oils, although over a period of hours the heavier pyrolytic oil settles out from the mixture because there is little mutual solubility. Pyrolytic oil may be soluble to a limited extent in water.

There is a possibility that the pyrolytic oil may also have use as an asphalt. Most asphalts are currently obtained from the heavy, viscous fractions of selected petroleum crudes. The pyrolytic oil may have similar properties."

This completes the summary discussion of the state of the art of the principal conversion processes used to produce synthetic fuels from biomass. By way of a brief review, the principle conversion processes and their products are shown in Table 16.

TABLE 16

SUMMARY OF THE PRINCIPAL PROCESSES AVAILABLE TO CONVERT BIOMASS FEEDSTOCKS TO SYNTHETIC FUELS

	Conversion Process	Products
A.	Fermentation	Ethanol
B.	Anaerobic Digestion	
	Controlled Processes for Methane Generation	Methane, CO_2
	Recovery of Methane from Sanitary Landfills	Methane, CO_2
C.	Pyrolysis	Char, Pyrolytic Oil, CO, H_2, CO_2, H_2O, CH_4, etc.
D.	Gasification	CO, H_2, CO_2, H_2O, CH_4, etc.
E.	Liquefaction	
	High Temp., High Press., Catalytic Liquefaction	Heavy Oil
	Conversion to Synthesis Gas and Subsequent Manufacture of Desired End Products	Methanol, Ethanol, Ethylene, Gasoline, etc.
	Pyrolysis	Pyrolytic Oil, Char, Gases

The technologies summarized in Table 16 are thought to be the principal conversion processes available at this time for making synthetic fuels from biomass feedstocks. The reader should be aware that not all biomass feedstocks have been used with each of the technological approaches listed. Due to the broad range of feedstocks shown in Table 1 as biomass, it is highly unlikely that each of these feedstocks could be used with each of the conversion systems shown in Table 16.

In addition to the technologies summarized to this point, there are other systems and processes which may be applied to convert biomass feedstocks to synthetic fuels. These include:

A. Hydrogasification
B. Steam Gasification
C. Molten Salt Reactors
D. Liquid Oil Recovery from *Euphorbic* Type Plants
E. Densified Biomass Fuels

Each of these technologies is reviewed in Section 3.

III. OTHER CONVERSION PROCESSES USED TO PRODUCE SYNTHETIC FUELS FROM BIOMASS

A. HYDROGASIFICATION

The Solar Energy Research Institute (1979) has provided a brief summary of the state of the art of hydrogasification:

> Hydrogasification, in which H_2 gas is added under high pressure, is... being studied and has the potential for high, direct yields of methane.
>
> Pyrolysis is normally carried out in an inert gaseous environment. When the pyrolysis is conducted in hydrogen, with rapid heating (greater than 1000 $^oC/s$), it is possible to increase the devolitization of the feedstock and enhance the hydrocarbon yield.
>
> The hydrogasification reaction takes place in two stages. If the char is prepared and stabilized in an inert atmosphere, the rate of hydrogasification is very low. On the other hand, if hydrogen is in direct contact with the freshly formed char during pyrolysis, the rate of gasification is several orders of magnitude greater. For rapid heating, the gasification rate is almost 100 times faster than for the slow heating case.
>
> The rate of hydrogasification depends on temperature, hydrogen pressure, and time. The rate of reaction of stabilized char with hydrogen is negligible during normal residence times. The stabilization reaction is assumed to consume only a fraction of the char available for hydrogasification. Hydrocarbon production in atmospheric gasifiers by direct hydrogenation of char is slow.

> The only published hydrogasification studies on biomass have been conducted for a naturally occurring peat. Figure 10 shows the effect of hydrogen pressure on the rapid heating gasification of peat. At 60 atmospheres of hydrogen in 4 — 7 seconds at 1400 °F, the amount of carbon gasified increased by roughly 40% over that for pyrolysis in an inert atmosphere.
>
> Hydrogasification of other biomass feedstocks is presently being investigated and it is believed that a considerable quantity of hydrocarbons can be derived from such materials under high hydrogen pressures and rapid heating.

Note from this SERI review that hydrogasification of biomass feedstocks is in the very early stages of laboratory investigation. Further, it requires that significant quantities of hydrogen be available in order to carry out such a conversion process on a large continuous scale. Thus, it is likely to be many years before such a system could be seriously considered for production of synthetic fuels from biomass feedstocks.

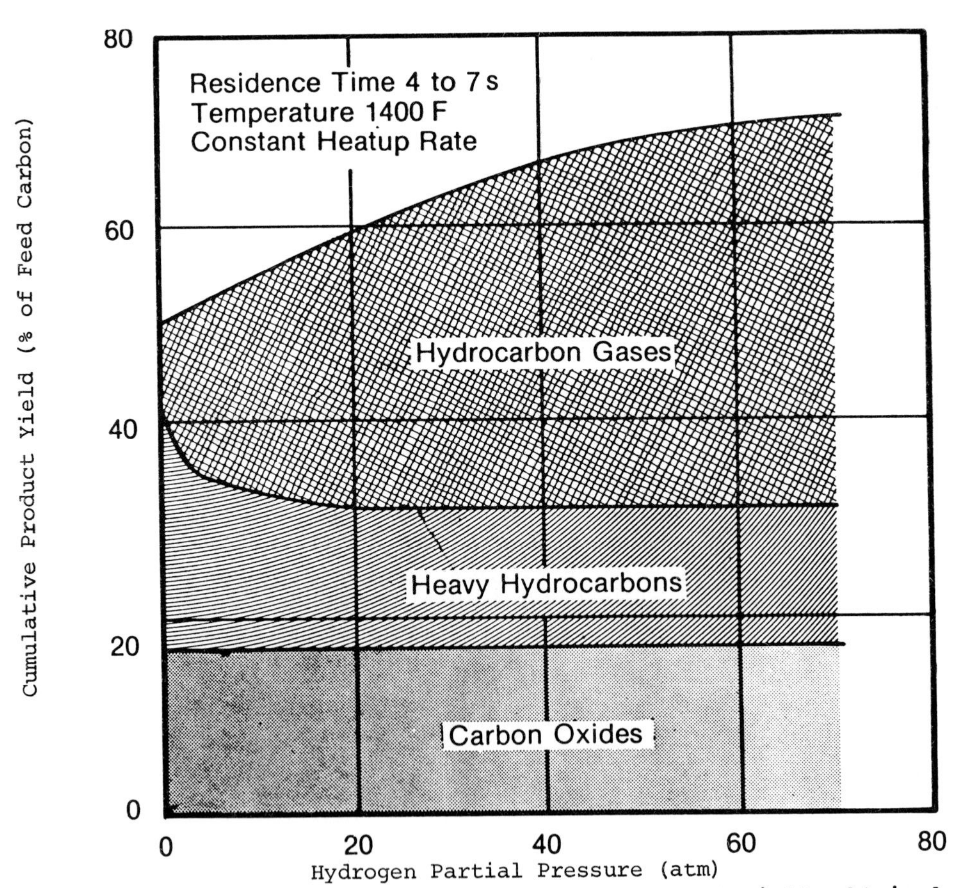

Figure 10. Effect of hydrogen partial pressure on product yields obtained during peat gasification (From SERI, 1979)

B. STEAM GASIFICATION

With sponsorship from a U. S. Government research contract, steam gasification of biomass is being studied by Wright-Malta Corporation. The project is summarized by Coffman (1979):

> The project purpose is development of a process and equipment for conversion of as-harvested green wood chips into medium BTU fuel gas. The chemistry that has proven to be most satisfactory is wood ash-catalyzed steam gasification (1150 ^{o}F, 300 PSI, with 1 hour dwell time). The equipment that has been evolved is a slender, jacketed rotary kiln, heated regeneratively and by the wood decomposition exotherm.
>
> Process chemistry has been defined in two bench-scale (10 lb/hr) facilities. The "mini-kiln", a 1 ft by 3 ft rotating autoclave, is a batch reactor, and provides data as a function of temperature. The "biogaser", a 2 inch by 10 ft auger reactor, is continuous and provides integrated gas composition data. The "mini-kiln" has also given useful information on the exotherm and heat transfer.
>
> Calculations have shown that the product gas, if raised to 1500 ^{o}F and 200 PSI, shifts to a H_2/CO ratio of 2/1, with only a small residual methane content. Condensation of this syngas will yield methanol at an overall energy efficiency (green wood to methane) of 70 - 75%.
>
> The process development unit (6 green tons/day), presently under construction, will be highly instrumented for complete process definition. Its mechanical features include a cylinder/lock valve feed, lock valve discharge, small diameter stuffing-box seals, and provision for steam as well as regenerative gas heating.
>
> With operation of the process development unit in the second half of 1980, development of the fuel gas process will have been completed. Syngas shifting and methanol systhesis will be project tasks in 1981.

Steam gasification, like hydrogasification , is in the early developmental stages. It will likely require 10 years of efforts to bring this technology to the point where it can be considered as commercially viable and proven technology. But like other technologies currently being investigated, it holds promise for relatively high conversion efficiencies.

C. MOLTEN SALT REACTOR TECHNOLOGY

In the preceeding discussions of pyrolysis and gasification technology, fluid bed systems were reviewed. Fluidized beds have become popular during the past 35 years because they provide reasonably good mixing of the reactants, uniform temperature profiles, and an opportunity to carry out a reaction in the presence of a catalyst which can be regenerated on a continuous basis. A parallel technology revolves around molten salt reactors. Yosim and Barclay (1977) provide a summary of molten salt reactor technology:

The concept of molten salt gasification is shown schematically in Figure 11. Shredded combustible waste and air are continuously introduced beneath the surface of a sodium carbonate-containing melt at about 1000 ^{o}C. The waste is added in such a manner that any gas formed during combustion is forced to pass through the melt. Any acidic gases, such as HCl (produced from chlorinated organic compounds) and H_2S (from organic sulfur compounds), are neutralized and absorbed by the alkaline Na_2CO_3. The ash introduced with the combustible waste is also retained in the melt. Any char from the fixed carbon is completely consumed in the salt.

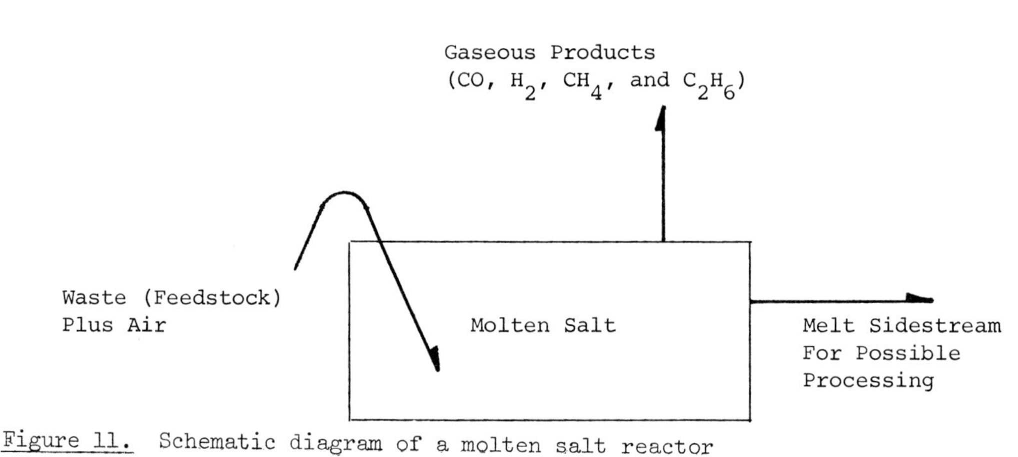

Figure 11. Schematic diagram of a molten salt reactor

Gasification of the waste is accomplished by using deficient air, i.e., less than the amount of air required to oxidize the waste completely to CO_2 and H_2O. Thus, the waste is partially oxidized and completely gasified in the molten salt furnace. The product gas flows to a conventional gas-fired boiler in which it is combusted with secondary air.

As a possible option, a sidestream of sodium carbonate melt can be withdrawn continuously from the molten salt furnace, quenched, and processed in an aqueous regeneration system which removes the ash and inorganic combustion products retained in the melt and returns the regenerated sodium carbonate to the molten salt furnace. The ash must be removed to preserve the fluidity of the melt at an ash concentration of about 20% by weight. The inorganic combustion products must be removed at some point to prevent complete conversion of the melt to the salts, with an eventual loss of the acid pollutant removal capability.

The advantages of molten salt gasification are as follows:

1. Intimate contact of the hot melt, air, and waste provides for complete and immediate destruction of the waste.

2. No HCl from plastics such as PVC, or H_2S from sulfur containing wastes is emitted.

3. No tars or liquids are produced during gasification.

4. Combustion products are sterile and odor-free.

Yosim and Barclay (1977) indicate that a pilot plant molten salt reactor has been constructed and operated in California on a variety of feedstocks including rubber, wood, coal, and others. The pilot plant is 3 ft in diameter with a height of 10 ft. Apparently from the data presented, the feed rates for feedstocks are extremely limited compared to other technologies. For example, the wood feedstock was fed at only 2.08 lbs/hr. Now, with a reactor of 3 ft in diameter (i.d.) that suggests that the feed rate was actually 0.3 lbs of feed per square foot of reactor surface per hour. Comparable rates for fluid bed reactors are 160 lbs/hr-ft^2. It is not clear from the summary by Yosim and Barclay as to whether the feed rates shown are actually realistic operating limits for molten salt reactors. If they are, then the future for molten salt reactors will be very limited because of their feed rate limitations. In comparison to fluid bed reactors, molten salt reactors would not be competitive due to large size requirements for a given throughput of feedstock.

Based on a review of the literature, there is apparently little interest in molten salt reactor technology for conversion of biomass feedstocks to synthetic fuels.

D. LIQUID OIL RECOVERY FROM *EUPHORBIA* TYPE PLANTS

There are a variety of green plants categorized as *Euphorbia* type plants which have some potential to supply synthetic oil or possibly be processed to yield methanol, ammonia, SNG, etc.

> At present, *Euphorbia* is undergoing extractive analysis at the University of California's Lawrence Berkeley Laboratory to characterize the nature of the hydrocarbon liquid contained in the plants. It is also anticipated that the chemical composition, heating value, and similar properties of the oil-free cellulosic plant material will be characterized in the future. (SRI, Int'l, 1979)[4]

SRI, International has investigated the possibility of using *Euphorbia* as a feedstock for processing of oil and cellulose. The hydrocarbon liquid content of the assumed feedstock is 8.7% by weight, with an assumed heating value of 17,000 BTU/lb. Based on these assumption, the processing facility envisioned would have a feedstock rate of 3000 dry tons/day and a product of high BTU oil equivalent to 261 tons/day. At this time, it is unknown as to whether the plant would actually operate since no process demonstration units have been constructed and operated to prove the technical feasibility of the proposed system. In any event, the product oil from such a plant would be only one of several important products in the plant output. The primary product of economic value would be cellulose.

In summary, *Euphorbia* type plants may have an extractable oil content which could then be used as a fuel supplement. The yield would likely be less than 10% of the dry weight of the plant feedstock (substantially less than yields of pyrolytic oils, chars, or combustible gases which might be generated from biomass feedstocks). Thus, while the technology may be feasible in terms of the equipment requirements or processes involved, it probably will not be of great significance as a synthetic fuel production technology.

E. DENSIFICATION OF BIOMASS

The production of synthetic fuels from biomass feedstocks can reasonably include the densification of biomass where such densification serves to expand the utility of the feedstock. For example, there are several installations in the United States which are presently using densified wood and bark residue fuels as a coal supplement. Without the densification process, the biomass fuels would not be used as coal supplements in some of the facilities.

There are five means of densifying biomass feedstocks as outlined by Reed (1979):

1. Pelleting — a die perforated with 1/4 in. to 1/2 in. holes rotates against pressure rollers, forcing feedstock through the holes at high pressure and densifying the feedstock;

2. Cubing — a modified form of pelleting producing a large size product 1 in. to 2 in.;

3. Briquetting - feed is compacted between rollers containing cavities; the product looks like charcoal briquettes;

4. Extrusion — a screw forces a feedstock under high pressure into a die, forming 1 in. to 4 in. diameter cylinders; and

5. Rolling / compressing — employs a rotating shaft to wrap fibrous material and produce high density rolls of 5 in. to 7 in. diameter.

The densification process takes advantage of the physical properties of two of the major components of biomass materials, cellulose and lignin. Cellulose is stable to 250 °C, while lignin begins to soften at temperatures as low as 100 °C. Densification is carried out at temperatures that ensure that the cellulosic material remains stable but that soften the lignin fraction, making it act as a "self-bonding" agent that gives the final [product] its mechanical strength. The water content of the feedstock must be controlled in the range from 10% - 25% to minimize pressure requirements for densification.

Densification proceeds by heating a biomass material [of proper moisture content] to 50 °C to 100 °C to soften the lignin, followed by mechanical densification that increases the biomass density to a maximum of 1.5 g/cm^3 and heats the material another 20 °C to 50 °C. The additional temperature increase liquefies waxes that act as additional binders when the product is cooled.

A typical biomass compaction plant is [described as follows]. The first step in the process is separation — stones and sand must be removed from forest or agricultural wastes and inorganics from municipal waste. The remaining biomass portion is then pulverized with hammer mills or ball mills to a size somewhat smaller than the minimum dimension of the pellets to be formed. This fraction is then dried in a rotary kiln or convection dryer. Finally, dried biomass is fed into the compactor which delivers pellets for storage or use.

During the past 4 years there has been a great deal of controversy over the benefits and problems related to densification. The principal benefits appear to be:

1. Densified biomass fuels are of low moisture content and high bulk density. Therefore, they can be shipped at substantial cost savings compared to low density, high moisture biomass fuels.

2. Densified biomass fuels are more uniform in both size and moisture content than non-densified, typical biomass fuels. As such, the densified fuels are more desireable for direct combustion purposes and for feedstock to gasification reactors. They avoid combustion upsets (due to size or moisture content variations) and they avoid plugging and bridging problems in fuel feed systems and in gasification reactors.

3. The relatively low moisture content of the densified fuels means that direct combustion systems using such fuels will operate at higher effiency than for wetter, non-densified fuels. Similarly, in gasification systems, the product gases will have lower moisture levels (using low moisture content densified feedstocks) and thus will have higher heating values than for gases generated using high moisture content feedstocks. Note, however, that densification has nothing to do with the higher combustion efficiency or the higher gas BTU value. The key is the moisture content of the feedstocks. It may be desireable to dry feedstocks but not necessarily to densify them for use in direct combustion of biomass or in gasification of biomass.

These benefits of densified biomass fuels are offset by the cost of preparing the fuels, including material handling, size reduction, drying, and densifying. Present costs for the complete densification process in Oregon are approximately $22/ton. On an as delivered basis, this densification process adds roughly $1.30/million BTU to the cost of the fuel. By comparison, the cost of non-densified fuel is currently approximately $0.50/million BTU. So, if densified biomass fuels are to be purchased, it is necessary to justify the added cost for the fuel.

In many installations the added cost of the fuel can be justified on the basis of (1) the competitive cost for coal; or (2) the lower sulfur content of the biomass fuel which may permit operation of an existing plant without the installation of sulfur removing flue gas scrubbers.

An interesting recent development has been growing interest in densifying biomass fuels (particularly wood and bark residues), transporting them to existing coal fired steam generating facilities, pulverizing the densified fuels, and then burning the fuels in suspension burners as a coal supplement. From an energy standpoint, this makes little sense. From a practical standpoint, it may offer some promise of reducing sulfur emissions to the atmosphere. But if suspension burning of wood and bark fuels is economically and environmentally desireable, then it would be more efficient to pulverize the fuel, dry it, and burn it, without any involvement of densification and repulverization.

There are three common misconceptions about densified biomass fuels which deserve comment:

1. The heating value of densified fuels (expressed as BTU/dry lb) is no greater than the heating value of the biomass feedstock (also expressed as BTU/dry lb).

The point is that the drying and densification processes do not add to the heating value of the fuel on a dry basis. Stated another way, the process of compressing a relatively dry biomass feedstock does not add to its inherent heating value.

2. Densified biomass fuels are often promoted as "clean" burning fuels which do not pollute the environment. Such statements are inherently false since practically speaking all combustion systems emit pollutants to the environment including CO, NOx, and particulate. The process of drying and densifying biomass feedstocks does not remove the inherent inorganic ash component of the feedstock, nor does it change the fixed carbon content of the feedstock. Thus, densified biomass fuels still have the potential to emit CO, and particulate to the atmosphere. Further, because of their lower moisture levels (compared to the more moist biomass feedstocks usually used for direct combustion) the temperatures in the combustion zones are apt to be sufficiently high to generate significant quantities of NOx.

3. It is occasionally suggested that densified biomass fuels can resist moisture pickup better than unprepared fuels. Experience with pelletized fuels has been to the contrary. Pelletized bark and wood fuels, when exposed to rain, tend to become rapidly unpelletized leaving a pile of wet "sawdust" remaining. However, it is understood that once the biomass is pelletized, it may be treated with a water resistant layer of material to prevent moisture pickup.

In summary of the state of the art of densification of biomass, it is a well developed, commercially proven process which can be accomplished by a variety of techniques. Densified biomass fuels may be very useful as fossil fuel supplements but the costs of densification are high compared to the cost of typically available non-densified biomass feedstocks. Essentially the use of densified biomass fuels is a matter of economic analysis on a case by case basi

In review of the technologies discussed in Section 3, the following conclusions are reached:

1. Hydrogasification is in the very early stages of laboratory investigation. It requires that significant quantities of hydrogen be available in order to carry out the conversion process on a continuous, large-scale basis. As such, it is not likely to be seriously considered for the production of synthetic fuels from biomass resources.

2. Steam gasification holds the promise of high conversion efficiencies of biomass feedstocks to methanol. But the technology is in the early stages of development and will likely require 10 years to become commercially viable.

3. Molten salt reactor technology has been proven in small pilot scale and has been used effectively for some specific operations. But the technology does not appear to hold much promise for synthetic fuels production from biomass feedstocks due primarily to the low solids feeding rates quoted in the literature. Other technologies such as, fluidized beds, entrained flow reactors, and even moving beds appear to have much higher conversion rates than the molten salt systems. Thus, the enconomics of molten salt systems will probably preclude their extensive use.

4. Liquid oil recovery from *Euphorbia* type plants may have some interest in the future, but it will likely be based principally on cellulose recovery rather than on oil recovery. The liquid oil content of these plants is not apt to exceed 10% on a mass basis. This technology is not likely to advance rapidly on a commercial scale for liquid oil recovery. It is too early in the developmental stages to predict what impact it may hold in the future.

5. Densification of biomass is commercially proven technology which is gaining in popularity at a fairly rapid pace in the United States. The primary reason appears to be that densified fuels can be used successfully as fossil fuel substitutes with attractive economic returns. Whether this trend will continue is difficult to predict. Obviously, it will depend to a large extent on economic considerations and to a lesser extent on environmental considerations.

IV. ECONOMIC COMPARISON OF SYNTHETIC FUELS PRODUCED FROM BIOMASS FEEDSTOCKS

Up to this point, the most prominent and promising technologies for conversion of biomass feedstocks to synthetic fuels have been reviewed from the perspective of determining the state of the art of each technology. In Section II, the major technology categories of fermentation, anaerobic digestion, pyrolysis, gasification, and liquefaction were explored. In Section III, additional techologies were discussed. Not all of the approaches in conversion are at the stage of commercial application or proven technology. One of the reasons for technological lag may be economic restraints. The economics of conversion technologies are reviewed in this section.

The first technology reviewed in Section II is fermentation for the production of ethanol. Until recently, prodcution of ethanol for fuel purposes was not considered to be economically attractive. However, with a variety of tax incentives, with some improvements in distillation efficiency, and with the opportunity to sell by-products from the operation as cattle feed supplements or as protein sources for other markets, the economics of fermentation for alcohol fuel production are now improving. This is evidenced by the development during the past year of significant numbers of production facilities in the United States. Brazil has, of course, received substantial publicity for their efforts to supplement fossil fuels with ethanol. However, there are differences in the economics of the two countries and only when gasoline prices in the United States approached $1/gal did the economics of fermentation for fuel production become attractive. With the prospect of continually rising fuel prices in the United States, the economic advantage of ethanol production should continue to improve.

The second technology area reviewed in Section II is anaerobic digestion (both on a controlled process basis and in sanitary landfills where the methane gas generated may be collected). Obviously the economics of this technology have been good for many decades as evidenced by the use of methane gas on-site in sewage treatment plants across the country. It is recognized that most of the process gas produced through anaerobic digestion is roughly 65% methane with 35% carbon dioxide. This means that the heating value of the gas mixture is well below that of pipeline grade gas. Thus, on-site use of the gas may well be economic but to treat the gas to remove CO_2 and use it for pipeline grade gas may be uneconomic. Further, the energy balance of most sewage treatment plants is such that there is not enough methane left over after meeting the plant energy requirements to justify a large capital expenditure for cleanup and utilization in other locations.

With ever increasing prices of pipeline gas, the economics of gas generation facilities using anaerobic digestion are also changing. Several facilities have been constructed during the past three years to utilize animal manure from feedlots as a feedstock for methane generation. It is likely that the trend will continue. The cost to upgrade the medium BTU gas mixture of CH_4 and CO_2 to pipeline quality remains prohibitive without substantial tax incentive but the economics of that technology are apt to change rapidly in the next 5 years. Of the various biomass feedstocks that could be used in an anaerobic digestion system, animal manure appears to be the most promising from the standpoint of economics.

Methane collection from sanitary landfills has several unknowns at this time from the viewpoint of economics. Apparently, based on the projects planned and underway now in California, the recovery of methane is attractive on an overal basis. But there may be a substantial element of public relations involved in the economics as well as a need to safely remove the methane generated by the landfills. The point is this: landfills are not capable of generating large quantities of methane gas - certainly not large enough to meet the needs of a typical community using the landfill. The production rates are slow on a continuing basis but there are some landfills where substantial quantities of methane gas have collected in a reservoir. In such facilities, it may be very beneficial to collect the stored gas, upgrade it using molecular sieve technology, and distribute it as pipeline grade gas. This apparently is the case for most of the collection facilities currently in use.

The economics of pyrolysis processes appear to favor the production of char at this time. There are many units in operation in the United States and in other countries where wood and bark residues are used for the production of char in pyrolysis reactors. In some, but not all of these reactors, the low BTU gases generated are burned in waste heat boilers for on-site steam generation. Full scale commercial applications of pyrolysis technology for the production of oils is not yet underway in the United States and has not been proven on a continuing basis. The economics of pyrolytic oil production as a replacement or supplement for No. 6 fuel oil are improving rapidly with rising crude prices but it may be several years before there is sufficient economic incentive to promote this technology on a large scale.

Of the three categories of thermochemical processes (pyrolysis, gasification, and liquefaction), gasification is currently receiving the greatest interest for full scale commercial application in the United States. Low BTU gases from gasification facilities have great potential for replacing high priced fossil fuels in veneer dryers, lime kilns, cement kilns, lumber kilns, and a variety of other systems where direct combustion of biomass would not be satisfactory. The economics of using such systems are attractive. The greatest hindrance to the implementation of the technology is the lack of proven equipment in large commercial size ranges. However, within a matter of 1 or 2 years, this situation is expected to change dramatically as full scale commercial systems are installed and get some operating experience to prove their capabilities.

Liquefaction facilities for biomass are not particularly attractive at this time from an economic standpoint. Liquefaction via high pressure, high temperature catalysis is very expensive and not proven even in the pilot plant stages as being technically feasible. No doubt, the process development unit in Oregon has produced liquefied fuels from biomass but only on a limited basis and with great difficulty in the operation of the plant. Liquefaction of biomass feedstocks via the route of synthesis gas formation and subsequent processing of the synthesis gas to form desired end products is also unproven on any commercial scale at this time and will likely be quite expensive by today's standards. However, with the increasing cost of petroleum feedstocks, the approach may hold some reasonable economic opportunities within the next few years. In the event of a major shortage of liquid fuels (methanol, gasoline, ethylene, etc.) synthesis gas plants based on biomass feedstocks can become a reality within a matter of only a few years — the time controlled by the delays in design and construction of the plants and the period required for startup and debugging. As noted above, liquefaction via pyrolysis techniques is not economically attractive at this time but may become more attractive as crude oil prices rise.

In Section II it was noted that SRI, International has studied a variety of paths for the conversion of biomass feedstocks to synthetic fuels. Based on their evaluation, the most promising paths (those with the greatest potential for market penetration) were selected for specific scrutiny.

Table 17 summarizes the results of the SRI, International analysis. The table has been arranged in order of the ascending projected costs for the synthetic fuels produced.

TABLE 17

SUMMARY OF SRI, INTERNATIONAL'S FINDINGS ON THE MOST PROMISING ROUTES FOR PRODUCTION OF SYNTHETIC FUELS FROM BIOMASS FEEDSTOCKS

	Route	Conversion Process	Base Case Cost ($/MMBTU)
1.	Wood to Oil and Char (Total Product Basis)	Pyrolysis	2.70
2.	Wood to Intermediate BTU Gas (Via Low Pressure Process)	Pyrolysis	3.40
3.	Wood to Intermediate BTU Gas (Via High Pressure Process)	Pyrolysis	4.00
4.	Wood to Oil and Char Byproduct (Char at $1.25/MMBTU)	Pyrolysis	4.50
5.	Cattle Manure to Intermediate BTU Gas	Anaerobic Digestion	4.90
6.	Wood to Heavy Fuel Oil	Catalytic Liquefaction	5.40
7.	Wood to Synthetic Natural Gas	Oxygen-Blown Gasification	6.50
8.	Cattle Manure to Synthetic Natural Gas (100,000 Head Envir. Feedlot)	Anaerobic Digestion	7.00
9.	Wood to Methanol	Oxygen-Blown Gasification	7.80
10.	Cattle Manure to Synthetic Natural Gas (10,000 Head Envir. Feedlot)	Anaerobic Digestion	14.40
11.	Kelp to Synthetic Natural Gas (Speculative Design Case)	Anaerobic Digestion	21.20
12.	Wheat Straw to Intermediate BTU Gas	Anaerobic Digestion	22.10
13.	Corn Stover to Ethanol	Fermentation — Acid Hydrolysis	25.55
14.	Algae to Ethanol	Fermentation — Acid Hydrolysis	25.90
15.	Sugar Cane to Ethanol	Fermentation — Enzymatic Hydrolysis	27.00
16.	Wheat Straw to Ethanol	Fermentation — Enzymatic Hydrolysis	52.60

Note from Table 17 that the order of increasing costs for technology appears to be: (1) Pyrolysis; (2) Anaerobic Digestion; (3) Oxygen-Blown Gasification; (4) Fermentation — Acid Hydrolysis; and (5) Fermentation — Enzymatic Hydrolysis.

It should also be pointed out regarding Table 17 that the data indicated for projected costs of synthetic fuels is based on a variety of assumptions concerning the cost of the feedstock, costs for purchase, and erection of equipment, operating manpower requirements, etc. The costs are not to be interpreted as absolute costs but will serve to give the reader some relative idea of the projected order of magnitude of costs for synthetic fuels from biomass.

The costs projected in Table 17 may serve as a guide as to which technologies are the most economically attractive for future conversions. However, it should be remembered that economics can quickly be forgotten in the event of major shortages of critical fuels, regardless of whether they are in the gaseous, liquid, or solid states. The point is this: if gasoline becomes in very short supply, then tax incentives plus increased costs will serve to implement additional facilities for the fermentation of various products to ethanol even though it is a very expensive liquid fuel to produce. Similar events can easily happen regarding other synthetic fuels from biomass.

V. BARRIERS TO THE COMMERCIALIZATION OF SYNTHETIC FUELS PRODUCTION FROM BIOMASS FEEDSTOCKS

Biomass fuels are known to have a significant potential on a worldwide basis for replacement and supplement of fossil fuels. Yet there has been relatively little activity in the area of manufacturing synthetic fuels from biomass feedstocks. Klass (1979) has summarized some of the major barriers to the commercialization of biomass energy technology:

> "Many factors must be examined in depth to choose and develop systems that are technically feasible, energetically and economically practical, and environmentally acceptable. These factors are particularly important for large-scale biomass energy farms where continuity and efficiency of operation and synfuel production are paramount. The problem is not so intractable though that it defies solution. But there are several major barriers that must be overcome or at least reduced in size to facilitate commercial use of biomass energy technology on a scale that will satisfy a significant portion of our energy demand. These barriers, none of which is insurmountable, are briefly described as follows:
>
> Barrier No. 1: Excessive Cost of Biomass-Derived Fuels
>
> At the present time, the commercialization of biomass energy is proceeding at the proverbial snail's pace. The excessive cost of synfuels from biomass in integrated growth, harvesting, and conversion systems, and from integrated waste collection and conversion systems is the prime reason for the low commercialization rate. Although synfuel production capacity (plant size) and financing conditions impact directly on synfuel costs, the estimated and actual manufacturing costs of most biomass-derived synfuels are not presently competitive with fossil fuels.
>
> The major factor that seems to influence synfuel costs from biomass is biomass cost itself; conversion and other associated costs are often a

smaller part of the total cost. Tax incentives and other forms of subsidy have already been suggested to reduce synfuel costs and thereby stimulate the investment of private capital. Whether this approach can be effective remains to be established.

Barrier No. 2: Negative Net Energy Production Efficiency

Considerable disagreement exists among the developers of biomass energy technology as to whether certain synfuels can be manufactured at positive net energy production efficiencies. The technology for the manufacture especially of ethanol has been criticized by several experts as having negative net energy production efficiencies. Careful analysis of these claims often reveals serious technical flaws.... For example, some advocates of negative net energy production efficiencies of ethanol plants are referring only to the use of a small portion of the biomass feed such as the grain; the remainder of the feed is ignored. Others often take no energy credits for the by-products produced along with the alcohol, and still others consider only the manufacture of pure alcohol which is not necessary or perhaps even desireable in many fuel applications.

Improvements in operating efficiency can be incorporated in most biomass synfuel systems; however, an easily understood net energy analysis should be delineated as early as possible to assure all interested parties that the system in question can deliver net energy in the marketplace.

Barrier No. 3: The All or Nothing at All Syndrome

Several analysts have concluded that biomass will never serve as a large source of energy supply. A variety of reasons has been presented to support this position. One analyst has concluded that any long-term appraisal of fuel farming looks unfavorable because more agricultural land will be needed for foodstuffs, and that our land, water, and inorganic nutrients must be conserved for long-term production. Another analyst takes the position that since agricultural land is as fixed in some respects as petroleum reserves, agricultural production might not actually be a source of renewable raw materials because of the continued population increase and the resulting food shortages. Another analyst concluded that forest biomass can never be used on a large scale as an energy source because of the growing demand and the projected shortages for lumber.

Many of these arguments are based on the premise that all of the energy needs of the U.S.A. would be met by land-based biomass in any biomass energy scenario. Little if any consideration is given to biomass energy that meets only part of the demand for specific chemicals, organic fuels, or electric power needs, and since so much land is needed to meet total demand, the concept is not feasible. This rationale is often related to existing methodologies and does not take any new developments into account.

Barrier No. 4: Governmental Regulations and Resistance to Change

The limitations of governmental regulations and the generally unfavorable or neutral response of the public to change present a large barrier to commercialization of biomass energy. The fact that large areas of land and water might be involved will undoubtedly bring organized opposition and protest groups to bear on the various agencies in control of issuing permits and regulations. Cooperation between all involved parties is essential.

Barrier No. 5: Multi-Product Plants are Low in Profitability

Almost all large-scale chemical plants produce marketable by-products or co-products and most agricultural operations produce by-products along with the main product. Biomass energy systems with very few exceptions will be operated in the same manner; multiple product plants will predominate. Some biomass strategists feel this creates an extra burden that tends to make biomass energy less profitable. On the contrary, multiple product plants will make biomass energy more profitable. Flexibility can be designed into many processes to make it possible to match product distribution with market demand. In fact, several analysts have found that a multiple product plant is often more profitable than single product plants designed to manufacture the same products.

Other barriers to commercialization of biomass energy, such as those concerned with the transfer of technology to the owner-operator, the availability of equipment and instrumentation, specific environmental and safety problems, acquisition of sufficient land for biomass production, acquisition of sufficient volumes of organic wastes for conversion, financing, and obtaining advance approvals from utility commissions ... could be added to this list. None is large enough to block the growth of biomass energy. I suggest that commercialization can be achieved at a higher rate by concentrating on the elimination of these barriers, especially those that are intrinsically characteristic of biomass energy."

From the technical summary presented in this work, it is apparent that many of the barriers viewed by Klass are indeed real. However, in view of the perspective gained from this review, it appears that economic barriers are balanced by technological barriers at this point in time. As an example, the economics of making gasoline from biomass fuels appear to be attractive but the technology is not demonstrated and many difficulties lie ahead in making the process commercial in scale and application. On the other hand, the technology for making methanol from wood is probably ready for commercialization, but the economics are unattractive compared to the present methanol industry based on the availability of natural gas as the feedstock.

One consideration that has not been mentioned relative to biomass fuels, is that the most profitable use of biomass fuels is in direct combustion rather than in conversion to synthetic fuels. As of 1980, it is estimated that the use of biomass fuels in industry through direct combustion represents about 1.4 Quads annually. Roughly 30 million tons of wood and bark residue fuels are burned in industrial boilers for steam generation producing 0.5 quads annually. The balance is found in recovery furnaces in the pulp and paper industry.

Additional energy is gained through direct combustion of wood fuels for home heating, for cooking, etc. Estimates of the use of biomass energy for home use range up to 1 Quad annually. (These estimates pertain to the United States only).

The point is this: the economics of manufacturing synthetic fuels from biomass depend to a large extent on the yield of desired product from the feedstock. Direct combustion gives the greatest energy yield of all of the alternative approaches. Furthermore, it is well proven on a commercial basis and is well accepted by both the public and industry. Any plans to utilize biomass fuels in systems other than direct combustion will be in competition for the feedstock (potentially) that is presently used for combustion. Taking the yields into account, that places synthetic fuels production from biomass in a potentially unfavorable light.

The reader is cautioned that the comments above are major generalizations but may be valid in some instances — particularly with regard to the use of wood feedstocks for synthetic fuels. But there are other feedstocks where direct combustion will probably not be in competition such as with kelp, animal manure, cheese whey, and many agricultural crop residues.

VI. SUMMARY AND CONCLUSIONS

This paper has reviewed the state of the art of manufacturing synthetic fuels from biomass feedstocks. Information for the summary was gained from literature references, from site visits to research facilities, and from the author's personal experience in the field.

Five major technological areas are reviewed:

1) Fermentation
2) Anaerobic Digestion
3) Pyrolysis
4) Gasification
5) Liquefaction

as well as some related technologies, including:

1) Hydrogasification
2) Steam Gasification
3) Molten Salt Gasification
4) Liquid Oil Recovery from *Euphorbia* type plants
5) Densified Biomass Fuels

From the technological standpoint, fermentation is well developed, particularl in the "tail-end" technologies. The preparation of the feedstocks on the "front-end" of the system has received considerable attention in recent years with work in the areas of enzymatic hydrolysis and acid hydrolysis. This has made it technically feasible to utilize cellulose base feedstocks for the production of ethanol but the system is not well demonstrated on a commercial scale. Other feedstocks offer attractive economic returns in fermentation processes. Technical advances are being made in improving the efficiency of the distillation process to lower the energy requirements for separation of the alcohol product from the water solution.

Anaerobic digestion has been well developed for many years but is now being expanded to a variety of feedstocks with good success. Methane generated by the process can now be upgraded by use of molecular sieves which are a recent development. This has improved not only the interest in methane generation by controlled processes but also the interest in methane recovery from sanitary landfills. Several large demonstration projects are in operation and in the planning and construction stages.

Pyrolysis technologies for the production of char, pyrolytic oils, and gases have received a great deal of interest across the United States as a means of generating a variety of synthetic products from biomass fuels. The data base on the operation of specific systems on specific feedstocks is expanding but not at a rapid pace. The technology could be classified as in the developmental stages on a broad basis with only a few commercially proven systems on the market. Most of these are manufacturing char for the coke and charcoal industries. There is some indication that pyrolysis systems may be used economically for the production of oils in the future, but at present that is somewhat speculative.

Gasification systems have been the most promising of the thermochemical approaches to converting biomass feedstocks to synthetic fuels. The technology is fairly well proven on pilot scales and it has been shown that it can be used as a replacement for fossil fuel in many installations. The difficulty at present is that there are no large-scale commercially proven systems on the market with long time operating experience. This has limited the expansion of the technology in the United States. However, it is likely that this will change rapidly in the coming years, particularly in view of the escalating oil prices.

Liquefaction technologies are more complex than gasification approaches (with the possible exception of pyrolysis oil formation) and they appear to be behind in development and demonstration. High temperature, high pressure catalytic formation of liquids from biomass fuels is in the early developmental stages only. Production of synthetic liquid fuels via the synthesis gas route is similarly not proven on a commercial scale from biomass feedstocks and may be a long way from that point. Technologically there are many processes that need to be proven before the systems can be put into commercial operation and even then the economics may not favor the technology.

As for the other technological approaches (hydrogasification , steam gasification, molten salt reactors, and liquid oil recovery from *Euphorbia* type plants) these are all in the early developmental stages and are many years away from commercial application on a significant scale. They hold some promise for the future but are not yet ready to make a large impact. On the other hand,densified biomass fuels as a substitute for fossil fuels is well under way in isolated locations.

An economic analysis of some of the alternative technologies for biomass conversion to synthetic fuels indicates that,in order of increasing cost, the technologies might be arranged as follows:

1. Pyrolysis
2. Anaerobic Digestion
3. Oxygen-Blown Gasification
4. Fermentation — Acid Hydrolysis
5. Fermentation — Enzymatic Hydrolysis

This list, of course, leaves out many of the technological alternatives for conversion, but it lists the most promising in terms of market penetration for the United States in the future. It does not necessarily indicate that these specific technologies will be appropriate for other countries of the world.

Many of the barriers to the development of synthetic fuels are reviewed. In brief they include:

1. Excessive Cost of Biomass-Derived Synfuels
2. Negative Net Energy Production Efficiency
3. The All or Nothing at All Syndrome
4. Governmental Regulations and Resistance to Change
5. Multi-Product Plants Are Low in Profitability
6. Technology Transfer

and others. It is suggested that none of these barriers is large enough to block the growth of synthetic fuels production if sufficient effort is made to remove the barriers.

Footnotes

1. Sladek, T. A., 1978, "Ethanol Motor Fuels and Gasohol," Mineral Industries Bulletin, Colorado School of Mines, 0010-1745/78/2103-0001.

2. Footnote (1): The summary of fermentation technology is taken from Ref. 1 by Sladek, pages 2 — 5 of the reference text.

3. Wang, D. I. C., Biocic, I., Fang, H.Y., and Want, S.D., 1979, "Direct Microbiological Conversion of Cellulosic Biomass to Ethanol", 3rd Annual Biomass Energy Systems Conference Proceedings, Sponsored by the Solar Energy Research Institute, Golden, Colorado, June 5, 6, 7, 1979.

4. Pye, K.E., and Humphrey, A.E., 1979, "Production of Liquid Fuels from Cellulosic Biomass", 3rd Annual Biomass Energy Systems Conference Proceedings, Sponsored by the Solar Energy Research Institute, Golden, Colorado, June 5, 6, 7, 1979.

5. Tsao, G.T., 1979, "Selective Solvent Extraction in Utilization of Stored Solar Energy in Cellulosic Biomass", 3rd Annual Biomass Energy Systems Conference Proceedings, Sponsored by the Solar Energy Research Institute, Golden, Colorado. June 5, 6, 7, 1979.

6. Bente, P.F., 1979, Bio-Energy Directory, 2nd Ed. Edited by Bente, P.F., Published by The Bio-Energy Council, Washington, D.C.

7. Bungay, H.R., 1979, "Fuels From Fermentation of Biomass", 3rd Annual Biomass Energy Systems Conference Proceedings, Sponsored by the Solar Energy Research Institute, Golden, Colorado, June 5, 6, 7, 1979.

8. Footnote (2): Rensselaer Polytechnic Institute has a contract to "Co-ordinate, analyze, report, and plan the research on fermentation of biomass to produce fuels and petrochemicals". Prof. Bungay is the Principal Investigator for the project. Because of his work in this area, significantly large exerpts were taken from his presentation at the 3rd Annual Biomass Energy Systems Conf. for inclusion in this paper.

9. SRI, Int'l, 1978 "Mission Analysis for The Federal Fuels From Biomass Program." Proj. Leader Schooley, F.A. Prepared for the U. S. DOE Div. of Distributed Solar Technology, Fuels from Biomass Systems Branch. Contract No. EY-76-C-03-0115PA-131, Volume 1: Summary and Conclusions.

10. Footnote (3): The economic information is taken from Table III-3, p. 74 of The Report of the Alcohol Fuels Policy Review, Pub. by the U.S. Dept. of Energy, Assistant Secretary for Policy Evaluation. June 1979, Washington, D.C., DOE/PE-0012.

11. Eckenfelder, W.W., 1966, Industrial Water Pollution Control, Pub. by McGraw-Hill Book Company, New York. Note: This is from the McGraw Hill Series in Sanitary Science and Water Resources Engineering.

12. Varani, F.T. and Burford, J.J., 1977, "The Conversion of Feedlot Wastes into Pipeline Gas", Chapter VI.— Fuels from Wastes, Edited by Anderson, L.L. and Tillman, D.A., Published by Academic Press New York, 1977.

13. Jones, J.L. and Fong, W.S., 1978, "Biochemical Conversion of Biomass to Fuels and Chemicals". Volume V of Mission Analysis for the Federal Fuels from Biomass Program, Prepared for the U.S. DOE, Division of Distributed Solar Technology, Fuels from Biomass Systems Branch, Contract No. EY-76-C-03-0115PA-131.

14. Cheremisinoff, P.N. and Morresi, A.C., 1976, Energy From Solid Wastes, Chapter 14, Published by Marcel Dekker, Inc, New York.

15. Calden, D.L., 1976, "Methane Recovery from Landfills", Published in a Monograph on Alternative Fuel Resources, edited by Hendel, F.J. The Monograph is based on papers presented at the Symposium on Alternate Fuel Resources held in Santa Maria, California March 25-27, 1976, Sponsored by the American Institute of Aeronautics and Astronautics, Vandenberg Section.

16. Lewis, F.M. and Ablow, C.M., 1976, "Pyrogas From Biomass", A paper presented at a Conference on Capturing the Sun Through Bioconversion March 10-12, 1976, Washington, D.C. Note: The paper was published as an in-house paper by Stanford Research Institute, Menlo Park, CA.

17. Milne, T., 1979, "Pyrolysis - The Thermal Behavior of Biomass Below 600 C", From Volumes I & II of A Survey of Biomass Gasification published by the Solar Energy Research Institute, SERI/TR-33-239, July 1979. Note: Milne is credited with authoring Chapter 5 of the Vol. II. It was assumed that he also authored the synopsis of Chapter 5 in Vol. 1, however it is not specifically stated to be so.

18. Fisher, T.F., Kasbohm, M.L, and Rivero, J.R., 1975, "The 'PUROX' System", A paper presented at the AIChE 80th National Meeting, Boston, Mass. Sept. 9, 1975.

19. Solar Energy Research Institute, 1979. Note: This is the same reference as 17 (above). No specific author is shown in the reference by SERI. The information is from Volume I, ppgs I-24 through I-27.

20. Reed, T.B., 1979, "Survey of Gasifier Types"- Chapter (8) Summary of Volume I.— A Survey of Biomass Gasification published by the Solar Energy Research Institute, SERI/TR-33-239, July 1979. Note: Reed is credited with authoring Chapter 8 of Volume III. It was assumed that he also authored the synopsis of Chapter 8 in Vol. I, however, it is not specifically stated to be so.

21. Keene, A.G. and Nyce, A.C., 1979,"Coal Gasification Technology: Wood Retrofit Potential", 3rd Annual Biomass Energy Systems Conference Proceedings, Sponsored by the Solar Energy Research Institute, Golden, Colorado, June 5, 6, 7, 1979.

22. Ergun, S., 1979,"An Overview of Biomass Liquefaction", 3rd Annual Biomass Energy Systems Conference Proceedings, Sponsored by the Solar Energy Research Institute, Golden, Colorado, June 5, 6, 7, 1979.

23. Berry, W. L., "Operations of the Biomass Liquefaction Facility", Albany, OR., Wheelabrator Cleanfuel Corp., July 1, 1978 - June 30, 1980, DOE Contract ET-78-C-06-1092, From 3rd Annual Biomass Energy Systems Conference Proceedings, Sponsored by the Solar Energy Research Institute, Golden, Colorado, June 5, 6, 7, 1979.

24. Kohan, S.M. and Barkhordar, P.M., "Mission Analysis for the Federal Fuels From Biomass Program", Vol. IV — Thermochemical Conversion of Biomass to Fuels and Chemicals, Prepared for the U. S. DOE, Div. of Distributed Solar Technology, Fuels from Biomass Systems Branch, Contract No. EY-76-C-03-0115PA-131. Note: This is part of the six volume report by SRI, International (1978-1979).

25. Diebold, J. P. and Smith, G. D., 1979, "Thermochemical Conversion of Biomass to Gasoline", 3rd Annual Biomass Energy Systems Conference Proceedings, Sponsored by the Solar Energy Research Institute, Golden, Colorado, June 5, 6, 7, 1979

26. ERCO, 1977, "Pilot Scale Pyrolytic Conversion of Mixed Wastes to Fuel", Volume I, A report by Energy Resources Company, Inc. to the Industrial Environmental Research Laboratory, U.S. Environmental Protection Agency, Cincinnati, Ohio, under Contract No. 68-03-2340.

27. Knight, J. A., Hurst, D. R., and Elston, L. W., 1977,"Wood Oil from Pyrolysis of Pine Bark - Sawdust Mixture", A Paper Presented at the Symposium on Fuels and Energy from Renewable Resources, Chicago, 1977.

28. Preston, G. T., 1976,"Resource Recovery and Flash Pyrolysis of Municipal Refuse", A Paper Presented at the Symposium on Clean Fuels From Biomass, Sewage, Urban Refuse, and Agricultural Wastes, Sponsored by the Institute of Gas Technology, Orlando, Fla., January 27-30, 1976.

29. North American Combustion Handbook, 1st Ed., North American Mfg. Co., Cleveland, Ohio 1952.

30. ERCO, 1979, "Mobile Pyrolysis System Mechanical Design Report", A report by Energy Resources Company, Inc. under Contract No. 68-03-2349. Date of Report Dec. 28, 1979, Principal Author Dr. H. M. Kosstrin, Note: This report is considered to be a preliminary report.

31. Coffman, J. A., 1979, "Steam Gasification of Biomass", 3rd Annual Biomass Energy Systems Conference Proceedings, Sponsored by the Solar Energy Research Institute, Golden, Colorado, June 5, 6, 7, 1979.

32. Yosim, S. J. and Barclay, K. M., "Production of Low-BTU Gas from Wastes Using Molten Salts", Chapter III of Fuels From Waste ed. by Anderson, L. L. and Tillman, D. A., Pub. by Academic Press, 1977.

33. Footnote (4): This information is found in Volume VI: Mission Addendum of the SRI, International Report entitled "Mission Analysis for the Federal Fuels from Biomass Program", pages 51 - 74, (See ref. 9).

34. Klass, D. K., 1979, "Barriers to Commercialization of Biomass Energy Technology," Conference Proceedings from the Mid-American Biomass Energy Workshop, Bloomington, Minnesota, May 21-23, 1979.

ETHANOL FROM BIOMASS

GEORGE F. HUFF

Gulf Oil Science and Technology

ABSTRACT

Biomass represents stored solar energy. The light energy of the sun is converted by green plants into chemical energy which is stored in chemical compounds such as starch and cellulose. Both materials can be used as raw materials for the production of ethanol which can be used as gasoline additive, for improving the octane number, or even as a gasoline replacement. The fermentation of starch as contained in grains, such as, maize and wheat, is well known and is widely practiced.

This paper defines the fermentation of cellulose, a much more abundant material found in municipal and agricultural wastes, but which is not yet utilized in commercial processes. However, research to date indicates that commercialization will be achieved within the next few years.

Economic projections for the commercialization of this process are presented. Preliminary results suggest that ethanol produced from cellulose will be price-competitive with non-subsidized gasoline.

ISBN 0-12-467101-2

INTRODUCTION

The subject of alcohol fuels from biomass is receiving increasing interest worldwide. Alcohol fuels are by no means a complete or even principal solution to the coming shortages in liquid energy, but they can make a contribution to the solution as they take their place beside other alternate liquid fuels. Alternate fuels must be high on the agenda of the scientific and technical community in the closing decades of this century if civilization as we know it is to be preserved, and parts of mankind are not to be doomed to return to savagery.

Biomass represents stored solar energy. This form of energy is stored by living plants containing chlorophyll. Green plants receive sunlight and in combination with water, atmospheric carbon dioxide, and soil nutrients convert light energy to chemical energy in a process called photosynthesis. This chemical energy is expressed in the form of certain chemical compounds containing carbon. The existance of nearly all living organisms depends, either directly or indirectly, on the photosynthetic process. Those organisms which are incapable of using sunlight to synthesize their own substance are nevertheless able to use the substance of green plants as food. One might reflect that the dominant liquid fuel that we enjoy today, petroleum, represents the stored energy of sunlight which fell on the earth hundreds of millions of years ago. The material of the ancient plants which grew at that time was subsequently subjected to geologic forces which converted it to the oil we know today. Thus, the direct conversion of biomass to liquid fuels can be viewed as a short-cut which eliminates the eons of time required to make petroleum.

Two of the principal chemical materials produced by the photosynthetic process are cellulose and starch, which are depicted in Figure 1. Cellulose is basically the structural material of living plants and is the most abundant organic chemical. Starch is less abundant and is generally found in the fruits of the plants. Both are polyanhydrides (also known as polysaccharides) of the six carbon sugar, glucose, and differ only in the nature of the ether linkage between the units. The starch polymer has the alpha 1, 4 configuration and the cellulose polymer, the beta 1, 4. This difference arises, of course, because of the assymetry of the carbon atoms in the glucose ring.

Glucose is the principal energy source or food for living organisms. All higher organisms have the ability to break down starch through hydrolysis to its constituent glucose units which are then assimilated as food. Only a few lower organisms can treat cellulose in this way. Thus, starch is generally thought of as a food material, while cellulose is not. Ruminant animals can subsist on cellulose, but they do not have this capability in their own bodies. The conversion of cellulose to glucose and of glucose to protein is accomplished by microorganisms which live symbiotically in the first

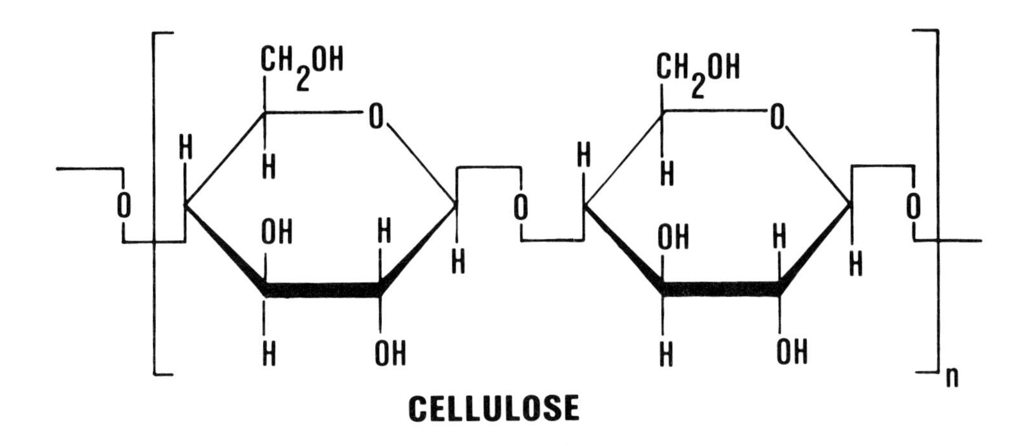

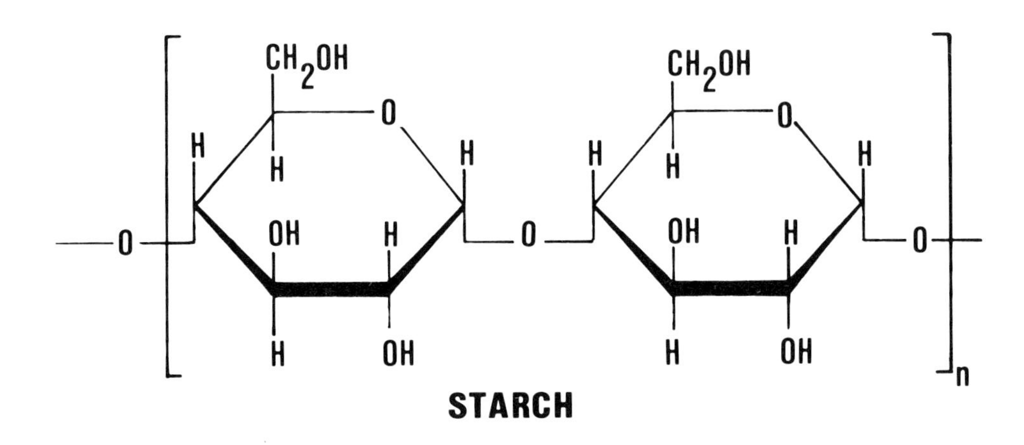

Figure 1

stomach or rumen of the animal. Hence, the lower organisms do the work for the higher organism.

The fermentation of sugars such as glucose to ethanol has been known and practiced since the beginning of recorded history, but this process was fully understood only through the research of Louis Pasteur in the late nineteenth century. Sugar combined with a common yeast, such as *Saccharomyces cervisiae*, is converted to alcohol and carbon dioxide. The enzymes of the yeast, collectively known as zymase, catalyze this familiar reaction.

Both starch and cellulose can be used as raw materials for ethanol production if these substances are first hydrolyzed to glucose by the insertion of a water molecule between the glucoside units of the polymer chain. Either acid or the appropriate enzyme (amylase in the case of starch, and cellulase in the case of cellulose) may be used to catalyze this hydrolysis. Acid hydrolysis has the disadvantage that the proton not only catalyzes the hydrolysis of the polysaccharide, but also catalyzes other unwanted reactions as well, including the degradation of glucose. Enzyme catalysis, on the other hand, is highly specific. Enzymes are said to "mediate" the particular reaction in which they are involved.

Starch hydrolysis is relatively easy. Older industrial practice employed a light acid treatment, but in modern practice the amylase enzyme is preferred since it eliminates not only undesirable side reactions, but also the inconvenience of handling and later neutralizing the acid. For cellulose, however, hydrolysis is much more difficult as will be discussed in a later section.

Ethanol from Starch

Common sources of starch for ethanol fermentation are grains, such as, corn or wheat. The industrial technology for the production of ethanol from grain is widely known and well developed (1). In this technology grain (corn) is first ground to the consistency of meal. The meal with other ingredients is mixed with water to produce a mash which is "cooked" for about 10 minutes at temperatures ranging between 145°F (63°C) and 350°F (177°C). The cooked mash is then combined with amylase enzyme and held for about 2 minutes at 100°F (38°C) before being introduced to the fermentors. The amylase is produced in a separate fermentation by the action of the fungus *Aspergillus niger* feeding on a portion (approximately 1.5%) of the cooked mash.

Yeast (*Saccharomyces cervisiae*) is added to the saccharified mash in the fermentors where the production of alcohol takes place. At the end of the fermentation cycle the "beer", containing 7 to 10% alcohol, is fed to the distillation train. The distillation train consists essentially of two towers. The first is a stripper-rectifier to which the beer is introduced. Alcohol at a concentration of 95% by volume (190° proof) is taken overhead and the column bottoms, containing the distillers dark grains (DDG), are evaporated to dryness. The dry grains, which are produced at rate of about 6 pounds per gallon of alcohol, are sold as an animal feed and provide an important credit to the economics of the process.

The 190° proof alcohol is the composition of the azeotrope of the binary system, ethanol-water. The alcohol cannot be further enriched by more distillation of the two component system. To break the azeotrope in order that anhydrous alcohol may be obtained, a hydrocarbon (often benzene) is added to form a three component system. This ternary mixture is then sent to the second tower — the dehydration tower. The tower bottoms stream is 99.5% alcohol (199° proof). The overhead stream is a mixture of water, hydrocarbon, and some alcohol. The hydrocarbon and alcohol values in this stream are then recovered by further processing.

The economics for this process are indicated in Table I.

TABLE I

50 Million Gallons per Year
Grain Alcohol Plant
Operating Costs (1980 dollars)

Fixed Investment $58.0 million

Working Capital $5.6 million

Cost Item	Annual $MM	$/Gal.
Fixed Charges	9.28	0.186
Raw Materials		
Corn @$2.50/bu.	48.57	0.971
Other	3.28	0.066
Utilities (power @4¢/kwh)	2.36	0.053
Labor	2.20	0.044
Total Production Cost	65.96	1.320
By-Product Credits	(20.00)	(0.400)
Freight, Sales, G&AO	5.74	0.115
Total Operating Cost	48.01	1.035
Taxes	9.54	0.191
Net Profit after Tax	9.54	0.191
Selling Price for 15% Return after Tax	--	1.417

All values are expressed in terms of 1980 dollars. A 50 million gallon per year plant, 100% equity financed with 10 year depreciation, is assumed. The price of corn is taken to be $2.50 per bushel (56 pounds per bushel) and 19.45 million bushels are required per year. The price of electric power is assumed to be $0.04 per killowatt hour and 65.84 million kWh are consumed per year. With these assumptions, a selling price which will return 15% of total investment after tax is calculated to be $1.42 per gallon. Since this is only slightly higher than the retail price of gasoline in early 1980, the economic viability of "Gasohol"—a blend of 90% gasoline and 10% alcohol—is indicated.

It may be observed that an outstanding characteristic of corn to ethanol economics is the sensitivity to the price of corn. The cost of raw material is 70% of the final selling price. On the other hand, capital charges including gross profit are less than two-thirds of the raw material cost. As will be seen later this is exactly the reverse of the situation in the economics of ethanol from cellulose.

The energy balance for a 50 million gallon grain alcohol plant such as described above, is displayed in Table II (2). Values are given in millions of BTU's per hour.

TABLE II

Energy Balance for Ethanol from Corn
50 Million Gallons per Year
(MM BTU/hour)

Energy Produced		Energy Consumed	
Ethanol	535.0	Agriculture	233.5
By-products	74.5	Alcohol Processing	346.4
Total	609.5	Total	579.9

Net gain is 29.6 or 5%

In this balance the energy required to grow the corn feedstock is included as an energy input because modern agriculture is an energy-intensive activity. As seen, if one considers only total energy, the manufacture of ethanol from corn is scarcely more than a break-even proposition. If all the energy inputs represented petroleum energy, there would be little point in manufacturing alcohol fuel from corn. However, in the United States there is not a total energy crisis, there is a liquid energy crisis. If the energy input for alcohol processsing were to be provided by coal, and the import of the equivalent amount of petroleum were not required, the net gain on a liquid energy basis would be 301.5MM BTU/hour or 229%. Furthermore, the major component of the agricultrual input is the energy represented by ammonia based fertilizers. The predominant feedstock for ammonia is natural gas. If the ammonia were to made from coal, as it inevitably will be, the net liquid energy gain will be even higher. To the extent that coal is used as an energy source, the alcohol process can be considered as an indirect method of coal liquifaction.

The availability of grain for alcohol manufacture in the United States is difficult to estimate for it is highly dependent on future government policies regarding land use and exports. The Department of Energy (2) has suggested that by the year 2000 there will be an availability of food crops (including sugar crops - cane, beets, sweet sorghum) sufficient to produce 12 billion gallons of ethanol. This is more than enough alcohol to convert the entire U.S. gasoline pool to "Gasohol".

Ethanol from Cellulose

Ethanol from starch or grains is an existing industrial technology. Ethanol from cellulose is not generally industrially practiced, but the technology is undergoing development in the United States and elsewhere. The major thrust of this work is to improve the hydrolyzability of cellulose.

The difficulty in cellulose hydrolysis was alluded to earlier. This subject is reviewed extensively in a recent paper by Lipinsky (3). The problem is that natural cellulose is associated with hemicellulose and lignin. The former is a polysaccharide made up of glucoside units and xylan units. Xylanose is a five carbon sugar, whereas glucose contains six carbon atoms. In hemicellulose, xylan is the predominant saccharide. Lignin is a highly aromatic hydrocarbon polymer which in living plants serves as the "glue" which holds cellulose fibers together and lends structural integrity to the stems or stalks. The presence of both hemicellulose and lignin tends to protect cellulose from hydrolytic attack. In addition, the cellulose itself is in a crystalline form which also inhibits reaction with water. Any practical method of reducing cellulose to glucose must include some form of pretreatment which will, either chemically or mechanically, lessen the inhibiting effect of lignin and hemicellulose and which will destroy the crystallinity of the cellulose.

As with starch, cellulose can be hydrolyzed by acid or enzymes. Acid hydrolysis has been studied for many years and was practiced by both the Germans and the Japanese during World War II. It is said that this technology is still being employed in the U.S.S.R. About 1947 a demonstration plant was built and operated on the west coast of the United States by the U.S. Forest Service, but was shortly shut down because of unfavorable economics.

The acid treatment of cellulose is much more severe than in the case of starch. Typically, a water slurry of a cellulosic material, containing about 1% sulfuric acid, is subjected to a temperature of 450°F (230°C) and a pressure of 500 psig for a short residence time (20 seconds). This treatment serves both to overcome the effects of lignin, hemicellulose, and cellulose crystallinity, and to hydrolyze the cellulose. The disadvantages of this method are high corrosion and numerous by-products stemming from the further degradation of glucose. Yields of glucose in excess of 60% of theoretical are rarely reported. Current research in the United States, notably at Dartmouth College and New York University, is devoted to specialized equipment design with a view to mitigating these problems.

Enzyme hydrolysis does not involve corrosion and high yields of glucose are possible. However, severe pretreatment is necessary to expose the cellulose to enzyme attack. In Figure 2 are indicated the four principal steps in the enzymatic process for the production of ethanol from cellulose. These are: (a) pretreatment of feedstock, (b) enzyme production, (c) simultaneous saccharification and fermentation, and (d) product recovery or distillation. The important ingredient, yeast, is recycled by techniques which are already well established in the fermentation industry.

Feed pretreatment, which is very important to the success of enzyme hydrolysis, can be either chemical or mechanical. Research is proceeding in different laboratories on a number approaches. Some approaches involve an acid pretreatment which dissolves the hemicellulose and loosens the lignin to expose the cellulose (University of California at Berkeley). This method has the disadvantage of corrosion and the inconvenience of acid handling and neutralization, and it does not attack the problem of cellulose crystallinity. Other approaches involve the use of solvents to dissolve lignin (University of Pennsylvania), or even to dissolve the cellulose itself so that it can later be precipitated in amorphous form (Purdue University). All of the solvent methods involve complex processing steps and, if even a modest solvent loss is incurred, they will be clearly uneconomical.

Mechanical pretreatment or fine grinding appears at this time to be the simplest and most practical pretreatment. It has long been known that grinding to particles of micron dimensions, because of the attendant increase in surface area, renders cellulose more susceptible to enzyme attack. Also, the crystallinity of cellulose is essentially destroyed. A recent paper by Millet, Effland, and Caulfield (4) discusses this phenomenon.

A problem with comminution, or fine grinding, is the energy requirement. Careful equipment selection is necessary for an economical operation. Investigators at the U.S. Army Natick Laboratory found good results with a laboratory ball mill, but energy consumption was excessive. They are now experimenting with a roller mill which promises greater efficiency. The Gulf Oil Corporation process, which appears to be the most advanced from a technical and economic standpoint, has adopted what is believed to be the most energy efficient grinder currently commercially available. It is a horizontal attrition mill consisting essentially of a cylindrical chamber filled with small steel balls and containing an internal agitator. A water slurry of the cellulosic feed at about 6% consistency is ground continuously in this device to extremely fine particle size. The energy requirement is about one-tenth of the requirement for a conventional ball mill. Attrition mills such as this are currently used for grinding paint pigments and certain food materials.

Enzyme production is accomplished by the aerobic fermentation of a mold called *Trichoderma reesei*. The parent strains of this organism are ubiquitious soil fungi and are responsible for much of the rotting of wood and other plant materials. In nature *T. reesei* receives its food by secreting a protein substace called cellulase. This is the enzyme that hydrolyses cellulose to glucose. After the glucose is formed it is absorbed by the mold to provide energy and growth. In the ethanol process the enzyme is prepared in a deep fermentation which contains cellulose, the *T. reesei* mold, inorganic nutrients, and to which air is supplied. Conditions are adjusted so that the mold produces the maximum amount of enzyme which is finally harvested and used for the

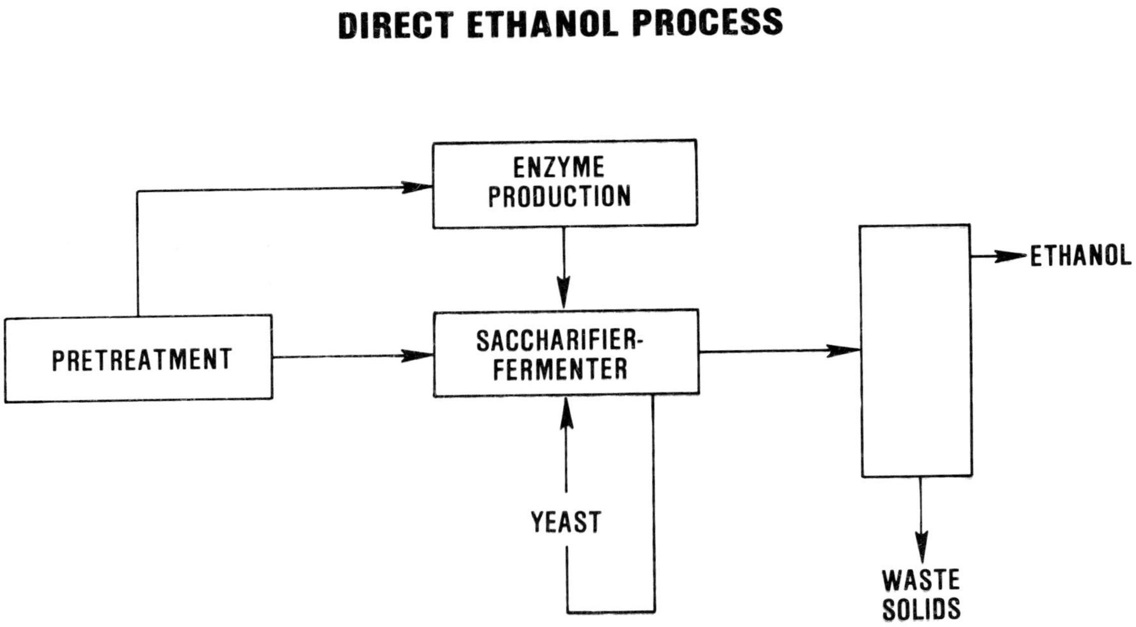

DIRECT ETHANOL PROCESS

ENZYME PRODUCTION

PRETREATMENT

SACCHARIFIER-FERMENTER

ETHANOL

YEAST

WASTE SOLIDS

Figure 2

hydrolysis of the main cellulose feedstock. In the Gulf program, a continuous fermentation process was developed which has been refined so that enzyme production can be accomplished in one-senventh the time required for conventional batch fermentation. Ten percent for the feedstock stream is diverted to the enzyme fermentors for this purpose. Not all of this cellulose is consumed, however. An excess is maintained, since if all cellulose were to disappear, the mold would cease producing enzyme.

The effluent from the reactor (containing enzyme, mold bodies or mycelia, unconverted cellulose, and nutrients) is then joined with the main feedstock stream. The enzyme is not isolated and purified as attempted by other investigators (5). Such procedures will only add to costs and may degrade the activity of the enzyme. The development of this continuous process is a significant advance in that it permits the economical production of large quantities of enzyme which greatly facilitate hydrolysis by overwhelming the cellulosic substrate with the enzyme. In the conceptual design of a 50 million gallon per year alcohol plant, the economics of which will be presented later, it is anticipated that the enzyme production will be 30 tons per day or 20 million pounds per year.

The enzyme so produced can then be used to catalyze the conversion of cellulose to glucose as the raw material for alcohol fermentation. Figure 3 depicts this reaction. The three arrows proceeding from cellulose to glucose represent the triple mode of action of the cellulase enzyme. Cellulase is a catch-all term for this enzyme which is actually at least three enzymes performing three different tasks in concert. First is endoglucanase which randomly attacks a cellulose polymer along its length to form lower molecular weight fragments. The second is cellobiohydrolase which attacks these fragments at the ends and breaks off two glucoside units at a time to form cellobiose or the dimer of glucose. The third is beta glucosidase which converts one molecule of cellobiose to two molecules of glucose.

The next major step in the process is simultaneous saccharification and fermentation where enzyme, yeast, and the cellulosic feedstock are combined in the same vessel. This marks another departure of the Gulf technology from that of others who formerly sought only to produce glucose which would subsequently be fermented to alcohol (6). The simultaneous method cuts down on reactor volume in that there is no need for one reactor for saccharification (hydrolysis) and another for fermentation. Instead, one reactor serves for both functions. Also, there is a synergism whose benefit should be obvious. That is that hydrolysis is aided by the continuous removal of glucose by the action of the yeast. Since the hydrolysis is a reversible reaction, the buildup of glucose concetration tends to inhibit further production of glucose. If glucose is converted to alcohol as fast as it is formed, this inhibition will be removed.

This synergism is graphically indicated in Figure 4. What is represented are the results of experiments with a standard prepared cellulose (Avicel). This material is almost pure cellulose and is highly crystalline. No pretreatment of this substrate was performed. In the control experiment, a 6% slurry of cellulose was inoculated with a fixed amount of enzyme to bring the enzyme concentration to 1 mg/ml, and the percent conversion of cellulose to glucose was measured. In the second experiment, the procedure and amounts were identical except that yeast was added to the mixture, and the

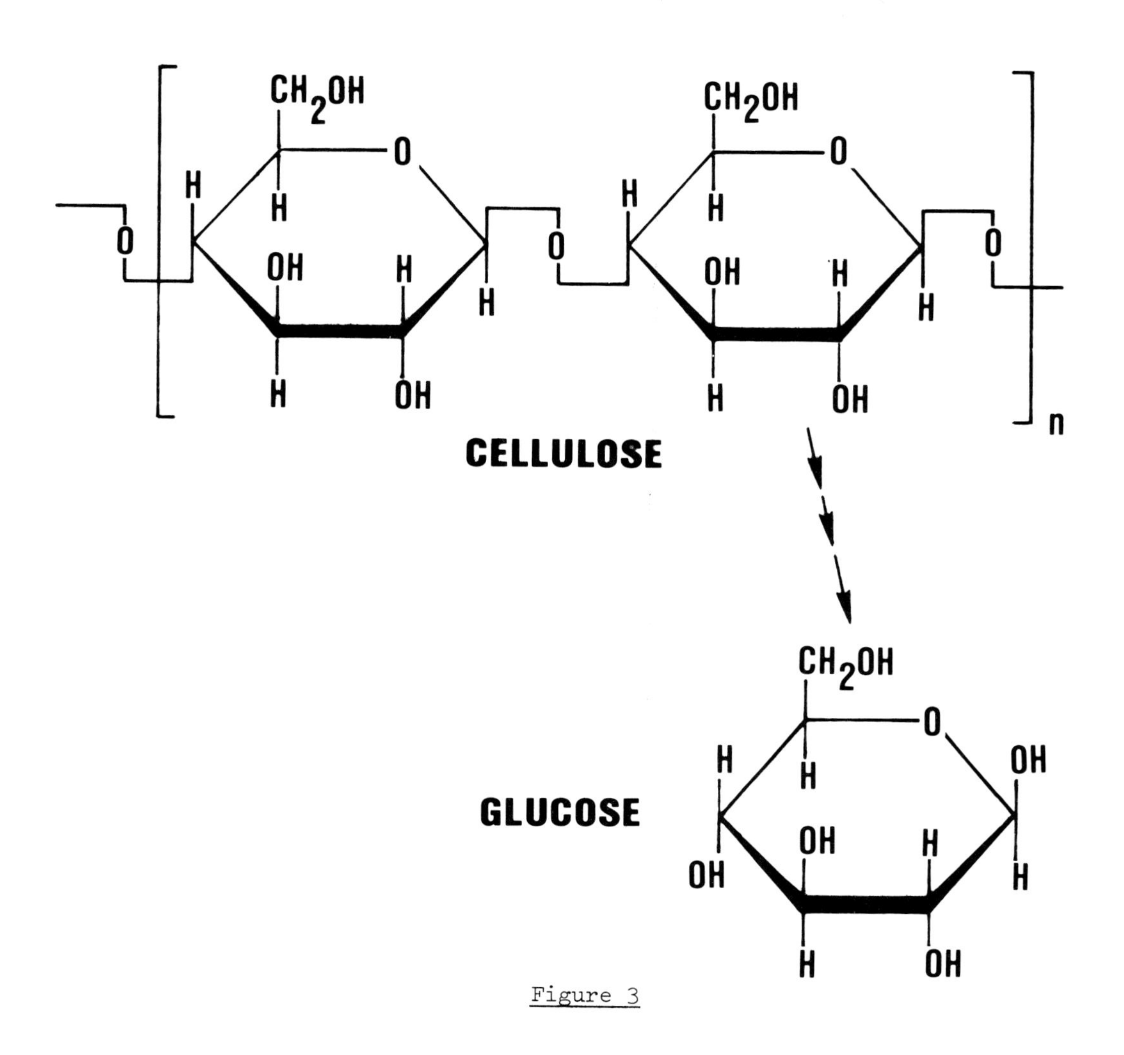

Figure 3

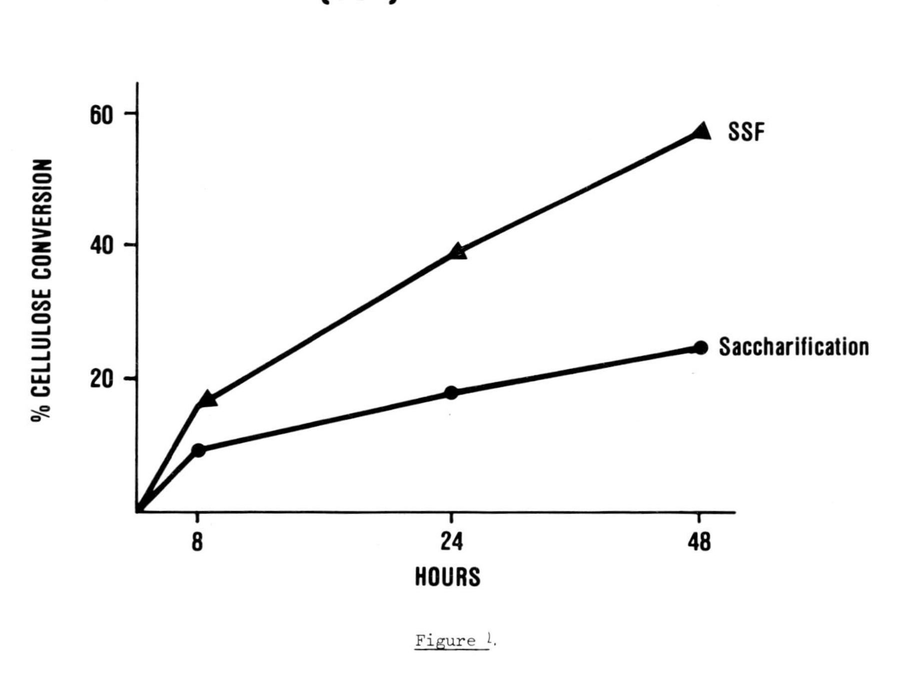
COMPARISON OF SIMULTANEOUS SACCHARIFICATION/
FERMENTATION (SSF) AND SACCHARIFICATION
% CELLULOSE CONVERSION
60
40
20
8
24
48
HOURS
SSF
Saccharification

Figure 1.

percent conversion of cellulose to ethanol was measured over time. As can be seen the rate of conversion of cellulose in the latter case is more than double the rate in the case where cellulose is merely hydrolysed to glucose.

Under actual process conditions, the ligno-cellulosic feedstocks are subjected to severe mechanical pretreatment which destroys all cellulose crystallinity, and the enzyme dosage is three times that used in these experiments. Under these conditions, 90% conversion of cellulose is experienced in 24 hours. The overall ethanol yield is 81% (90% conversion cellulose to glucose and 90% conversion of glucose to ethanol).

Thus, the strategy of the above described approach to the enzymatic conversion of cellulosic materials to alcohol involves three principal elements:

1. Severe mechanical pretreatment to increase susceptibility of feedstocks.

2. The use of massive inoculations of enzyme to obtain highest rates of hydrolysis.

3. Simultaneous saccharification and fermentation to minimize reactor volume.

Actually, these elements are interrelated and must be optimized among each other. For example, increasing mechanical treatment would allow for lower enzyme dosages. Lowering either or both the severity of pretreatment and enzyme dosage would require larger reactor volumes because of the longer time required for conversion. This economic optimization has not yet been made precisely.

The validity of this process strategy has a theoretical basis in that it is predicted, at least qualitatively, by a kinetic model developed by Humphrey (7). Humphrey assumed that the enzymatic hydrolysis of cellulose proceeds in two steps, cellulose to cellobiose and cellobiose to glucose. He further assumed that the cellulase (at least the endoglucanase and cellobiohydrolase portion) was adsorbed at specific sites on the surface of the cellulose and that the partition of the enzyme between that which is adsorbed and that which is free in solution could be described by the Langmuir adsorption isotherm. The following rate equation was derived to describe the disappearance of cellulose in enzyme hydrolysis:

$$\frac{dC_x}{dt} = -K_x C_x \left[\frac{E_x}{\alpha + E_x}\right] \left[\frac{I_2}{I_2 + C_2}\right]$$

Where

C_x = concentration of cellulose, grams/liter

K_x = a term which varies directly with the cube root of the number of cellulose particles per gram of cellulose

E_x = concentration of unadsorbed enzyme in solution, grams/liter

α = a constant (empirically equal to 0.30 grams/liter)

C_2 = concentration of cellobiose, grams/liter

I_2 = inhibiting concentration of cellobiose, grams/liter (That concentratio which will prevent further conversion of cellulose)

The derivation of this expression will not be repeated here since for this th reader is referred to Humphrey's paper.

This rate equation expresses all three elements of the process strategy:

1. Mechanical pretreatment (fine grinding)

   As indicated, the term K_x varies directly as the cube root of the number o particles in a gram of cellulose. As the particle size of a given mass o material is reduced, the number of particles will vary inversely as the cub of the mean particle diameter. Thus, if the particle size is reduced b grinding by a factor of 10, the number of particles will be increased by factor of 1,000. Therefore, K_x and the reaction rate will be increased ten fold.

2. High enzyme concentration

   The second term, in brackets, is derived from the Langmuir isotherm. Sinc it is generally agreed that the enzyme is tightly bound to the adsorptio sites, E_x will have a very small value until enough enzyme is present t saturate the available sites. After this the value of E_x will increas rapidly. At very low values of E_x the second term approaches zero, at hig values it approaches unity.

3. Simultaneous saccharification and fermentation

   The third term, in brackets, deals with product inhibition, specifically th inhibition by cellobiose toward the hydrolysis of cellulose. There is simi larly an inhibition by glucose toward the hydrolysis of cellobiose. Simul taneous saccharification and fermentation removes glucose as fast as it i formed and therefore cellobiose concentration is kept to a very low level At low values of C_2 the third term approaches its maximum which is unity.

The development of this enzymatic process has proceeded from laboratory throug bench scale to a small (1 ton per day of feedstock) pilot plant which operated betwee 1975 and 1979. Data obtained from this facility was used to design a demonstratio plant (50 tons per day) where commercial-size equipment can be evaulated. It is fro the operating results of such a demonstration plant that the confidence to design an build the first commercial plant will be derived. This next step in the developmen has not yet been taken.

Throughout the foregoing discussion of the enzymatic process it should be remem bered that we are dealing with fermentation processes. Fermentation processes hav four principal characteristics — two of them good and two of them bad. On the positive side are the opportunity to operate at ordinary pressures and temperatures and the fact that corrosion is almost non-existent. This means that high-pressure construction and specialty alloys are not required, a circumstance that tends to lower capital costs. On the other hand, one must work with low concentrations and low rates, which lead to large reactor volumes and tend to increase capital costs. The degree to which these latter variables can be controlled will determine the economic viability of any fermentation process. The information available to date on the Gulf enzyme process indicates that it will be economically viable, although final judgement on this point must be withheld until the demonstration plant is built and operated. However, as with any fledgling technology, improvements can be expected to arise in the future which will make the process even better than now foreseen.

To evaluate the process in the light of existing information, a conceptual design of a commerical plant was made and the economics were estimated for the year 1983, which is the earliest date that such a plant could come on stream given the need for a prior demonstration plant program. It was estimated that such a plant, with construction beginning in 1981, would require an investment of $112 million. The plant would produce 50 million gallons of alcohol per year and would consume 2,000 Tons (oven dry) of cellulosic feedstock per day. (8)

In Figure 5, estimated alcohol selling prices in current dollars which will return 15% on investment after tax (30% before tax) for cellulose alcohol, corn alcohol, and synthetic alcohol from the hydration of ethylene are presented. In these estimates, 7% inflation was assumed throughout the period. Corn was assumed to escalate at only 5%, an assumption supported by both history and projections by the U.S. Department of Agriculture. Ethylene was assumed to increase at 9%, a rate higher than general inflation.

The 1983 price for cellulose feedstock was assumed to be $14.00 per dry ton. The feedstock contemplated was the combustible portion of municipal solid waste and the $14.00 price was derived from the alternate use for this material as a fuel. The price for corn was assumed to be $3.00 per bushel and the price of ethylene was assumed to be 18 cents per pound — actually ethylene is over 20 cents per pound today.

Also shown in this figure are projected market prices for ethanol — a high value reflecting 7% escalation in price (with general inflation) and a low value reflecting 5% escalation. The actual 1980 price of alcohol is in excess of $1.60 per gallon so that these estimates are obviously low.

The three upper solid price curves are based on 100% equity financed plants with 0 year depreciation. The lowest curve is based on a leveraged situation with 80% debt and 20% equity where the debt is financed through tax free bonds.

Inspection of the chart indicates that fermentation alcohol will eventually drive ut synthetic ethanol. There are signs that this is already happening. No new thylene hydration facility has been built in the United States for the past ten years and several major producers (those with older plants) have dropped out of the market.

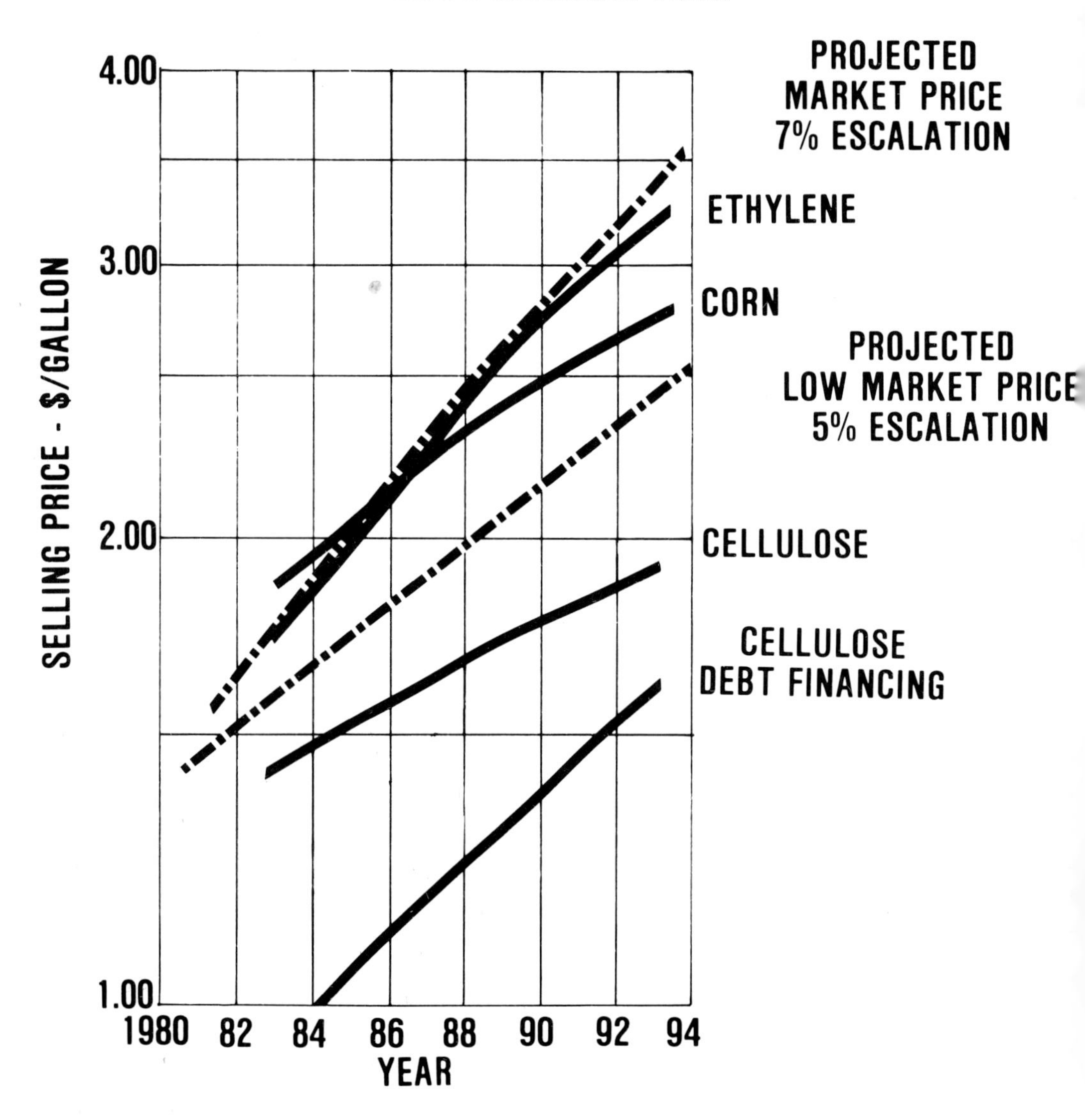
ETHANOL ECONOMICS
SELLING PRICE FOR 15% ROI AFTER TAX
50 MILLION GALLONS/YEAR
PROJECTED MARKET PRICE 7% ESCALATION
ETHYLENE
CORN
PROJECTED LOW MARKET PRICE 5% ESCALATION
CELLULOSE
CELLULOSE DEBT FINANCING
SELLING PRICE - $/GALLON
4.00
3.00
2.00
1.00
1980 82 84 86 88 90 92 94
YEAR

Figure 5

Cellulose alcohol will hurt corn alcohol if our projections are correct. However, if the projections are optimistic, even by a wide margin, cellulose will still be competitive with corn. Notice that the 1983, 15% return price is $1.43 per gallon. Gasoline price will certainly be much higher by then.

TABLE III

50 Million Gallons per Year

Cellulose Alcohol Plant

Operating Costs (1980 dollars)

Fixed Investment $91.59 Million

Working Capital $8.00 Million

Cost Item	Annual $ MM	$/Gal.
Fixed charges	14.65	0.293
Raw Materials		
RDF @ $11.30/O.D.T.	7.45	0.149
Other (nutrients, chemicals)	8.30	0.166
Utilities	10.25	0.205
Labor	2.20	0.044
Total Production cost	42.85	0.857
By-Product Credits	(21.00)	(0.420)
Freight, Sales, GA&O	5.74	0.115
Total Operating Cost	27.59	0.552
axes	14.95	0.299
et Profit after Tax	14.95	0.299
elling Price for 15% Return after Tax	--	1.150

These economics are also cast in 1980 dollars in Table III so that they can be ompared with the economics of corn based alcohol set forth Table I. If the costs are xpressed in 1980 dollars it must be assumed that construction on the plant began in

1978. Adjustment for inflation gives a fixed investment of $91.59 million and working capital requirement of $8.00 million.

Inspection of this table reveals that sensitivity to feedstock price is low. Th dominant factor is the cost of capital, both fixed charges and gross profit. Furthe development of this process must certainly concentrate on reduction of the capita requirement. As with the corn alcohol case there is a heavy dependence on the anima feed by-product credit. One circumstance that may arise is that it may not be possibl to sell the animal feed by-product. This may be true if municipal waste is used as feedstock because of the presence of a wide variety of contaminants and poisons. I this case this product would be burned in the plant to generate electricity and n purchased power would be needed. In such a case, a cost penalty of 30 cents per gallo would be incurred and the 15% return selling price in 1980 dollars would rise fro $1.15 per gallon to $1.45 per gallon. Even so, the alcohol would still be competitiv

Of equal importance to the economics of the process is its energy efficiency. Table IV an energy balance similar to the one in Table II for corn ethanol is presented There is an important difference, however. In the corn case the energy input contri buted by the feedstock is taken to be the agricultrual energy required to produce th corn. In the cellulose case the energy input represents the heat of combustion of th feedstock as if it had an alternate use as a boiler fuel. In the case of cellulosi wastes the energy required to produce the waste material should be assigned by an realistic energy accounting system to the primary use of this material and should no enter into the calculus of the energetics of the cellulose to ethanol process.

TABLE IV

Energy Balance for Ethanol from Cellulose
50 Million Gallons per Year
(MM BTU/hour)

Energy Produced		Energy Consumed	
Ethanol	535.0	Cellulosic Feedstock	1257.3
Animal Feed	256.7	Fuel Oil (combustion aid)	49.5
		Purchased Electricity	277.0
Total	791.7		1583.8

Gross efficiency - 50%

Net efficiency (Corrected for boiler efficiency differences) - 54%

Net efficiency (If feedstock valued at zero) - 192%

Table IV indicates that the gross efficiency of this process is 50%. If one corrects the inputs and outputs for combustion efficiency (for example ethanol burns with an 85% efficiency and cellulosic material with 67% efficiency), the net energy efficiency is 54%. This approaches the energy efficiency of most coal conversion processes; the most efficient among these is the SRC II process at 72%. If the cellulosic feedstock is waste material which is mostly disposed of by burying in sanitary landfills or by incineration without recovery of energy, the energy value of the feedstock can be considered to be zero. If so, the energy efficiency will be 192%.

The information in Table IV is based on the requirement to purchase 27,700 kW of electric power, 2,000 oven dry tons of feedstock per day, and 194 barrels of fuel oil per day. The outputs are 150,000 gallons of ethanol and 534 tons of animal feed per day. The purpose of the fuel oil is to serve as a combustion aid to burn the solid or insoluble residues from the fermentation step in a boiler which, through cogeneration, supplies all the steam requirements of the plant as well as 3,400 kW of electric power. The soluble residue from the fermentation step is the animal feed. The energy associated with the feed is its heat of combustion. The energy associated with electric power is the amount of coal required to produce this amount of power — 10,000 BTU per kWh.

If, as suggested above, the animal feed cannot be sold as such and must be burned as fuel in the plant, the plant will be nearly self-sufficient in electricity. In this case the energy efficiency is 70% where the feedstock is valued for its heat of combustion, and 1266% where it is valued at zero. Thus, the use of the by-product as fuel improves the energetics of the plant even though the economics suffer.

Another matter of great importance is the availability of cellulose or cellulosic materials for feedstocks for alcohol production. The most desirable, of course, are waste materials which have little or no other value. Since wastes are widely scattered the cost of picking them up is a significant item. Therefore wastes which are already collected for other reasons are naturally preferred. In Table V are set forth the major sources of cellulosic materials in the United States. It is interesting to note that the total annual waste production, both collected and uncollected, represents about 9 quads of energy — about one-ninth of total energy demand in the U.S. Thus, as was said at the outset, biomass is not a complete answer to the energy problem, but it can make a contribution to the ultimate solution.

TABLE V

Availability of Cellulose Waste in the United States

(Tons per Year)

Collected	
Municipal Solid Waste	140,000,000
Pulp and Paper Mill Waste	3,000,000
Selected Agricultural Waste	1,000,000
Uncollected	
Forest Residues	145,000,000
Wood Processing	20,000,000
Agricultural Waste	300,000,000

Of the already collected materials, the most abundant by far is municipal solid waste. The 140 million tons indicated would theoretically provide 10 billion gallons of alcohol annually. The portion of MSW which is desired as a feedstock is the organic or combustible fraction. A number of communities in the U.S. have refuse recycling plants which separate this material, and other valuable materials as well, from trash as collected. The separation of the organic fraction is accomplished through air classification and the product is generally known as refuse derived fuel (RDF). RDF is seeking a market as a boiler fuel, but it is certainly not as convenient as more conventional fuels. Its price per BTU therefore is sharply discounted from the BTU price of coal.

In Table VI is given the average composition (dry basis) of municipal solid waste in the United States. The source of these figures is the U.S. Environmental Protection Agency (9). About 75% of the total weight is the air-classified RDF fraction. This fraction is typically 53% cellulose and it is this composition on which the alcohol plant design and economics are based.

TABLE VI

Average MSW Composition in the U.S.

Weight percent (dry basis)	
Paper	35
Yard Wastes	15
Food Wastes	15
Misc. Organics	11
Glass	9
Ferrous metals	7
Rubber/Leather	3
Plastics	2
Textiles	2
Non-ferrous metals	1

Ethanol Markets

In the United States there are two distinct markets for ethanol: industrial alcohol and alcohol fuels.

The industrial demand for ethanol is 250 million gallons per year, with a growth rate of 2% to 3%. It is used principally as a solvent although it has other specialized uses. Ethylene-based alcohol satisfies about 80% of this demand. As the prices of ethylene and other petrochemicals rise, reflecting the increased cost of crude oil, biomass derived ethanol will regain its former place (pre-petrochemical industry) as a building block chemical. Many chemicals such as acetaldehyde and acetic acid will be more economically produced from ethanol than from ethylene. Consequently, the growth rate of industrial ethanol should far exceed its present 2 to 3%.

The alcohol fuel market is now somewhat less than 100 million gallons annually. This market is limited only by the availability of alcohol, and the potential demand is only a matter of conjecture. The present "Gasohol" movement in the United States promises to provide a demand that will not be satisfied until many large biomass-based ethanol plants are built. One of the intriguing properties of alcohol blended with gasoline is the octane improvement it brings. The research octane blending value of ethanol is 135, the road octane blending value is 120. Thus, the admixture of 10% of ethanol to a 90 octane gasoline would be expected to increase road octane by 3 points to 93. This fact makes alcohol worth more than merely a one-for-one replacement of gasoline (2). Since the consumption of gasoline in the U.S. is more than 100 billion gallons per year, an eventual market for alcohol fuels in the vicinity of 10 billion gallons is not inconceivable.

Research Recommendations

In conclusion, mention might be made of several areas of research investigation which would have a high pay-out in the area of ethanol from biomass. First, organisms should be discovered or developed, through modern genetic engineering techniques, which operate efficiently at higher temperatures. The use of such creatures would have the dual effect of increasing rates, thus overcoming one of the drawbacks of fermentation processes, and of discouraging the existence of contaminating organisms which pose a constant hazard to any fermentation process. Second, means or organisms should be sought which convert xylans or five carbon sugars to either gluscose or to ethanol directly. Such a capability would increase the yield of ethanol from wood or municipal solid waste by 50%. In the case of farm residues such as corn stalks, wheat straw, sugar cane bagasse, because of the high hemicellulose content of these grasses, yields of ethanol would be multiplied two and a half times. Third, research in the area of the use of lignin for its valuable chemical values should be encouraged. Lignin can be recovered as a separate entity in most of the processes which are being studied, but is merely being thought of as a plant boiler fuel. The recovery of higher grade chemical values would substantially improve ethanol process economics.

REFERENCES

1. Katzen, R., et al., "Technical and Economic Assessment - Motor Fuel Alcohol from Grain and Other Biomass," Proceedings of the Alcohol Fuels Technology Third International Symposium - Volume 11, Asilomar, California, May, 1979.

2. U.S. Department of Energy, "The Report of the Alcohol Fuels Policy Review", DOE/PE-0012, June, 1979.

3. Lipinsky, E. S., "Perspectives on Preparation of Cellulose for Hydrolysis," Advances in Chemistry Series 181, pp. 1-23, American Chemical Society, Washington, D.C., 1979.

4. Millett, A. M., et al., "Influence of Fine Grinding on the Hydrolysis of Cellulosic Materials—Acid vs. Enzymatic," Advances in Chemistry Series 181, pp. 71-89, American Chemical Society, Washington, D.C. 1979.

5. Huff, G. F., et al., "Enzymatic Hydrolysis of Cellulose", U.S. Patent 3 990 945, assigned to Bio Research Center Co., Ltd., Tokyo, Japan, 1976.

6. Gauss, W. F., et al., "Manufacture of Alcohol from Cellulosic Materials Using Plural Ferments," U.S. Patent 3 990 944, assigned to the Bio Research Co., Ltd., Tokyo, Japan, 1976.

7. Humphrey, A. E., "The Hydrolysis of Cellulosic Materials to Useful Products," Advances in Chemistry Series 181, pp. 24-53, American Chemical Society, Washington, D.C., 1979.

8. Emert, G. H., et al., "Economic Update of the Bioconversion of Cellulose to Ethanol," AICHE Meeting, San Francisco, CA., Nov., 1979.

9. Environmental Protection Agency, Office of Solid Waste, Fourth Report to Congress, "Resource Recovery and Waste Reduction: Chapter 2, Post-Consumer Solid Waste Generation and Resource Recovery Estimates," (SW-600) 1977.

PROSPECTS FOR PHOTOVOLTAIC CONVERSION OF SOLAR ENERGY

SAMIR A. AHMED
City College of New York

ABSTRACT

This paper examines the status of developments in the use of photovoltaics for the conversion of solar energy into practically usable electric power. The fundamental features and limitations of photovoltaic solar conversion are discussed and developments in cell efficiencies, costs, and manufacturing processes are examined for single crystal and thin film cells. The use of concentrators to improve the overall efficiency of photovoltaic solar conversion is explored, including recent developments in this area with multicolor cells, flat plate dye concentrators, and thermophotovoltaic converters.

The different aspects of photovoltaic conversion are then combined and examined as systems from the economic viewpoint, for their potential to compete as alternate sources of usable energy. It is then seen, that photovoltaic solar conversion, while attractive in principle—as it's energy source is abundant, inexhaustible, and clean—has many hurdles to overcome before it can become a significant viable alternate energy source. Basic problems are efficiency and cost. The developments required to improve them make it unlikely that photovoltaic solar conversion will play a significant role as an energy source before the year 2000. Beyond that period it is possible that photovoltaic plants will start to be competitive with coal and nuclear plants for central electricity generation, which will be the market they must penetrate if they are to make a significant contribution as a viable alternate energy source.

ISBN 0-12-467101-2

1.0 INTRODUCTION

In the search for alternative energy sources increasing attention has been focused on solar energy for very basic reasons: it is plentiful, clean and inexhaustible. At noon, on clear days about 1000 W/M^2 irradiates the earth's surface. The sunlight falling on the U.S., for instance, represents about 5×10^4 Quads per year (1 Quads = 10^{15} Btu,and 1 Btu = 1054 Jouls). The total annual U.S. energy usage is about 80 Quads. Thus, if solar energy could be economically collected and converted to enable energy forms with a conversion efficiency of 10 percent, (a level readily attainable in photovoltaic conversion devices), the solar energy falling on 1 percent of the land surface of the U.S. would suffice to meet national energy requirements.

There are a variety of technologies which can derive end-use energy from solar insolation. Hydroelectric generation for example is well established and widely used, domestic solar heating presents important opportunities for energy substitution, wind power, important in the past, may see a revival as energy costs goes up.

Among the several alternative technologies for solar energy conversion, however, photovoltaic electric systems have some unique and attractive features. Photovoltaic systems convert light falling on a flat semiconductor material to electricity directly. They therefore do not require intermediate stages of conversion to thermal and mechanical energy prior to conversion to electrical energy. With no moving parts involved, photovoltaic components and systems should potentially have long lifes and possibley enjoy lower maintenance costs.

Furthermore, in a photovoltaic electricity generation system, the actual photovoltaic conversion device, a semiconductor cell or an array of cells, is inherently modular, and will convert solar power to electricity whether it is a few watts or megawatts. As a consequence, a wide variety of applications can be considered, ranging from small remote systems for isolated communities or use to large central generating stations (where some economies of scale would apply).

In spite of these advantages the use of photovoltaics remains very limited. The primary reason for these limitations is cost, though considerable cost reductions have been achieved in recent years, particularly in the area of cost of the photovoltaic cell modules themselves that are at the heart of a photovoltaic conversion system.

Thus, costs of a peak kilowatt of photovoltaic generating capacity are down from about $200,000 in 1959, to a few thousand dollars at present. [Peak kilowatt output (kWp) for a photovoltaic system is defined as the output of a photovoltaic system with its collector cell modules perpendicularly oriented to the sun's rays at noon on a clear day]. While a considerable improvement on earlier costs, present costs are still about an order or magnitude higher than the $300-600/kWp typical for conventionally generated electric power.

With present costs, photovoltaic solar conversion remains restricted to space use, and some remote applications where difficult access for maintenance or fuel, or cost of connection to an electricity grid makes photovoltaics a viable alternative. There are, for instance, estimates of over one million primarily rural communities in developing countries for whom photovoltaic systems would present a viable means for meeting electricity needs.

Examination of photovoltaic system costs shows that production costs for the photovoltaic cell modules themselves, are still high. There are also additional non-negligible costs for the associated power conditioning and storage required to make the electricity produced by the photovoltaic cells practically usuable and reliable, and less susceptible to the variations of available sunlight.

Cell module costs are also high, since it is expensive to process and refine silicon, the primary material presently used, to the degree of purity necessary, and from it to grow the high quality crystals needed for present generation photovoltaic devices. Furthermore, both these processes, which are very energy—intensive, and the complementary slicing process required for producing the silicon crystal wafers required for the modules, are wasteful of material.

Traditionally, the electronic semiconductor industry has been able to cut costs through miniturization. Unfortunately this route is not available for photovoltaic devices. The electrical output generated by a photovoltaic solar cell is directly proportional to the amount of sunlight intercepted and converted by the cell. Reducing the cell area, therefore,directly reduces its capacity to intercept and convert light, i.e., it reduces its generating capacity.

There are several approaches possible to reducing costs and making photovoltaic systems more attractive:

(i) There is still much development work that can be done to improve production techniques and reduce costs for silicon and other single crystal solar cells that are the dominant type at present.

(ii) Efficiencies of single crystal cells still have a considerable potential for improvement. Presently, 12 percent is readily attainable in commercial single crystal silicon cells, with goals set at 22 percent (versus a 29 percent theoretical limit). GaAs cells, which are more expensive, have already experimentally achieved 23 percent (versus a theoretical maximum of 37 percent).

(iii) Thin film solar cells are also attracing great interest because because of their potentially low costs of production by vapor deposition, plasma deposition, and similar techniques. These "thin film" cells are to be differentiated from single crystal cells not by their thicknees(since both are thin), but rather by their material of fabrication and their polycrystalline or disordered characteristics. Efficiencies attainable with thin film cells are still presently rather low, with much research and development work being devoted to their improvement.

(iv) Concentrator systems appear to offer the greater potential for realizing cost - effective photovoltaic systems. These systems collect light (e.g., with a heliostat) from a large area and concentrate it onto a solar cell for conversion.

A concentrator system therefore makes more effective use of the photovoltaic conversion cell, and reduces its relative contribution to the overall system costs. It also puts a premium on the photovoltaic conversion cell efficiency.

For these applications it then becomes more viable to consider more complex cells yielding higher efficiencies. These include multilayered, multifunction cells, that can, in principle, make fuller use of the solar spectrum than single junction cells, which are better suited to convert limited parts of the spectrum while wasting others. Theoretical efficiencies of multijunction cells are over so percent(versuse 36 percent for single) crystal GaAs, which has best match for utilizing the solar spectrum).

Thermophotovoltaic devices are another approach that can be used with concentrator systems to absorb the total light received and remit it at a wavelength that optionally matches the characteristics of a single-junction cell, while recirculating radiation that is not initially absorbed and converting it to a wavelength that matches the cell bandgap[1].

In addition to work on the development of cells to be used with concentrator systems, there is much work going on concentrator research and development. Particularly interesting ongoing work for instance, is the investigation of flat plate concentrators with active absorption and reemission.

Large-Scale Photovoltaic Electricity Generation

While there are other limited uses and markets for it, the widespread use of photovoltaic energy conversion ultimately depends on it becoming competitive with conventional central power generating systems such as coal and nuclear. On-site photovoltaic conversion, even in conjunction with cogeneration for heating does not appear too viable for residential application, though on-site photovoltaic systems with cogeneration appear to have some economic potential for commercial sites. (Fig. 1.1 shows[3] the performance chain that would be required for a typical solar photovoltaic power plant).

Analysis of photovoltaic systems and potential development (Section 5), show that two approaches may lead to economically competitive large-scale utility applications by the year 2000. In particular, thin film cells used in flat plate collector systems without concentrators appear as potentially viable[2], if conversion efficiencies of over 10 percent are attained at flat plat cell module costs of no less than $\$25/m^2$, and total capital investments of less than $1000/kWh.

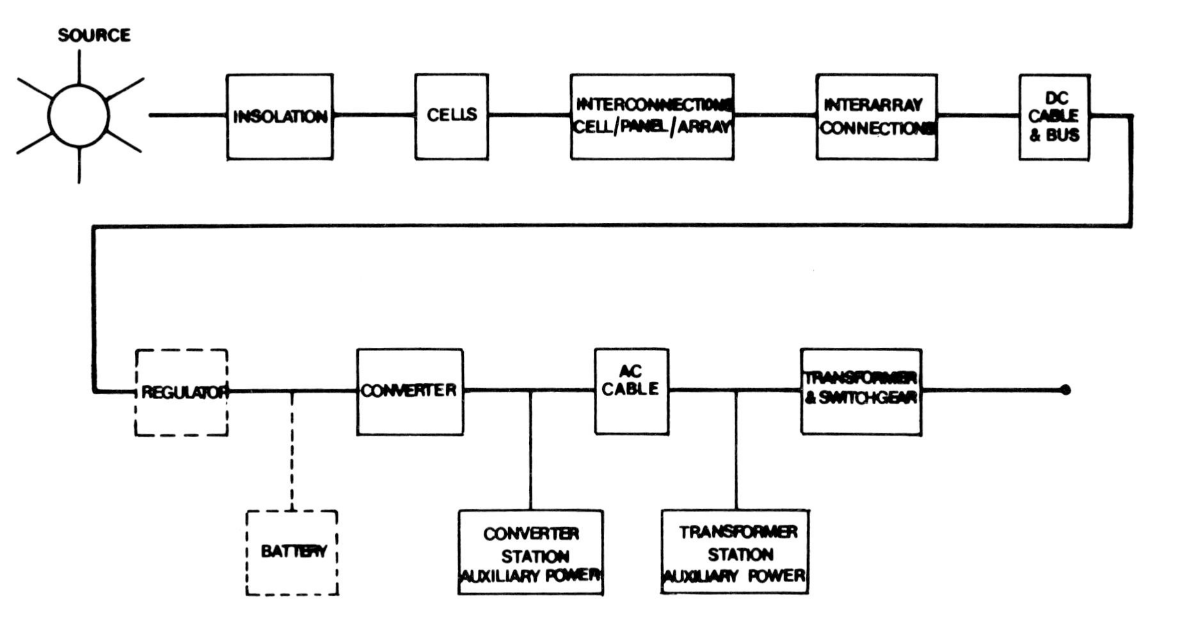

Figure 1.1. PV plant performance chain Ref. 3

The other approach that appears to have potential for economically competitive large-scale power generation in the future, is based on the use of high concentration systems, which concentrate the solar radiation by a factor of 50-500 onto high efficiency photovoltaic converters. Since, under these circumstances, the photovoltaic device itself will represent a smaller fraction of the system cost (as compared with a flat plate system), the use of higher quality, more elaborate, and expensive photovoltaic cell converters becomes viable. This would be the case with single crystal silicon modules, if costs can be kept below $\$50/m^2$, and efficiencies improved to 20 percent; GaAs cells, at module costs of $\$60-70/m^2$ and efficiencies of 23-29 percent; or multijunction cells, at $\$70-90/m^2$ and efficiencies of 27-33 percent, with capital investments[2] for the three types in the vicinity of $800/kWp.

It should be noted that it will be quite some time before photovoltaics can have a significant impact on the energy picture. It is expected that it is only by 1990 that the technology required for large-scale economical photovoltaic systems will have been largely developed. Based on that assumption, and the requirements for building up productive capacity, necessary raw materials, and rates of production, it appears unlikely that significant penetration of the central electricity generation market will occur before the end of the century.

Even a 1 percent penetration of the U.S. power generation market by photovoltaics by the year 2000 would place a strain on the necessary resources. Thus, it is estimated that $2\times10^{10} W_p$ of photovoltaic capacity would be required to generate about 1 percent of estimated U.S. electricity requirements in the year 2000 (approximately 5×10^{12} kWh). If production were started in 1990 this would require an annual increase in photovoltaic capacity of $2\times10^9 W_p$.

Establishing this capacity would require more than doubling present annual silicon production, if a silicon-based technology were utilized. Even more common material requirements would be substantial. For a flat plate system for instance, total requirements for 1 percent penetration of U.S. electricity production were estimated at approximately 5 percent of annual comsumption for steel, and 17 percent for cement.

Because of their diurnal variations, solar plants could be expected to replace intermediate conventional generation plants. Fortunately, the daily solar radiation cycle does correspond generally with demand on intermediate load generating plants. However, if solar power does make significant inroads into central electricity generation, storage capacity or auxiliary generation capacity would have to be built up in order to ensure reliability for cloudy periods.

Given the time-frame required for the development of solar technology, production facilities, and necessary cost reductions for photovoltiac solar power to be competitive with conventional electricty generation, it appears likely that if photovoltaics are to significantly penetrate the electricity generation market, it would be in the 2000-2020 period. At that time, central electricity production in the U.S. is expected to be very largely free of any dependence on liquid fuels. Penetration by photovoltaics at that time would therefore be primarily by the displacement of coal or nuclear capacity.

2.0 BASIC FEATURES OF PHOTOVOLTAIC SOLAR CELLS

2.1 INTRODUCTION

Solar cells are photovoltaic devises in which photons with energies above a threshold (bandgap) value are absorbed to produce, via electronic excitations, electron-hole pairs of charge carriers (Fig. 2.1). By establishing an internal electric field in the device, these optically generated charge carriers can be separated, collected by electrodes, and made to flow as a useful current in an external circuit, thereby directly achieving light to electricity conversion.

The internal electric field in a photovoltaic device can be established in several ways, including p-n junctions, Scholtky barriers, and semiconductor-liquid-electrolyte interfaces.

In a p-n junction, solar cell energy conversion starts with the penetration and absorption of a photon of sufficient energy in the p (and/or n) material creating an electron-hole pair. The freed electron then moves across the junction, into the n material; then, via the contact electrode on it, through the external circuit, constituting an electric current, back through the other electrode into the p material to recombine with a hole and complete the circuit.

In a Scholtky barrier device, the electric field is at the inerface of a high work-function metal and the semiconductor. Thus, illuminating a metal semiconductor sandwich with a Scholtky barrier in it, will cause electrons to flow into the semiconductor and holes to flow into the metal, producing a current in an external circuit.

In a semiconductor-liquid-electrolyte inerface, the electrolytes play the same role that metals do in Scholtky barrier devices. In this case, however, electrons and holes entering the electrolyte can be used to produce chemical reactions which produce fuels directly, making these interesting systems with potential for producing liquid fuels directly from sunlight.

2.2 FUNDAMENTAL FACTORS LIMITING EFFICIENCY OF PHOTOVOLTAIC CELLS

With even an ideal photovoltaic cell much of the solar spectrum falling on it is wasted. Photons with energies below the threshold (bandgap) value are not absorbed and hence cannot be used for the generation of photovoltaic currents. Photons with energies equal to or higher than the threshold, bandgap energy, are absorbed. However, only the fraction of their energy that is equal to the bandgap energy is usefully utilized for the creation of electron-hole pairs, and hence for the useful production of photovoltaic currents. The balance of the energy, beyond the bandgap energy, is dissipated as heat. (See Fig. 2.2 and 2.3)

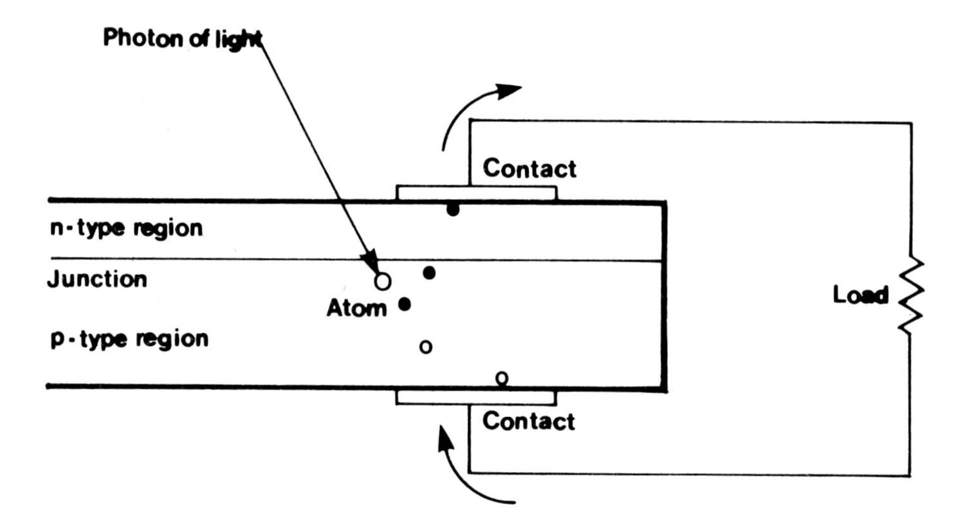

Figure 2.1 Ref. 6

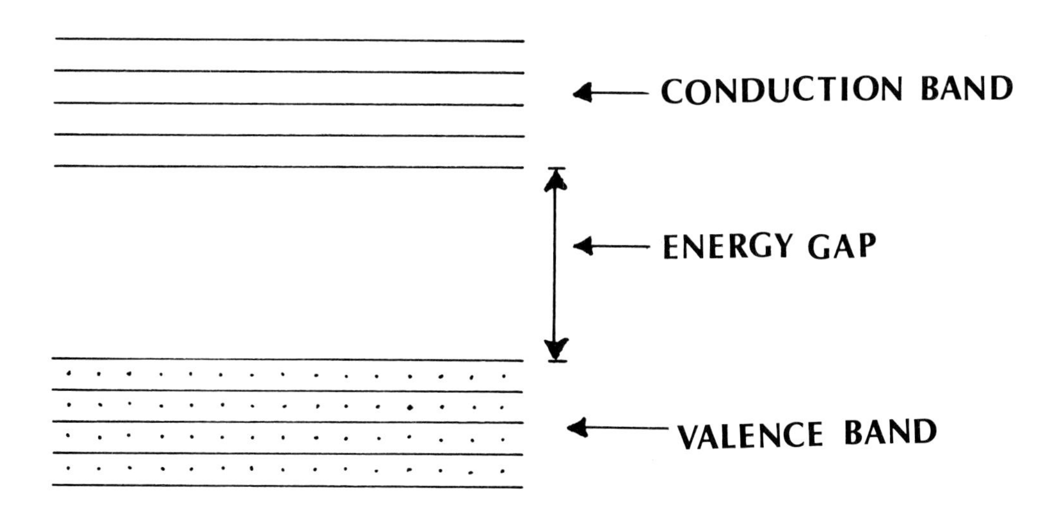

Fig. 2.2 Energy-levels diagram for a semi-conductor

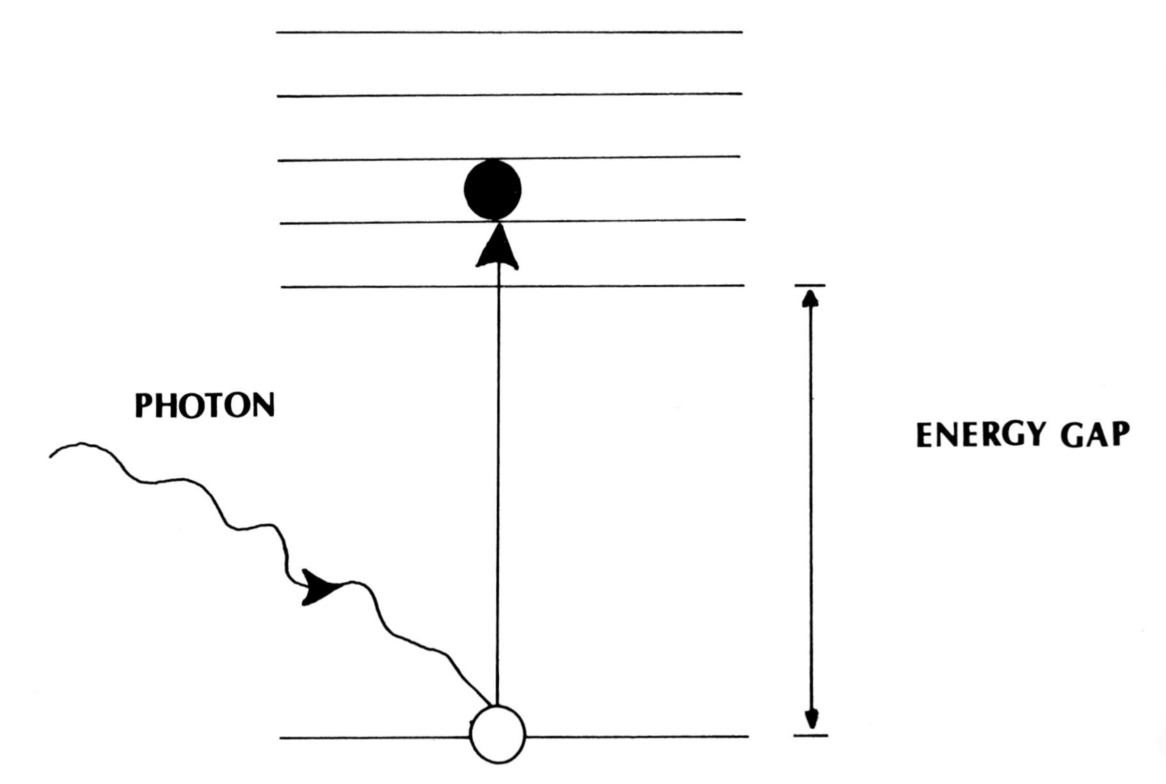

Fig. 2.3 A photon-electron collision can activate an electron across the energy gap. [an electron-hole pair is created if photon energy > E_g]

Consideration of these loss factors, and their relationship to the wavelength or energy distribution of the solar spectrum, permits definition of an optimum bandgap of aaproximately 1.5 ev. Thus, an ideal cell with a bandgap of 1.5 ev would be best matched to utilize the solar spectrum to a maximum. For instance, for gallium arsenide, with a bandgap of 1.4 ev, the theoretical ideal cell efficiency is 36 percent. For silicon, with a bandgap of 1.1 ev, it is 29 percent. Fig. 2.4 shows the relationship between bandgap and ideal theoretical efficiency for conversion of the solar spectrum.

Some of the above limitations on efficiency arising from incomplete spectrum utilization may be overcome by using more complex multilayered cells, discussed in more detail in Section 3.

Multilayered cells would consist of several layers of cells, each with a different bandgap. These cells would be connected in series, with the cell layer with the largest bandgap intercepting the incident light first, with the succeeding layers of cells absorbing and converting the light which was not absorbed in the first layers. With a sufficient number of layers much fuller use can be made of the incident solar spectrum. Efficiency goals for multilayers cells are nearly 40 percent[2], considerably higher than those for single layer cells.

In addition to the theoretical limitations on single cell efficiency due to mismatch between bandgap energy and solar spectrum energies, there are additional factors which further limit efficiencies in practical cells. Two of these factors are: undesirable front surface recombination on the one hand, and incomplete absorption of the radiating photon flux on the other.

In particular, photons with energy much higher than the bandgap are often absorbed so near the front surface of the photovoltaic cell that the resulting electron-hole pairs recombine near the surface without flowing through the junction and contributing to the cell current.

Photons with energies just above the conduction band, on the other hand, are often only weakly absorbed, and thus a fraction of them can pass right through the cell material, effectively without contributing to the creation of electron-hole pairs and cell current. Both these factors further reduce the amount of the solar spectrum that can be usefully converted by a photovoltaic cell.

Still other factors limiting practical cell efficiencies well below the theoretically attainable limits include reflection of radiation at the cell front surface and the interception of some of the radiation by the network of metallic electrodes normally on the surface to collect the cell current.

The losses from front surface reflections can be greatly reduced through the use of anti-reflex coatings and textured surfaces, while some cell designs also eliminate the front surface gridding completely, though at a cost in cell complexity.

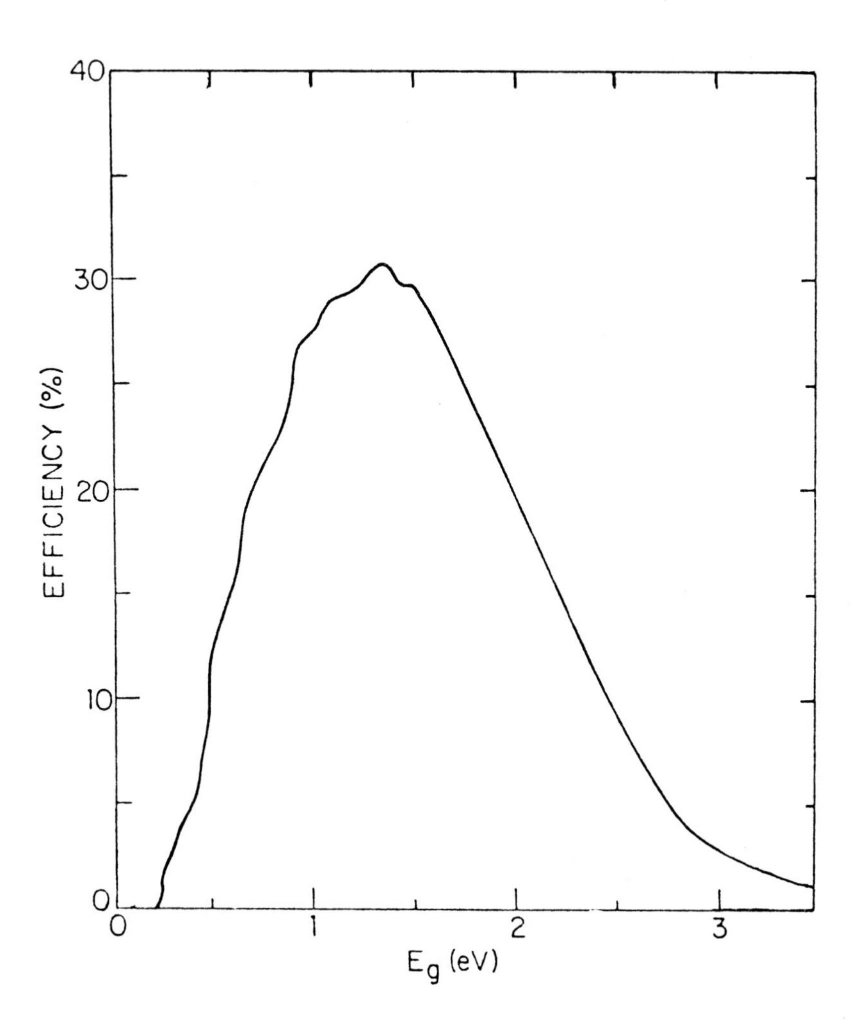

Figure 2.4. Ideal solar cell efficiencies at 300 K (oscillations are due to atmospheric absorptions)

Ref. 2

3.0 PRINCIPAL TYPES OF PHOTOVOLTAIC SOLAR CELLS AND THEIR CHARACTERISTICS

3.1 INTRODUCTION

Photovoltaic cells can be placed in two broad categories: (a) cells made from single crystals and (b) so-called thin film cells. Since both categories are made in thin layers, it is important to understand that in the context of photovoltaics the term "thin film" is more associated with the method of production and the characteristics of the cells produced rather than with the thickness of the cells.

Thus, "thin film" refers to cells where the active layers are polycrystalline or disordered films which have been deposited or formed on substrate using techniques such as evaporation, spultering, spray pyrolisis, plasma deposition, and electrolysis. (It should be noted, however, that some of these techniques, such as evaporation and molecular beam epilaxy, can and are also used to produce thin films of high crystalline quality, e.g., for the highest efficiency GaAs - GaAlAs cells though these are not technically called "thin film cells").

Single crystal cells are characterized by higher efficiencies than thin film cells. It is expected that research on thin film cells will gradually improve their efficiencies, though not to the level of single crystal cells. It is the potential for low production costs that motivates much of the effort in thin films.

With economic considerations in mind, it appears likely as will be seen in Section 4, that thin film cells will have the edge for use in flat plate collectors, while single crystal cells with their higher efficiencies will be used in concentrator systems, where the costs of the photovoltaic solar cell is only a small component of the total system cost.

3.2 SINGLE CRYSTAL PHOTOVOLTAIC CELLS

3.2.1 Single Crystal Silicon

Single crystal silicon represents the most highly developed material for photovoltaic conversion at present. The raw material for it is abundant, and the processed material is readily available commercially, with a well established manufacturing technology and well understood physical properties. Furthermore, with a 1.1 ev bandgap, situated fairly well with respect to the solar spectrum, silicon-based photovoltaics have a relatively high theoretical efficiency, as discussed in Section 2, of 29 percent. In practice, reproducible conversion efficiencies of approximately 12 percent are readily attainable, while experimental efficiencies of 17 percent have been attained. It is estimated that efficiencies of 18 percent and ultimately 22 percent may be eventually attained in quantity production[2].

The most serious obstacle to the economic utilization of single crystal silicon is, as with all photovoltaics, its high production costs. Considerable effort is being put into development of lower cost fabricating techniques, with the optimum approach still to be identified. U.S. DOE cost goals are $ 40 per square meter by the mid-1980s.

Other lesser problems relate to the protection of the surfaces by encapsulation, in tolerance to temperature, and long life.

To produce single crystal silicon cheaply it is first necessary to have low cost, solar grade polycrystalline silicon, with enough purity to be used in the production of single crystal cells. There are several processes being investigated for producing solar grade Si at less than $10/kg in quantity production by the late 1980s.

There are presently two main approaches being followed for obtaining low cost single crystal or semicrystalline silicon cells. They are: (a) selecting single crystal wafers from large diameter single crystal ingots grown in a costly Cyochraleki growth procedure, and (b) ribbon growth processes which produce near-crystalline silicon in a continuous ribbon.

A third alternative also being examined is to cast polycrystalline silicon into rectangular bricks, which are then sawed into wafers.

The first approach produces the highest efficiency cells, while the other two have potential of lower production costs with, however, reduced efficiency. Table 3.1 from Ref. 2 shows estimates of costs that might be attainable in the late 1980s for single crystal Si cell modules made from slicing single cell ingots produced by mutiple pulls from the same crucible. Multiblade slurry sawing is assumed. Multiblade saws under development are expected to attain the low slicing costs indicated by cutting up to 1000 wafers simultaneously. However, the wafer slicing process is inherently wasteful because of material wastage[4].

Ribbon technology[2] is capable of continuous production of ribbons typically 8 cm wide and yielding 0.121m^2 per hour. Cell efficiencies of Si ribbon material are also considerably lower, 8-12 percent, compared to the 14-18 percent attainalbe with single crystal devices. There are problems in maintaining ribbon quality in continuous operation. Multiple-ribbon machines are being developed and are expected to reduce costs. Estimates in Ref. 2 put costs of modules produced from ribbon grown material at $0.73/$W_p$ (in 1975 dollars), compared to the $0.62/$W_p$ for the multiblade sawing of single crystals(Table 3.2).

Cast polycrystalline Si in rectangular brick form produces low cost Si sheets when sliced into thin wafers. However, cell efficiencies obtained with this material are considerably lower, and the cost of the finished module is dominated by the fabrication costs. Estimates for polycrystalline silicon cost[2] are $0.75/$W_p$, again higher than the $0.63/Wp estimated for cells produced from multiblade slicing of single Czochralski grown crystal ingots.

Further possibilities for cost reduction require investigation of manufacturing procedures. Float-zone growth of single Si crystals is a well established process that may eventually offer advantages over the Czochralski process.

TABLE 3.1

Module Cost Components for Cells Made by Multiblade Sawing of Continuous-Czochralski Single Crystals

Process	Yield	Process Cost	Yielded Cost Contribution per Unit Module Area	Cost Contribution per Peak Watt
Poly	1.0	8 \$/kg	13.0 $\$/m^2$	0.081 $\$/W_p$
Crystal	0.8	10 \$/kg	16.3	0.102
Slice	0.95	18 $\$/m^2$	19.9	0.125
Cell	0.9	20 $\$/m^2$	21.1	0.132
Module	0.95	28 $\$/m^2$	29.5	0.184
Total			100 $\$/m^2$	0.62 $\$/W_p$

(1975 dollars)

The following assumptions are made in this table:

16% module efficiency
Cells pack 90% of the module area
Crystal silicon yields cell wafers at .85 m^2/kg, or 20 slices/cm of crystal.

Ref. 2

TABLE 3.2

Module Cost Components for Cells
Made by a Multiple-Ribbon Process

Process	Yield	Process Cost	Yielded Cost Contribution per Unit Module Area	Cost Contribution per Peak Watt
Poly	1.0	8 \$/kg	5.5 $\$/m^2$	0.046 $\$/W_p$
Ribbon	1.0	25 $\$/m^2$	29.2	0.244
Cell	0.9	20 $\$/m^2$	23.4	0.195
Module	0.95	28 $\$/m^2$	29.5	0.246
Total			88 $\$/m^2$	0.73 $\$/W_p$

(Note: This cost estimate is based upon ribbon technology process cost which is judged to be significantly lower than can currently be anticipated. See text.)

(1975 dollars)

The following assumptions are made in this table:

12% module efficiency
Cells pack 100% of the module area.
Polycrystalline silicon is converted to 250 micron ribbon, giving a conversion factor of 1.7 m^2/kg.

Ref. 2

Module component assembly, fabrication, and encapsulation also are important contriburors to cost, and possibilities exist for lowering these, such as, encapsulation of cells in tubular glass enclosures of the type used for fluorescent lighting. Other improvements can also be made in cell qualities. Light reflection at the surface can be reduced or eliminated by antireflection coatings or textured front surfaces which effectively trap the light, thereby reducing reflection.

Because of its well established place in photovoltaic, as well as the relative abundance of raw materials silicon-based technology is likely to continue to dominate the solar cell field for some time to come, for both flat cell and concentrator systems. Significant penetration of electricity generation at the turn of the century, if it comes, is likely to be based on silicon photovoltaics.

There are, however, photovoltaic cells based on other semiconductor materials which are capable of higher efficiencies than silicon-based cells. The relatively high cost, however, makes it unlikely that they will displace silicon easily, except perhaps in high concentration systems where cell efficiency is at a premium, and cell costs relatively unimportant. GaAs-based photovoltaic cells, which have a high theoretical efficiency, are discussed briefly below.

3.2.2 Galium Arsenide

With a bandgap of 1.4 ev, near the theoretical optimum for the response of a semiconductor to the solar spectrum, GaAs photovoltaic cells have a high ideal theoretical conversion efficiency of 36 percent. As a result, GaAs photovoltaics have received considerable attention.

High photovoltaic conversion efficiencies over 20 perecnt, have been demonstrated in GaAs test cells[5]. As well as its potential for high efficiency, GaAs has another advantage that is important particularly in conjunction with concentration systems. It has a low electrical resistance in the junction region which permits the cell to retain its efficiency at higher temperatures.

Major problems of GaAs cells are likely to be costs, manufacturing reproducibility, lifetime, and reliability factors. Costs, however, are the more serious drawback, with estimates for eventual costs still one to three orders of magnitude above those of silicon-based photovoltaics. These considerations would appear to restrict the application of GaAs to concentrator systems where the conversion module efficiency is at a premium and costs of the module itself are only a relatively small part of the total.

3.3 THIN FILM PHOTOVOLTAIC CELLS

3.3.1 Introduction

As discussed earlier in Section 3, thin films cells, in the context of photovoltaic, refers to cells where the active layers are polycrystalline or disordered films. The active layers can be deposited in several forms to obtain p-n junctions, Scholtky barriers, or MIS structures where metal and semiconductor are separated by a thin insulating barrier.

The driving force behind the interest in polycrystalline thin films is their potential for cost reduction. Thus, cost goals an order of magnitude lower than those for single crystal silicon appear feasible ultimately.

However, much lower demonstrated conversion efficiencies, less than 5 percent, for polycrystalline silicon to a large extent offset potential cost reductions. Considerable research is required on basic factors, such as, grain size, boundary effects, etc., before optimum manufacturing processes can be developed to take advantage of the potential low costs. Some of the thin film cells which have been recently attracting attention are discussed below.

3.3.2 Amorphous Hydrogenated Silicon Cells (a-SiHx)

a-SiHx has properties which in principle are well matched to solar cell requirements. The bandgap can be adjusted, by varying hydrogen concentration, so that it is at or close to the ideal for solar spectrum utilization. Cells can be produced with either p or n doping and have a high optical absorption coefficient which can be optimized by the addition of C or G during the fabrication process. Production techniques are low cost and amenable to mass production, and include planner deposition and other vapor deposition techniques.

Efficiencies for a-SiHx cells are still low however. About 6 percent has been obtained from MIS cells made for a-SiHx. Scientifically, there are still many open questions on the potential for a-SiHx cells. These include reasons for the low efficiency, and possible lack of stability, which is still too low for solar applications. Much fundamental research remains to be carried out on these types of cells before these questions can be settled.

3.3.3 Polycrystalline Silicon

Another type of production process for photovoltaic cells that has potential for low cost is the vapor deposition of pure polycrystalline silicon thin films on metallurgical grade silicon substrates, with the utilization of acid for the extraction of impurities. Efficiencies of nearly 10 percent have been obtained with such cells and further improvements appear possible. Possible ultimate production costs for this process are still not defined.

3.3.4 Cadmium Sulfide/Copper Sulfide

Cells made from this combination of materials have received attention both because manufacturing techniques for these are relatively inexpensive, and because the high absorbance of the solar spectrum by copper sulfide permits the use of thinner films cells (than silicon) with consequently lower demands on manufacturing costs and on scarce material resources.

Disadvantages of this type of cell are relatively low efficiencies, typically 6 percent, and limited life. Limitations on life arise, it appears, from contaminants, including moisture, which cause chemical reactions and instabilities, which it appears are aggravated at higher operating temperatures. Mid-1980s goals are $10 per square meter and conversion efficiencies of 10 percent.

3.3.5 Cadmium Telluride

With an absorption bandgap of 1.5 ev, close to the theoretical optimum for solar spectrum conversion, cadmium telluride appears attractive as a solar cell material. Furthermore, it has a large optical absorption coefficient and it can be fabricated using low cost vapor deposition techniques into thin film cells.

To date, however, efficiencies attained with CdTe have been dissappointingly low, typically 8 percent, for reasons not fully understood, but possibly related to impurities and defects in the material resulting from the fabrication process.

3.4 MULTILAYERED CELLS

As discussed in Section 2, the relationship between bandgap and solar energy spectrum limits attainable cell efficiencies. Multilayered stacks of cells, with a downward progression of bandgaps, offers the potenial for overcoming some of these limitations to produce compound cells of considerably higher efficiencies.

In these configurations, the cell with the largest bandgap is on top and intercepts the incident light first, converting the higher energy photons, while the remainder pass through to be progressively converted by succeeding layers, with losses due to excess photon energies greatly reduced.

Thus, whereas a single layer cell with an optimally situated bandgap would have an ideal efficiency of 37 percent, a two layer (two bandgap) stack would have ideal efficiency of 50 percent, and a three layer arrangement an efficiency of 56 percent, while eventual practical goals might be 29, 35, and 39 percent respectively.

There are several ways of making stacked cell configurations. Epitaxical deposition of different layers for the different semiconductors may be used. Two photovoltaic junctions have been integrated in series using a tunnel diode. Epitaxical growth also offers some possibilities.

4.0 CONCENTRATOR SYSTEMS FOR PHOTOVOLTAIC SOLAR POWER CONVERSION

4.1 INTRODUCTION

With the high costs associated both with the photovoltaic cells themselves, and with collecting the current generated from large flat plate systems, the use of concentrators in conjunction with high efficiency cells offers attractive alternatives. Analysis of well designed concentrator systems and comparison with flat plate systems show the former are likely to be more cost—effective, with particularly substantial reduction in terms of structural requirements, materials, and costs. Furthermore, because concentrator systems collect and direct the light more effectively onto the photovoltaic cell, overall system conversion efficiencies are higher for systems with some concentration.

In this context, however, there is another competing factor. Cell efficiency, which is constant for lower radiation levels, decreases at the higher intensities attained as the concentration ratio is further increased beyond the optimum. Figs. 4.1 and 4.2 show this effect. (Concentration ratio (CR) is defined as: Aperture Width/Absorber Width). As a consequence, there is an optimum concentration which will depend on the type of cell involved, and its configuration, gridding, etc.

Another factor to be taken ito account in concentrator systems is the increase in temperature of the cell at high insolation intensities, and the drop in efficiency associated with it. For many concentrator systems this will necessitate the incorporation of cooling systems.

Cell efficiency at a given temperature, T, is given by[3]:

$$T = \text{Cell efficiency at } T_o \times \left[1-k(T-T_o)\right]$$

where:

T = operating temperature in °C

T_o= reference temperature, usually 28 °C

k = decrease in efficiency per unit temperature increase.

Silicon has a k of 0.004°C, GaAs 0.0025, Cds 0.003. At 28°C a commercial silicon cell, for instance, would have an efficiency of 12 percent. At 200°C it would have only 8.5 percent.

Some other negative factors that are enhanced by increased concentration ratios include: higher losses because of greater separation between solar arrays and consequently longer connections; more complexity of design and hence higher probability of mechanical problems.

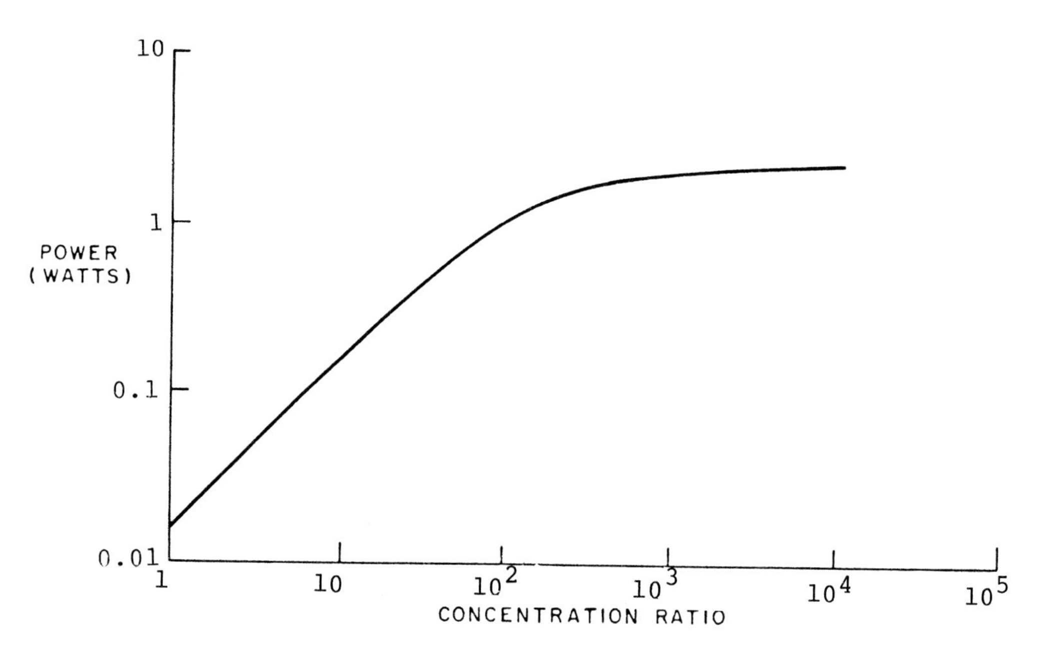

Ref. 3

Figure 4.1. Output power vs. concentration ratio for Si p-n junction

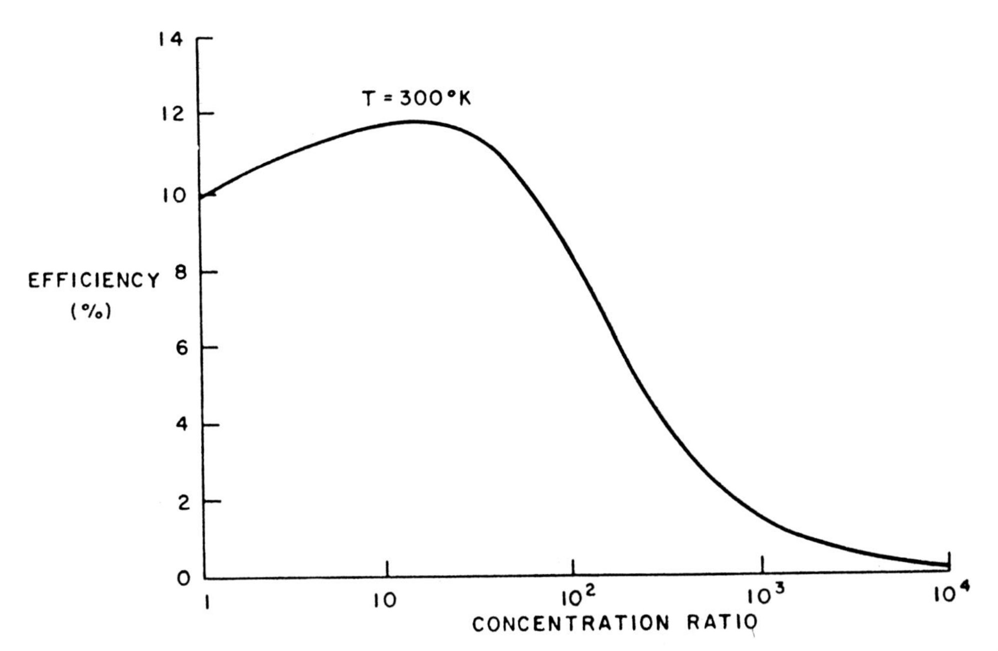

Figure 4.2. Solar power conversion efficiency vs. concentration

Ref. 3

4.2 Concentrator Systems - Overview

The planar wedge (Fig. 4.3), the Fresnal wedge (Fig. 4.4), and the compound parabolic concentrator are three typical low cost, low CR concentrators. While they can be used in a stationary management, with appropriate orientations for the solar insloation, thier performance can be improved by tracking.

Medium concentration systems are the compound parabolic concentrator (Fig. 4.5) and the cylindrical parabolic trough (Fig. 4.6), with concentration ratios of 3-10 and 20-50, and estimated costs[8] of \$3-10/W_p and and \$20-30/W_p, respectively.

Fresnal lens and paraboloid dish concentrators with CR = 20-500 and 100-1000 respectively, represent high concentration systems that require tracking of the sun on two axis, and warrant the use of high efficiency cells. Cost estimates are \$2.50/W_p for the Fresnel systems and \$2.70/W_p for the paraboloid dish.

Characteristics for a low concentration system and a medium concentration system are compared in Tables 4.1 and 4.2, while Figs. 4.7 and 4.8 compare the typical contributions to losses in flat plate fixed-tilt planar systems and those in parabolic dish systems.

More recently, work has been carried out on heliostats where the reflecting element is an aluminized polyester film supported on a light framework within an inflated dome to provide protection from the weather. Concentration ratios range from 50-500 for those lightweight dome-enclosed designs which have estimated[2] costs of \$1.20/Wp.

4.3 Dye Concentrators

The approach used in dye concentrators is based on planar solar collectors in which the vertically incident radiation is absorbed by a luminescent material (e.g., a dye) close between two lying planar surfaces. The luminescent material reemits the light at longer wavelengths. Trapping of the light is then achieved since it is radiated isotropically, with much of it reemitted at angles more grazing to the enclossng surface than the critical angle for total internal reflection, ensuring its trapping between the surfaces.

The trapped light can then be used to irradiate a photovoltaic cell at the edges of the luminescent medium. The advantage of this approach is that luminescent material can be produced at low cost embedded in a plastic material in a planar form, in which it can be used to collect and concentrate the light on an efficient and more expensive photovoltaic cell.

Suitable luminescent planar configurations can be achieved in a variety of ways. A thin plastic sheet with high refractive index and with dye molecules in it is suitable. Much of the light absorbed and reemitted by the dye molecules is reemitted isotropically. Since the sheet is thin, much of this light will be at less than the critical angle for the plastic surfaces. It is therefore trapped and propagated along by multiple, totally internal reflections from both top and bottom surfaces to the edges of the sheet. Some of the edges of the sheet could be made reflective, while others would have efficient solar cells.

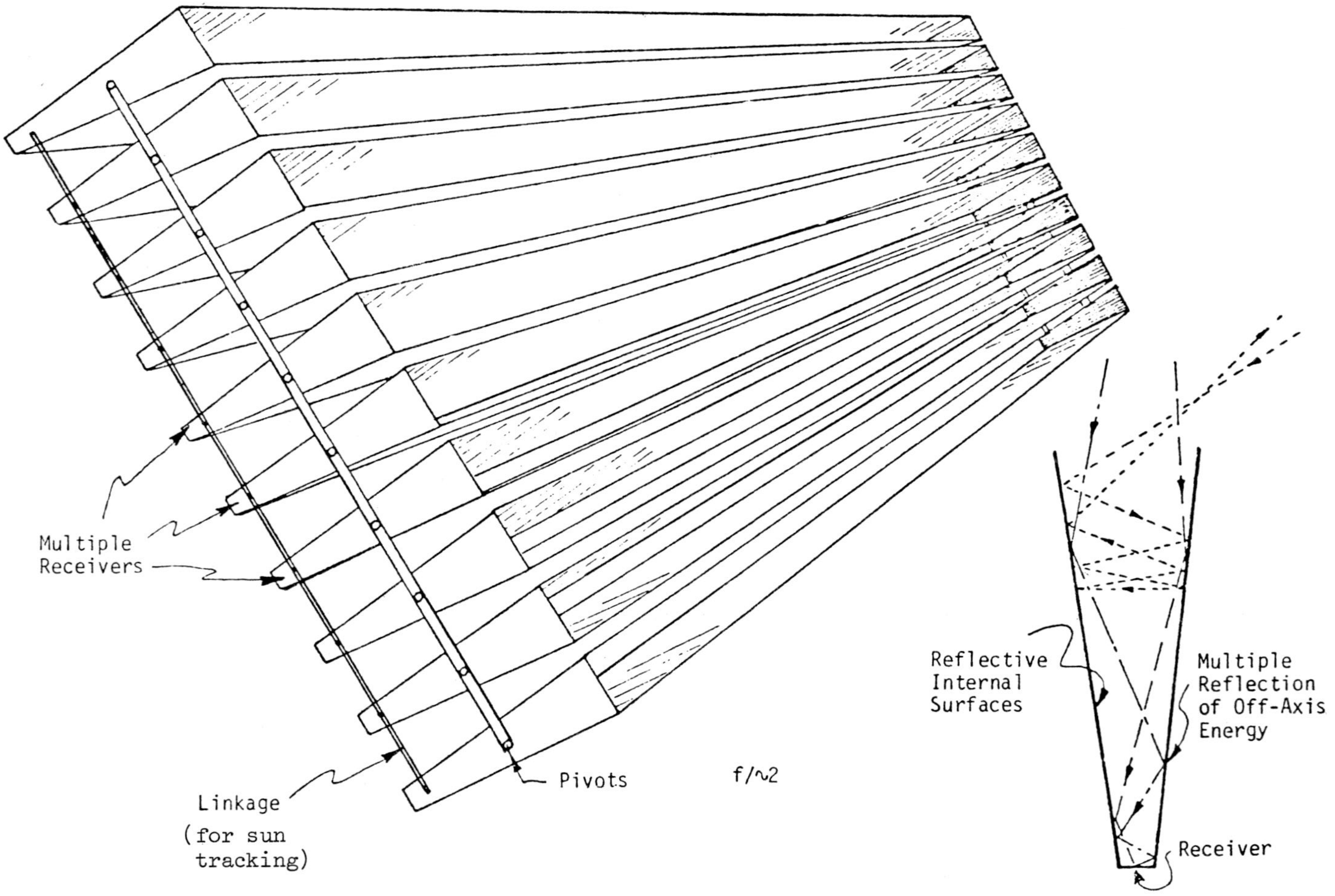

Multiple Receivers

Linkage (for sun tracking)

Pivots

$f/\sim 2$

Reflective Internal Surfaces

Multiple Reflection of Off-Axis Energy

Receiver

Figure 4.3. Planar wedge array

Ref. 3

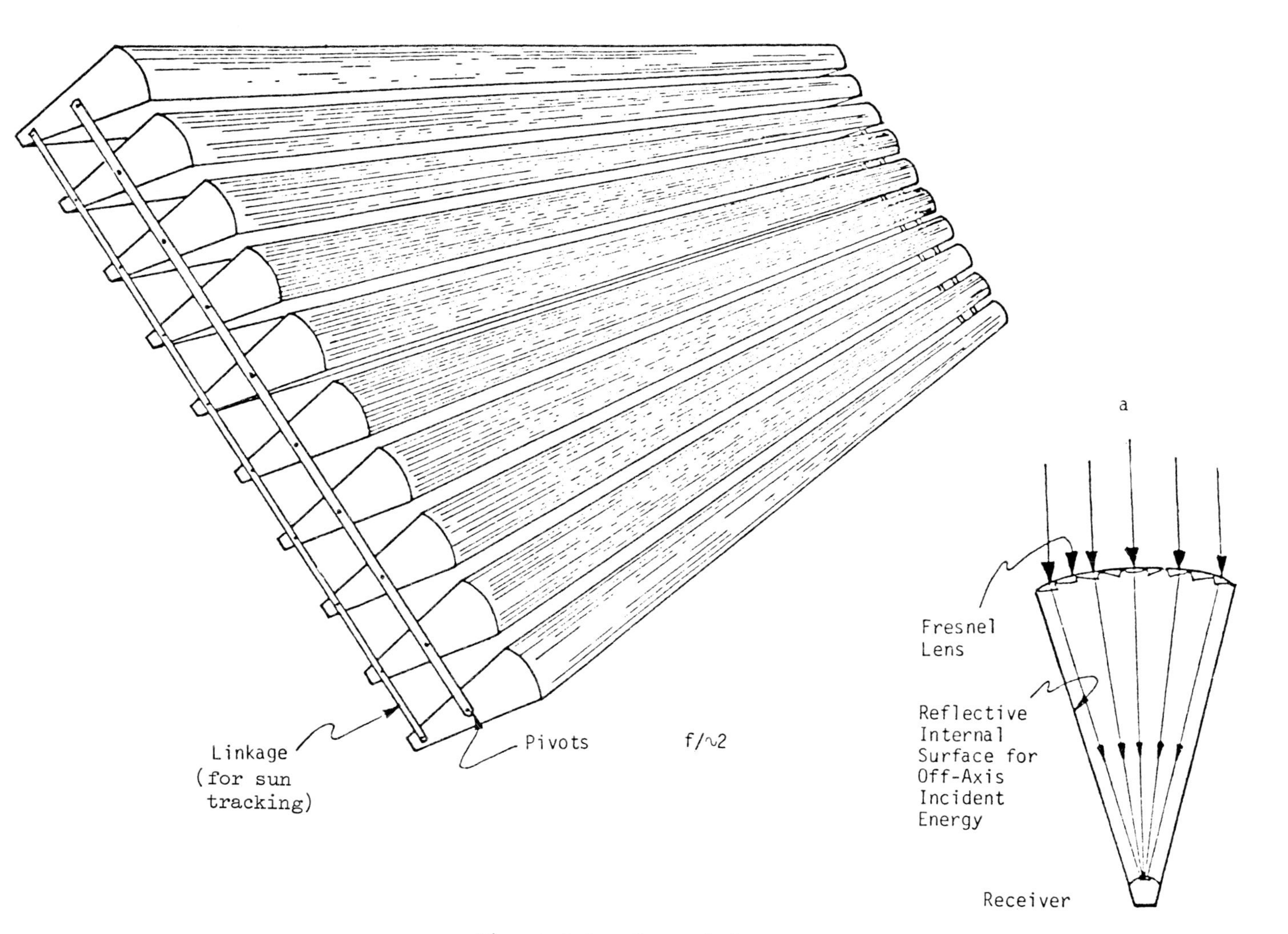

Figure 4.4. Fresnal lens array

Ref. 3

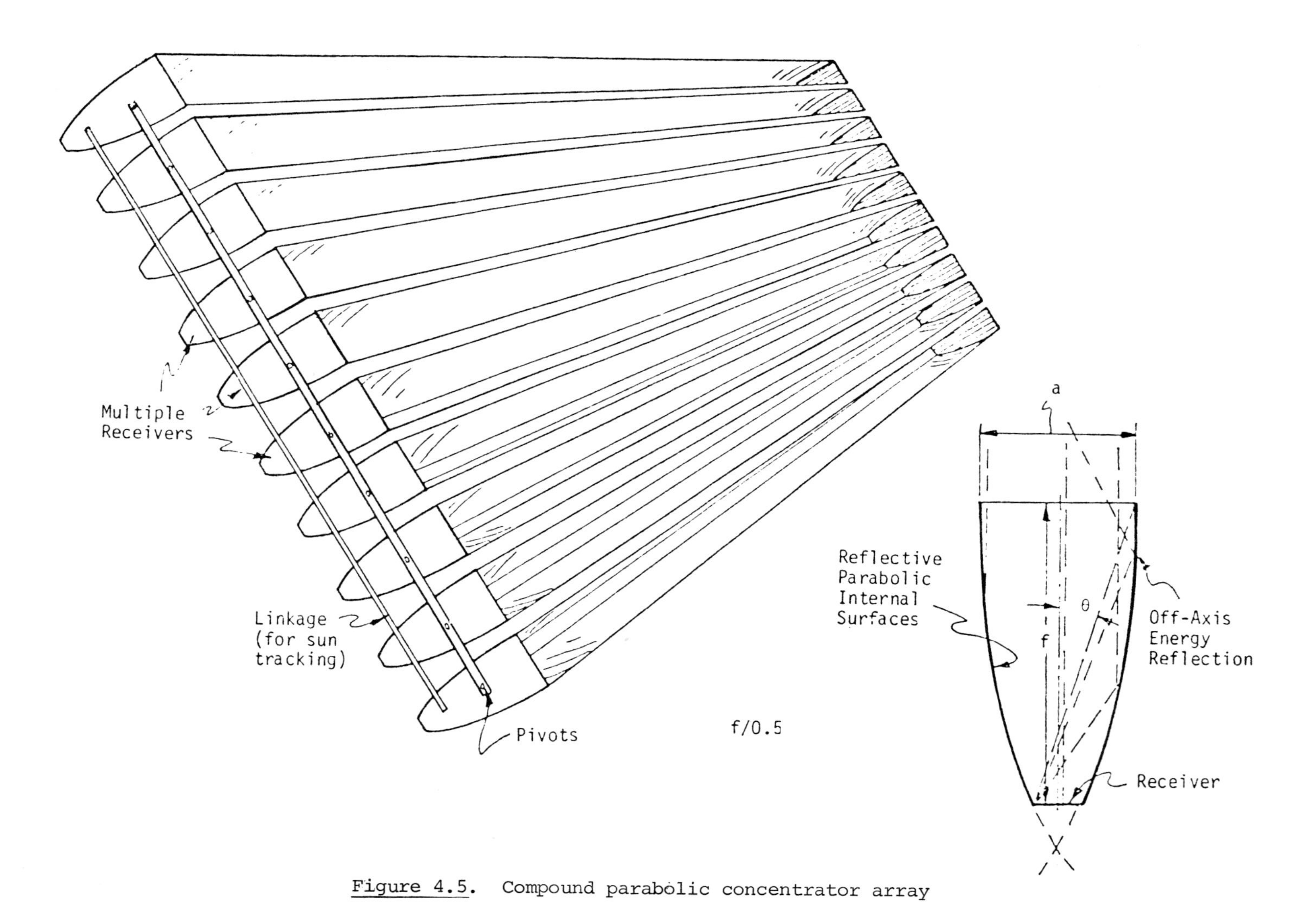

Figure 4.5. Compound parabolic concentrator array

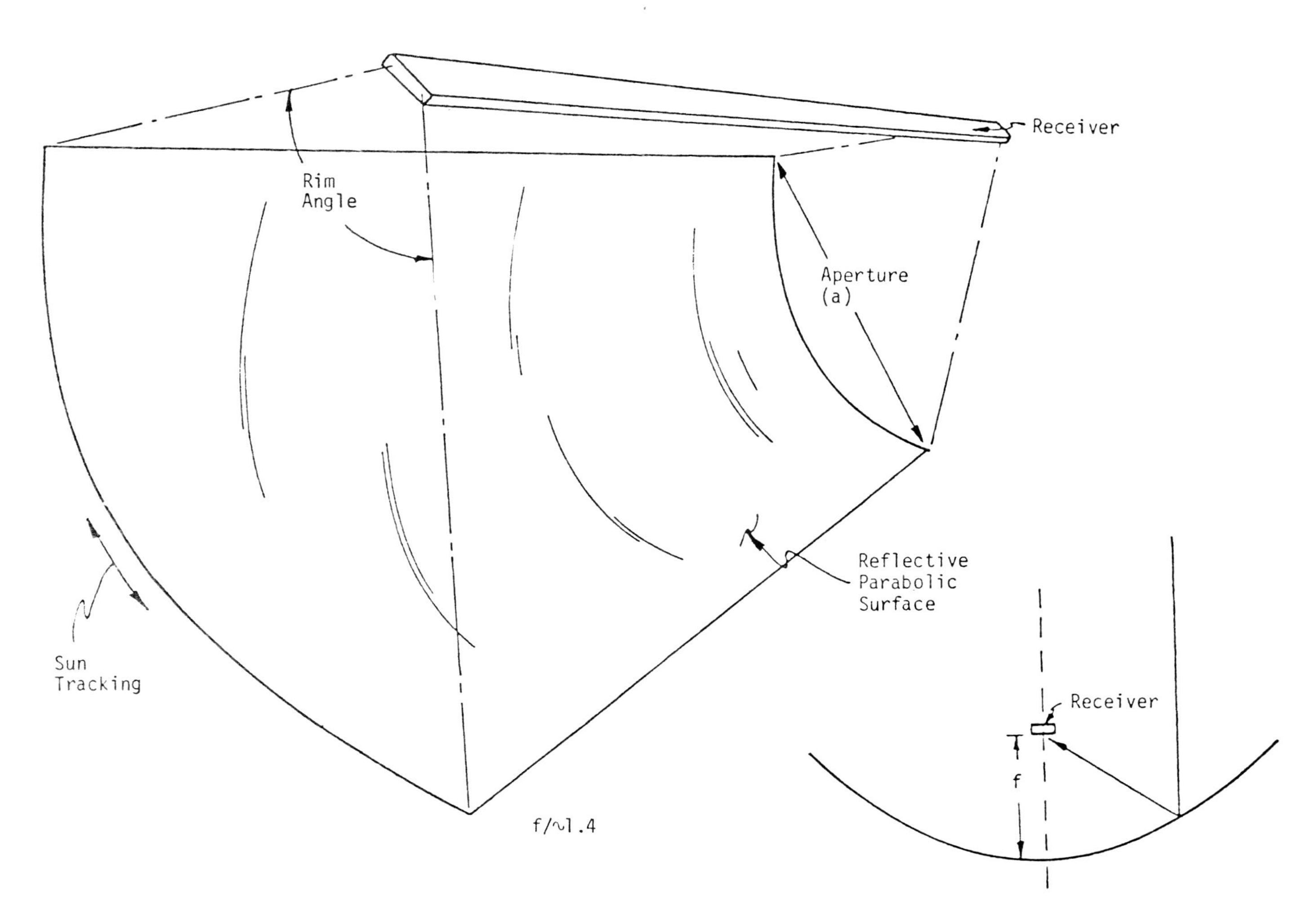

Figure 4.6. Cylindrical parabola (Trough)

Ref. 3

TABLE 4.1

Characteristics of Planar Wedge Collector

Operational Mode	Optical Characteristics	Advantages	Disadvantages
- Stationary or sun tracking	- Concentration ratio <8	- Simple in concept, particularly for non-tracking mode	- Low CR
- E-W or N-S orientation (E-W if stationary)	- Reflective surface area several times aperture area	- May be arrayed in parallel segments fashion, if sun tracking can be driven with common linkage	- Multiple reflections increase losses (absorptive) and much energy incident at large angles to receiver.
- Inclined for latitude	- Multiple internal reflections	- Will collect some diffuse solar radiation	- Multiple receivers required, one for each segment.
	- Effective CR depends on solar disc, specular reflectance, surface flatness, and alignment and drive errors if tracking.		- Morning and afternoon end losses in stationary E-W orientation
			- Profile to wind and extreme weather undesirable (collect dust, snow, rain unless aperture covered) requires structure to maintain flatness.
			- Sun tracking requires rotating elements, drive, and command system

Ref. 3

TABLE 4.2

Characteristics of Compound Parabolic Concentrator (CPC)

Operational Mode	Optical Characteristics	Advantages	Disadvantages
- Stationary or sun tracking - E-W or N-S orientation (E-W if stationary) - Inclined for latitude	- Concentration ratio about 4 stationary, about 10 sun tracking Reflective surface area 4-5 times (or more) aperture area Effective CR depends on solar disc, specular reflectance, contour accuracy alignment, and drive system errors if sun tracking	- May be used in non-tracking configuration May be arrayed in parallel segmented fashion. If sun tracking, can be driven with common drive system Will collect some diffuse solar radiation	- Low CR - Reflector area 4-5 times aperture area - Some energy incident at large angles to receiver - Multiple receivers required-one for each segment - Morning/afternoon end losses for stationary E-W orientation - Large profile to wind & extreme weather - Requires significant structure to maintain contour - Difficult to fabricate large parabolic section accurately - Requires rotating elements, drive, and command, if sun tracking

Ref. 3

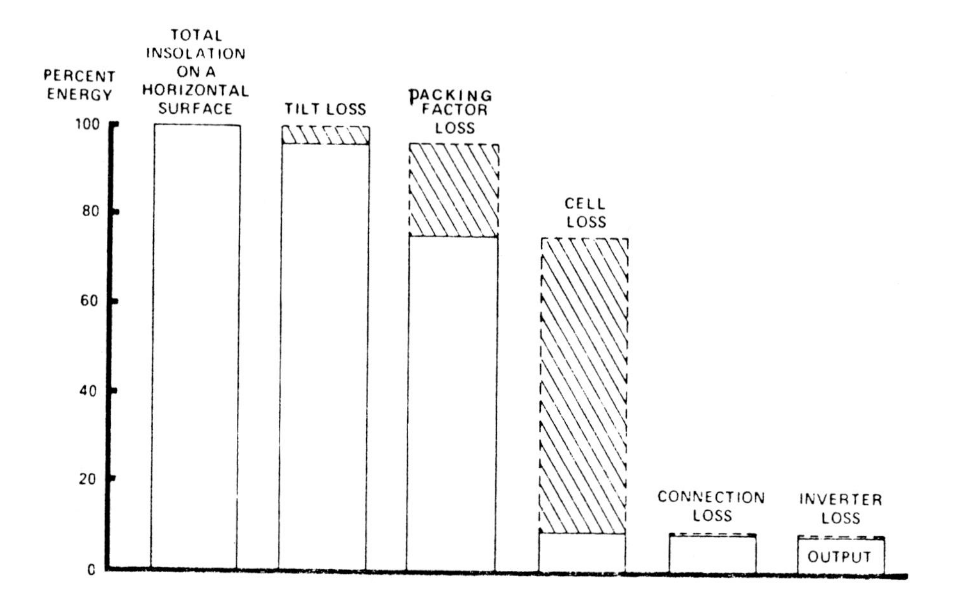

Figure 4.7. Flat panel — fixed tilt — performance train

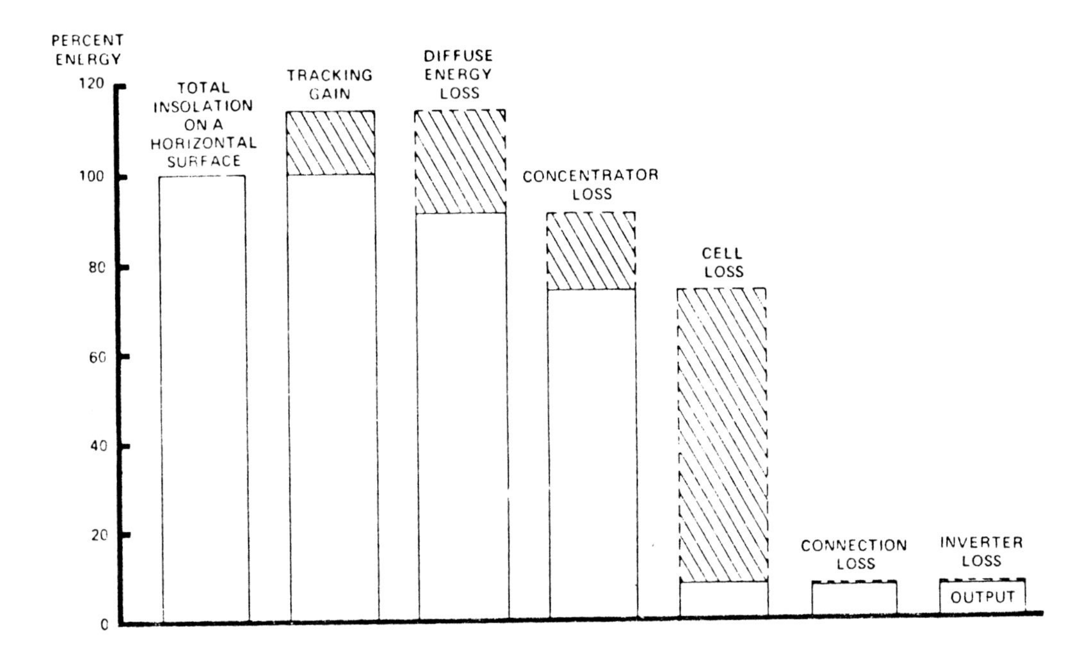

Figure 4.8 Parabolic dish — performance train

There are several advantages to this type of concentrator configuration. As a flat plate collector, it can absorb diffused as well as directed sunlight and requires no tracking, yet, it is estimated to be capable of achieving concentration ratios of 50.

There are many suitable fluorescent dyes for use in this application. They have high quantum yields and, interestingly, they can also be combined in mixtures which can capture all the solar spectrum and cascade it by interdye energy transfer processes to one larger wavelength, relatively narrow band where it can be efficiently converted by a single bandgap photovoltaic cell.

While there are many potential advantages to dye concentrators, including possibly low cost, there are also problems that need to be overcome, in particular degradation due to solar exposure of the plastic sheet and the dyes situated in it.

4.4 Thermophotovoltaic Conversion

Thermophotovoltaic devices are an approach which seeks to recycle energy that would normally be wasted in photovoltaic conversion and thereby increase efficiency. As discussed in Section 2, in a single bandgap cell, much of the energy of incident solar radiation is wasted because the photons have more energy than is strictly required to produce an electron hole pair, or not enough energy. In either case there is wasted energy winding up as heat.

The thermophotovoltaic concept seeks to make use of this heat. Fig. 4.9 shows the configuration for a thermovoltaic converter. Highly concentrated sunlight is focused, in two stages, first from an external paraboloid onto the entry window of the unit, then through a secondary focusing lens onto a black body radiator in an evacuated cavity. Under these conditions the black body and the cavity can rise to temperatures in excess of 2000°K. The black body radiator reemits the energy at wavelengths characteristic for 2000°K, which are of course at longer wavelengths than the incident solar spectrum.

This longer wavelength emission irradiates Si solar cells which are thermally insolated by a vacuum. Photons which are absorbed are converted to electricity, while those that are too low in energy are transmitted through the silicon (which is transparent at these longer infrared wavelength) onto a mirror, and back through the silicon cell onto the black body radiator.

High concentration (over 1000)efficiencies of over 50 percent are theoretically possible, but in practice they are likely to be limited by parasitic losses, such as, imperfect mirror surfaces and free carrier light absorption, to about 40 percent. With the requirement for very high concentration and relatively close mechnaical tolerances the cost-effictiveness of this type of system is an open question that remains to be determined.

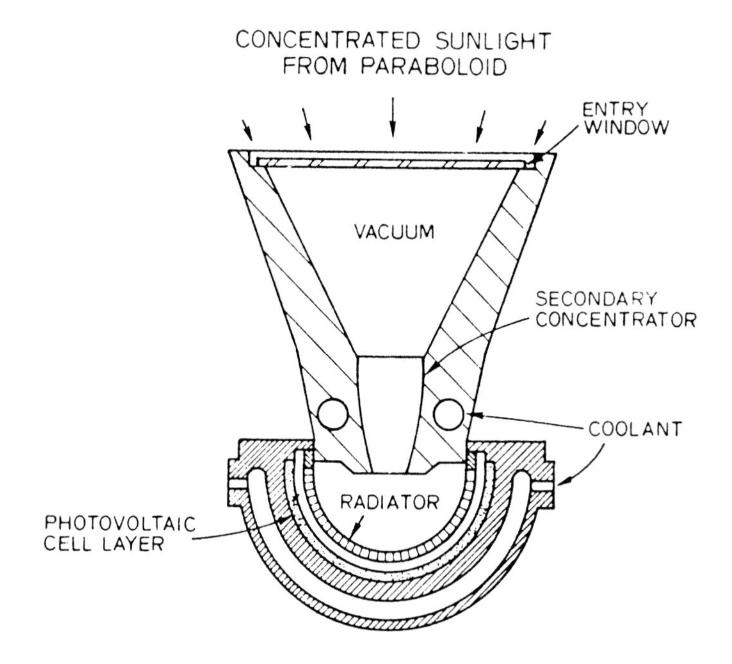

Figure 4.9. Thermophotovoltaic converter

5.0 TECHNO-ECONOMIC FEATURES OF PHOTOVOLTAIC CONVERSION SYSTEMS

5.1 INTRODUCTION

The large-scale use of photovoltaic devices depends, essentially, on photovoltaic electric power generation becoming competitive with conventional systems for central power generation. There are other markets that can be served by higher cost photovoltaic systems, these include space applications, and nongrid-connected remote applications. Compared to the requirements for central electrical power generation in industrialized countries, however, these markets are very small and of limited potential.

However, it should be noted that the market for remote nongrid-connected applications can serve as a useful impetus for the development and testing of photovoltaic systems. This market, which for the U.S. would be primarily an export market, has been studied by the Solar Energy Research Institute and others[2 3]. It is a market which could range anywhere from 50 MW_p/year to over 500 MW_p/year.

Typical applications in this market would be for remote water-pumping and electrification of remote rural or desert communitites, for whom it would not be economical to extend national grid systems. It is, for instance, estimated that there are over one million villages of 500 inhabitants in developing countries in nongrid-connected areas which could usefully use 5 kW of power each. Providing power to these villages represents a potential market of over 5000 MW average for isolated and independent generating system. Supplying these with photovoltaic power would require, along with some energy storage, an installed photovoltaic capacity of approximately 30,000 MWp. Meeting this potential demand, as system costs are reduced, would provide an added useful incentive to the development of cost-effective photovoltaic systems, as well as achieving a socially worthwhile goal.

However, since any really significant impact of photovoltaics on the overall energy picture will occur only if and when photovoltaic electric power generation becomes competitive with conventional systems of central power generation, the examination of the economic features of photovoltaic conversion will be carried out in the context of comparisions to conventional systems of central electric power generations. Furthermore, the quantitive estimates of possible trends and changes will be presented in the context of the U.S. energy situation, with existing and projected conventional power generating capacities, costs, etc., serving as the reference base.

As far as the time—frame to be discussed, it appears reasonable to assume that it will take another ten years or so, to 1990, before device and manufacturing technologies presently being examined would have been developed to the point where reasonably large-scale implementation of photovoltaic systems might start. At that time, if economic and policy reasons warrant it, it could be feasible, in terms of productive capacities, technologies, etc., to start fairly large-scale installation of photovoltaic generating capacity so that by the end of the century, it would represent significant amount (1-2 percent, say) of the approximately 5×10^{12} kWh electricity production expected for the year 2000, (Fig. 5.1).

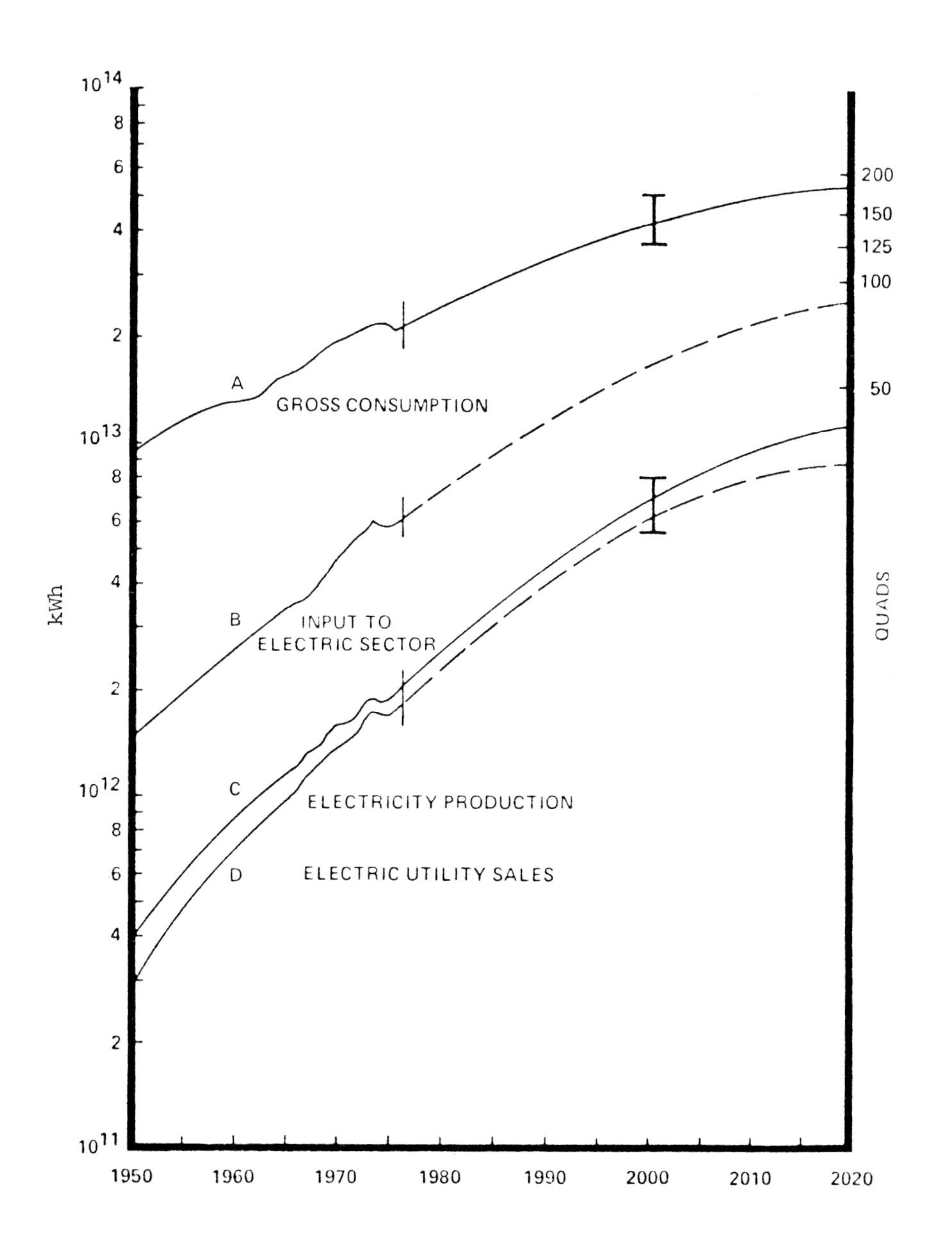

Figure 5.1. U.S. National energy history and projection

Ref. 3

Even to achieve the 1 percent level (approximately 5×10^{10} kWh) in the 10 years from 1990-2000 would require the installation of 20,000 MWp photovoltaic capacity in the sunbelt (where 1 Wp of installed capacity can produce 2.5 kWh/year). Installed uniformly this would represent 2000 MWp/year of capacity installation. This is approximately 1,000 times total present annual production. With expected capital investment requirements at 70-140 ¢/Wp installed in 1979 dollars, the above installation rates represent investments of 1.4-2.8 billion dollars per year and array areas of 200 km^2 (at 10 percent conversion efficiency) in sunbelt regions.

Realistically, therefore, the comparisons and evaluations of a significant role for photovoltaics in the overall power generation picture are concerned with the period beyond the year 2000.

The examination in this paper of the economic features of photovoltaic power generation concentrates on systems for central power generation. It is particularly in the area of intermediate load generation that photovoltaics might be expected to penetrate the central power generation picture, with the daylight hours generally matching the load requirements.

Potentially significant usage of photovoltaic conversion exists also for on-site system. These can range from systems mounted on residential roofs to those on larger industrial and commercial sites. The on-site applications have some savings associated with use of roofs for system support. However, inappropriate orientation would, in many cases reduce this advantage The most serious drawback of on-site systems would be associated with their relatively small scale and the consequently higher relative cost of installation and components and power conditioning systems. In addition, either backup connection to utility grid is required with on-site systems, or expensive storage and/or auxiliary generating capacity.

A potentially important advantage of on-site systems is the possibility of cogeneration of heat and electricity from the same system. For residential systems, however, where economics would rule out the use of concentrators and tracking, the advantages of cogeneration do not appear significant. On the other hand, cogeneration for commercial and industrial sites appear to offer significant advantages. This is logical, since at large industrial and commercial sites, the same economies of scale that apply to central power generation applications would start to come into play, in addition to the advantage of making effective use of the thermal waste produced. The requirements for power conditioning, backup generation capacity, or connection to the grid remain, however.

At this stage it is still too early to say whether on-site photovoltaic conversion with cogeneration at large industrial or commercial sites will have advantages over photovoltaic conversion for central power generarion. It remains safe to say, however, that if photovoltaics are to become significant contributors to the overall U.S. energy picture, they would first have to become competitive in the area of central power generation.

5.2 Conventional Power System Costs

In examining the competitiveness of photovoltaic electrical power generation, the basic questions to be answered are: what is the cost per unit of electric energy produced (e.g., kWh) over the lifetime of the system, and how does that compare with similar costs for other generating systems. The unit used in the electric utility industry for comparisons of this type is the levelized busbar energy cost (BBEC).

The requirement for a levelized value concept arises from the fact that the revenue requirement for a system (e.g., an electric utility) to meet its financial obligations will normally involve a series of annual fixed and variable costs over the lifetime of the plant, which are difficult to compare when different alternatives are being considered. To make these comparisions, present value arithmetic is used, which gives a present value of future revenue requirements. The present value can then be levelized over the time period of interest to give a monetary value equivalent to the year by year revenue requirements.

Thus, the levelized value of the stream of revenue requirements is constant over the period of interest so the sum of the present worth of the equal annual levelized revenue requirements is equal to the sum of the present worth of actual (and varying) revenue requirements over the same time period.

In terms of electrical power generation systems, the levelized busbar energy cost (BBEC) is a fixed equivalent price of unit energy (in constant dollars) at which energy could be sold throughout the life of the plant, to provide the revenues needed to meet all revenue requirements (or total costs of the system) throughout its life, including capital, interest taxes, operating and fuel costs.

The BBEC for a generating system can be determined[2] by the following equation:

$$\text{BBEC (mill/kWh)} = \frac{10^4 \text{ R.C.}(\text{¢/Wp})}{\text{H}} + \text{OM(Mill/kWh)} + \text{Fl(mill/kWh)}$$

where:

R = equivalent fixed charge rate over period of interest that cover repayment of principle, return on equity, taxes, and insurance.

C = initial capital investment per peak watt of capacity;

$$\text{H} = \frac{\text{annual plant output (in kWh)}}{\text{rated plant capacity}} = \text{affecting hours of operation}$$

OM = levelized operation and maintenance cost

FL = levelized fuel cost

As discussed earlier, the intermediate load electrical generation market represents the most plausible area for significant large-scale penetration by photovoltaics, so it is the economics of intermediate load generation plants that can be taken to represent goals for photovoltaics if penetration is to occur.

Using the above equation, levelized busbar costs estimates for the period 2000-2030 are obtained for intermediate coal-fired electric generation systems. Taking H = 4280 h/yr as typical for an intermediate load system, (equivalent to 50 percent utilization), and C = 67+9 ¢/Wp for coal plants; gives a cost of 23 + 3 mill/kWh for the first, capital related term in the equation. With an estimate of OM = mill/kWh, and calculations of FL based on DOE and EPRI estimates of coal, fuel costs, and 34 percent plant thermal efficiency, the range of BBEC for intermediate coal electric generation in the period 2000-2030 is given below[2]:

COST ESTIMATES FOR GENERATION OF INTERMEDIATE LOAD ELECTRIC POWER BY COAL

(1975 DOLLARS)

	DOE (high)	DOE (moderate)	EPRI
Fl (mill/kWh)	41	21	20-26
BBEC (mill/kWh)	65-71	45-44	44-56

This table effectively represents the range of values with which photovoltaic conversion would have to compete, to penertrate the intermediate load market in the 2000-2030 period. The range of BBEC is reflected in Figs. 5.2 and 5.3(from Ref. 2) which are discussed later to examine the range of costs in which photovoltaic systems must fall to be competitive

The capital requirement number[2] used in these estimates (67-93¢/Wp) may be compared with Table 5.1 for 1976 capital cost estimates for conventional plants, and Table 5.2 for 1979 start-up estimates[9] for coal electric plants, where the equivalent numbers range between 60-88¢/W. Furthermore, present information suggests that capital cost reuirements are escalating more rapidly than inflation.

5.3 Photovoltaic System Costs

The costs of photovoltaic systems can be divided into two basic components: module costs, and the remainder of the system or non-module costs. The module is the specific unit or part of the system that receives the light and converts it dircetly into electricity, and can consist of flat plate solar panels or concentrator/cell comination units.

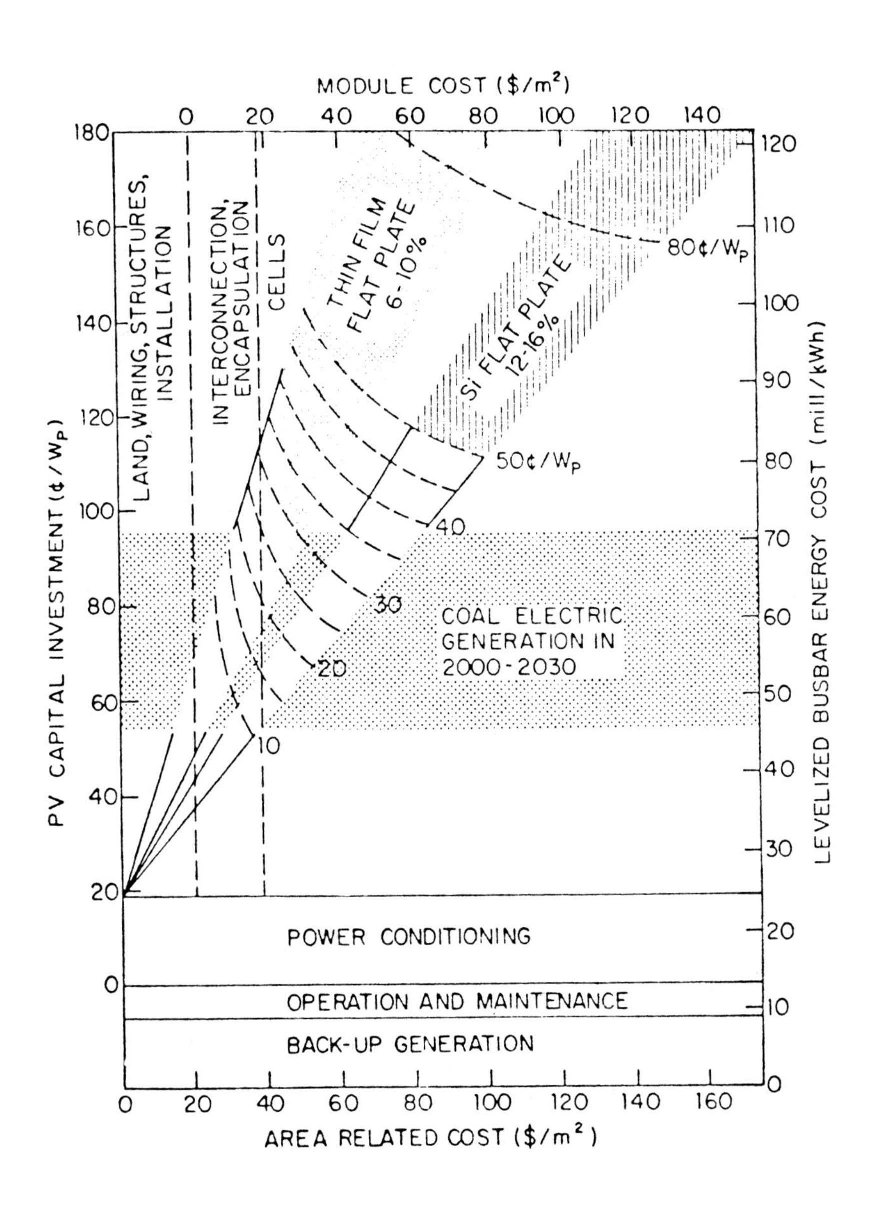

Figure 5.2

Ref. 2

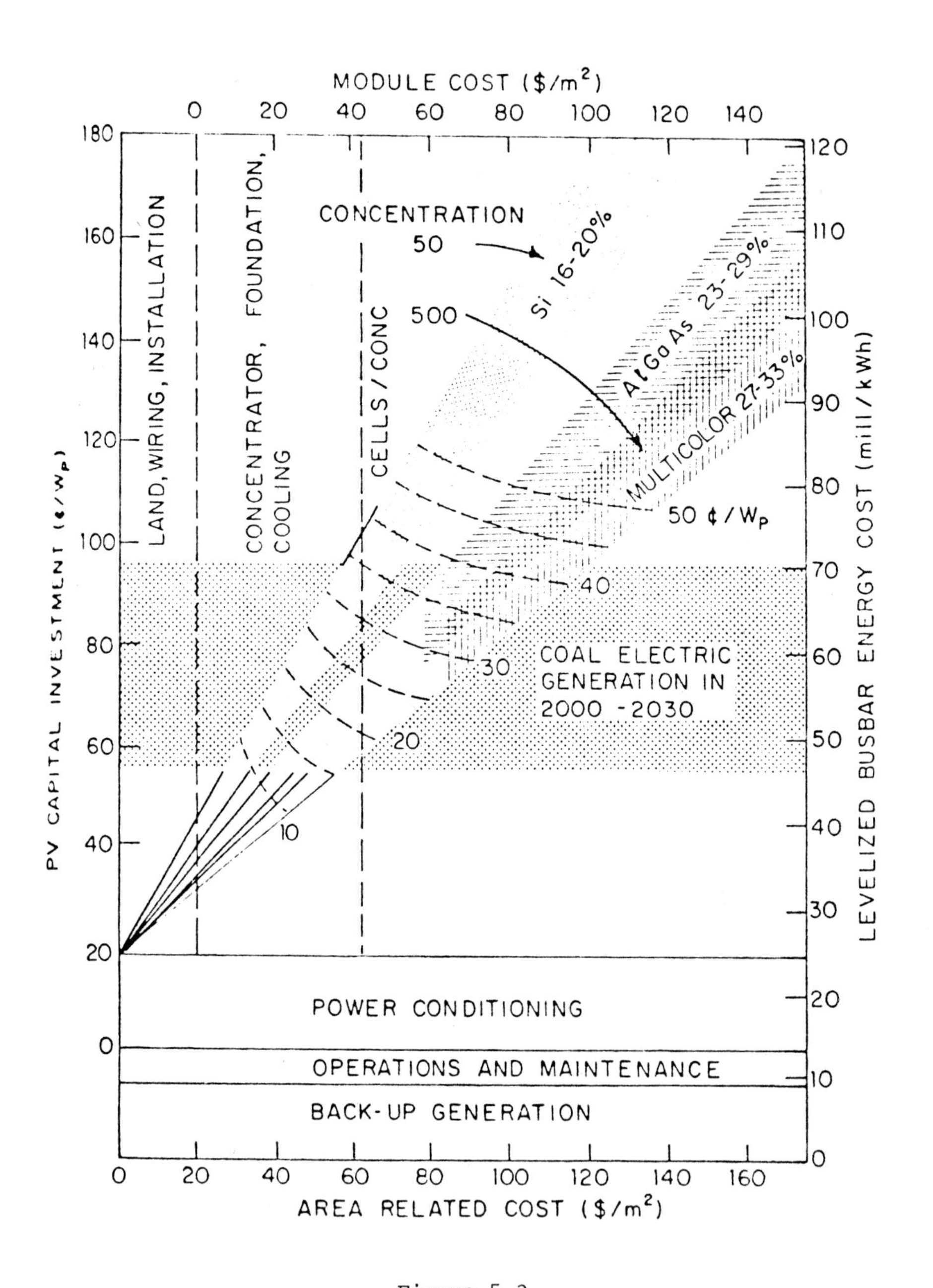

Figure 5.3

Ref. 2

TABLE 5.1

Conventional Thermal Plant Capital Cost Estimates
(Costs in Current 1976 $ per net kW)

Plant Type	Unit Size MW	Typical Value $/kW	Range $/kW
Nuclear	1000	835	670-1000
Coal	1000	640	480-800
Oil	1000	575	410-740
Simple Cycle Gas Turbine	100	160	130-190
Combined GT -ST Cycle	400	295	230-360

Source: Projects Engineering Operation, General Electric Company

Ref. 3

TABLE 5.2

Regional Coal Power Plant Characteristics*

(Costs for 1978 End-of-Year Plant Startup—2-1000 MW Units)

Region	Total Capital Requirement** $/kW	Operations & Maintenance Fixed $/kW/yr	Operations & Maintenance Variable Mills/kWh	Operations & Maintenance Consumable*** Mills/kWh	Full Load Heat Rate Btu/kWh	Average Annual Heat Rate Btu/kWh	Sulfur Removal %
Northeast	765 685-885	13.50	0.94	2.10	9395	9670	90
Southeast	675 605-785	11.95	0.83	1.81	9400	9680	90
East Central	730 655-845	12.90	0.90	2.60	9450	9735	90
West Central	735 660-850	10.85	0.76	0.76	9830	10125	70
South Central	765 685-885	11.30	0.79	1.16	9910	10205	78
West	830 745-960	12.30	0.86	1.11	9490	9770	70

* Based on promulgated 1979 NSPS Standard and supercritical steam conditions.

** The most likely range and the expected value of capital costs are shown.

*** Lime/Limestone plus sludge and ash disposal.

The non-module costs include:

(i) land, physical structure, and system installation costs, estimated at \$140/kW;

(ii) power conditioning equipment and switchgear to convert the variable insolation-dependent output generated by photovoltaic cells into a fixed voltage and frequency a.c. output and connect it to the grid, with total costs estimated at \$20/m^2;

(iii) indirect costs, including engineering contingencies, etc., were estimated at 35 percent of total non-module costs,

(iv) operation and maintenance costs, estimated at 4 mill/kWh;

(v) backup capacity costs

In the comparison of photovoltaic and conventional systems, the capital investment range for a photovoltaic plant to be competitive with coal is calculated back from the range of levelized busbar costs estimated in Table 5.2 for coal electric generation. For the photovoltaic, the folfowing is assumed: the fixed charge rate is 15 percent again, 4 mill/kWh for operations and maintenance costs, and a nominal insolation equivalent to H = 2500 h/yr, in the U.S. Southwest. These calculations, with the results of backup generation requirements added to them, yield allowable capital investment costs for photovoltaic plants in the 52-98¢/Wp range, as is depicted in Fig. 5.2 on the left-hand vertical scale, and corresponding to the estimated range of 45-72 mill/kWh for coal BBEC on the right-hand vertical scale.

While the above discussion has defined allowable limits for photovoltaic system capital costs, if they are to be competitive with coal, it remains to relate these capital costs to the other costs and parameters of photovoltaic systems.

5.4 Flat Plate Photovoltaic Systems

For a flat plate system, the capital investment cost, C, in ¢/Wp is given by:[2]

$$C = \frac{C_s + C_a}{N_p\ N_m\ I_p} + C_{pc}$$

where:

C_S = cost of site, structure, installation (assumed \$20/m^2)

C_A = area cost of modules (variable, dependent on type of module)

N_P = plant efficiency, module output busbar a.c. (Tapically 90)

N_m = module efficiency (light to module output, variable depending on type of module)

I_p = peak insolation (assumed 1 kW/m^2)

C_{pc} = power conditioning capital cost (assumed \$140/kW)

The module cost per peak watt, C_W, is given by:

$$C_W = C_A/N_m I_p$$

which can be used in the previous expression above to give a capital investment cost, C, in ¢/Wp by :

$$C = \frac{C_W}{N_p}\left(\frac{C_S}{C_A} + 1\right) + C_{pc}$$

In the total capital investment cost, an indirect cost factor of 1.35 will be assumed.

These relations are shown in Fig. 5.2, where the effects of different module costs and different efficiencies are related to investment costs for two types fo flat plate photovoltaic convertors: thin film plate, and single crystal (Si) flat plate.

The range covered by the silicon (Si) flat plate, shown by the shaded region, is bounded by the two sloping lines relating to 12 and 16 perecnt efficiencies which represent the range for Si flat plate cells, and by the curved 50 ¢/Wp lower limit on Si flat plate costs which is the 1986 cost goal of the DOE program.

As can be seen even at those costs, the shaded region for single crystal Si flat plate does not overlap the shaded region for costs of coal electric generation in Fig. 5.2. It is thus expected that single crystal silicon flat plate will not be quite competitive with coal electric generation, at least according to the assumptions made for Fig. 5.2, in the 2000-2030 period.

For thin film flat plate cells with efficiencies ranging from 6 to 10 percent, minimum costs for the module are estimated at $\$25/m^2$. This is made up of $\$7.5/m^2$ for the photovoltaic cell ($\$6.5/m^2$ for metalization $\$1/m^2$ for antireflection coatings) and $\$17.5/m^2$ for non-cell module costs ($\$3.5/m^2$ for interconnection and $\$14/m^2$ for encapsulation).

The total $\$25/m^2$ module cost determines the cut-off of the shaded region Fig. 5.2 for thin film flat plates. As can be seen, there is a small region overlapping the shaded region for the range of coal electric costs. This indicates that according to the numbers assumed, thin film flat plate photovoltaic systems might be competitive with coal electric costs, if module conversion efficiency, N_m is close to 10 percent.

5.5 Concentrator and Multilayered Photovoltaic Systems

The use of concentrators to collect light over a large area and concentrate it onto a photovoltaic device warrants the use of higher efficiency (though more expensive) cells, such as, single crystal GaAs or multilayered cells. Through the use of low-cost heliostats as well as other methods, concentrator systems are found to be potentially considerably more cost-effective than flat plate systems.

Non-module costs associated with concentrator systems are estimated at $\$20/m^2$ ($12.5 for site preparation, $\$7.5/m^2$ for heliostat installation). Module costs are estimated at $\$42.5/m^2$ (heliostat $\$17.5/m^2$, added costs for parabolic-shaped reflectors $\$15/m^2$, coding $\$8/m^2$). Costs of cells to be used with the concentrators are estimated[2] as follows:

Si cells, $\$250/m^2$ at concentration of 50 = $\$5/m^2$

GaAs cells, $\$7500/m^2$ at concentration of 500 = $\$15/m^2$

Multilayered cells, $10,000 m^2 at concentration of 500 = $\$20/m^2$

Fig. 5.3 shows the results of parametric cost calculations carried out for concentrator systems (similar to those depicted in Fig. 5.2 for flat plate systems). As can be seen, there are now more regions of overlap where photovoltaic systems might be expected to be competitive with coal electric generation in the 2000-2030 period.

5.6 Material Requirements for Photovoltaic Electrical Generation

Even to achieve one percent penetration of the electricity generation market by 2000 would require considerable diversion of resources into a photovoltaic program. Table 5.3 shows estimates of the requirements associated with both flat plate and concentrator systems if these were to supply one percent of estimated U.S. electricity generation in the year 2000 (5×10^{12} kWh). For the purposes of Table 5.3, efficiencies from light to busbar are assumed to be 10 percent for flat plate systems and 12 percent for concentrator systems. As can be seen, large amounts of material are required for supporting structures. The material requirements for plastic bubble enclosed concentrators clearly show the merits of lightweight systems.

TABLE 5.3

Structural Materials to Support and Illuminate PV Cells. Materials Requirements to Add $4 \cdot 10^{10}$ kWh ($2 \cdot 10^{10}$ W_p) of Generation, Compared with 1974 U.S. Production or Consumption*

Array Type	Material	PV Requirement	1974 U.S. Annual Consumption (Production)	Requirement as a Percentage of Cons. or Prod.
Flat Plate	steel	$5.2 \cdot 10^6$ T	$(1.1 \cdot 10^8)$ T	5 %
	cement	$1.4 \cdot 10^7$	$(8.3 \cdot 10^7)$	17
Concentrator (steel)	steel	$1.4 \cdot 10^7$		13
	cement	$6.0 \cdot 10^6$		7
	aluminum	$8.6 \cdot 10^5$	$6.0 \cdot 10^6$	14
Concentrator (plastic)	steel	$2.8 \cdot 10^6$		3
	cement	$7.0 \cdot 10^5$		1
	aluminum	$3.3 \cdot 10^5$		6
	oil	$2.5 \cdot 10^6$ Bbl	$6.0 \cdot 10^9$ Bbl	.04

$1 \text{ T} = 10^3$ kg

The requirements shown would allow approximately 1 % of the year 2000 electrical generation from PV.

Ref. 2

The requirements for cell materials are also considerable if significant use is to be made of photovoltaics for electricity generation. Table 5.4 shows estimates of cell material requirements associated with a 1 percent penetration of the electricity generation market in the U.S. In the estimate of Table 5.4, it is assumed that Cd and Si would be used for flat plate sheets, while Ga and Ge would be used as substrate material for stacked multilayer concentrator cells with In and Sb as constituents. As can be seen from the table, a significant photovoltaic program could strain existing material resources considerably, particularly the requirements for Cd, Ga, and Ge.

TABLE 5.4

Cell Materials Requirements[a]

Element	$2 \cdot 10^{10}$ W_p Requirements	Production (metric tons)	Present Annual Production	Comments
Ge	250		76 (world) 13 (US)	World reserves of Ge are estimated at 2000 tons.
Ga	120 T		7 T	Up to 140T/yr could be recovered from Zn and Al refining.
Sb	4		$2 \cdot 10^4$	U.S. use of primary Sb
In	4		76 (world)	
Si	$1.2 \cdot 10^5$		$1 \cdot 10^5$	Metallurgical grade; 1974 production of high-purity Si was about 300T/yr. Estimated 1978 production is 1000 T.
Cd	$2.1 \cdot 10^4$		$6 \cdot 10^3$	U.S. use.

(a) Because the numbers in this table are based on schematic rather than actual designs, they should be regarded as giving only order of magnitude estimates.

Ref. 2

6.0 CONCLUSION

General conclusions can be drawn from the preceding discussions. The conversion of solar energy into electricity by photovoltaic devices is attractive in principle: the energy source is abundant at sufficient densities, and is clean and inexhaustible; devices and technology exist to convert it directly into electricity in usable forms. However, in spite of these attractive features, photovoltaic solar conversion must overcome many hurdles before it can make significant contributions as an energy source.

The primary difficulties arise from the low efficiencies attainable by the simpler photovoltaic conversion devices and their relatively high costs. The methods that exist, e.g., with concentrators, to obtain higher overall system efficiency, succeed in improving efficiency but by greatly increasing costs in associated equipment, mirrors, tracking systems, etc.,so that trade-offs result in some, though not very great, advantages for the more complex systems. Because of these factors, and the development required to improve them, it appears unlikely that photovoltaics solar conversion will play a significant role as an energy source before the year 2000.

Beyond that period it is possible that photovoltaic plants could start to be competitive with coal and nuclear plants for central electricity generation, which is the market they must penetrate if they are to make a significant contribution as a viable alternate energy source.

REFERENCES

1. Ehrenreich, H. and J. Martin, "Solar Photovoltaic Energy," Physics Today, September 1979.

2. American Physical Society, "Principle Conclusions of the American Physical Society Study Group on Solar Photovoltaic Energy Conversion," H. Ehrenreich, Chairman, 1979.

3. Electric Power Research Institute, "Requirements Assessment of Photovoltaic Power Plants in Electric Utility Systems," EPRI ER 685, Project 65 1-1, General Electric Company, Technical Report Vol. 2, June 1978 (Palo Alto, California).

4. Holden, S.C. et al., Varian Associates, "Slicing Silicon into Sheet Material," Quarterly Report, ERDA/JPL 954374-77/5, April 1978.

5. Woodall, and J.M. and H.J. Hovel, Applied Physics Letter. Vol. 30, p. 492,1977.

6. Electric Power Research Institute, EPRI Journal, March 1978, (Palo Alto, California).

7. Usmani, I.H. International Symposium on Science, Inventions, and Social Change, Albany, New York, 1978.

8. Cole, R.L. et al., "Applications of Compound Parabolic Concentrators to Solar Photovoltaic Conversion," Argonne National Laboratory Report, ANL-77-42 ERDA Contract W-31-109-Eng-38.

9. Electric Power Research Institute, 1979 Technical Assessment Guide, Special Report, PS-1201-SR, (Palo Alto, California).

THE WIND ENERGY PROGRAM IN THE UNITED STATES OF AMERICA

CARL ASPLIDEN, TERRY HEALY, EDWARD JOHANSON, THEODORE KORNREICH, RICHARD KOTTLER, WILLIAM ROBINS, RONALD THOMAS, IRWIN VAS, LARRY WENDELL, AND RICHARD WILLIAMS

ABSTRACT

An historical account of the deployment of wind energy conversion systems (WECS) in different applications is given. This shows that WECS have always been competitive with regard to other energy sources, although this is dependent on application and needs.

The United States' Wind Energy Program is presented as an example of the great efforts now underway in many parts of the world in today's renaissance of WECS. In all such programs the major objective and goal is to make WECS on every scale, cost-competitive again. Great progress has already been made, both in machine development and technology, as well as in the field of resource assessment and other associated areas. This progress is exemplified and substantiated by new generations of machines, which within a very short time have become simpler in design, more rugged and reliable, and more economical.

A renewal of interest in the deployment of WECS has now developed in several potential market areas, such as, agricultural and central electric production industries.

---

*
Carl Aspliden, Battelle Memorial Institute

Terry Healy, Rocky Flats Test Facility, *"Small Wind Energy Conversion Systems"*

Edward Johansen, JBF Scientific Corporation, *"Economic Analysis"*

Theodore Kornreich, JBF Scientific Corporation, *"Environmental Issues Assessment"*

Richard Kottler, JBF Scientific Corporation, *"Environmental Issues Assessment"*

William Robins, NASA-Lewis Research Center, *"Large Wind Energy Conversion Systems"*

Ronald Thomas, NASA-Lewis Research Center, *"Large Wind Energy Conversion Systems"*

Irvin Vas, Solar Energy Research Institute, *"Wind Energy Innovative Systems Program"*

Larry Wendell, Battelle Memorial Institute, *"Wind Energy Characteristics"*

Richard Williams, Rocky Flats Test Facility, *"Small Wind Energy Conversion Systems"*

ISBN 0-12-467101-2

HISTORY AND BACKGROUND

Wind energy has been utilized for several thousands of years. Probably the oldest preserved documentation of such a deployment is a 5,000 year old drawing of a Nile river boat equipped with a sail to produce translational motion. The first wind turbines were probably very simple, vertical axis designs similar to those used in ancient Persia as early as 300 to 200 B.C. for grinding grain. These turbines used bundles of reed for sails. The utilization of such turbines spread throughout the Mediterranean along the coast of Africa and Asia with the expansion of the Arab empire and Islam. Primitive horizontal axis wind turbines were developed much later. These consisted of wooden booms and jib sails and are still found in use on many islands in the Mediterranean (Figure 1). Toward the end of the 12th century and during the 13th century, windmills were introduced into western and northern Europe by returning crusaders.

During the Middle Ages the Dutch made radical improvements in the basic design, and the windmill was extensively employed in water pumping to drain shallow coastal lakes and marshes for the purpose of expanding arable land. Several such large projects were executed during the early part of the 17th century (Figure 2). Typically, several dozens windmills were engaged in such operations. They were producing 40 to 50 horsepower each and pumped water at the rate of a thousand cubic meters per minute. With the invention of the printing press in 1560, the wind powered paper mill was constructed a few years later to meet the quickly growing demand for paper. Also, during the 16th century, wind operated sawmills were fabricated to process imported timber for internal use as well as for an expanding export market. At about 1850, more than 9,000 windmills were in use in the Netherlands for a wide variety of purposes. This was possible mainly due to the fact that several refinements were introduced, particularly in the case of the rotors. These were four-bladed, consisted of a stock with wooden bars on both sides which were covered with sails. To improve on the aerodynamic design, the bars were later moved off the center to the trailing edge. The gear train consisted of several large wooden-toothed wheels which were greased to reduce friction and noise.

It is interesting to note that even during this time legal and environmental problems existed. For example, it was not allowed to put up any construction or to plant trees in the vicinity of windmills and, as a rule, landlords did not permit their tenants to build their own windmills to grind grain.

After the middle of the 19th century, when the industrial revolution got a stronger and stronger foothold in Europe ,the use of wind power declined sharply. By the turn of the century, most wind turbines had been replaced with steam engines throughout Europe.

The renaissance of the windmill occurred between 1930 and 1960 in several countries in Europe. The Rusians built and installed a hundred kilowatt turbine at Yalta on the Black Sea in 1931 (Figure 3). This turbine was tied in with a steam power plant some 30 km away. It had a three-bladed rotor with pitch control and its annual output was about 250,000 to 300,000 kWh per year.

<u>Fig. 1.</u> Mediterranean sail wing windmill

<u>Fig. 2.</u> Old Dutch windmill

Fig. 3. The Russian wind turbine at Yalta

In Denmark, as much as 200 MW installed capacity was divided between a few thousand large and about 30,000 small wind machines at the beginning of the century. New developments and designs took place after the World War II, especially the famous Gedser machine, which had a three-bladed fixed pitch and a rated output of 200 kW in a 16 m/s wind, and operated for many years and delivered thousands of kilowatt hours into the Danish utility grid. The Gedser machine has served as a design base for many recently developed machines (Figure 4).

During the 1940s and '50s, 200 kW turbines rated at 16 to 17 m/s wind speeds were built in England. The first one was interconnected with a diesel-powered utility network but was in operation only for a short period and was shut down because of operational problems. The second machine, known as the Enfield turbine, was designed by a Frenchman and had some unique features (Figure 5). It had a hollow tower 27m high, and a hollow two-bladed rotor with openings in the blade tips. During rotation, a pressure gradient was established which forced the air through openings in the tower near the base through a small turbine in the tower (which drove an alternator below the turbine), and out through the blades. Although it had a very innovative design, it was found to have a lower efficiency than conventional horizontal axis turbines.

Similarly, in France and Germany, several large wind-powered electric turbines were built in the late 1950s. One 800 kW unit rated at 18 m/s with a 30m diameter three-bladed rotor, operated intermittently at a site outside Paris for several years. This unit was operated at a constant speed of 47 rpm and was coupled to a 50Hz, 60 kW utility grid using a synchronous alternator which operated 1000 rpm and generated 1000 V. A smaller unit with a 22m diameter rotor and operated at 56 rpm had an asynchronous generator producing 130 kW at rated wind speed of 13.5 m/s . The capital cost of these units was around $1,000 to $1,150 per kW in 1960 dollars.

Professor Hutter in Stuttgart, Germany, introduced a number of new ideas during the 1940s and '50s, such as, lightweight constant-speed rotors controlled by variable pitch propeller blades (Figure 6). The blades were made from fiberglass and plastic. The tower was a small diameter hollow pipe supported by guy wires. The largest machine built was a 100 kW machine in an 8.5 m/s wind, and it operated for more than 4,000 hrs.

In the United States, the windmill situation was somewhat different from that developed in Europe. Due to the large distances between farms in rural America, centralized electric power was introduced relatively late in remote parts of the country. Therefore, there was a strong emphasis on small machines principally for electric generation, water pumping, or sawmill operations. Some 6,000,000 small units had been built for such purposes by 1930. It is estimated that more than 150,000 are still in operation. By and large, all these machines were small and had metal multi-blades of the order of 3-5 meters diameter. The only attempt to build a large turbine in the United States was made in the 1930s when a team of top-notch scientists designed the so-called Smith-Putnam machine (Figure 7). Until very recently it was the largest machine ever built. It had a two-bladed steel rotor with a diameter of 57m and weighed 116 tons. At a rated wind speed of 14.5 m/s, it produced 1.25 MW of power while operating at a rated speed of 28 rpm. In 1945, after intermittent operation over four years on the top of a 600m hill in Vermont

Fig. 4. The Danish Gedser machine

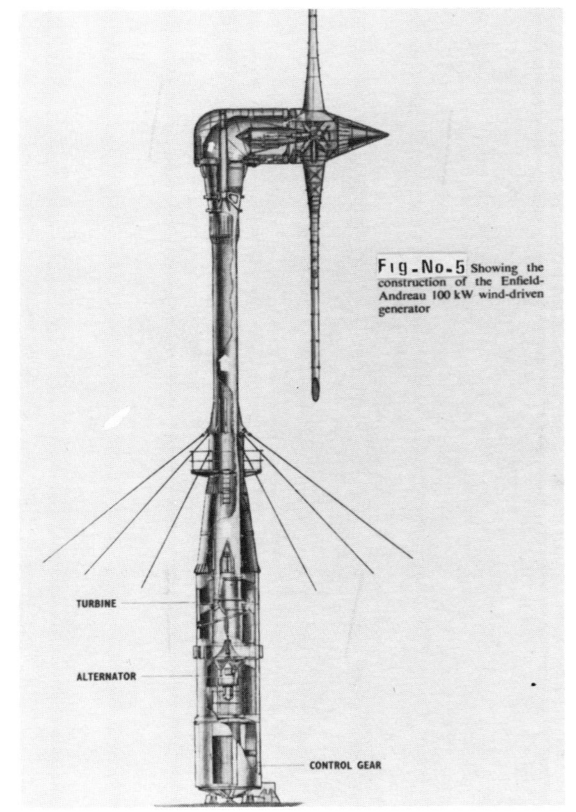

Fig. 5. The unique Enfield-Andreau 100 kw wind-driven generator.

<u>Fig. 6.</u> Machines based on designs by Hutter

<u>Fig. 7.</u> The megawatt Smith-Putnam machine

in northeastern United States, one of the blades broke off near the hub. The failure was due to material fatigue, which had been identified prior to the failure but, because of wartime material shortages, had not been corrected. The project was abandoned because at that time the machine was not cost-effective and could not compete with conventional coal and oil generating plants.

These examples, selected through the history of the deployment of wind energy extracting systems, clearly indicate two major keypoints: (1) the state of the art of wind energy technology depends on the application and on the competition; further, it depends on the local needs and on the local availability of other energies and materials; (2) the size of a wind machine may range from less than 1m to more than 100m, corresponding to a fraction of a horsepower to several thousands of horsepower respectively. Both of these circumstances have contributed to the fact that wind energy extracting systems always in the past have been able to favorably compete with other systems. It is now, therefore, our responsibility to once again bring this technology back and to update it so that it becomes commercially viable on a large scale. This is our objective and goal.

WIND ENERGY CHARACTERISTICS

Wind energy is air in motion relative to the surface of the earth. The air moves because of uneven heating of the surface. By and large, the atmosphere is not heated directly by the incoming radiation but the solar radiation first is absorbed by the surface and is then transferred in various forms back into the overlying air. Since the surface of the earth is not homogeneous (land, water, desert, forest, etc.) the amount of energy that is absorbed varies both in space and time. This creates temperature-density-pressure differences which in turn create forces which move air from one place to another. For example, land and water along a coastline absorb differently, so do valleys and mountains. This creates breezes. However, the atmosphere operates on many space and time scales ranging from a fraction of a centimeter and a fraction of a second to tens of thousands of kilometers and several months. For example, in a year, the tropical regions receive an excess of energy while the polar regions have a deficit. But the tropics don't get hotter from year to year nor do the poles get colder with time. Obviously, due to this differential heating, there is an exchange of energy across latitudes. For this reason, as well as because the earth rotates around it's axis, certain semi-permanent planetary-scale circulation patterns are established in the atmosphere.

From these examples, and general sketch, it is understood that some areas would be preferable to others for kinetic energy extraction in the boundary layer. In addition to these major forcing agents, there are others, such as, topographical features, which may alter the energy distribution considerably so that many exceptions from this general picture exist, especially on a local scale.

Preliminary wind energy resource assessments have already been made in several countries. As a matter of fact, due to efforts mainly made by the

World Meteorological Organization, wind information in the boundary layer is now available from most parts of the world. Therefore, at this time, an effort is now underway to produce a preliminary wind energy resource map on a global scale. There are no doubt some areas from which no observations are available. Most of these are, however, inaccessible and of little economic significance. Also, there are many areas in which data are available but not processed into formats such that they are of direct use in a wind energy assessment. Some limited resources in funds, personnel, and equipment will be required to obtain these data for processing. The general plan covering this work indicates that a preliminary global map of wind energy resources can be made available for distribution several months prior to the planned United Nations Conference on New and Renewable Energies in August, 1981, in Nairobi, Kenya.

The resource assessment program discussed in this paper is almost entirely based on the work that has been undertaken in the United States. This is for two reasons. Firstly, the author is most familiar with this program and secondly, which is very important, the United States geographical spread encompasses a large variety of topographical and meteorological characteristics which may not be found in another single country. Therefore, the methods and techniques which are employed in the United States must be applicable to polar and tropical conditions, to deserts and forest areas, to coastal areas, flatlands, and mountainous regions. Thus, it is believed that the techniques discussed here are covering most of the applications that may present themselves in a resource assessment program in almost any part of the world.

RESOURCE ASSESSMENT

The activities of resource assessment and the development of siting methods have been consolidated under the program area called Wind Energy Prospecting. Work has been progressing simultaneously in these two activities as shown in Figure 8. National and regional resource assessment activities are described below. Siting methods for large and small wind energy conversion systems (WECS) are described in a following section.

National Assessments

Three national wind energy assessments were performed early in the Federal Wind Energy Program. The first assessment was published by Sandia Laboratories (Reed, 1975). The other two assessments were performed as integral parts of separate mission analysis studies and were based on the same National Climatic Center (NCC) data as the first assessment. Because of discrepancies in the amount and distribution of the wind power density in the three national assessments, a synthesis of these assessments was completed by Elliot (1977). This synthesis analyzed the causes of the discrepancies and included the most plausible results in a single assessment (see Figure 8).

These initial assessments contributed to the preliminary wind energy scoping exercises; however, with increased interest in wind energy utilization, more refined and reliable assessments were required on a regional scale. For example, the wind power density values shown in Figure 8 are representative of terrain with good exposure to the wind. The amount of land in a given area with good exposure can vary from about 80% in flat terrain to less than 2%

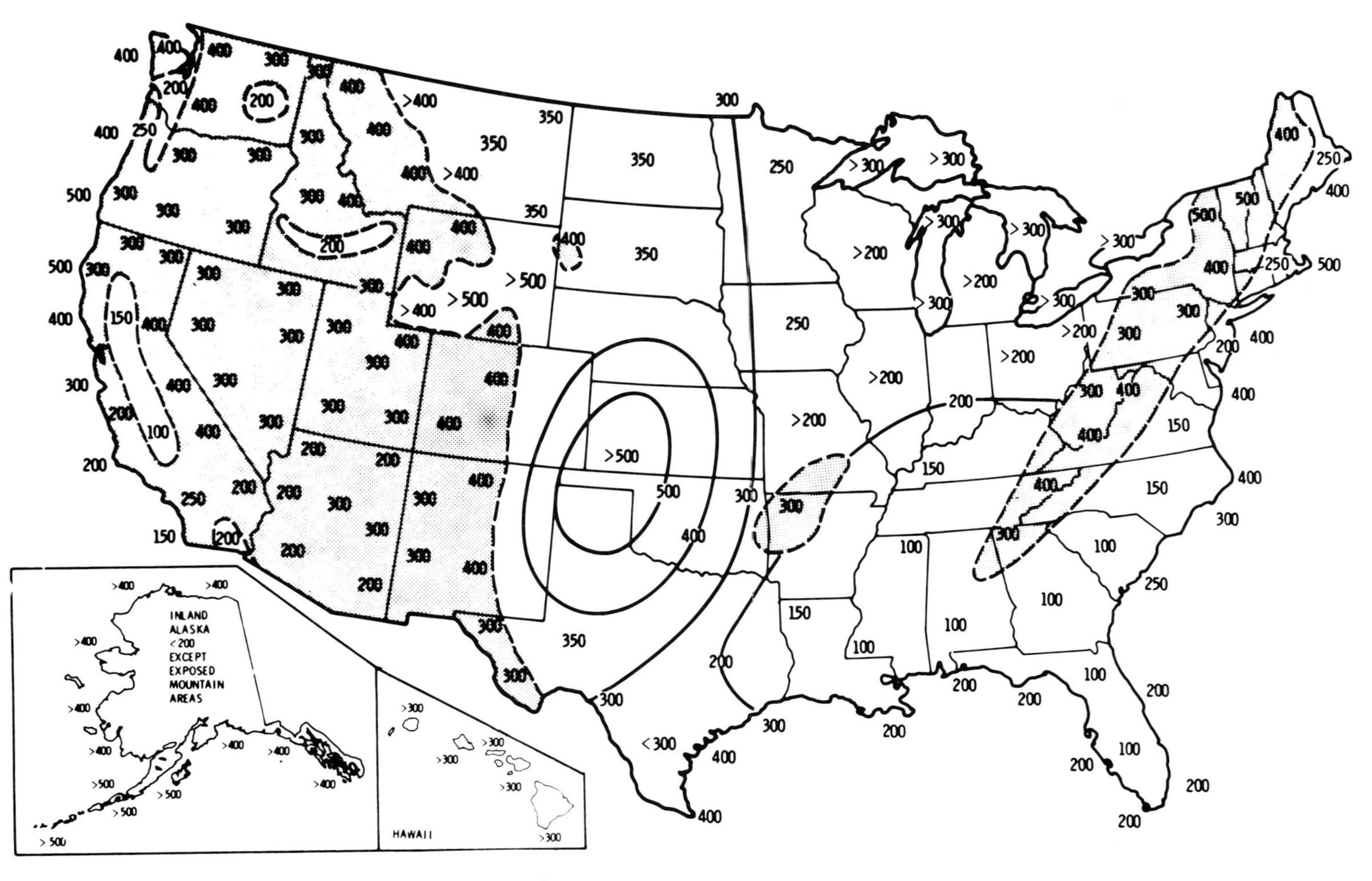

Fig. 8. Annual average wind power (watts/m^2) at 50m

in mountainous terrain. This variation can produce unreliable estimates of wind power potential in given areas within a region.

Regional Assessments: Techniques Development

The first objective of the Wind Characteristics Program Element's (WCPE) regional assessment program was to develop and test prototype techniques for the analysis of wind energy potential and distribution over a large area (Renne and Elliot, 1978). These techniques involved the utilization of a data set much larger than that used for the national assessment, the application of meteorological and topographic factors in the analysis, and the use of indirect methods of wind power estimation in areas where no wind measurements existed. Five states in the Pacific Northwest were selected as a test of these techniques.

Existing wind data provided the basis for assessing the Northwest's wind energy resource. Techniques were developed for identifying screening and analyzing these data. Table 1 indicates the sources and format of wind data that are available in the Northwest. However, not all of these data need to, or should, be used in a wind resource assessment.

TABLE 1

Number of Stations with Wind Data in the Northwest Region and Peripheral Area

Source and Type	Available	Used in Assessment
National Climatic Center (NCC)		
Summarized	205	173
Digitized	97	61
Unsummarized	207	80
U.S. Forest Service	685	63
Canadian Summarized	34	34
Other	200	51
Total	1428	462

Screening procedures were developed to identify stations with the most useful data and to eliminate stations that would not significantly contribute information on the distribution of the wind resource. Even though the screening procedures eliminated almost two out of three data stations available, the data coverage for the Northwest was increased from 80 stations used in the national assessment to 462 stations.

The increase in data coverage by almost a factor of six certainly contributed to a more reliable analysis, but many areas of the region still lack adequate data for wind energy estimates. In many of these areas combinations of meteorological and topographic features were employed in the analysis to enhance the reliability of the overall assessment. A few features that indicate high wind energy potential are:

- o gaps, passes, and gorges in area of frequent strong pressure gradients;
- o long valleys extending parallel to prevailing wind directions;
- o high elevation plains and plateaus;
- o exposed ridges and mountain summits in areas of strong geostrophic winds.

Some features that indicate low wind energy potential are:

- o valleys perpendicular to the prevailing wind direction;
- o sheltered basins;
- o short and/or narrow valleys and canyons.

The features listed above can be located in a region by careful observation of relief maps and typical surface pressure maps.

Other indirect methods of estimating mean wind speed include observations of deformed vegetation (Hewson and Wade, 1977) and landforms affected by the winds (Marrs and Kopriva, 1978). These techniques were tested to a small degree in the Northwest region. The major portion of the assessment was accomplished with the enhanced data set and the topographic-meteorological analyses. The topography of the Northwest region is indicated by the raised relief map in Figure 9.

Regional Assessment for the Northwest

The techniques described above were applied in the Northwest region, which is defined as the states of Washington, Oregon, Idaho, Montana, and Wyoming (Elliot and Barchet, 1979). The goal of this assessment was to produce a wind power atlas for the region, which would serve immediate needs in the commercialization of wind as an alternative energy source.

The wind power density was analyzed on a seasonal and annual average basis for each state in the region. The results of the analysis of annual average wind power for the Northwest region are shown in Figure 10. The increase in in resolution can be seen by comparing the results in Figure 10 with the results for the five Northwest states in the national analysis shown in Figure 8. The effect of the analysis of meteorological and topographical features is particularly apparent in the long valley in southwestern Montana and the large east-west valley in central Washington.

As in the national analysis (see Figure 8), the stippled areas indicate that the wind power densities apply only to well-exposed terrain. However, in the regional analysis there is significantly more definition in the number and distrubution of the values shown in the stippled area than was possible to show in the national assessment.

A seasonal contrast in wind power density for the State of Washington may be seen in Figure 11. Wind power is significantly higher in the mountainous

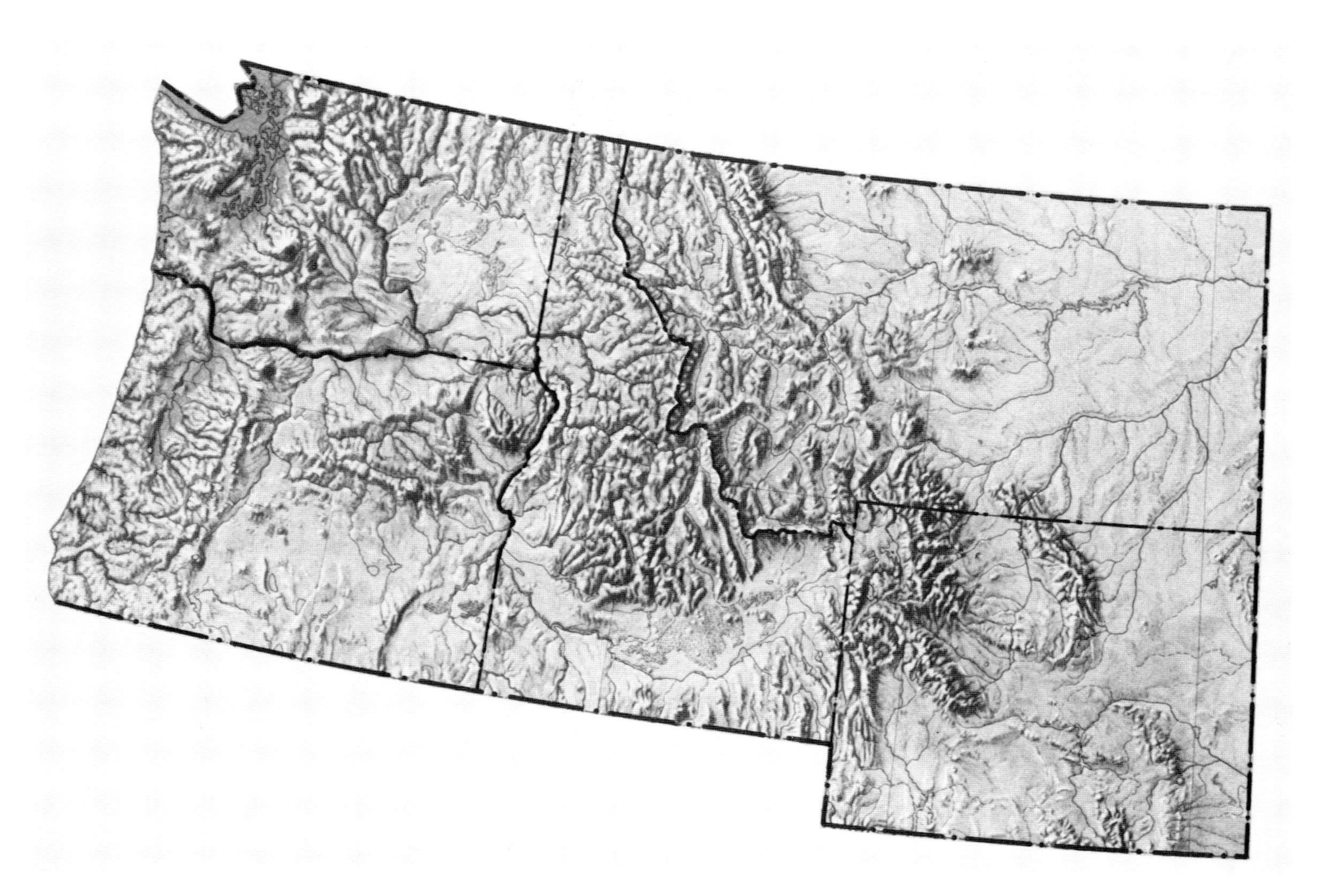

Fig. 9. Relief map of Northwest region

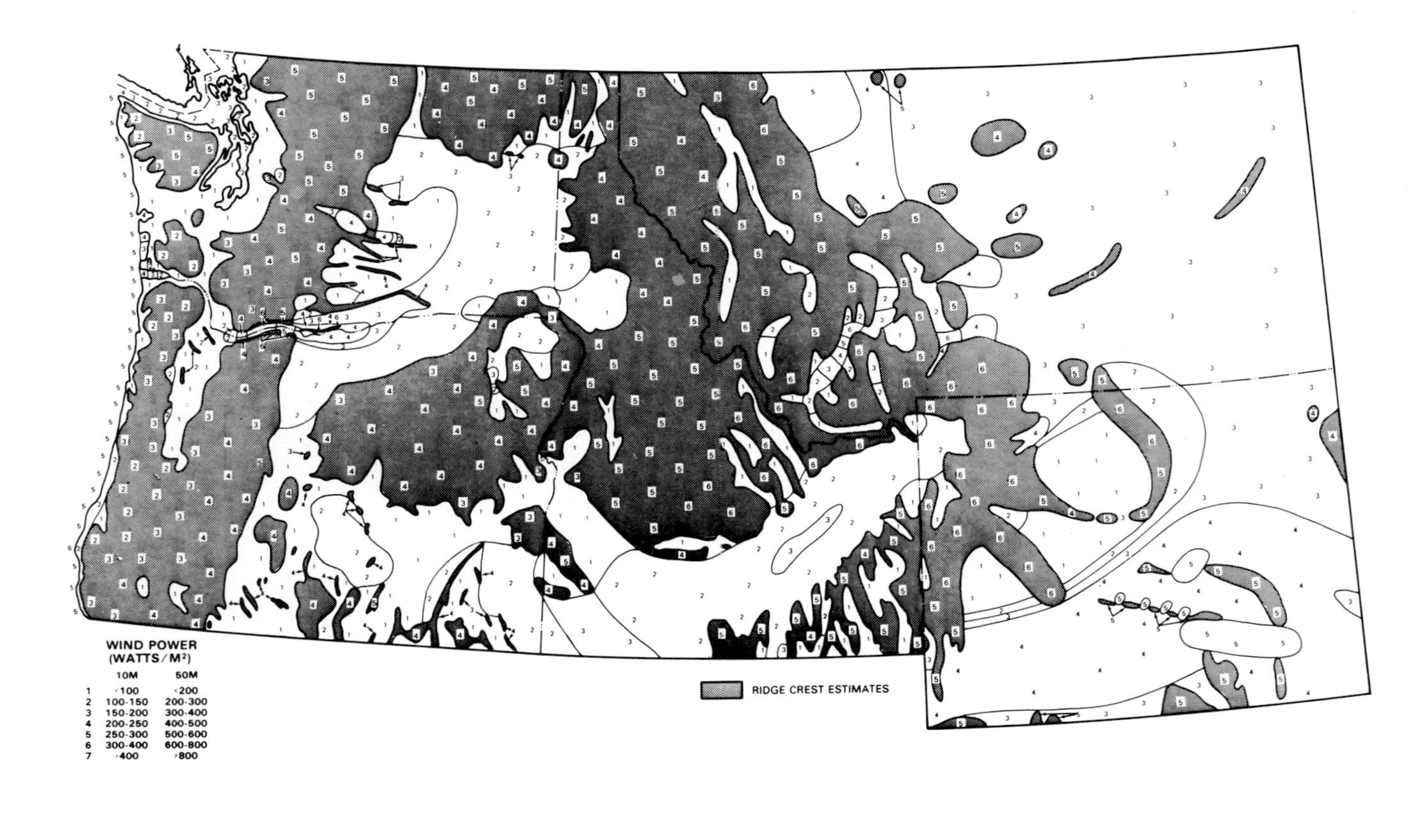

Fig. 10. Northwest anuual average wind power (watts/m^2)

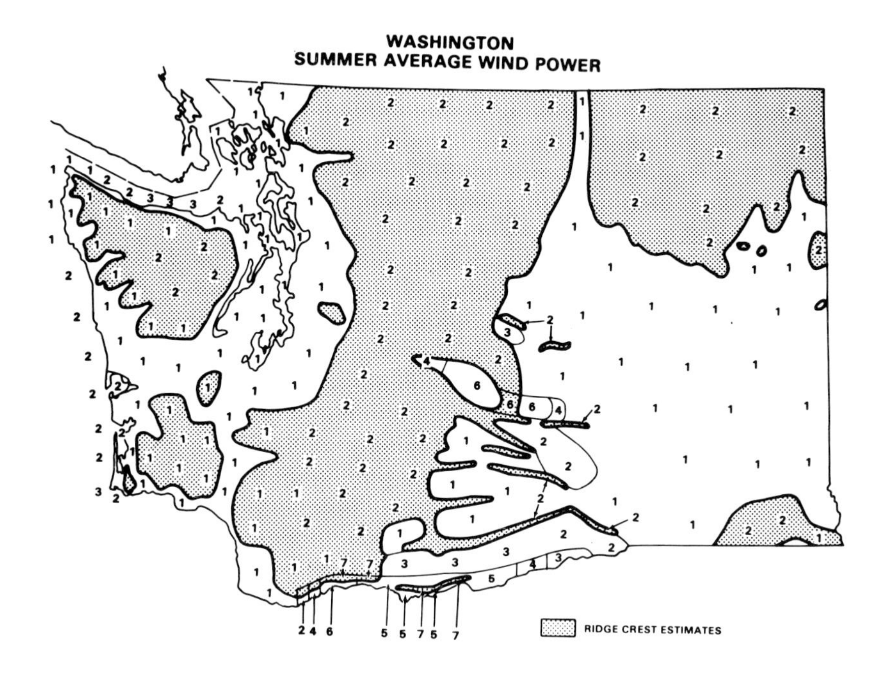

Fig. 11(a). Washington summer average wind power

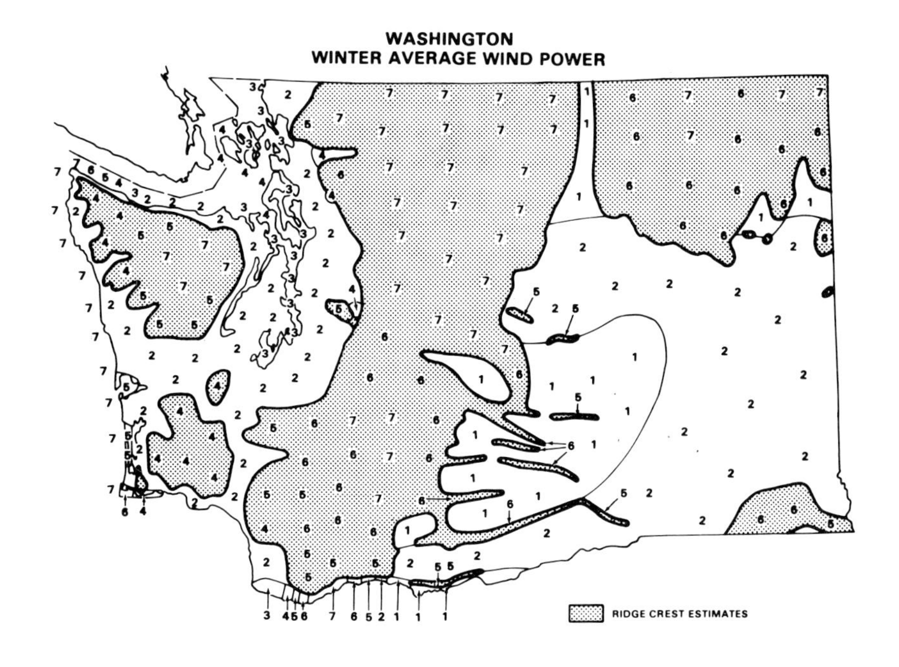

Fig. 11(b). Washington winter average wind power

regions in the winter. A summer maximum exists, however, in the valley extending eastward from the mountains in the central part of the state. Seasonal shifts are evident in the maximum wind power density along the Columbia River Gorge at the southern border of the state.

The analysis of the wind resource maps on a state and regional basis depends on information from individual stations. For those stations with hourly or three-hourly data on NCC magnetic tapes, a very detailed presentation of the temporal variation and character of the wind resource can be obtained. However, the geographical area represented by a station may well be very small and, in many cases, not representative of the windiest or best-exposed sites in a particular area.

Figure 12 presents selected examples of the wind characteristics at one station. Between 10 and 18 graphs of the same type are plotted on one page for each state, allowing convenient station-to-station comparison. The interannual graph illustrates the range of annual wind power density that can be expected. The seasonal variation of the wind resource is shown in the graph of monthly average wind power density. The variation of the wind resource during the course of an average day is given by the diurnal graphs for the four seasons. A Raleigh wind speed distribution based on the mean speed is shown along with the observed wind speed frequency distribution. The coincidence of the two peaks in the speed direction graph indicates that the highest wind speeds occur from the prevailing wind directions. The percent of time a selected level of wind power occurs at the station is also shown.

The analyses discussed above provide information on the distribution of wind power over a geographical region. Another concern in the regional assessment is the wind power potential over all of a region or any selected portion of a region may be obtained by digitizing the power density values, from maps as shown in Figures 11(a) and (b), and the landform classifications (USDI, 1970) for the region as shown in Figure 13. The grid box size for the Northwest assessment is one-third degree longtitude by one-fourth degree latitude and is shown for the State of Washington in Figure 14. By assigning fractions of the power density to different percentages of the area in the grid box, according to landform classification, the wind energy potential for the grid box can be estimated. The analysis can be performed for any collection of grid boxes within the region to determine the power available as a function of the land area. An example for the State of Washington is shown in Figure 15. The grid approach provides a convenient way to assess the effect of existing and proposed wilderness exclusion areas in which turbines may not be installed.

Table 2 summarizes similar analyses for the states of the Northwest region by giving estimates of the percentage land area over which the wind power density at 50m equals or exceeds 400 watts/m^2.

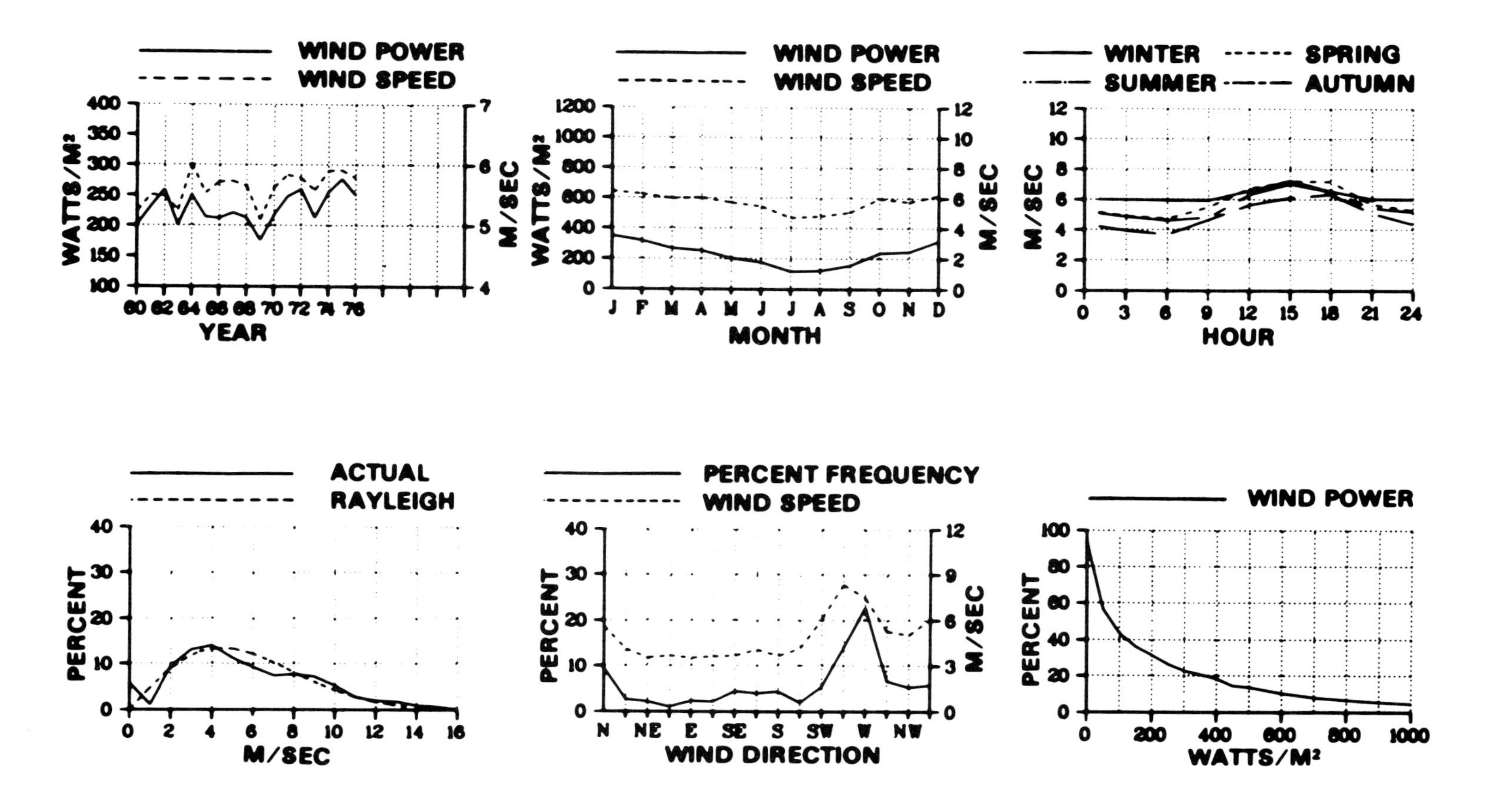

Fig. 12. Wind characteristics for Cutbank, Montana; 10/59-12/76; anemometer height=6m; average speed=5.7 m/s; average power = 230 watts/m^2

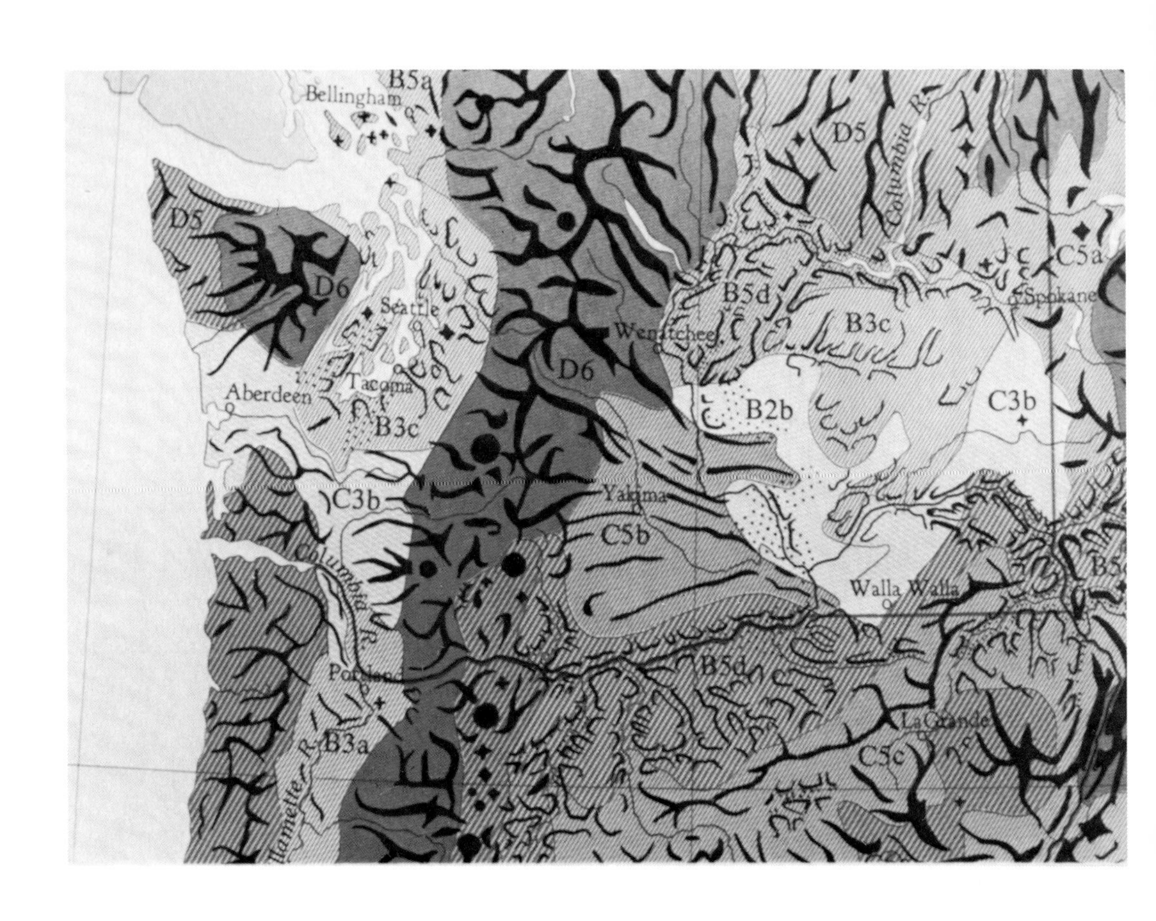

Fig. 13. Land form classification for the State of Washington

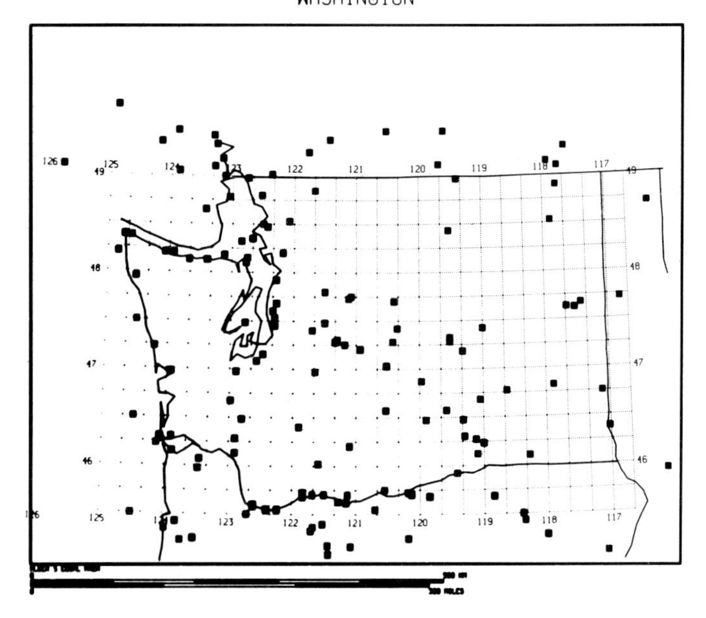

Fig. 14. Latitude and longitude grid used to digitize wind power analysis and land form classification. The symbols represent the wind data stations used in the wind power analysis.

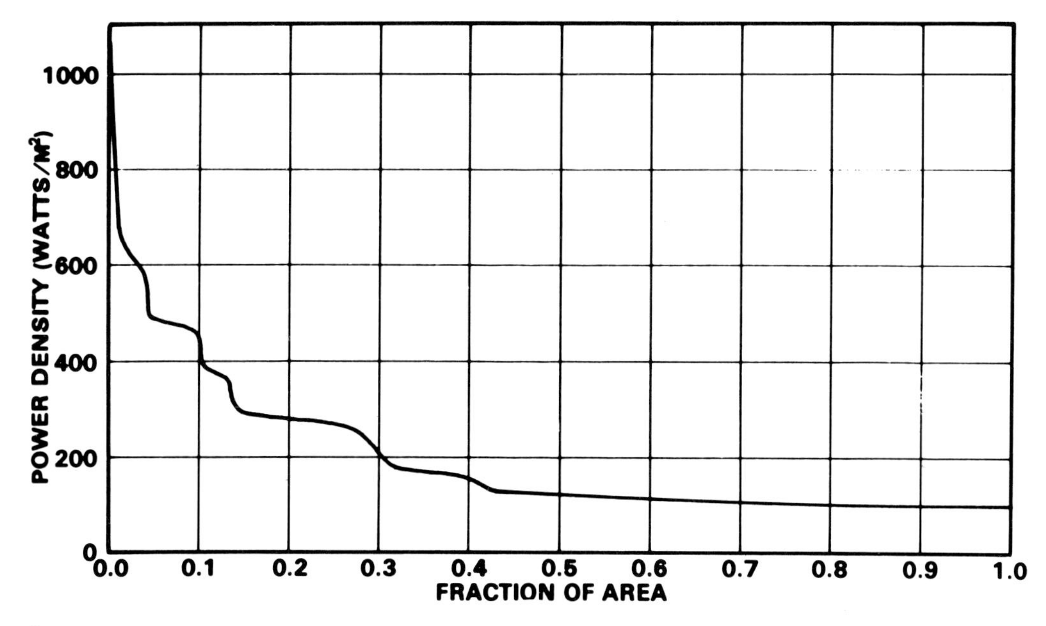

Fig. 15. Areal distrubution of annual wind power for the State of Washington

TABLE 2

Land Area with 400 watts/m^2 at 50m

State	Percentage	Area (km^2)
Idaho	7.7	11900
Montana	9.5	24700
Oregon	5.7	10300
Washington	3.8	4600
Wyoming	26.6	49100
Northwest	11.2	100600

Other Regional Assessments

The wind power atlas for the Northwest region will serve as a guide for the analyses of other regions of the country. The Department of Energy has released a request for proposals to conduct wind energy assessments for the eleven other regions, shown in Figure 16, which cover all the other states and territories of the United States. The present goal is to complete all the regional assessments by October 1980. In the process of completing these regional assessments, a wind energy data base will be established in a form convenient for updating, as usable wind data becomes available through new measurements or through analyzing unsummarized historical data.

SITING METHODS

Along with resource assessment activities, the Wind Energy Prospecting program area includes the development of siting methods for both small (less than 100 kW) and large WECS. Different approaches to siting small and large machines are necessary because of differences in the way these machines are used and differences in their costs.

Siting Methods for Small Machines

Small machines are used for generating power at its points of use; this restricts the location of the machine to a small area. The small capital cost of the machine limits the magnitude and sophistication of the siting effort that can be conducted by the individual small WECS customer.

Proper siting procedures for small WECS must attack two basic problems: finding the best or , at least , an acceptable location for the turbine within a given area; and accurately estimating the wind characteristics at the site. Locating a site is the simplest problem, because basic features of flow over obstacles and many terrain features are fairly well understood. Guidelines can be formulated that will enable a person siting a machine to avoid an unwise choice of location. Such guidelines are provided in a Siting Handbook for Small WECS (Wegley et al., 1978).

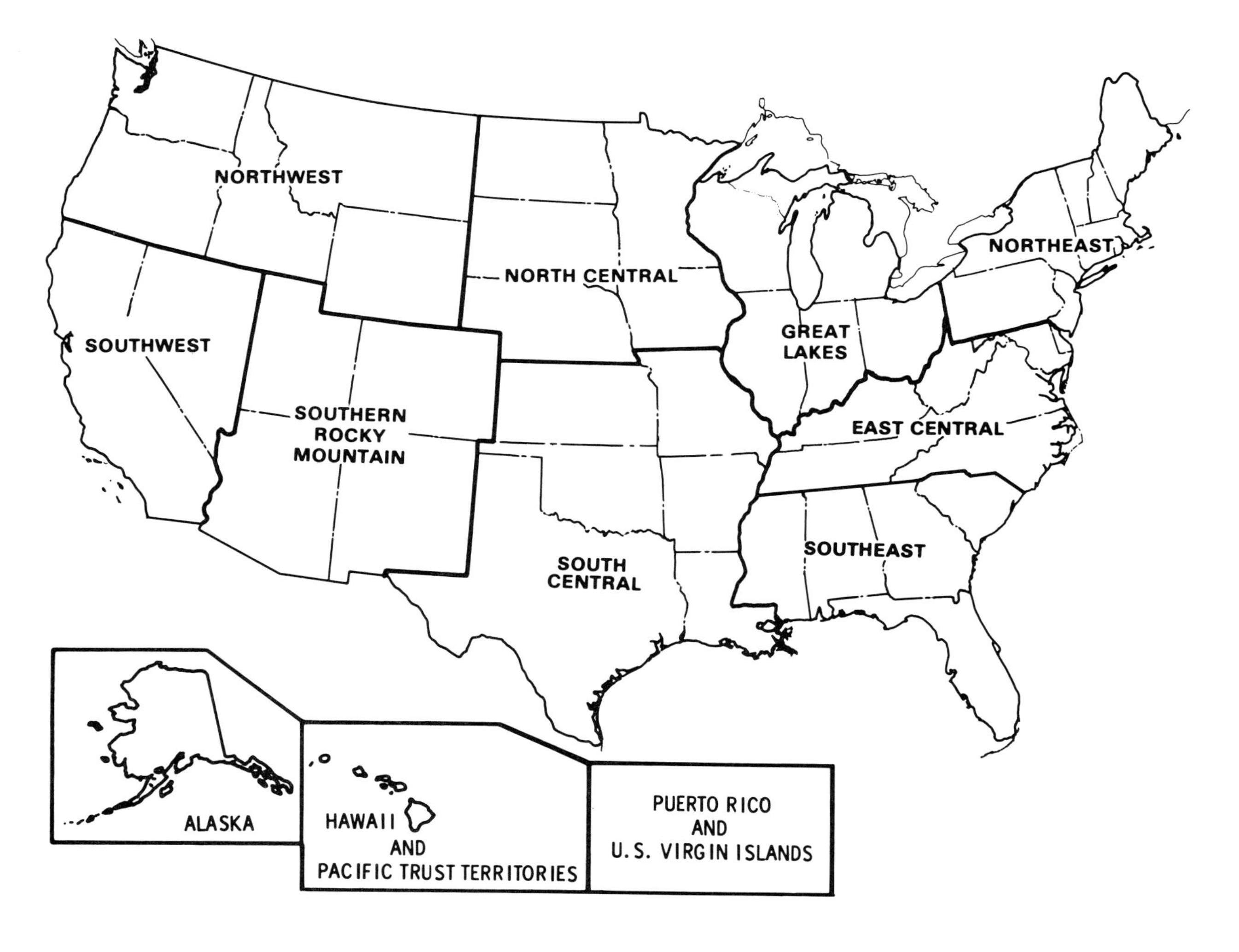

Fig. 16. Geographic division for regional wind energy resource assessment

Accurately estimating the wind characteristics at a site is more difficult. Even in the case where data from a nearby weather station can confidently be applied, there are hidden pitfalls. For example, the elevation , location, or exposure of weather station anemometer may have changed over the period of data collection, thus affecting the mean wind speed and other wind characteristics. Although changes in anemometer location and exposure may be noted in the wind records, the data user must be prepared to look for these changes. This level of expertise or commitment to wind resouce analysis may not be possessed by potential users of small WECS. In most cases, potential users will need assistance in turbine siting and economic or performance analysis, at least until there is widespread experience with wind machines in the user's immediate vicinity.

To assist in this area, the Wind Characteristics Program has developed and tested a short course for siting small wind turbines. A version of this course is now being prepared for wide dissemination.

<u>Siting Methods for Large Machines</u>

When wind turbine generators are used to produce electricity for a central grid, the geographic area in which they can be located may be very large. There is a great deal more freedom in site selection, since the objective is to locate machines in places where their performance will minimize the cost of energy production for the entire grid. Finding such locations requires a fairly extensive program of analysis. Fortunately, the siting of large machines is less restricted by cost because the capital investment is much larger than that for small machines.

The process for finding the locations mentioned above has been divided into the following stages:

- large area analysis,
- mesoscale evaluation,
- candidate site screening,
- candidate site evaluation,
- site development.

During the course of the Wind Characteristics Program,various techniques have been developed for use as tools in the different stages of the site selection process.
These techniques include:

- numerical modeling of the flow (Traci et al., 1978);
- physical modeling of the flow (Meroney et al., 1978);
- biological indicators (Hewson et al., 1979);
- geological indicators (Kolm and Marrs, 1977);
- topographical indicators (Elliot, 1979);
- social and cultural indicators;
- measurements (Walker and Zambrano, 1979).

The applicability of these techniques in the different stages of the site selection process is indicated in Figure 17. A detailed discussion of the techniques and their use is provided in the handbook for siting large WECS. The work is presented in two volumes and the handbook will be completed in March 1980.

	LARGE-AREA ANALYSIS	MESOSCALE EVALUATION	CANDIDATE SITE SCREENING	CANDIDATE SITE EVALUATION	SITE DEVELOPMENT
NUMERICAL MODELING		X			X
PHYSICAL MODELING					X
BIOLOGICAL INDICATORS	?	X	X		X
GEOLOGICAL INDICATORS	X	X	?		
TOPOGRAPHICAL INDICATORS	X	X	X		X
SOCIAL & CULTURAL INDICATORS	X	X	X		
MEASUREMENTS	X	X	X	X	X

Fig. 17. Applicability of techniques to stages of wind turbine siting

A brief description of the stages put forth in the handbook for finding the locations for large machines will illustrate the process of siting a large WECS. The goal of the strategy presented is to identify several WECS farms that could be developed sequentially. In choosing these sites, such issues as land use, accessibility, public acceptance, and proximity to existing transmission lines must be considered in addition to wind power potential. Seasonal and diurnal characteristics of the wind are also important. A proper combination of sites is one in which the net power output best matche the load and generating characteristics of the utility. A flow chart outlining a decision process for developing a system of WECS farms is shown in Figure 18.

Large Area Analysis

The strategy shown in Figure 18 begins with a large area analysis, which includes an examination of pertinent wind resource assessments along with an other available wind related information in the service area. Figure 19 shows a distribution of transmission lines superimposed over a map containin the Washington wind power density from the Northwest regional assessment. One or more potential high wind areas are identified (e.g., dashed box in the central part of the state) and selected for analysis in the next stage of the site selection process.

Mesoscale Evaluation

The second step of the strategy is the mesoscale evaluation. This involves analysis of the flow field in the selected area to find candidate locations for potential clusters of turbines. If strong terrain forcing is apparent and the seasonal and diurnal variations in the forcing of the flow are large at least a full year of wind data is required for the analysis of flow characteristics in the area. Existing data or new measurements should be taken from those key locations that provide the most information on the mass flow through the area. Figure 20 depicts flow affected by the topography in the area. Numerical flow models may be used to assist in this stage of the analysis as well as most of the other tools referred to in Figure 17.

Candidate Site Screening

After a number of sites for multiple WECS installation have been identified, the potential sites are further screened. This screening is accomplished by site visits. During these visits, the surrounding terrain is examined and any small-scale terrain features or obstacles that could affect wind characteristics at the site are noted. Soil conditions are also examined. Site screening is not a time consuming process and, depending on the number of sites, can be completed in a few weeks or months.

Valuable information on wind characteristics at the site are obtained by examining the vegetation and by measuring the wind profile, using such inexpensive systems as wind sensing kites or instrumented tethered balloons. Profile measurements made during site visits are primarily used to identify obvious potential hazards, such as, flow separation and the turbulence and high wind shear that accompany them. One cannot, however, expect to obtain a meaningful wind profile climatology through such a small set of observations.

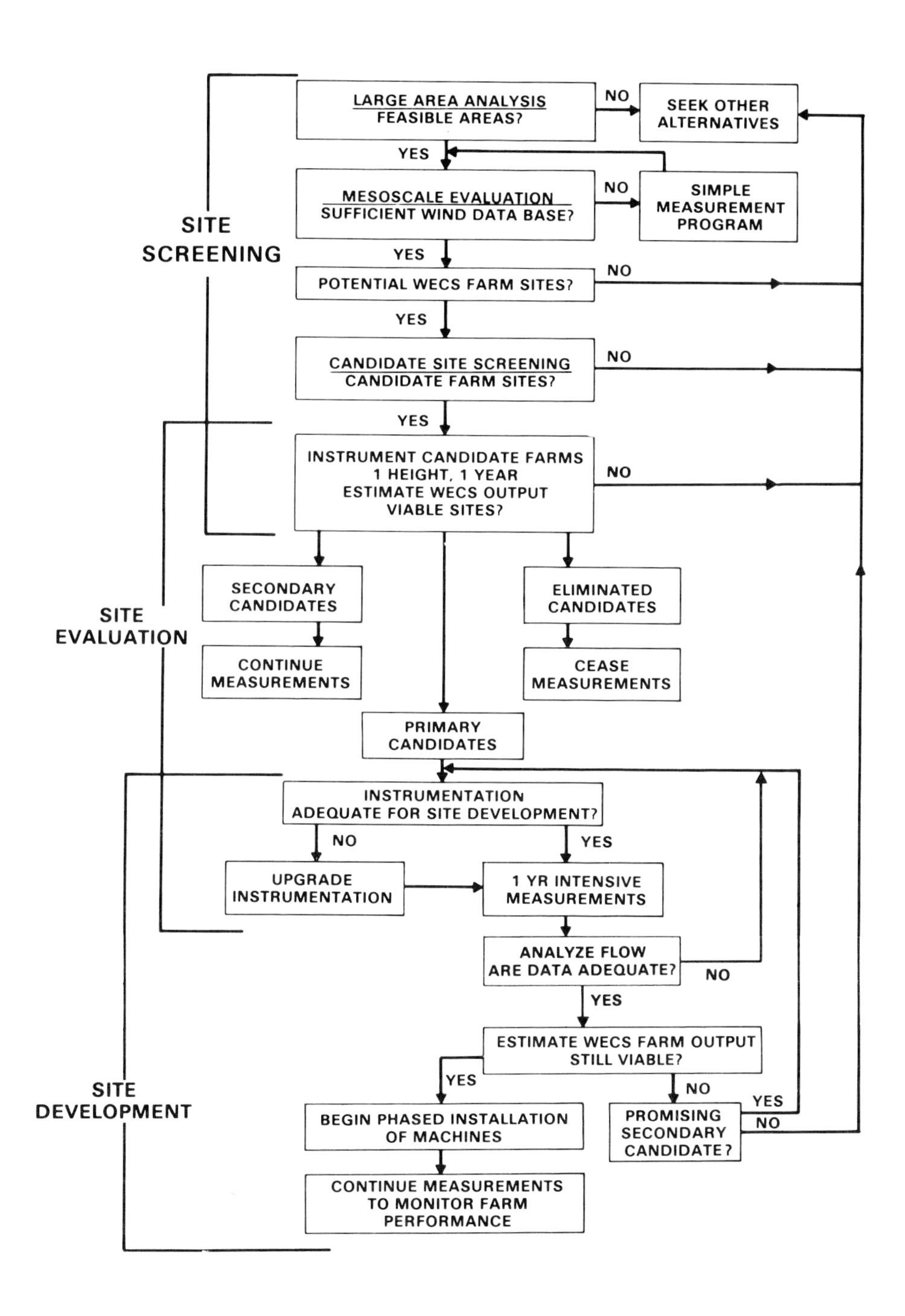

Fig. 18. Decision process for developing WECS farms

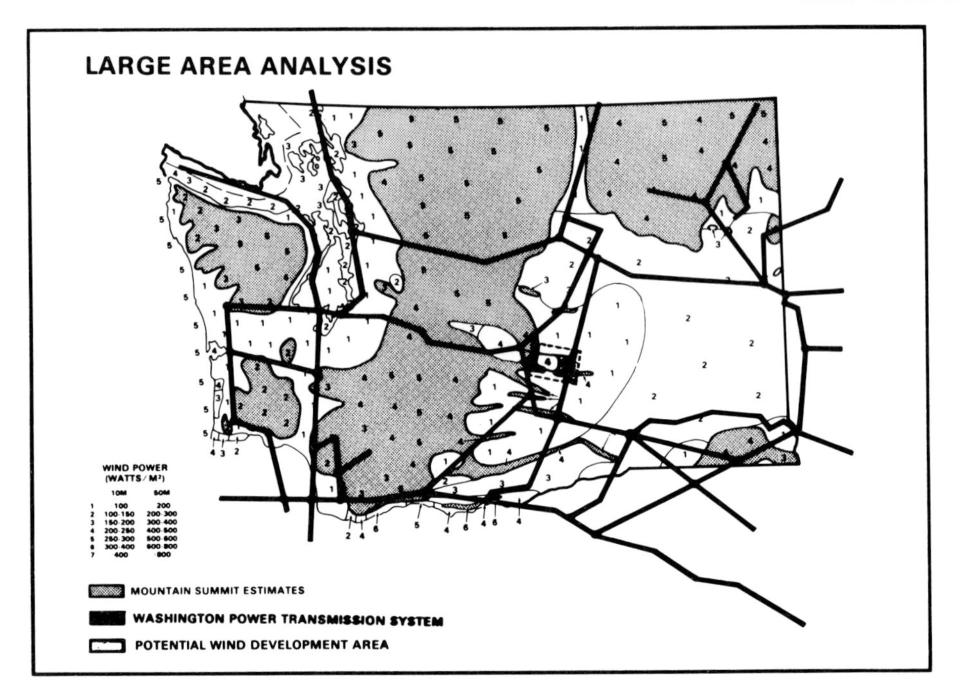

Fig. 19. Transmission lines and wind power densities in the State of Washington

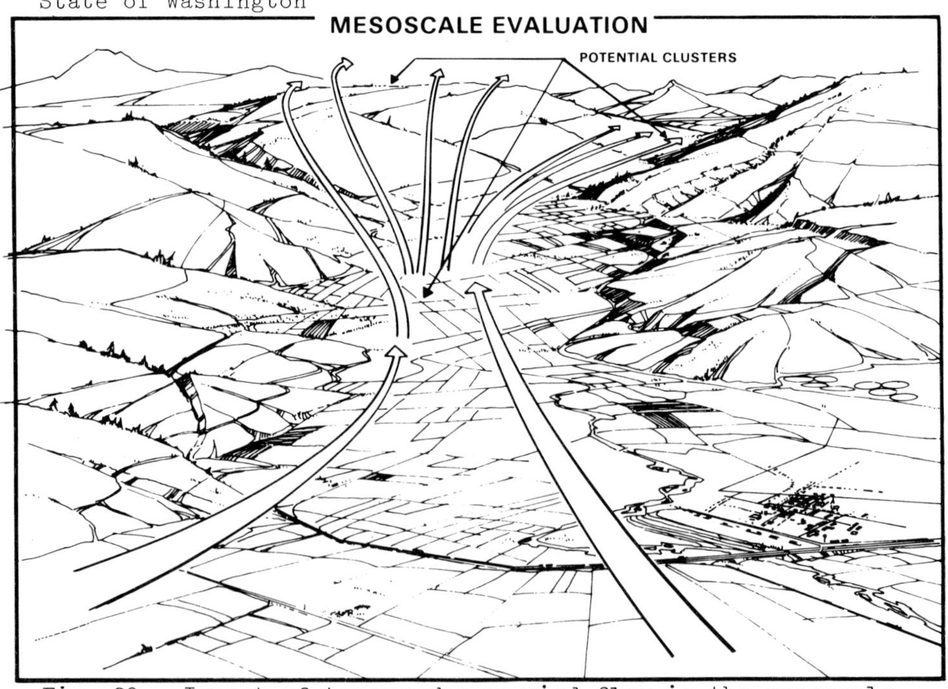

Fig. 20. Impact of topography on wind flow in the mesoscale evaluation process

Candidate Site Evaluation

Candidate site evaluations are used to determine more accurately how wind machines would perform at each site and to ascertain the combination of sites that would result in power output characteristics that best match the needs of the utility.

At this stage of the site selection process, wind data are needed at only one level, i.e., as near hub height as possible. Extrapolating wind speed data from the measurement level to hub height always results in some error, especially if the terrain in and around the site is complex. Wind measurements should be made at enough locations to properly gauge the wind resource. In flat terrain, a single measurement location is sufficient; in more complex terrain, measurements at several places may be required.

Although the length of time data must be collected is uncertain at present, two years of data are probably the maximum period needed for establishing the optimum mix of WECS farms from a set of candidate sites. One year of data, however, should be enough for identifying the leading sites in terms of total annual energy production, unless the year was considered abnormal climatologically. Experience with the 17 candidate sites in DOE's MOD-OA demonstration program showed no significant changes in site ranking by annual energy production when a two-year record was used instead of a one-year record.

Site Development

After a year of collection at candidate sites, a site may be chosen as the prime candidate for a WECS farm; however, data collection will probably continue at the other promising candidate sites. At the primary farm site, a more intensive program of site evaluation then begins; this evaluation seeks to:

- locate each machine in the farm;
- estimate the power output characteristics of the machinery array;
- document those wind characteristics that affect WECS operation and service life.

Such an evaluation requires detailed knowledge of the flow through a section of the atmosphere having a vertical dimension of 100 to 150 meters and horizontal dimensions equal to the size of the farm. If existing instrumentation on the site is insufficient, a more extensive program of measurement and modeling is suggested. After a year of data collection from the expanded system, the wind field is analyzed and the distribution of all important wind characteristics over the area is determined. Likely analysis techniques include both subjective interpretation and modeling. Physical modeling may also be used in determining the layout of a WECS farm. Careful attention should be given to the appropriate spacing of the machines in the area, based on wind turbine wake considerations for all meteorological conditions observed in the area.

SUPPORT FOR DESIGN AND PERFORMANCE EVALUATION

It was strongly implied in the discussion of siting methodologies that it is essential to determine the wind characteristics at a candidate location that might affect the WECS performance or service life. The purpose of this task area is to provide designers with wind characteristics relevant to producing reliable cost-effective wind turbines and to effectively evaluate the performance of these turbines.

Gust Models for Wind Turbine Design

Anyone who has stood outside on a windy day has experienced fluctuations in the wind, called gusts. Wind engineers have long factored wind gustiness into the design structures exposed to the wind. Unfortunately, there is no universally acceptable definition of wind gustiness that can be easily factored into the design of a wind turbine. Figure 21 depicts four different models of the fluctuation of wind speed that may be used in WECS design. In Figure 22, two different gust definitions are shown for the same fluctuation. The differences in the amplitudes and duration in the description of the same fluctuation are quite striking. Another complicating factor is that wind turbines have to be designed for both extreme event considerations and fatigue life considerations.

Because of the proliferation of gust models over the past three years, the Wind Characteristics Program has conducted a review of these models (Powell and Connell, 1980). This report presents an objective comparison of the models and will form a basis for a second generation gust model that may be widely acceptable to wind turbine designers.

Wind Measurements Over the Disk of a Rotor Blade

In the application of the gust definitions in WECS design, the general assumption is that measurements by a single anemometer are representative of conditions encountered by a wind turbine blade. To test the validity of this assumption, measurements have been made with circular arrays of anemometers set in a vertical plane perpendicular to the prevailing wind (Verholek and Eckstrom, 1978). Figure 23 shows the current configuration of anemometers with the approximate blade dimensions of three large turbines superimposed for comparision purposes.

Data are recorded at each anemometer at the rate of 10 samples per second. Figure 24 shows an eight-minute data sample at each anemometer location of an earlier array configuration (12.2m radius). Obviously these traces are not identical, which means at any given time the wind speed is likely to be different at the various anemometer locations. To get a clearer impression of what a turbine blade rotating through this disk might encounter, successive samples may be selected from each time series to simulate the wind record encountered by a rotating blade. A record of this type is shown in Figure 25. The most striking features of this record are the fluctuations caused by the wind shear in the vertical, and the variation of the wind at the center anemometer with respect to the anemometer at the hypothetical rotating blade. The direct application of this type of data into wind turbine design is still evolving, but it does provide valuable information on instantaneous shear that might be considered in WECS design.

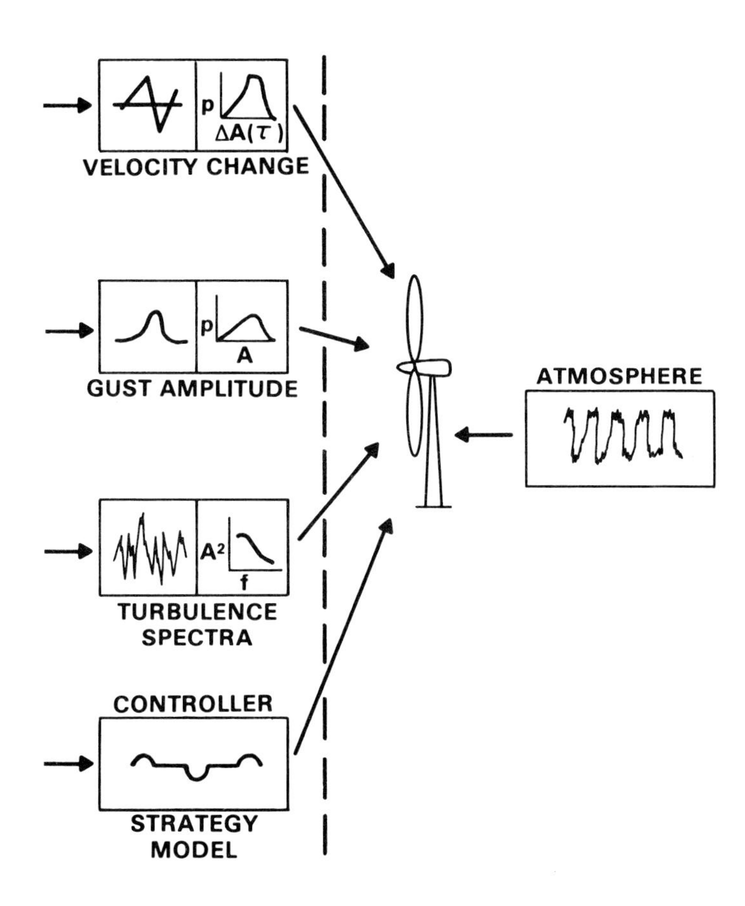

Fig. 21. Four different models of fluctuations of wind speed that may be chosen for criteria used in WECS design

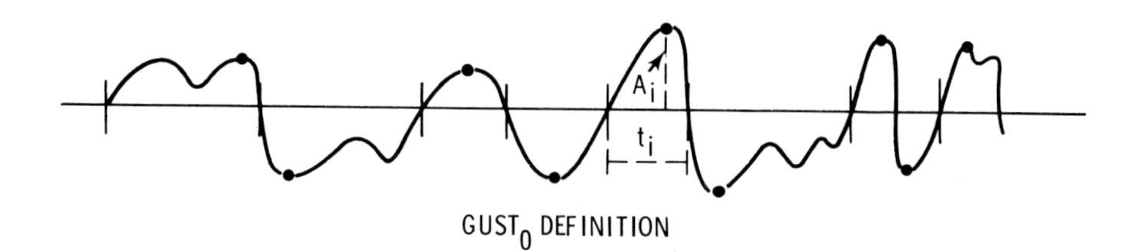

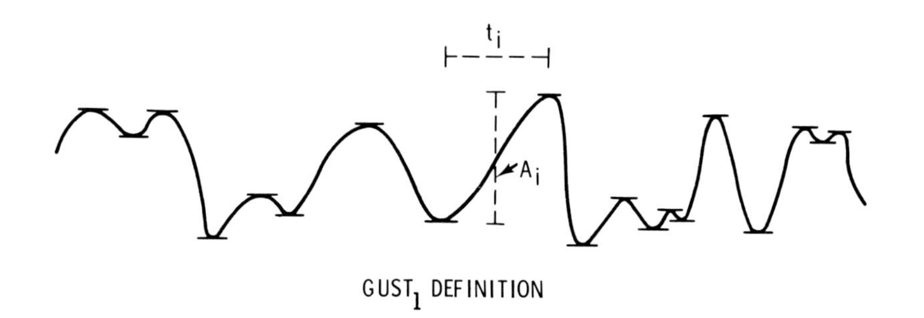

Fig. 22. GUST$_o$ and GUST$_1$ definition

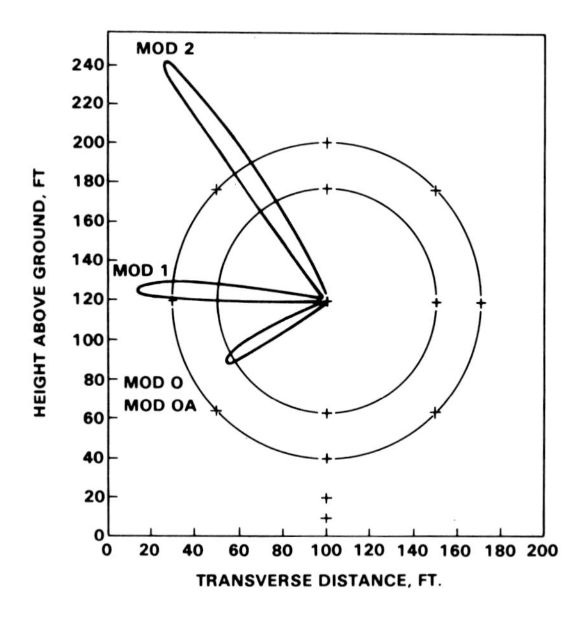

Fig. 23. A vertical anemometer array

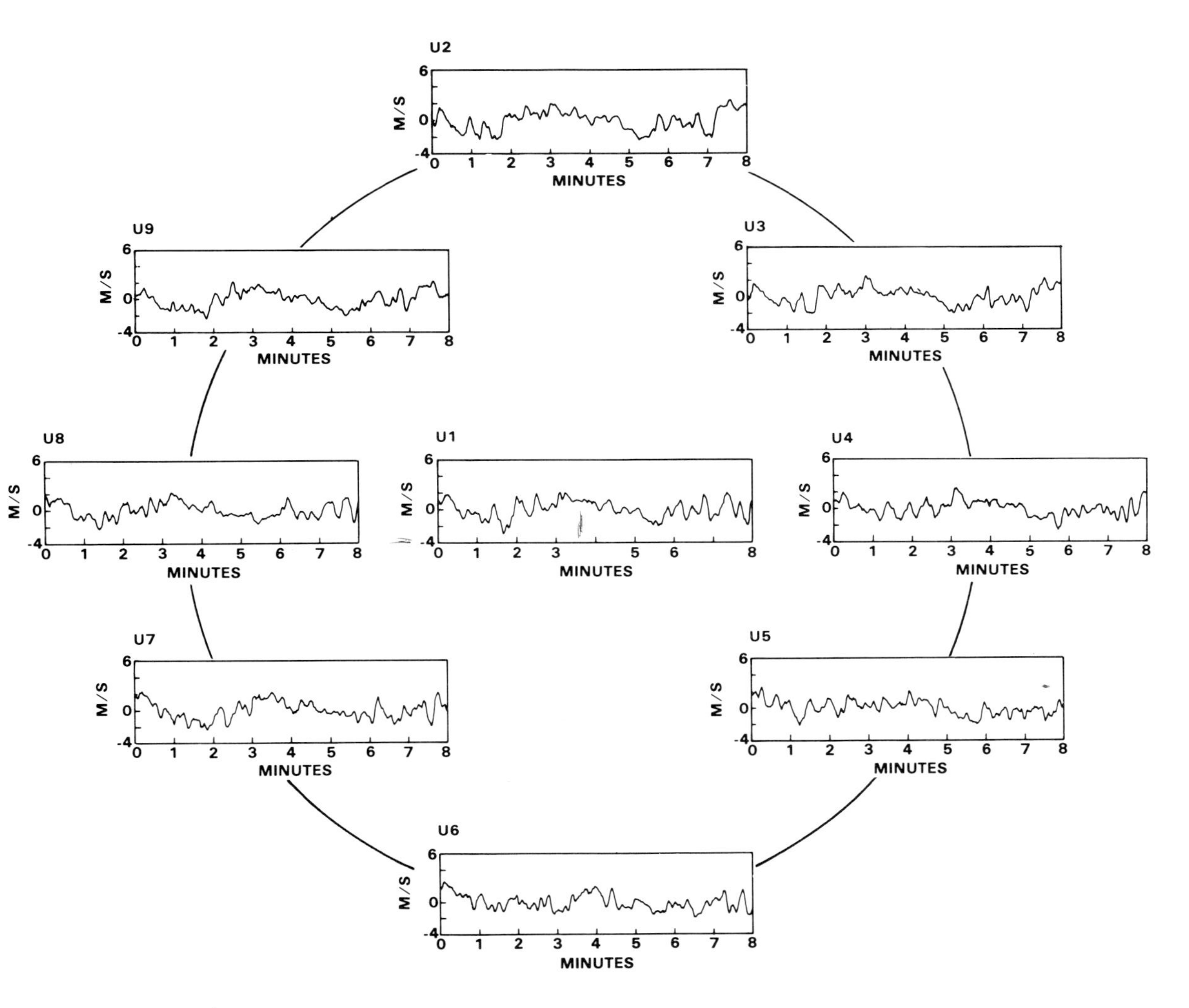

Fig. 24. Cross plane wind speed fluctuations at nine fixed points

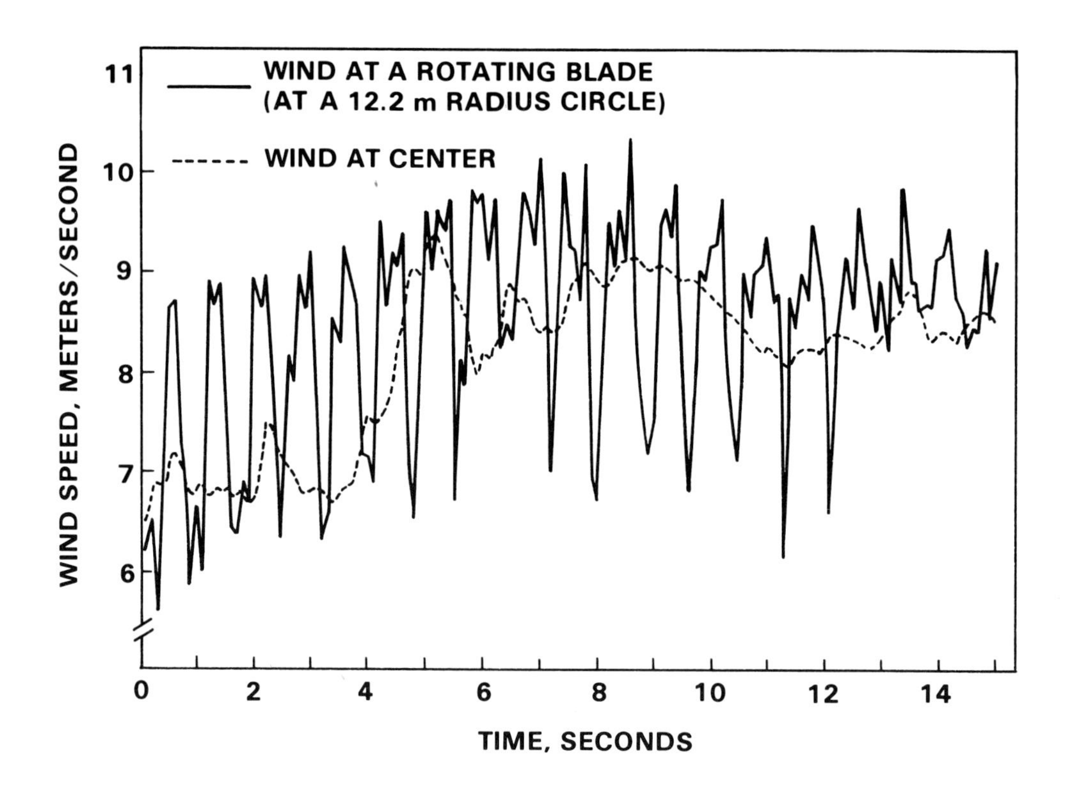

Fig. 25. Wind fluctuations seen by a rotating blade

Another use for data with a fast sampling rate is shown in Figure 26(a) and (b). In this case the measured data (a) is filtered (b) to simulate the wind fluctuations seen by a wind turbine. The turbine is assumed to have enough mass to respond to fluctuations with periods greater than five seconds. It is also assumed that the control system will handle fluctuations with periods greater than 50 seconds. The filtered data record may be analyzed to estimate the turbulence statistics actually experienced by the turbine.

SUPPORT FOR USER OPERATION OF WECS

As the possibility increases that WECS may achieve some degree of penetration into the electrical generation capacity of this country, the effect of a variable resource on their operation is receiving more attention.

Utility Survey

Approximately two years ago the Wind Characteristics Program surveyed eight utility companies ranging in size from a large power pool, down to a very small utility (Wendell et al., 1978). Both utility planners and operations personnel were asked what kinds of wind forecasts would they need with a 10% penetration of WECS. The types of forecasts they would like to have included:

- 0 to 1 hour for dispatching,
- 0 to 24 hours for load forecasting and unit commitments,
- 1 to 7 days for maintenance scheduling.

From the discussions, it was clear that these individuals felt they needed high reliability in the forecasts, but were skeptical about being able to get it. A need was apparent for a close liaison with utilities to better define their needs for forecast reliability and to keep abreast of their evolving needs as the probability of working with wind turbines in their generation mix increases.

Wind Forecaster Survey

During the same period of the utility survey, another survey was conducted to determine the availability of wind forecast products that might be appropriate to the needs of the utilities. Organizations contacted were the National Weather Service, the Navy and Air Force Weather services, and a subsequent working-group discussion between wind forecasters and utility personnel.

The primary conclusion was that available wind forecasts are not appropriate to the needs in wind energy forecasting. There are many wind forecasts being produced, but they are not specific enough to lend themselves well to estimating the power output of a wind turbine. It was also found that, with the exception of the National Weather Service, there was almost no wind forecast verification data being produced to establish reliability estimates for wind forecasts. These findings led to the establishment of wind-forecasting reliability studies within the Wind Characteristics Program.

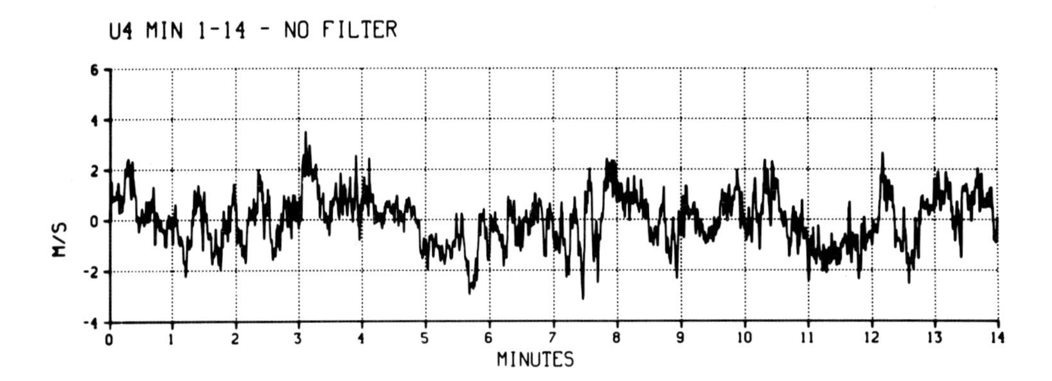

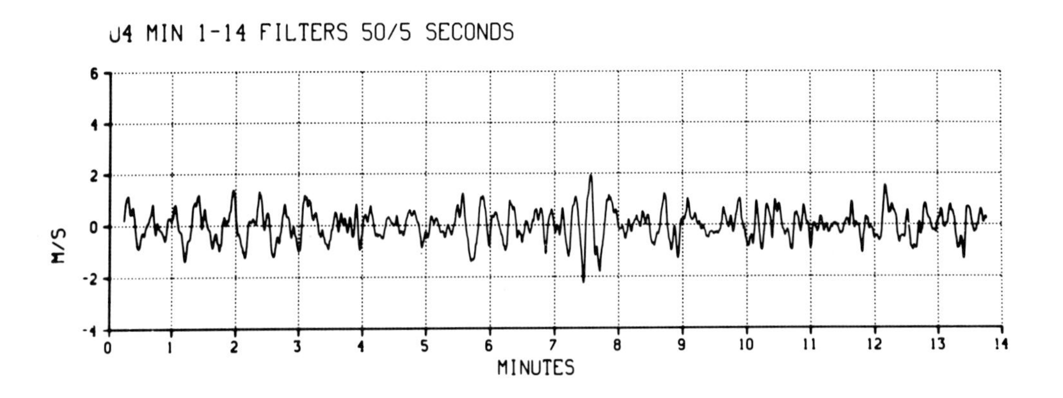

Fig. 26. Anemometer data filtered to simulate fluctuations seen by a rotor

Wind Forecasting Reliability Studies

The reliability studies encompass the two general forecasting methods in use, namely, subjective and objective. Subjective wind forecasts are generally made by an experienced forecaster using some guidance from numerical models. Objective forecasts are computer-produced using dynamic atmospheric models or statistical methods and sometimes a combination of the two. Reliability studies are being conducted on both types of wind forecasts. 24-hourly forecasts are produced and will be verified by PNL with actual data from the sites. The Wind Characteristics Program is also participating in the International Energy Agency's Forecast Verification Program,led by Sweden. PNL is providing assistance in the development of a standard verification code and providing data analysis.

Planned Research

From the preliminary results in the forecast verification effort, it is apparent that upgrading of wind forecasts is needed for (1) WECS arrays, (2) dispatching, and (3) planning. To determine the magnitude of the effort that should be applied to upgrading these forecasts, quantitative studies are planned to assess the impact of wind forecast reliability on WECS operations.

SITE EVALUATION

This program area is a FY 1980 addition to the Wind Characteristics Program. The responsibility for management of the measurement program at the candidate and turbine installation sites was assigned to PNL in late FY 1978 and was carried as a program, separate from the Wind Characteristics Program, and called the Meteorological Validation Program (MVP). The interaction between the two programs grew as data from the candidate sites became useful in other studies (e.g., forecasting verification) and information from the wind characteristics studies were useful in improving measurements and analyses reporting at the candidate sites. As a result, the MVP was merged into the Wind Characteristics Program as a program area with three major task areas: site selection support; site meteorological measurements and analysis; and large-machine site evaluation.

Site Selection Support

As DOE selects turbine installation sites or new candidate sites, PNL provides technical support to the Site Evaluation Boards, Site Selection Panels, and Site Selection Officials. This support includes providing analyses of site data or proposal data in a format which will facilitate the evaluation at hand. Support is also provided in the preparation of panel reports and documentation of proceedings with adequate technical back-up material.

Site Meteorological Measurements and Analysis

This activity includes selection and management of the subcontractor responsible for tower and wind instrument installation, data acquisition,and preprocessing of the data for transmittal to PNL. The meteorological data is then analyzed and a monthly data report prepared for dissemination to NASA and other requesters.

Recently, two turbine parameters, (1) power in kilowatts and (2) the discrepancy between the direction of the wind turbine and the wind direction, have been added to the wind data tapes at the sites where turbines have been installed. A map of the original DOE candidate sites, including those with turbines, is shown in Figure 27.

Large-Machine Site Evaluation

With the advent of the installation of clusters of wind turbines, a more sophisticated evaluation of an area will be required than a measurement and analysis program with a single tower. The effects of local terrain and the effects of wakes on spacing will have to be taken into consideration. Experience gained in this area should provide valuable documented information an procedures for other siting efforts involving clusters of WECS.

SMALL WIND ENERGY CONVERSION SYSTEMS

INTRODUCTION

Operated by Rockwell International, the Rocky Flats Wind Systems Program provides technical and management support to the Department of Energy (DOE) for the development and testing of small wind systems (machines with an output o less than 100 kW) designed primarily for farm,home,and rural use (figure 28)

The Federal Wind Energy Program (FWEP), under which the Rocky Flats (RF) program is funded, is designed to support the earliest possible commercialization of wind power to help meet the national energy requirements. This goal is being pursued by simultaneously developing advanced small wind systems, addressing the technical, economic, and institutional requirements for their use, and stimulating their commercial utilization.

Widespread commercialization of small wind energy conversion systems (SWECS) is dependent upon these systems achieving energy costs competitive with the cost of power obtained from conventional sources. But the achievement of competitive costs (by whatever means) will not in itself assure widespread SWECS use if consumer and institutional acceptance has not been achieved. A close coupling of SWECS testing and development activities with research projects oriented toward potential uses for wind power is required to develop the SWECS industry and create an institutional and consumer environment conducive to SWECS commercialization. The RF Program is designed to provide the comprehensive approach to achieve these goals.

Numerous small manufacturers are offering a variety of wind turbine generators (WTGs) for sale. However, the industry is still developing. Technical data relating to machine output, durability, and behavior under extreme and varying weather conditions must be made available, together with information on types of power output, the compatibility of various system components, and other questions that might arise when installing a wind turbine generator.

Widespread adoption of small wind energy conversion systems (SWECS) is dependent upon these systems being economically competitive. This means that the cost of power provided by a SWECS must offer definite economic advantage for consumers. Improvement in component and system design are two ways to meet this goal. And, while conventional energy costs are expected to esca-

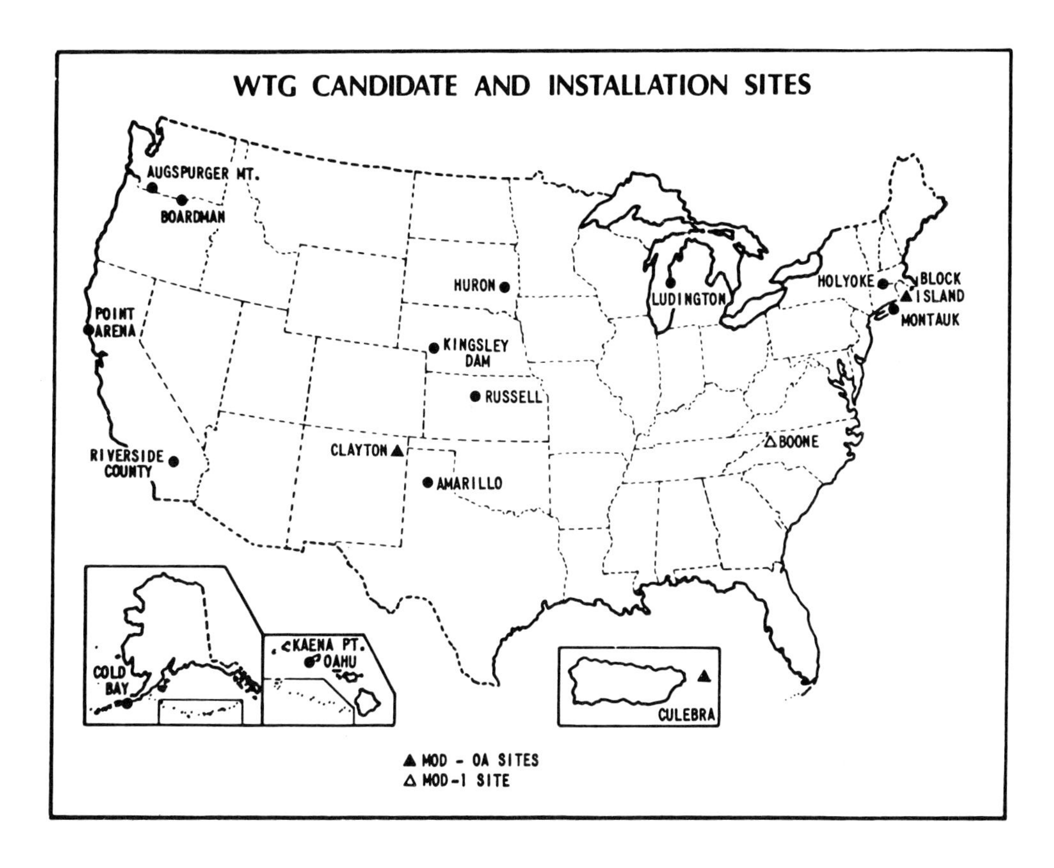

Fig. 27. Wind turbine candidate and installation sites

Fig. 28. Rocky Flats small machine test center

late in the near future, so too will manufacturing and installation costs associated with SWECS. Furthermore, the advent of competitive costs will in itself not necessarily assure widespread use. A close coupling of SWECS development with programs oriented towards potential use for wind power is planned to assure maximum stimulation of the SWECS industry.

OBJECTIVES

The primary objectives of the RF program are to stimulate the development and manufacture of SWECS by the private sector and accelerate the acceptance and use of SWECS by the public. To achieve these objectives, Rocky Flats has been assigned a series of specific functions by DOE. These include:

- Establish and operate a national facility where small wind systems are tested, thereby helping to assess the current state of the art technology and to identify required technology improvements.

- Subcontract activities in research and advanced systems development designed to reduce the cost and improve the reliability of SWECS.

- Provide appropriate technical support for the formation of standards to be used in the manufacture, product testing, data reporting, and installation of small wind systems.

- Disseminate the technical information generated by the program, so that it may be effectively used by potential SWECS users, researchers, manufacturers/distributors, and other DOE agencies/programs.

- Perform other required activities to assure that the program provides a basis for the widespread commercialization of SWECS.

A great many technical, economic, environmental, and social issues have to be resolved before significant benefits from wind power can be realized. Non-competitive costs are the greatest barrier to the use of wind systems. While much has been accomplished, the development of rugged, economical, small wind systems in partnership with the dedicated, progressive SWECS industry remains the program's primary challenge.

ROLE

Rocky Flats' position in the program is the focal point of DOE's Farm and Rural Use (Small) Systems Program Element. Inputs from the applied research and systems studies conducted, and special studies performed under contract to Rocky Flats, provide the basis for initiating development of new wind systems. Multiple contractors will be selected with an emphasis on small business participation whenever possible, to develop widespread industry technology capability and increase the competition for advanced small system designs. Each system will eventually be subjected to rigorous performance and reliability evaluation at the Rocky Flats Test Center followed by field evaluation at other locations. Data gathered during the evaluation process at both Rocky Flats and field sites will be used to determine priorities for further development.

The operation of the RF program requires the exchange of information and data with other FWEP participants. Primary interfaces are with DOE/HQ/Wind Systems Branch for overall program direction and specific project direction; with Battelle-Pacific Northwest Laboratories regarding wind characteristics; and with the U.S. Department of Agriculture Research Service on the machine requirements of farm applications.

TEST CENTER DEVELOPMENT

During FY 1976, DOE authorized the establishment of the Small Wind System Test Center. Since that time the primary activities have been establishing the site, installing towers and wind turbines, and initiating data collection The testing capacity being developed includes component tests, system tests, and environmental tests. Currently, the emphasis is on testing commercially available machines under natural wind conditions.

The test center is on the N.W. corner of the Rocky Flats plant site, approximately 20 km N.W. of Denver, between Golden and Boulder. Since the test center site is on a large flat plains area facing a mountain pass, the wind regime provides a large variety of conditions, from gentle breezes to hurricane-force wind storms.

Weather data is being taken at 10 and 40 meter heights on two meteorological towers and at several 15-meter high portable towers.

Wind speed and direction data are also taken at each SWECS tower. The anemometers are located on booms approximately 1-1/2 to 2 rotor diameters below the SWECS hub height, and provide input to the data collection system.

A well has been drilled at the site to provide water for normal operations as well as a mini irrigation test system. The well pump is used to fill a stock tank from which a 30 gpm (at 50 psi) pump feeds the irrigation to provide power for this test.

The data acquisition system consists of a generation and signal conditioning subsystem, an analog-to-digital conversion subsystem, and a central computer used for system control and permanent data storage. The electrical analog signals generated from sensors mounted on each SWECS are routed to microprocessors housed in an instrumentation and power control shed located at the base of each SWECS tower.

This microprocessor converts the analog signal to a digital signal, performs preliminary calculations, and temporarily stores the digitalized information. The central control and storage computer integrates each microprocessor at the test site on a scheduled basis, acquires the data, and permanently stores it on digital tape for subsequent analysis.

The SWECS power control and storage subsystems consists of a microprocessor control unit and a resistor bank located in the shed at the base of each tower. In addition, there is a 120 volt central power storage bank located in a shed near the computer trailer. The power storage bank consists of 60 deep discharge 2-volt batteries with a capacity of 1760 ampere hours.

TEST CENTER OPERATIONS

A major objective of the activities at Rocky Flats is to test small wind energy systems over a wide range of actual environmental conditions. The data collected from this testing will:

- o Provide consumer information for potential buyers of small wind systems;
- o Provide detailed engineering data for use by developers, manufacturers, and researchers of small wind systems.

Quantitative as well as qualitative data are being collected. The qualitative data includes information on the condition in which a machine is received from the manufacturer, usefulness of operating manuals, installation difficulties which were encountered, and operational experience. During the testing process, Rocky Flats personnel interface directly with the manufacturers of machines under test to discuss problems encountered and provide them with current operational data.

Long-Term Data

Long-term data results from continuous testing 365 days a year. The data collected is restricted to two areas of interest: (1) basic machine performance (rotor RPM, generator voltage and amperage, etc.); and (2) the general wind characteristics (velocity, direction) in which the machine operated during the test period.

Intensive Testing Data

Detailed information is also obtained relative to structural and/or performance characteristics of wind system components during periods of high winds and/or unusual environmental conditions. The high volume of data collected during intensive testing periods is obtained from such sensors as strain gauges, accelerometers, and thermocouples. Measurements are made in such critical components as blades, rotor shafts, gearboxes, and towers.

SWECS tested at Rocky Flats are classified into three broad categories. The first category is the "commercially available" machine. If the manufacturer has delivered at least three units and, of these three, at least one has been installed and made operational, the machine is considered to be commercially available. The second type is the "commercial prototype." This category requires that development has proceeded to the stage where at least one machine has been built and has been operated but is not in production. The third category encompasses those machines being developed under government contract. These machines include the 1 kW high reliability, 8 kW, and 40 kW developmental machines. Samples of machines now being tested are shown in Figure 29.

SYSTEMS DEVELOPMENT

Under the Systems Development task element of the RF Program, outside organizations are funded to develop advanced SWECS to achieve reduced wind-generated energy costs.

Needs have been identified for low-cost SWECS of approximately 8 kW output for home or farm use. Also, numerous applications exist in deep well irrigation, general farm/ranch applications, and for small isolated communities

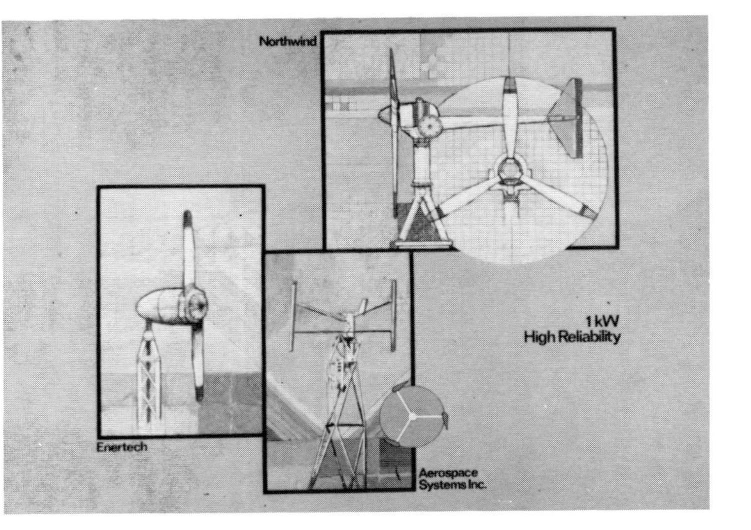

(a)

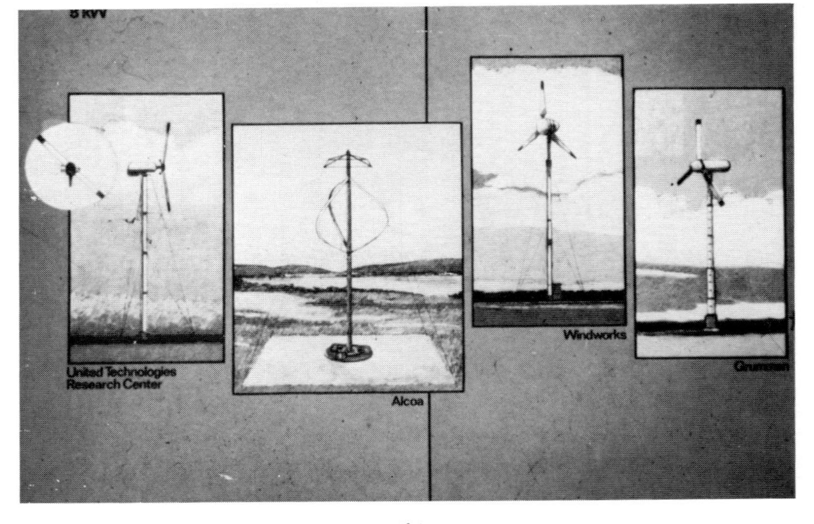

(b)

Fig. 29(a). Selection of 1 kW machines.

Fig. 29(b). Selection of 8 kW machines.

Fig. 29(c). Selection of 40 kW machines.

and industries for a machine of 40 kW output. Low system cost is particularly vital in each of these cases. The third major need was for a high-reliability system, developing approximately 1-2 kW, for remote locations where the cost of installing and maintaining conventional power systems is high. Possible applications in this category are communications repeater stations, pipeline glavanic protection, remote seismic monitoring stations, and off-shore navigational aids.

Contracts were awarded in 1977 for the 1-2 kW and the 8 kW projects. Two 40 kW contracts were awarded in late 1978. Multiple awards were made, with small business participation in both the 8 kW and High Reliability Programs.

FIELD EVALUATION PROGRAM

To enable the collection of data and operational experience under widely different environmental conditions, a Field Evaluation program is now underway. The major objectives of this program are: (1) to establish procedures for small wind systems interconnected with utilities, and accelerate the establishment of a standard rate structure for such applications, and (2) to stimulate the small wind systems industry. The program involves the installation and testing of commercially available and federally-developed wind turbines at selected user sites.

The program scope calls for the purchase of 120 SWECS to be installed in 56 states and territories. Multiple sites are now being selected in each state, 56 machines have already been delivered. By the end of the FY 80 (30 September 1980), SWECS installation will be complete in 35 states. By July 1981 installations will be complete in all states and territories.

STANDARDS DEVELOPMENT

The objective of this task is to provide technical support for the formation of consensus standards by private industry and their acceptance in the manufacture, product testing, and reporting of data on small wind systems. The development of data on these standards will aid in the acceptance of small wind systems by consumers, by the financial and insurance industries ,and by utilities with which SWECS will be interconnected.

RF developed a comprehensive plan for providing technical support to standards development. Under the plan, the American Wind Energy Association (AWEA) would be private industry's focal point for establishing voluntary consensus of institutional/consumer needs and conducting workshops attended by representatives from industry, utilities, financial institutions, consumer organization, and other groups affected by standards.

FUTURE DEVELOPMENTS

The near future will see an expansion of the advanced small machine development and test program managed by Rocky Flats. At the test site, atmospheric testing already underway will be supplemented by vibration testing of SWECS and subassemblies, dynamometer testing of SWECS generator/gearbox assembly, controlled testing, and environmental chamber testing.

The first series of 1-2 kW, 8 kW, and 40 kW advanced machines was delivered to Rocky Flats for testing in late 1979 and early 1980. It is anticipated that additional machines will be obtained through competitive design procurements as development needs and requirements are defined by marketing and application studies and the results of ongoing tests at Rocky Flats. Additions to the test site are already underway with an expansion of the data acquisition system and the number of test pads required to accomodate new machines.

A major effort in SWECS Supporting Research and Technology was initiated by Rocky Flats in FY 1979. Research on SWECS dynamics, mechanical and electrical characteristics, aerodynamics, and various components which have a significant influence on SWECS costs is now being conducted by several subcontractors.

SWECS market studies, institutional studies, and a study of the operational aspects of interconnecting large numbers of dispersed SWECS with utility systems began in FY 1979 under the new RF Special Studies task. Technical support to standards development has begun with analysis of the impact of standards, which will define the types of standards which would benefit the industry, consumers, various institutions, and utilities.

A comprehensive information dissemination plan is being developed to promote the use of small wind systems and provide pertinent product information generated by the test center and the field evaluation program. Annually updated guides to available systems, up-dated user's guides (designed for use with small WECS siting handbooks prepared by Battelle-Pacific Northwest Labs), and information and data packages for various classes of small wind turbines and related products will be made available to manufacturers, consumers, researchers, and others.

In coming years, the focus of the Rocky Flats Wind Systems Program will be on the development of cost-competitive hardware, the development of a solid technology base and a sound SWECS industry, the resolution of barriers to SWECS commercialization, and the dissemination of information vital to promoting the use of small wind systems.

LARGE WIND ENERGY CONVERSION SYSTEMS

INTRODUCTION

Since 1973, the United States Government has sponsored an expanding research and development program in wind energy in order to make wind turbines a viable technological alternative to existing electrical generating capacity. The current U. S. Wind Energy Program, under the sponsorship of the Department of Energy, is directed towards the development and production of safe, reliable, cost-effective machines which will generate significant amounts of electricity.

One element of the U. S. Wind Energy Program is Large Horizontal Axis Wind Turbine Development, which is being managed by the NASA Lewis Research Center. This activity consists of several ongoing wind system developments oriented primarily toward utility applications (Glasgow and Robbins, 1979). Four wind turbine projects designated the Mod-0, Mod-0A, and Mod-2 are part of the current development program for large, horizontal-axis wind turbines in the U. S. The machine configurations are illustrated in Figure 30.

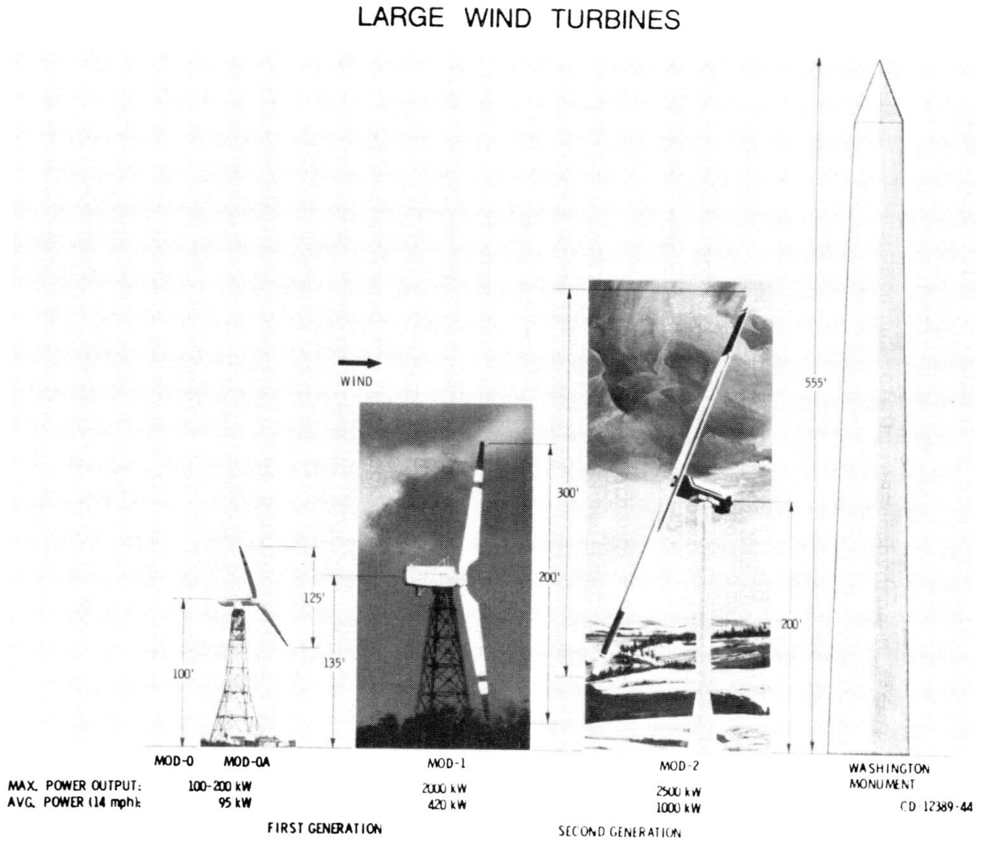

Fig. 30. Large wind turbines

In addition to the configurations currently under development for testing, efforts aimed at achieving lower machine costs have been initiated in 1979. These will include an advanced MW-class wind turbine project and an advanced 200 to 500 kW wind turbine project.

The machine design and technology development projects have been supported by substantial analysis and hardware/material testing. These include efforts to improve the methods of structural dynamic analysis, assessment of utility interface problems, testing of component materials and evaluation of new blade concepts by analysis, laboratory testing of blade sections, and operational testing of full-scale blades. This paper presents an overview of the NASA wind turbine activities. More detailed descriptions of the projects are presented in Lowe (1979) and Rambler and Donovan (1979).

DEVELOPMENT STATUS OF LARGE WIND TURBINES

Mod-0

The current program of research and technology development on large, horizontal-axis wind turbines was initiated with the Mod-0. The Mod-0 is a 2-bladed, 125-foot diameter, research wind turbine rated at 100 kW. This machine was designed by the Lewis Research Center. The Mod-0 has aluminium blades and the rotor is located downwind of the tower. However, Mod-0 has also operated with the rotor upwind of the tower to assess the effects on system structural loads and machine control requirements. The rotor speed is maintained at a constant rpm by rotating or pitching the blades about their lengthwise (spanwise) axes to control the aerodynamic torque imparted to the rotor as the wind speed varies. This type of speed control is referred to as full-span pitch control.

The nominal rotational speed of the Mod-0 is 40 rpm, but a belt drive incorporated in the drive train system (see Figure 31) has permitted the machine to be run at several different speeds for test purposes. Power is transmitted from the rotor through a speed-increasing gearbox to a synchronous generator operating at 1800 rpm to produce 60-hertz power.

The entire assembly illustrated in Figure 30 is mounted on a steel, open-truss tower. This assembly is oriented to the wind by a yaw control mechanism. With a change in wind direction, the yaw control system orients the entire assembly using a hydraulic yaw drive connected to a large diameter ring gear.

The Mod-0 is installed at NASA's Plum Brook facility near Sandusky, Ohio, and became operational in the Fall of 1975. It is being run in an automatic, unattended mode and synchronizes routinely with the Ohio Edison utility network. It has proved to be a valuable engineering test bed for evaluating advanced design concepts and validating the analytical methods and computer codes which are being used to design advanced machines.

Mod-0A

The Mod-0A Project will place four prototype units of the Mod-0 class into utilities to gain early in-service experience. The Mod-0A is essentially

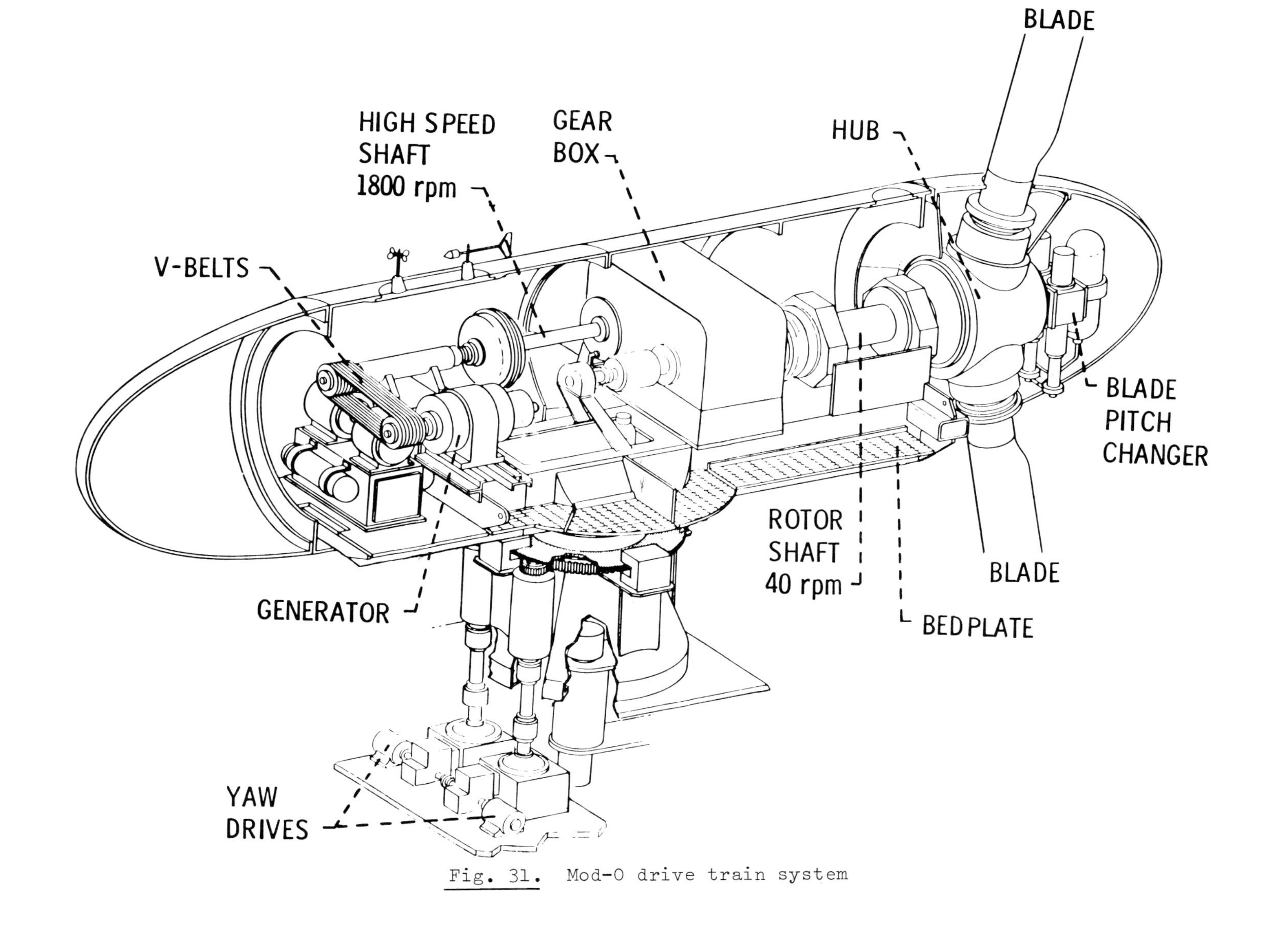

Fig. 31. Mod-0 drive train system

the same design as the Mod-0 except for a larger generator (200 kW) and larger gearbox. The Westinghouse Electric Corporation of Pittsburgh, Pennsylvania is the prime contractor responsible for assembly and installation. The blades are built by the Lockheed California Company.

The first Mod-0A was installed at Clayton, New Mexico; the first rotation occurred in November of 1977. Following a checkout period, Lewis turned the machine over to the City of Clayton in March 1978 to operate as an integral part of their utility system. The machine has operated successfully; it is operationally compatible with the utility grid and has generated 2 to 3 percent of the energy at Clayton since the machine was activated. As expected of the first machine in service, machine hardware problems have been encountered and have been corrected as they occur.

A second Mod-0A was installed at Culebra, Puerto Rico for the Puerto Rico Water Resources Authority and was activated in July 1978. A third Mod-0A is installed at Block Island, Rhode Island and first operated in May 1979 for the Block Island Power Company. Photographs of these three operating Mod-0A wind turbines are shown in Figure 32. A fourth Mod-0A is planned for Hawaii and will be activated in 1980 on the Island of Oahu for the Hawaiian Electric Company.

The Mod-0A project has demonstrated the technical feasibility of wind turbines in utility applications. It has provided valuable in-service testing of hardware and operations to help guide technology development.

Mod-1

The Mod-1 project was started in 1974. The Mod-1 is a 2-bladed, 200-foot diameter, wind turbine with a rated power of 2000 kW. The blades are steel and the rotor is located downwind of the tower. Full span pitch is used to control the rotor speed at a constant 35 rpm. The gearbox and generator are similar in design to the Mod-0A but, of course, are much larger. The tower is a steel, tubular truss design. The General Electric Company, Space Division, of Philadelphia, Pennsylvania is the prime contractor for designing, fabricating, and installing the Mod-1 (G.E., 1979). The Boeing Engineering and Construction Company of Seattle, Washington, manufactured the two steel blades. A single prototype is installed at Boone, North Carolina (Figure 33) The Mod-1 machine is undergoing a thorough checkout prior to being turned over to the local utility, the Blue Ridge Electric Membership Cooperative.

Mod-2

The Mod-2 project was initiated in 1976. Three machines are currently being fabricated. These are 2-bladed, 300-foot diameter, wind turbines with a 2500 kW rating. These machines are being designed with a new technology base developed as a result of research and development efforts on Mod-0, Mod-0A, and Mod-1. Because of this, the Mod-2 is referred to as a second generation machine. The rotor will be upwind of the tower. Rotor speed will be controlled at a constant 17.5 rpm. In order to simplify the configuration and achieve a lower weight and cost (the cost of these machines is closely tied to weight and complexity), the use of partial span pitch control is being incorporated rather than full span pitch. In this concept, only a portion

CLAYTON, NEW MEXICO

CULEBRA, PUERTO RICO

BLOCK ISLAND, RHODE ISLAND

Fig. 32. Mod-OA wind turbines

Fig. 33. Mod-1 wind turbine

of the blade near the tip (outer 30% of the span) is rotated or pitched to control rotor speed and power. To reduce the loads on the system caused by wind gusts and wind shear, the rotor is designed to allow teeter of up to 5 degrees in and out of the plane of rotation. This reduction in loads saves weight and, therefore, cost in the rotor, nacelle, and tower.

The Mod-2 tower is designed to be "soft" (flexible) rather than "stiff"(rigid). The softness of the tower refers to the first mode natural frequency of the tower in bending relative to the operating frequency of the system. For a two-bladed rotor the tower is "excited" twice per revolution (2P) of the rotor. If the resonant frequency of the tower is greather than 2P it is referred to as "stiff". Between 1P and 2P it is generally characterized as "soft" and below 1P as "very soft". The stiffer the tower, the heavier and more costly it will be. The tower's first mode natural frequency must be selected to be sufficiently displaced from the primary forcing frequency (2P) so as not to resonate. Care must also be taken to avoid higher mode resonances.

The tower is a welded steel, cylindrical shell design. This design is more cost-effective than the stiff, open-truss tower. The gearbox is a compact, epicyclic design which is lighter weight than a paralled-shaft gearbox such as used on Mod-1. The nacelle configuration of the Mod-2 is illustrated in Figure 34.

The Boeing Engineering and Construction Company is the prime contractor for designing, fabricating, and installing the Mod-2. Current plans call for 3 prototype units to be installed at a single site during 1980-81. The site has now been selected and is situated in the southern part of the State of Washington. An artist concept of the Mod-2 is shown in Figure 35.

WIND TURBINE COST-OF-ELECTRICITY

The cost of electricity (COE) in cents per kilowatt hour is plotted as a function of site mean wind speed (at 10m) in Figure 36 for the NASA wind turbines either operational or under development. This COE plot reflects the machine capital costs for the second units that are built and assumes that the machines will operate and generate electricity 90 percent of the time that the wind is in the proper speed range.

The cost of electricity (COE) produced by wind turbines is computed as follows:

$$\text{COE (cents/kWh)} = \frac{(\text{Capital Cost,\$})\ (\text{Fixed Charge Rate, \%})}{(\text{Annual Energy, kWh})} + \frac{(\text{Annual O\&M costs, \$})\ (\text{Levelizing Factor})\ (100)}{(\text{Annual Energy, kWh})}$$

The cost of electricity is taken to be at the output of the installation's step-up transformer. Capital cost, O&M costs, annual energy production, fixed charge rate, and levelizing factor, are briefly discussed in the following paragraphs.

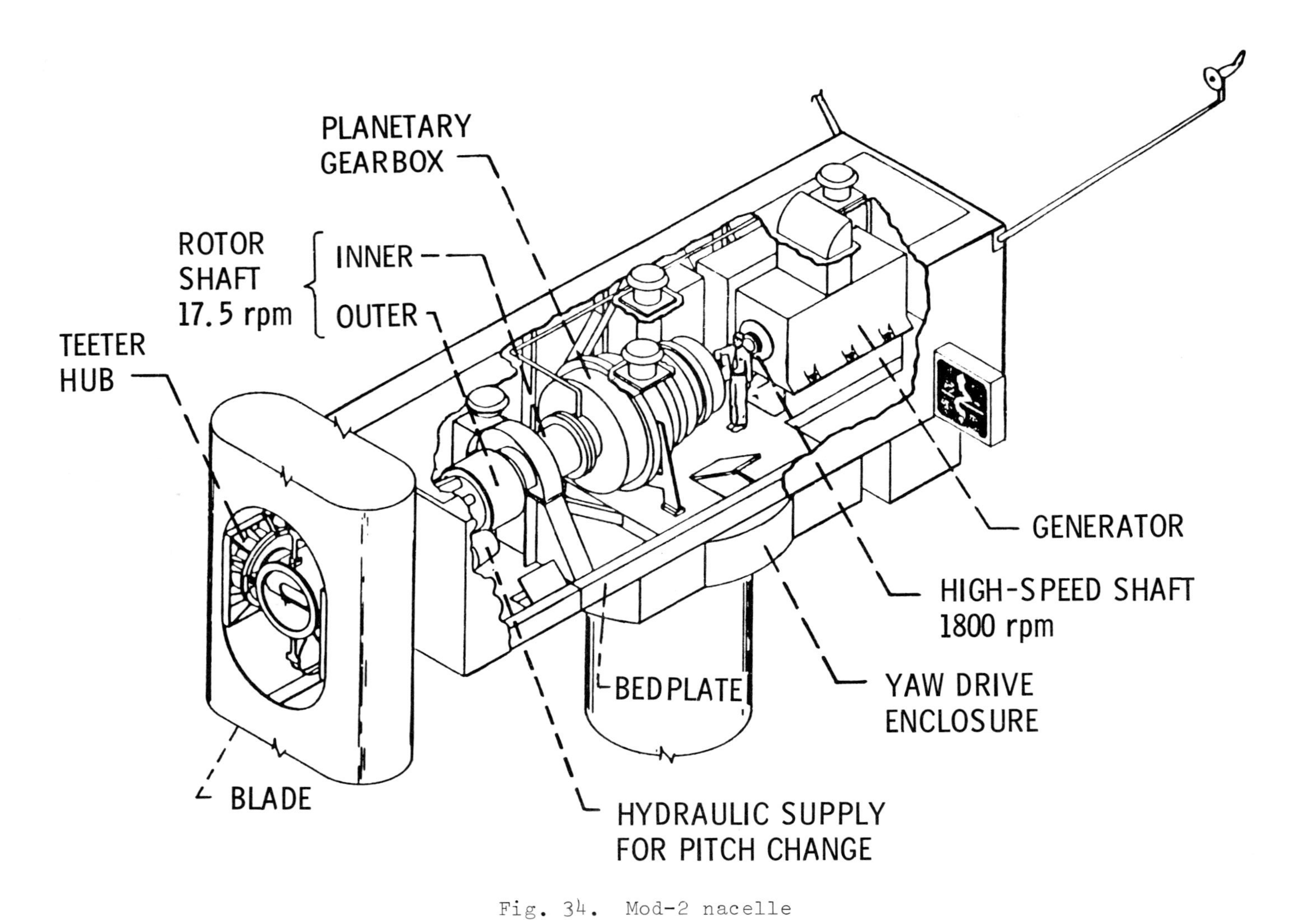

Fig. 34. Mod-2 nacelle

Fig.35. Artist concept of Mod-2

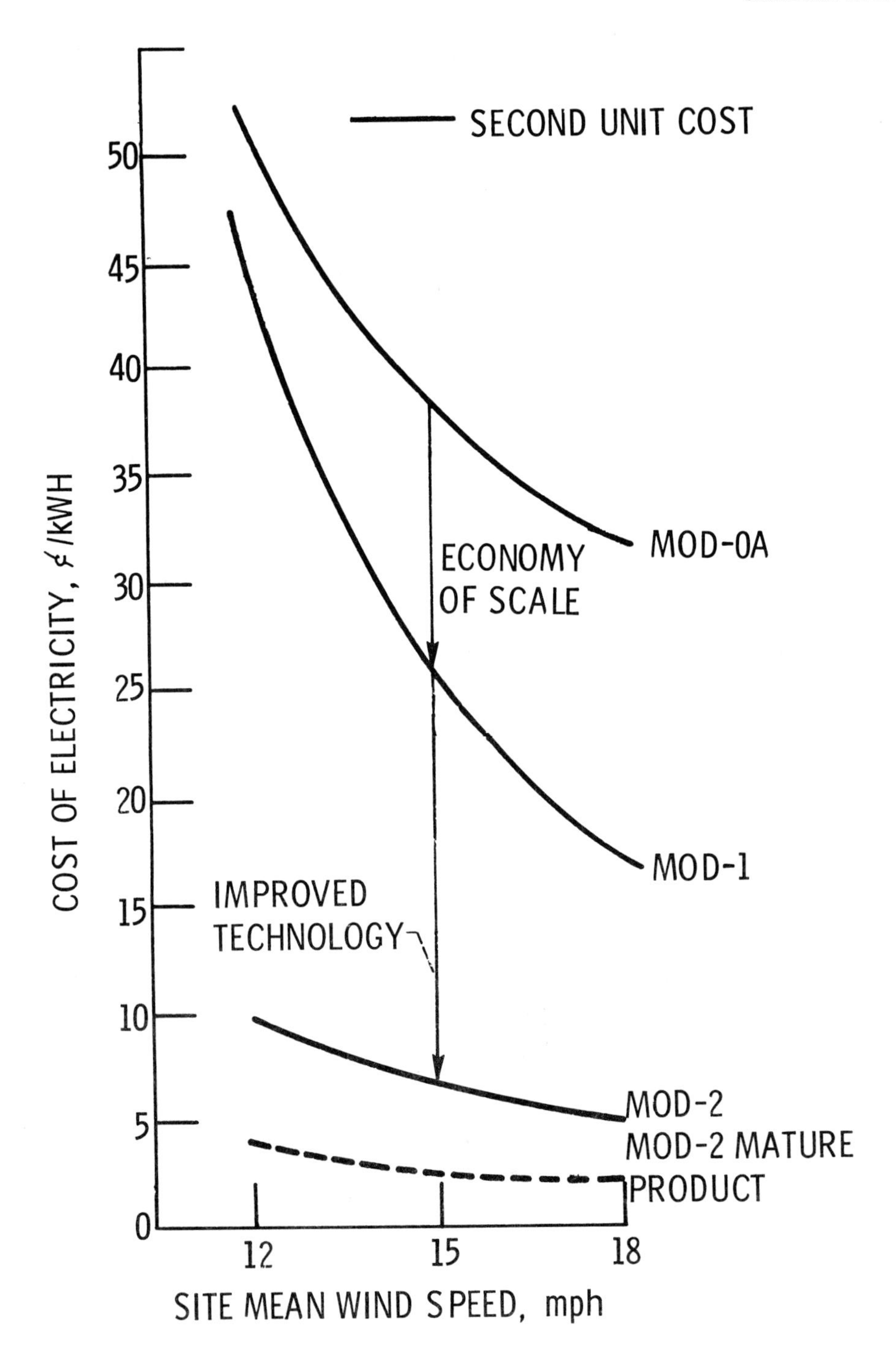

Fig. 36. Cost of electricity

Capital Costs

The installed equipment costs of the prototypes of the large, horizontal-axis wind turbines currently under development are (1977$):

	Mod-OA	Mod-1	Mod-2
Cost ($M)	1.61	5.40	3.37
($/kW)	8050	2700	1350

Second-unit costs are quoted so as not to include the nonrecurring costs associated with the first portotype unit.

The Mod-OA second-unit prototype cost represents NASA Lewis's estimate of the cost that would be required to build a second unit identical to the first unit at the Clayton, New Mexico site,and reflects the knowledge and experience gained as well as the actual costs incurred in that first installation. For Mod-1, the second-unit prototype cost is based upon an estimate made by the General Electric Space Division after they completed the fabrication and testing of the first Mod-1 prototype. The Mod-2 second-unit prototype cost is based upon the current estimate of the installed equipment for the second prototype unit to be built by the Boeing Engineering and Construction Company.

Fixed Charge Rate

The fixed charge rate (FCR) is a capital levelizing or annualizing factor which accounts for the return to investors, depreciations, allowance for retirement dispersion, income and other taxes, and other items,such as,insurance and working capital. It is a function of the design life of the unit, the general inflation rate, the debt/equity ratio of the utility, and other financial parameters, such as, the weighted average cost of capital.

A fixed charge rate of 18% has been assumed in computing the COE for large, horizontal-axis wind turbines. This is a representative value for investor-owned utilities, assuming a general inflation rate of 6%, no allowance for tax preferences, an after-tax weighted average cost of capital of 8.0% (10% before tax), and a 30-year life.

Levelizing Factor

In order to correctly compute the total levelized revenue requirements or COE of a wind turbine (or any utility power plant for that matter), expenses such as O&M costs, which will tend to increase with time due to inflation (and thus result in a variable stream of annual costs), must be levelized before adding them to the levelized capital investment.

Levelization of expenses can be accomplished by multiplying the first year's expense by a levelizing factor. The levelizing factor is a function of the general inflation rate, the cost of capital, and any real escalation (above inflation) to which the expense may be subject. Using the assumed values of economic parameters described above and a 0% real escalation rate on O&M costs, the corresponding levelization factor is 2%.

Annual Operations and Maintenance Costs

Sufficient operational data is not yet available to determine an appropriate O&M cost for the prototype units. Furthermore, because the prototype units are aimed at providing in-service testing and hardware qualifications, larger O&M costs will be experienced in these early prototypes than are expected from production units. The annual levelized O&M cost for the prototypes was assumed to equal 2% of the total capital investment. A total fixed charge rate of 20% (18% on capital plus 2%for O&M) was therfore applied to the total capital investment to compute the COEs in Figure 36.

During the Mod-2 design, Boeing performed detailed O&M estimates for production wind turbines operating in a 25 unit cluster. These O&M estimates are approximately 1% of the capital costs,thereby making the 2% estimates reasonable for the second unit COE estimates.

Annual Energy

The annual energy output for any horizontal-axis wind turbine can be computed for a specific wind speed duration curve by computing the power output at each wind speed and integrating it over the appropriate time duration for each wind speed. The power output as a function of wind speed for the Mod-OA, Mod-1, and Mod-2 wind turbines is shown in Figure 37.

The annual energy produced by the Mod-OA, Mod-1, and Mod-2 is illustrated in Figure 38 as a function of the mean wind speed at 30 feet above the ground. These annual electrical energy production curves account for the aerodynamic, electrical, and mechanical losses up to the busbar (output side of wind turbine's step-up transformer). They also include a 90% availability factor, i.e., the wind turbine is assumed to be available for service 90% of the time that the wind speed is in its operating range.

Estimated Cost-of-Electricity for Prototype Large Wind Turbines

The cost of electricity (COE) for the second prototype units of Mod-OA, Mod-1, and Mod-2 is shown in Figure 36 versus the site mean wind speed. The installed equipment costs, annual energy estimates, estimated O&M fixed charge rate, and levelizing factor discussed above were used in computing these COEs.

As noted in Figure 36, the significant reduction in COE from Mod-OA to Mod-1 is mainly attributable to economy of scale. The COE reduction from Mod-1 to Mod-2 is mainly the result of improved technology, i.e., moving from a rather stiff and heavy design to a relatively soft and lighter weight design. The COEs of the prototype units displayed in Figure 36 are not low enough to be generally attractive to utilities. However, in quantity production the capital costs for Mod-2 are expected to decrease substantially. The dotted curve shows the projected Mod-2 costs for the 100th production unit. The current estimate of the installed equipment cost (1977$) for the 100th production unit of Mod-2 is as follows, along with a current estimate of weights.

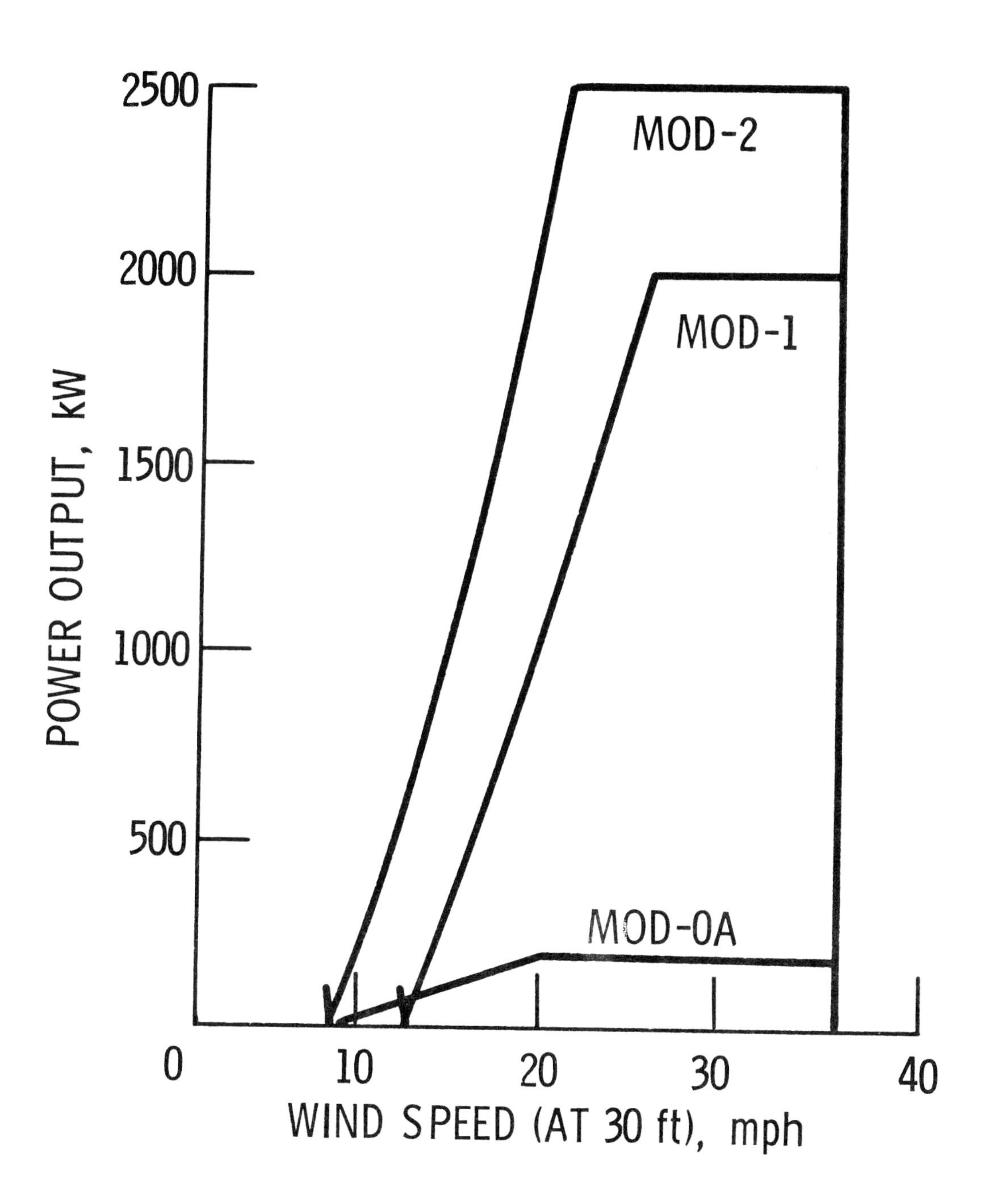

Fig. 37. Wind turbine operating range

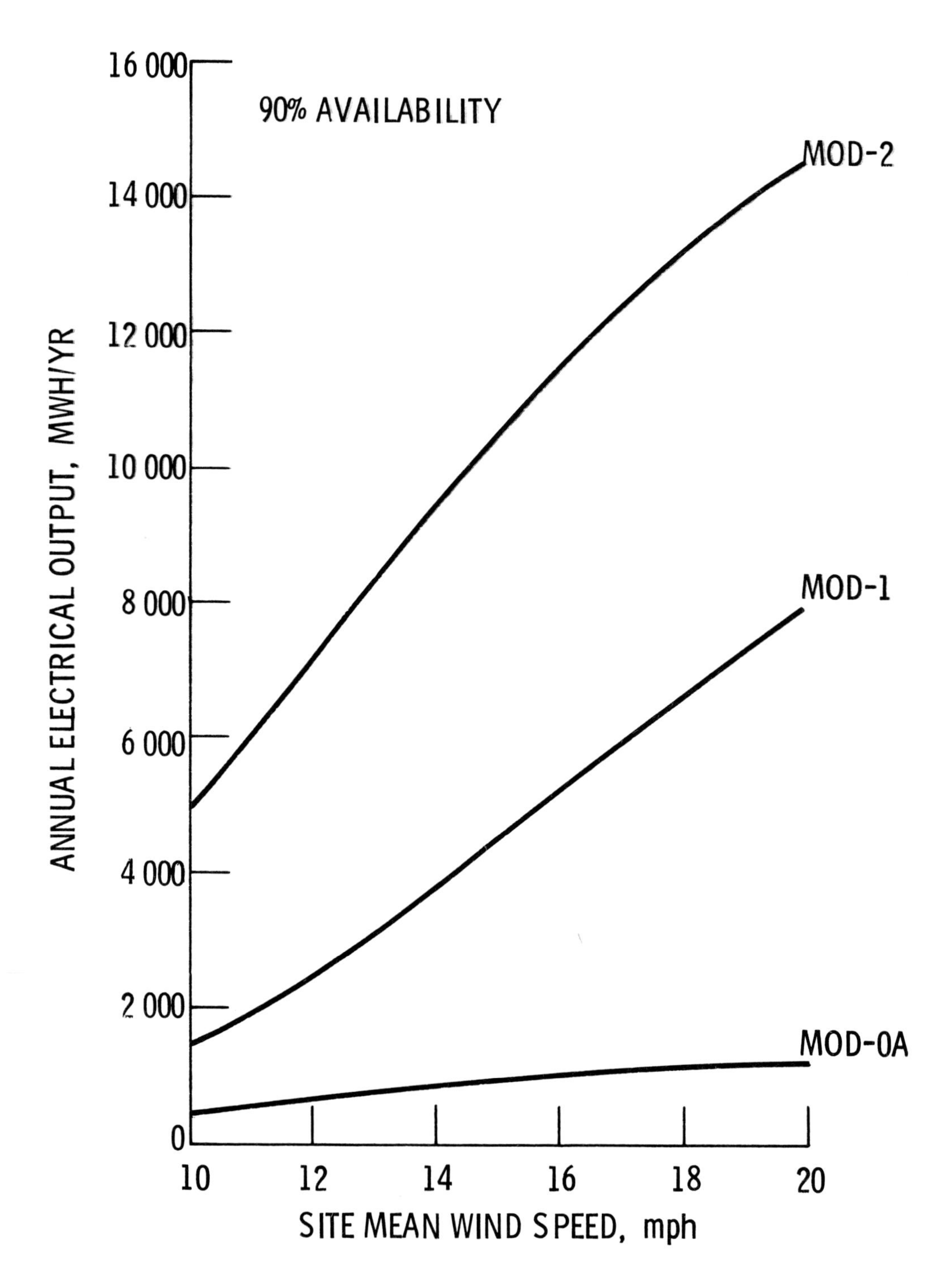

Fig. 38. Wind turbine annual output

	Weight on Foundation, lbs	100th Unit Cost, $K
Machine	(588204)	(1163)
Rotor Subassembly	169567	329
Drive Train Subassembly	103892	379
Nacelle Subassembly	63279	184
Tower Subassembly	251466	271
Transportation & Installation		(328)
Transportation		29
Site Preparation		162
Erection & Checkout		137
Initial Spares & Maintenance Equipment		(35)
Production Facility Depreciation		(35)
Subtotal		1561
FEE (10%)		156
Total Capital for Installed Equipment		1717

These costs are per-unit costs assuming a 25-unit cluster.

We anticipate that the Mod-2 will be cost-competitive in certain areas with attractive wind sites where current fossil fuel costs are high. However, before there is broad market penetration, further reductions in COE must be achieved. We believe that these reductions can be made. New system developments currently planned will incorporate new technology which will result in lighter weight, lower cost systems.

DEVELOPMENT ASSESSMENT

Technology

Major improvements have been made in wind turbine technology over the past five years. As a result, wind turbine system design is understood. The design tools are in an advanced state of development and are available to U. S. industry. Operational machines at utility sites have validated the basic system electrical, structural, and mechanical designs. The compatibility of the single unit wind turbine with utility interfaces has been successfully demonstrated. Further operational experience is required to address long-term reliability. Additional blade development is required and is underway to reduce cost and weight. Metal, fiberglass, and wood blades are all currently attractive candidates. A 50m fiberglass blade has been fabricated and tested by Kaman Aerospace and shows promise of providing a cost-effective design approach (Figure 39). Two 30m blades of the 50m design technology are being built for testing on the Mod-1 in 1980.

Environmental Issues

No serious environmental issues have been identified which would impede the development of large wind turbines. Experience to date has shown that large wind turbines can be designed to be safe, quiet, and clean; however, television interference is a siting consideration.

Fig.39. 50m fiberglass rotor blade

Economics and Market Potential

Large megawatt machines have the greatest potential for application in utility networks because of economics of scale, and are the only machines that will generate significant amounts of electricity. In applications, wind turbines must produce electricity at 2 to 3 cents per kilowatt hour to have wide applications as a utility fuel saver at current fuel prices. The Mod-2 machine approaches these costs, and moderate COE reductions and/or higher fuel costs will result in substantial market potential.

CONCLUDING REMARKS

First-generation-technology large wind turbines (Mod-OA and Mod-1) have been designed and are in operation at selected utility sites. Second-generation machines (Mod-2) are scheduled to begin operations on a utility site in 1980. These second-generation machines are estimated to generate electricity at less than 4 cents per kilowatt hour when manufactured at modest production rates. However, to make a significant energy impact, costs of 2 to 3 cents per kilowatt hour must be achieved. The U. S. program will continue to fund the development by industry of wind turbines which can meet the cost goals of 2 to 3 cents per kilowatt hour when operating at sites with modest winds (6 m/s at 10m). These lower costs will be achieved through the incorporation of new technology and innovative system design to reduce weight and increase energy capture. The challenge, however, is associated with acceptance by the utilities of wind turbines as part of their energy generating capability and the creation of competitive industry to produce wind turbines efficiently. The principals in government, industry, and the utilities are currently involved in meeting this challenge in the United States.

THE WIND ENERGY INNOVATIVE SYSTEMS PROGRAM

INTRODUCTION

The responsibility to provide program management for the Wind Energy Innovative Systems program rests with the Solar Energy Research Institute (SERI) in Golden, Colorado. The objective of the program is to determine the technical and economic feasibility of potential cost-competitive innovative systems. Studies performed in this program are subcontracted to small and large private companies as well as to universities. In FY '79, 17 studies were funded, 11 of which were R&D studies that address various types of innovative extraction, augmentation, and electro-fluid dynamic type systems. The remaining 6 were short-term studies assessing the value of innovative systems of generic order.

All projects include theoretical thermodynamic and aerodynamic studies to confirm energy viability, and performance characterictics and capabilities. These initial studies are followed by model design and test, economic evaluation, and finally, the establishment of concept cost-competitiveness. For those concepts that have the potential of being cost-effective, additional proof-of-concept tests are to be conducted. From these tests, system costs, performance, and engineering data will be developed to determine with reasonable accuracy the cost of energy of a manufactured system.

In early 1979, generic studies were solicited and six short-term studies were funded. Preliminary results from these studies were presented at the

WEIS Conference held in May 1979. A listing of these assessment studies is given in Table 4.

The present paper will review only a few of these projects, arbitrarily selected. Details of the studies may be found in the references provided at the end of the text.

THE MADARAS ROTOR POWER PLANT

A study has been carried out at the University of Dayton Research Institute on the Madaras Rotor power plant concept. The primary objective of the study was to evaluate the Madaras concept including wind tunnel tests of a rotating cylinder, an electromechanical design study, and a cost analysis for the system.

During the period 1929-34, analytical, wind tunnel, and full-scale aerodynamic studies were carried out on a wind powered Madaras rotor power plant, a system invented by Julius D. Madaras (Figure 40). The concept considered a circular closed track, 457m diameter, in which were mounted cylinders, 6.8m in diameter and 27m high. It was planned that 18 MW of power would be extracted from the system in a 13 m/s wind at a track speed of 8.9 m/s. At two points in each orbit the rotor spin changes direction. This schedule is required so that the driving force generated by the rotors act in the same direction. The reason for using a rotating cylinder instead of an airfoil is because a rotating cylinder generates a lift force about 10 times larger than that of an airfoil (Figure 41)

Experimental Studies

The performance of the Madaras rotor is complicated by the interaction of the atmospheric boundary layer with the flow generated by the spinning rotor. Wind tunnel tests were carried out to evaluate this complex phenomenon. These tests were performed under the direction of Professor Willmarth in the University of Michigan's Aerospace Engineering Department wind tunnel which has a cross-section of 2.1m x 1.5m. The purpose of the tests was to validate existing wind tunnel data of rotating cylinders, develop aerodynamic characteristics of rotating cylinders with end plates, obtain power requirements for rotating the cylinder, and evaluate cylinder performance in an atmospheric boundary layer. Models of 15 mm and 152 mm in diameter were tested. All cylinders were tested with internally housed motors capable of rotating the cylinder at speeds up to 2,000 rpm. The cylinders were rotated to high rpm to simulate a full-scale cylinder which could rotate at 168 rpm. Tests were made for various aspects ratios, end-plates, and spin speed. The power required to rotate the cylinder is shown in Figure 42. It is observed that there is a significant increase in power absorbed by the rotor for e/d greater than 2. Tests performed at different aspect ratios indicated that there was little difference in power absorbed as a function of aspect ratio.

The lift and drag coefficients have also been measured for different aspect ratios for U/V up to 6 (Figure 43). The C_L increases dramatically with increased end-plate size, especially for e/d between 1.25 and 2.0. For large values of aspect ratio the benefit of increasing e/d beyond 2.0 is not apparent. From these results, it appeared that a good design could be achieved with the rotor having an AR = 6 and e/d = 2. This combination would provide

TABLE 3

R&D PROJECTS

Project Title	Subcontractor
Innovative Wind Turbines (VAWT)	West Virginia University
Diffuser Augmented Wind Turbines (DAWT)	Grumman Aerospace
Tornado Type Wind Energy Systems (Tornado)	Grumman Aerospace
Tests and Devices for Wind/Electric Power Charged Aerosol Generator (EFD)	Marks Polarized
Electrofluid Dynamic Wind Driven Generator (EFD)	Univ. of Dayton Research Institute
Energy from Humid Air (Humid Air)	South Dakota School of Mines and Technology
The Madaras Rotor Power Plant Phase I (Madaras)	Univ. of Dayton Research Institute
Vortex Augmentors for Wind Energy Conversion (Vortex)	Polytechnic Institute of New York
The Yawing of Wind Turbine with Blade Cyclic Pitch	Washington University Technology Associates
Oscillating Vane Concept	United Technologies Corp.
Advanced and Innovative Wind Energy Concept Development - Dynamic Inducer	Aerovironment, Inc.

TABLE 4

GENERIC STUDIES

Project Title	Subcontractor
A Definitive Generic Study of Augmented Horizontal Axis Wind Energy Systems	Aerovironment, Inc.
A Definitive Generic Study of High Lift Device Wind Energy Systems	Aerovironment, Inc.
A Definitive Generic Study of Augmented Vertical Axis Wind Energy Systems	New York University
A Definitive Generic Study of Augmented Horizontal Wind Energy Systems	Tetra Tech, Inc.
A Definitive Generic Study of Sail Wing Wind Energy Systems	Washington University Tech. Associates, Inc.
A Definitive Generic Study of Vortex Extraction Wind Energy Systems	JBF Scientific Corp.

Fig. 40. Wind powered Madaras rotor power plant

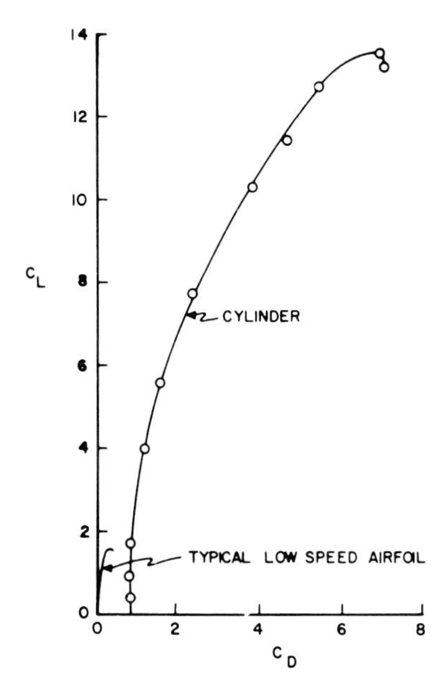

Fig. 41. Lift-drag polars of a rotating cylinder and airfoil

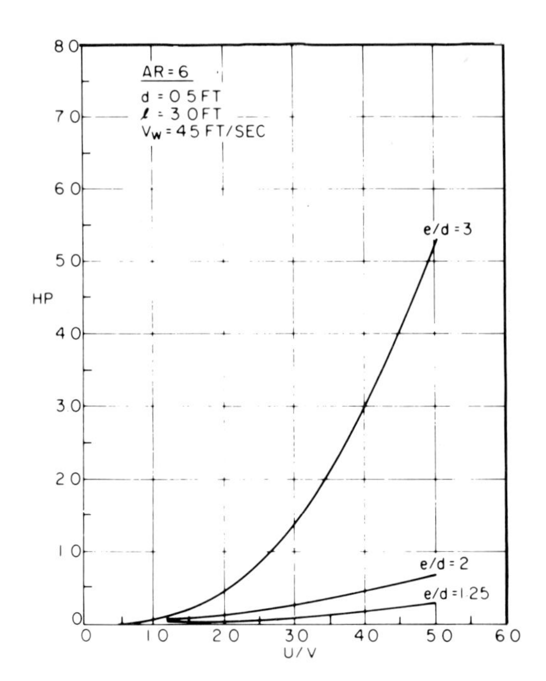

Fig. 42. Power required to rotate cylinder for various end-cap diameter ratios and two aspect ratios

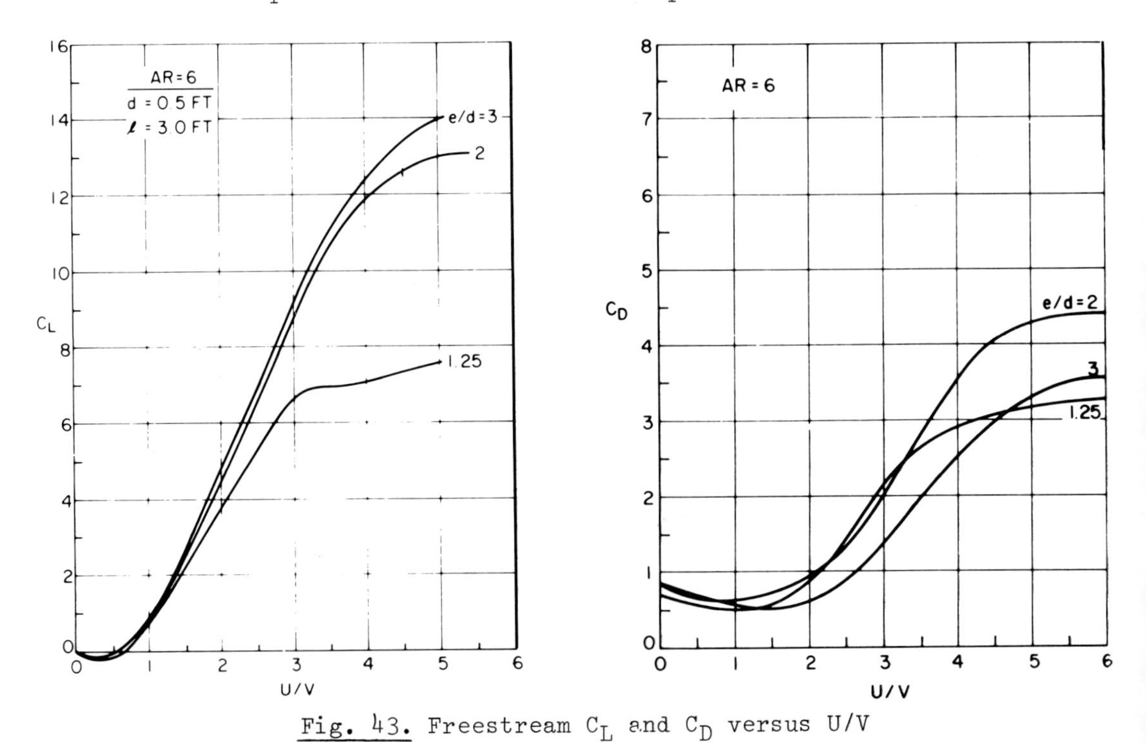

Fig. 43. Freestream C_L and C_D versus U/V

higher lift, lower drag, and reasonable power levels for spinning the rotor.

Measurements were also taken utilizing a boundary layer that was developed on the tunnel wall. Most of the observations or measurements conducted with this contrived boundary layer were similar to those obtained with the free-stream data. It was also concluded that the use of top and bottom end-plates instead of only top plates, would be beneficial to the performance.

Design Studies

A Madaras performance simulation program was developed which predicted the forces of the various components as a function of the angular position on the track. Design loads on the rotor, rotor support, tower car, and track were obtained from the program. Three wind conditions were evaluated, operational, operational plus gust, and static operations at hurricane conditions. The operational plus gust load was found to be the most severe condition. Load conditions were also analyzed due to aerodynamic loading on the endcap, acceleration caused by rotor track effects, centrifugal body forces, angular acceleration during rotor spin-up from 0 to 186 rpm, car weight, and wheel and lateral restraint loads. With the assumptions and conditions imposed, a rotor car design evolved.

aspect ratio	= 8	cylinder length	= 38.1m
e/d	= 2	end-cap diameter	= 9.8m
cylinder diameter	= 4.9m	car length	= 19.2m
track gauge	= 11m	car width	= 17.4m
car height	= 3.8m	gross car weight	= 328,000 kg

In addition, the structural configuration and materials for the rotor, track, support tower, and rotor car suspension system were all developed. Each rotor car developed approximately 1 MW. Power is transferred from the rotor car to an overhead triple—track rigid trolley rail. All rotor cars are joined together by two wire ropes, attached to the main frame.

The electrical design considered four components: the rotor spin system, the generator system, the controls and instrumentation system,and the electrical interface system. The rotor spin system consists of a 450 kW DC motor which is used to provide acceleration and dynamic braking to the rotor. Direction reversal is achieved by reversing polarity of the power leads to the armature, and speed control is achieved by balancing the input power level to the armature against the demand of the control function signal. One of the results of the spin motor study was to discover that there are three very significant sources of spin motor losses: power required to overcome the viscous friction of the rotor caused by spinning; power required to overcome rotor inertia while accelerating from 0 to 186 rpm and loss of energy during the rotor deceleration; and the power lost in heating the motor windings during the low speed acceleration stage at which time the motor is operating at a very low efficiency. Four 250 kW three-phase, 60 H_z AC induction generators were used for each car. The central system included a minicomputer to tie the rotor car to the central house and a monitoring of the wind sensor network to provide data for the spin schedule imposed on the rotor. The electrical interface system included the necessary feeder lines, synchronous reactors for the power feeder correction, utility feeder circuits and substations for the operation of the facility.

A performance analysis of the system was carried out to determine optimum characteristics for the rotor size and track. A circular and racetrack configuration was studied. For the racetrack configuration, the track speed was 13.4 m/s and the track size had an end diameter of 1372m with a straight section which varied from 3050m to 19,210m. The calculated net power output for circular and racetrack configurations is shown in Figure 44. It is observed that the racetrack configuration develops over twice as much net power as the circular track configuration. In general, track speeds less than twice wind speeds appear appropriate, particularly at high wind speeds.

In order to determine an optimum spacing for the rotor cars, a mutual interference study was conducted using the vortex analysis developed for the giromill. The analysis determines the effect of all vortices shed from each rotor in the plant on the vortices shed from all other rotors in the plant, and then determines the effect of this vortex field on the wind velocity vector at all points around the track. As rotor spacing decreases, mutual interference losses increase. As is shown in Figure 45, at constant wind speed, increasing the number of rotors decreases the performance of each rotor.

Based upon the rotor and track characteristics, cost estimates were developed for a system utilizing generalized equations which described plant geometry in terms of modular cost variables. The cost estimation was carried out by M.L. McClellen and Company, a professional engineering company specializing in cost estimation of large plants. The analysis indicated that the most efficient circular plants were not sufficiently economical for further considerations. Large racetrack plants appear to be more cost—effective but require sites that have predominant wind directions. The results of some of these studies are shown in Figure 46 and 47. The plots indicate the relationship of various parameters on the unit cost and energy related to rated power. The effects of interrotor spacing, the number of rotors, and length of straight track were evaluated for a mean wind speed of 8.1 m/s. The curves do not reflect the effect of learning curves or land costs, and is based on an annual cost equal to 16.5% of total plant costs and 30-year lifetime. Typically, for a 100 MW system with 75 rotor cars the system cost could vary between $2000 and $3000 per installed kW depending upon the track size. This results in an energy cost of between 7.5 to 9.0 ¢/kWh.

Additional calculations have been made utilizing a 90% learning curve on the major elements used in this system and locating a plant at Medicine Bow, Wyoming. Comparisions were made between these energy costs and conventional wind systems. Based upon the assumptions used in the current study, the energy costs for the Madaras system is somewhat higher than for a conventional system.

Summary Remarks

The results of the University of Dayton study indicate that a Madaras system utilizing a racetrack configuration would be preferred to one with a circular track. Sufficient information is available on the aerodynamics of a rotating cylinder to provide performance characteristics of the rotor.

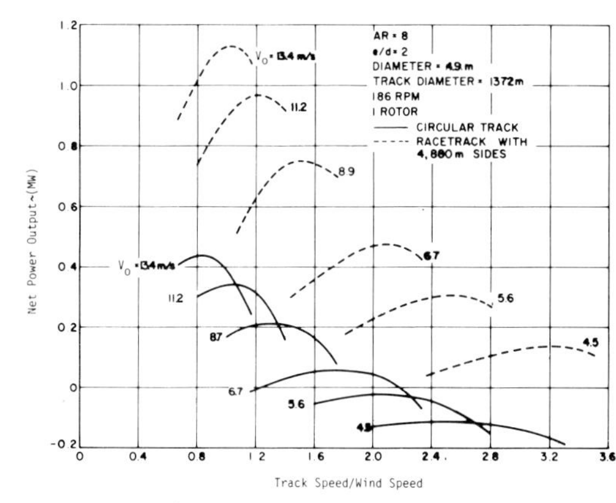

Fig. 44. Net power output for one rotor versus track speed/ wind speed

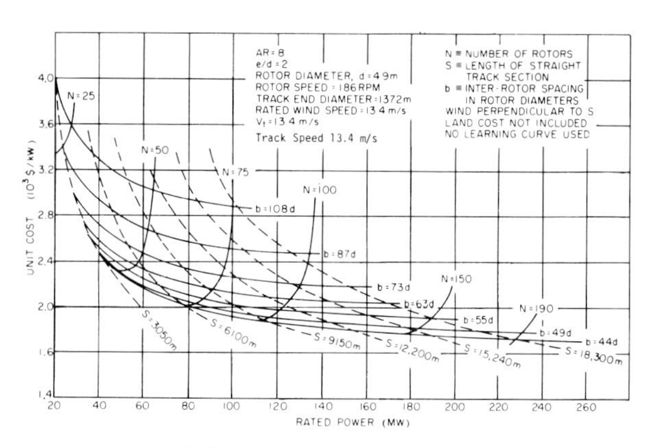

Fig. 46. Unit plant cost versus rated power for racetrack configuration

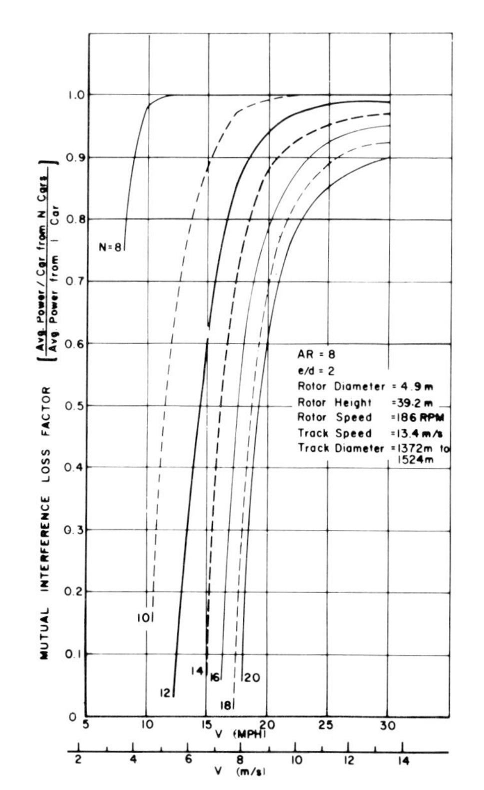

Fig. 45. Mutual interference loss factor versus wind speed for various numbers of rotors

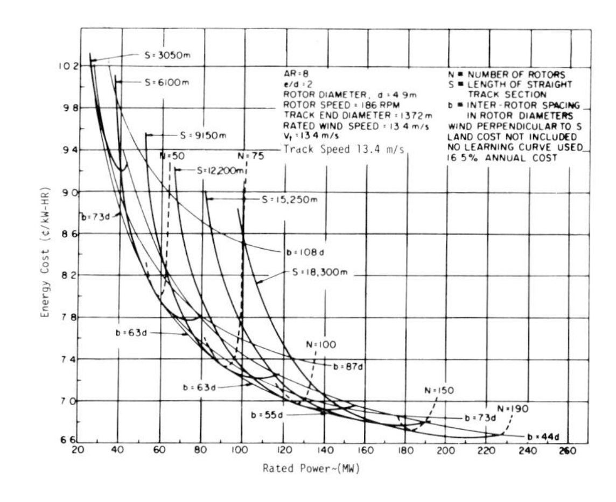

Fig. 47. Energy cost versus rated power for racetrack configuration

Optimization of the system has provided an indication of the rotor diameter, height, and end-cap geometry. It is clear that the system is quite complex, comprised of multiple cars each being fairly complex. It is anticipated that the operations and maintenance for the system, cars, track, and controls would probably be quite high. Calculations based upon costs developed for the system indicated that the energy costs are higher than a conventional system located at Medicine Bow.

DIFFUSER AUGMENTED WIND TURBINE

For the past three years, Grumman Corporation has been performing a study to evaluate the diffuser augmented wind turbine (DAWT) concept. The work has been both analytical and experimental with the major emphasis directed toward experimental verification of the concept. The objective of the study is to develop a cost-effective system by devising a means of increasing the power generating capability of a conventional wind turbine by increasing the mass flow through the rotor. Preliminary efforts of this concept have been carried out in England and Israel. Details of the Grumman Study are provided in the reference.

Theoretical Studies

For a conventional wind turbine, only 59.3% of the energy in the air can be extracted. In the case of a diffuser augmented wind turbine (Figure 48) with a turbine of the same area and wind speed, because of the pressure recovery of the diffuser, it is possible to sustain a reduced pressure downstream of the rotor. As a result, the air speed ahead of the DAWT is greater than the free speed by the ratio ε . Thus, the power extracted from the DAWT is given by

$$P_D = \varepsilon A_2 V_0 \left[\varepsilon^2 \tfrac{1}{2} \rho V_0^2 + P_2\right]$$

Utilizing one-dimensional theory, it is possible to define an augmentation ratio

$$R = \frac{C_{P_i}}{0.593} = \frac{C_T}{0.593} \left(\frac{q_2}{q_o}\right)^{3/2}$$

The maximum value of r can also be calculated and is given as

$$r_{max} = \frac{9}{8} (1 - \pi_4) \left[\frac{1 - \pi_4}{3(1 - C_{P_R})}\right]^{\frac{1}{2}}$$

where

$$C_{P_R} = \eta_D \left[1 - \frac{A_2}{A_4}\right]^2$$

The value of π_4 and η_D or ε are to be determined from experiment. One-dimensional theory implies that the diffuser area ratio between the base of the diffuser and the throat should be as high as possible. It is well known that separated flows in adverse pressure gradients are prevalent for angles in excess of 20°. Preliminary investigations by Grumman involved developing methods to demonstrate that area ratios between 2 and 4 were feasible using included angles greater than 20°. The purpose of using large angles was to make the diffuser section as short as possible with the possibility of reducing system costs.

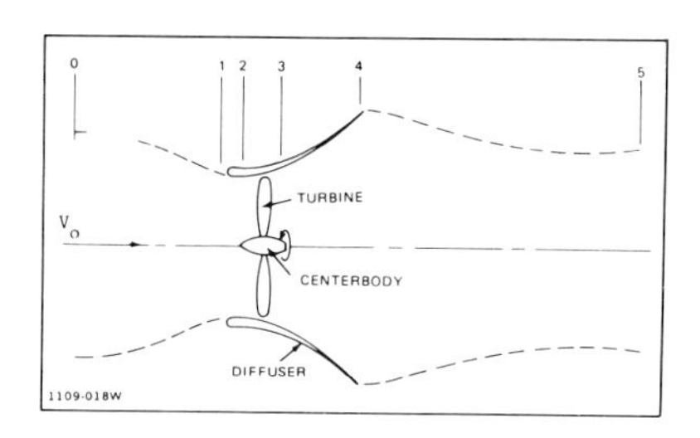

Fig. 48. Basic schematic of the DAWT with reference stations

Fig. 49. Baseline model installation in the 2.1 x 3 m wind tunnel with three-bladed, constand chord wind turbine 0.46 m diffuser and paraboloid centerbody end-pieces

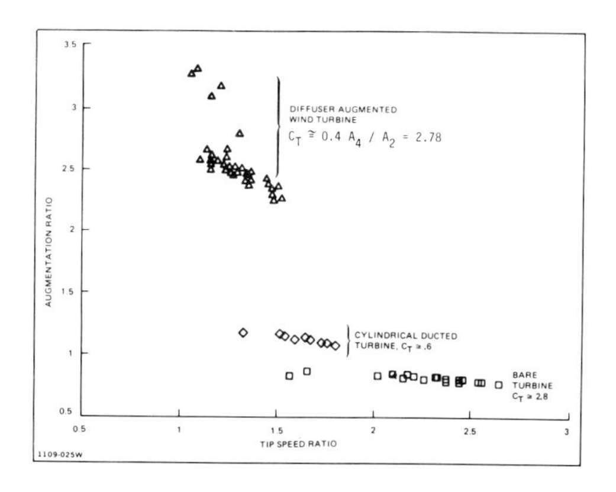

Fig. 50. Comparison of turbine performance in the wind tunnel for different augmentation systems

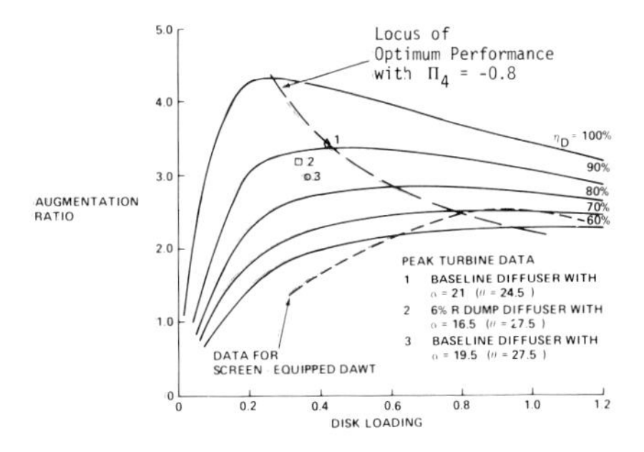

Fig. 51. Comparison of DAWT models test data with theoretical performance for $A_4/A_2 = 2.78$ and $\Pi_4 = -0.8$

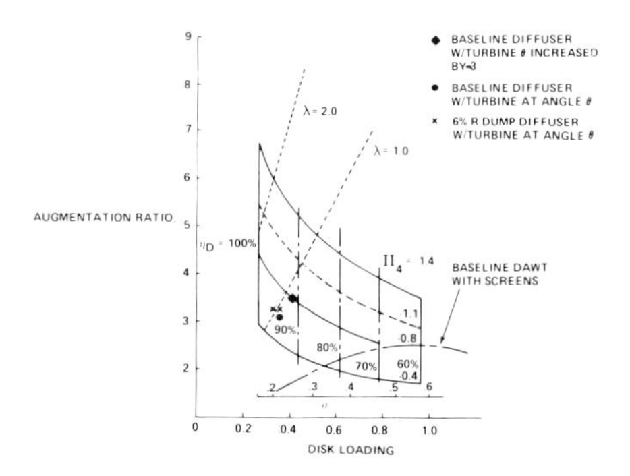

Fig. 52. Comparison of theoretically optimum performance with model test data, $A_4/A_2 = 2.78$

Experimental Studies

A multiphased study was carried out using various wind tunnel facilities and models. The first series of models were tested in an open jet. The model size was approximately 4.5 cm in diameter, which gave a Reynolds number of approximately 70 x 10^3 based on the model exit diameter. The models utilized screens of various solidities to simulate turbine disk loadings. Many designs were tested; for example, straight walls with slots to suppress flow separation, flapped, ring wings, etc. From these studies a baseline diffuser design was chosen for additional testing. This baseline had the following characteristics: area ratio of 2.78, included angle of 60°, length-to-diameter ratio of 0.5 with a two-staged walled slot-flow. Measurements with this model indicated that the exit pressure coefficient, π_4, was below atmospheric pressure with a value of -0.56.

The next phase used larger models with Reynolds numbers approaching 10^6. Tests were conducted only with the baseline geometry utilizing screens to simulate a turbine disk loading. The purpose of these tests was to confirm the low value of base pressure coefficient,which was found to be approximately -0.83.

During the next phase,the effect of flow in the diffuser using a turbine instead of a screen on the baseline diffuser model was considered. A photograph of the model in the wind tunnel is shown in Figure 49. The last phase of the work was to determine whether improvements could be made in π_4. Tests were carried out at Reynolds numbers of approximately 0.13 x 10^6, and it was observed that improvements in the diffuser performance were possible using a shorter diffuser, termed the "Dump diffuser."

Preliminary measurements to determine the augmentation ratio are shown in Figure 50. A short cylindrical duct improved the performance of the base turbine by a factor of about 2.0; the performance with a designed diffuser is approximately six times the value of the base turbine.

Using a one-dimensional analysis it is possible to calculate the augmentation ratio as a function of disk loading for various values of diffuser efficiency. Results of these calculations are shown in Figure 51. The locus of peak augmentation ratio is also shown on the figure for a pressure coefficient of -0.8 as well as measured data using screens and a turbine. The effect of disk loading is not strong for values of C_T occurring after r_{max}. The peak turbine data is also considerably higher than the screen-equipped data. The use of a turbine appears to have a major impact upon the efficiency of the diffuser. This could be because the turbine wake provides energy for the flow downstream of the turbine and aids the slot flow.

Further comparisions of the model tests with optimum performance data is shown in Figure 52 where the calculations are again for the one-dimensional configuration. Shown in the figure is the baseline diffuser with screen where the maximum level of augmentation ratio is approximately 2.5. The data taken from the baseline diffuser with a turbine varies between 3 and 3.5, and were taken at a tip speed ratio of approximately 1.0. The effect of tip speed ratio is shown on the figure and it is observed that an increase in the tip speed ratio could significantly affect the augmentation ratio.

A comparison of the current measurements made with an area ratio of 2.78, and "potential" performance of the systems are shown in Figure 53. For a larger system with an increase in tip speed ratio an augmentation factor of approximately twice current value appears feasible.

An engineering design for a prototype baseline DAWT was performed using an augmentation ratio of 2 (which was demonstrated at the time the study was initiated). The system used a 5.4m turbine (a modified Grumman Windstream 25), has a 5.6m diffuser entrance diameter, 9.1m exit diameter, and an overall length of 4m (Figure 54). Several modifications to this preliminary design were made. The major cost item is the diffuser and its construction. The current effort by Grumman is twofold: to consider various options in the manufacture of the diffuser to make it cost-effective(using, for example, steel, aluminum, fiberglass, or ferrocement as construction materials);and, to determine the cost of energy using the most recent performance estimates.

Summary Remarks

The diffuser augmented wind turbine offers a potential of generating significant amounts of energy from available wind. Current wind tunnel studies have shown that models using a turbine produced a higher augmentation than those using a disk to simulate the turbine. Augmentation ratios of 3.5 have been measured at a tip speed ratio of unity. The preliminary analysis indicates that this value could possibly be doubled if the tip speed ratio is increased. The first estimate for the system showed that the major component cost was attributed to the diffuser.

Current studies of the diffuser augmented wind turbine are concerned with confirming power coefficients of the baseline diffuser and the dump diffuser to provide an efficient system at power levels under 200 kW. In addition, the study is to include manufacturing techniques associated with different materials for the diffuser. It is anticipated that results from this study could provide direction and guidance of an engineering design for a potentially cost-effective diffuser augmented wind turbine.

ECONOMIC ANALYSES

DISPERSED RESIDENTIAL SYSTEMS

INTRODUCTION

This section is based upon the results of an investigation of the economic status of present day small wind systems (SWECS) for residential use (JBF, 1979). The residential cases examined were a remote home without utility backup power and a rural-suburban home with utility backup power.

For the remote residence two situations were examined. The first was a remote residence already owning a diesel generator and the SWECS was added as a fuel saver (the retrofit case). The second was a remote residence without an existing system, where the choices would be to either purchase a diesel generator or purchase a SWECS-battery-diesel generator combined system (new construction case). In addition, in the new construction case, a third

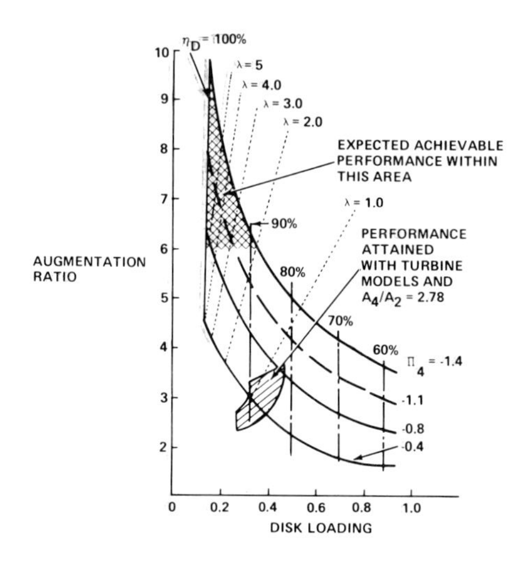

Fig. 53. Comparison of theoretically optimum performance with model test data, $A_4/A_2 = 4.0$

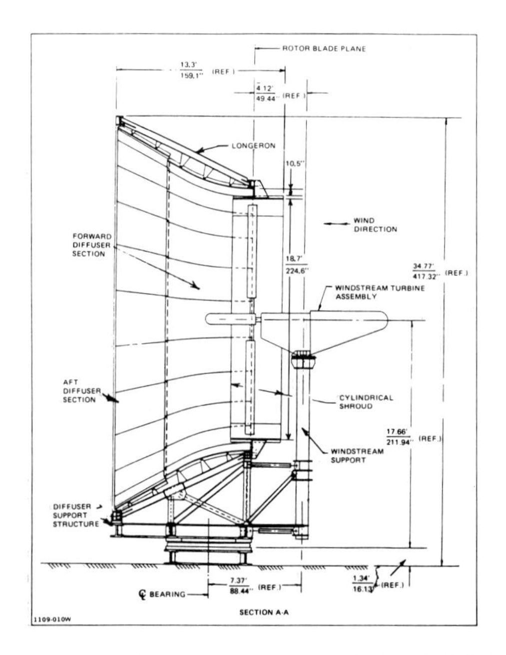

Fig. 54. Engineering design of DAWT field-test article

alternative considered was an "economy" system using a gasoline driven generator, inexpensive batteries, and a SWECS.

The wind data used for the analysis was hourly data from New England that displayed a wind probability distribution similar to large parts of the United States and exhibited a good diurnal wind-load hourly correlation as well as a good seasonal wind-space heating load correlation. Average annual wind speeds examined, with these wind data, were 4.5, 5.5, and 6.5 m/s.

The SWECS selected for analysis were existing 1.2, 3.2, and 4.0 kW machines, and an 8 kW machine representative of those in the prototype phase (DOE program at Rocky Flats). The SWECS costs used were representative of systems that can be purchased in 1979 and were not adjusted to account for future R&D related cost reductions.

Earlier distributed SWECS studies clearly indicated that economic analyses based upon annual mean wind speeds, or wind duration curves, are inadequate because they do not address the critical issues of hourly wind-load correlation, excess (or unusable) energy, and utility sellback of excess SWECS energy. Thus, the approach used was an hourly analysis that took into account the load existing at each hour and the wind energy available each hour. Excess energy, or SWECS energy generated at a time when the load was inadequate to absorb it, was either used for supplemental loads (hot water space heating), dumped, or sold back to the utility.

It was assumed that the residential user had a 25% federal income tax rate, paid a sales tax of 5%, experienced an annual inflation rate at 6% per year over the SWECS lifetime, obtained a 10% loan and used a 10% discount rate, and paid no property taxes on the system. It was assumed that diesel fuel (or gasoline) was 85¢/gallon in 1979 and would grow at zero, or 3%, above inflation. Utility rates were assumed to be 5¢/kWh and would grow at 0, or 1.5%, above the constant inflation of 6%. It was further assumed that the utility would neither penalize, nor reward, the SWECS user by changing his utility rate (e.g., demand charge).

Levelized energy costs were calculated over the lifetime of the energy systems. In addition, the life cycle break-even cost was calculated for each SWECS-user case examined. The break-even cost is the amount that the user could spend on the system and break even over its lifetime.

It should be noted that all users would not accept a life-cycle payback, but would require a shorter payback, perhaps three to five years. Thus, the case examined is conservative in that the SWECS break-even cost would be substantially lower for a short payback period. This translates into either a requirement for a higher SWECS system value or a lower SWECS system cost.

FINDINGS

Remote Residence:

1. Retrofit Case (fuel saver)

In the case where the SWECS is being used to displace diesel fuel only, the results shown in Figure 55 are obtained, for a lighting and appliance load.

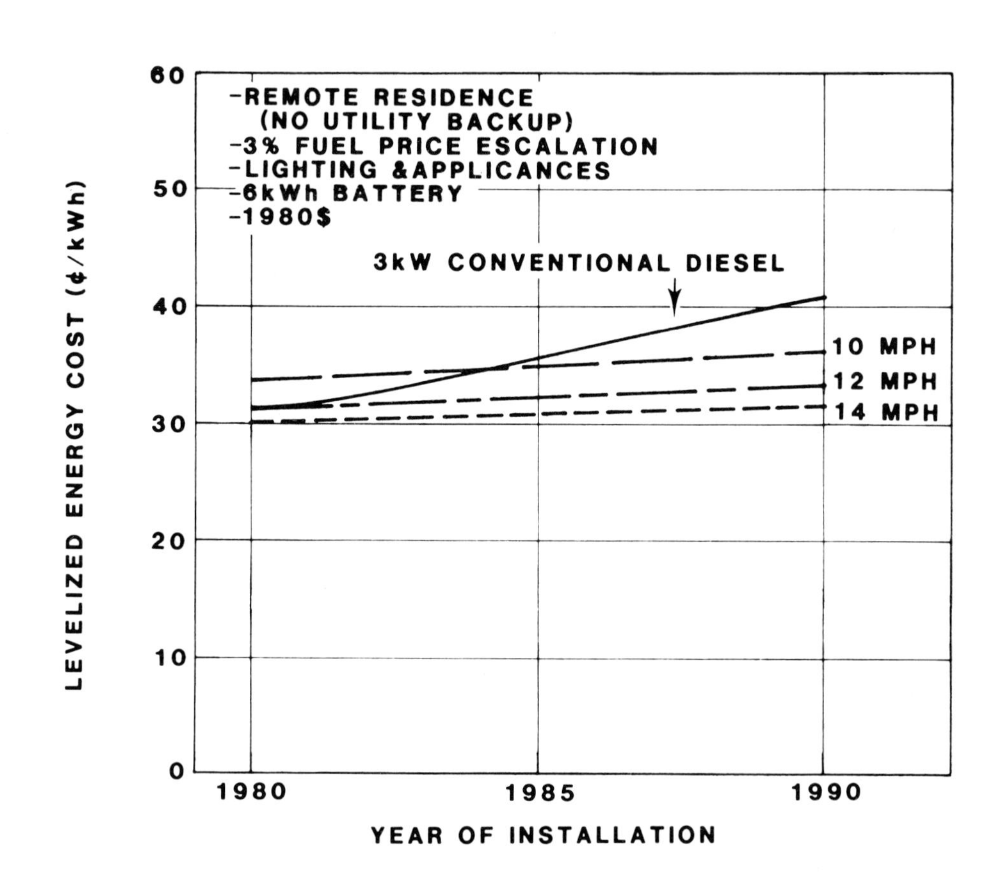

Fig. 55. Levelized energy costs for 1.2 kw SWECS system operating as fuel-saver

The SWECS is cost-effective over its lifetime for a 1980 installation at wind speeds of 12 mph, or greater, if fuel escalates at 3% above inflation. Similar results are obtained when a hot water load is added, except that the levelized energy costs for both the diesel and the SWECS system are lower. That is, it is more cost-effective to also use the system for hot water loads. If there is no fuel escalation above inflation, the SWECS is a poor competitor to the diesel.

2. New Construction

In this case the residential user has no system and would purchase either a 3 kW diesel generator or a SWECS-battery-diesel system. The latter would be utilized in a mode where the SWECS and batteries powered the load and the diesel would only be used to charge the batteries, thus allowing it to operate at peak efficiency and minimize fuel consumption.

The results obtained indicate that the SWECS-diesel system is "almost" competitive with the diesel-only system in 1980, and is cost-effective in 1985, at 3% fuel escalation. For the case of zero fuel escalation the SWECS-diesel is marginally effective for wind speeds 6.5 m/s.

An analysis was done using a larger SWECS (3.2 kW) for the residential load shape and wind characteristics used in this study, and the larger SWECS was less cost-effective. While there are undoubtedly remote applications with larger loads where this may not be true, the results suggest that the optimum size for a SWECS to serve a simple remote residential user, without utility backup, is on the order of 1 kW.

Residence with Utility Backup

In this case it was assumed that the SWECS would be used to power a typical rural-suburban residence and the load cases examined include lighting and appliance (L&A), hot water, and space heating.

Figure 56 shows the levelized energy costs for SWECS system sizes that were found to be near optimum size for the winds-loads-SWECS examined in this study. It can be clearly seen that existing SWECS for residences with utility backup do not quite demonstrate cost-effectiveness even on a life cycle basis, for utility escalation of 1.5% above inflation — by 1990. The only application that comes close is the larger SWECS for the equivalent of an all-electric home. This presumes no sellback of excess energy. It should be stressed that this conclusion is for existing SWECS, not the systems under development by DOE.

Figure 57 shows the effect of sellback, as a function of SWECS size, for the lighting and appliance load, for various sellback rates. The figure suggests with small sellback rates, small machines are closer to being cost-effective. For larger sellback rates, larger machines are more cost-effective. It appears as though sellback at about 3¢/kWh virtually eliminates the sensitivity of SWECS value to SWECS size — for the cases examined. It will take a sellback rate of 5¢/kWh, or higher, for SWECS to be cost-effective in 1980, for the cases examined in this report.

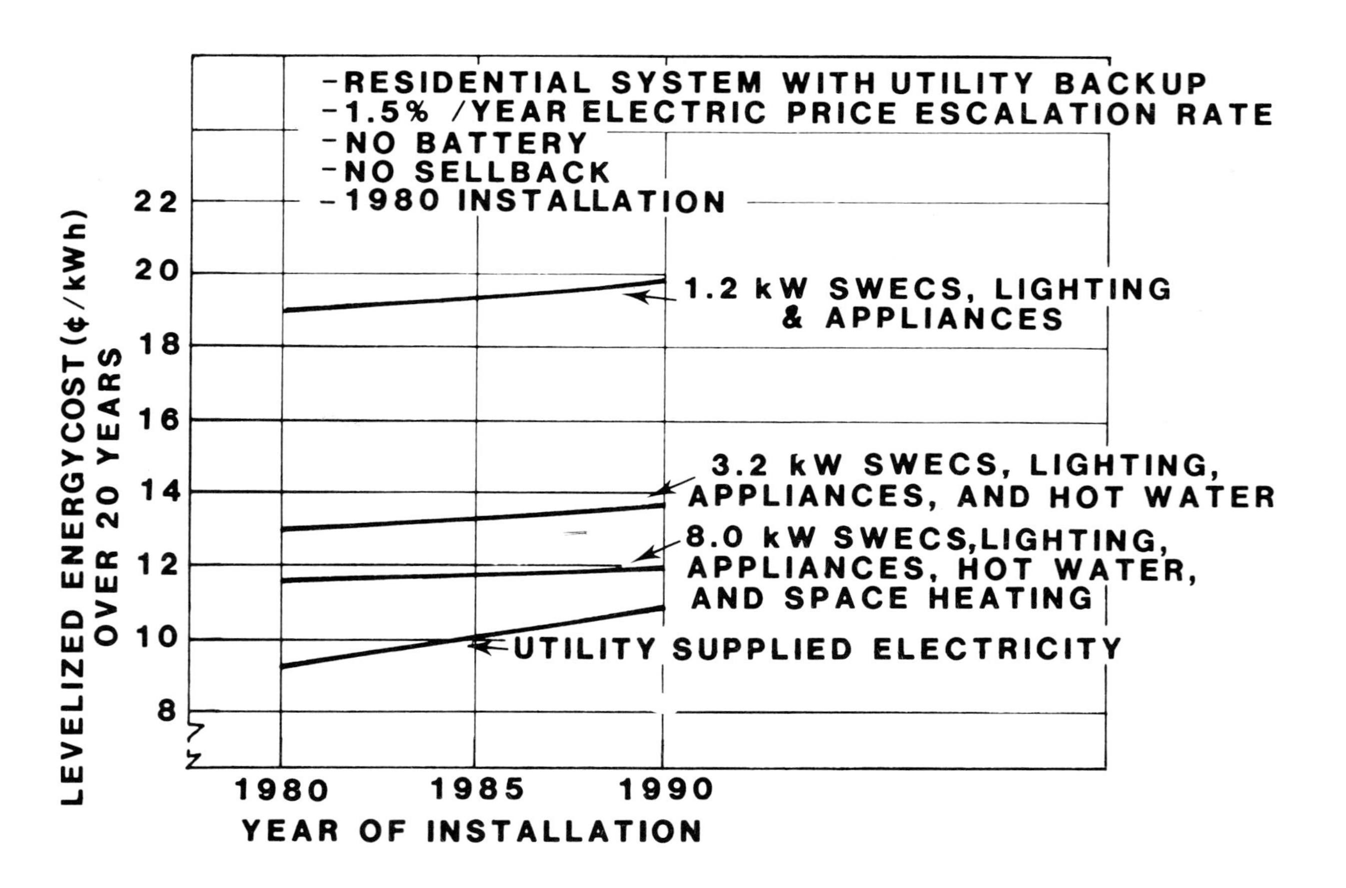

Fig.56. Levelized energy costs for SWECS systems and residential load applications with utility backup

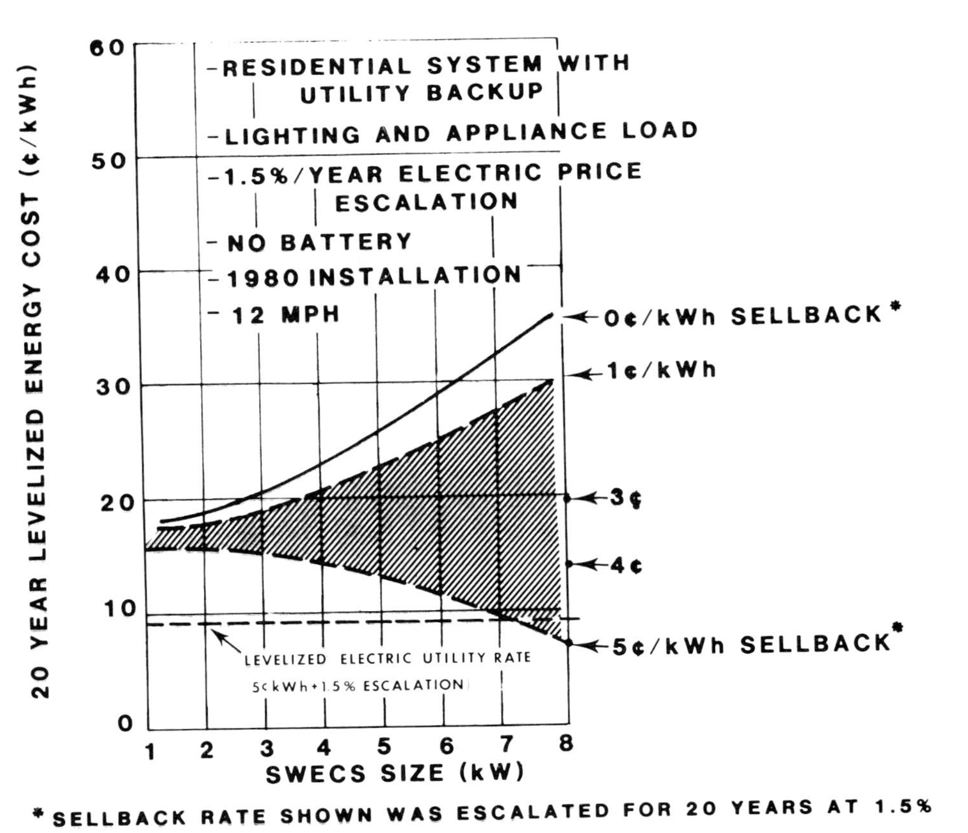

Fig. 57. Levelized energy cost, at various sellback rates, as a function of SWECS size

CONCLUSIONS

Remote Residence (no utility backup)

Fuel Saver (retrofit case)

When a remote residence already has a diesel engine, and the only value of the SWECS is to save fuel, the following results are obtained:

- In 1980, the SWECS is cost-effective (has the same, or lower levelized energy cost as the diesel fuel) for wind of 5.5m/s (annual mean), or greater — assuming a 3% fuel escalation above inflation for the system lifetime. For no fuel escalation, the SWECS is a poor contender, with its levelized energy cost being almost twice the diesel case. Adding a hot water load improves this but does not overcome it.

- A fundamental problem of the remote residential case is the disposal of excess SWECS energy, since the possibility of utility sellbacks does not exist. This leads to the selection of a relatively small SWECS (about 1 kW) for the load analyzed. However, other loads in addition to basic lighting, appliances, and hot water would justify a larger machine.

New Construction (no existing system in place)

In the new construction case, where the remote residence has the choice of simply purchasing a diesel or the choice of installing a SEWCS system (which also includes a diesel, or other backup source), the following results were obtained:

- The SWECS-diesel system would be cost-effective in the mid-1980s as compared to a 3 kW diesel alone. This assumes a 3% escalation of fuel above inflation. However, the SWECS system is "almost" competitive in 1980 with either zero or 3% fuel escalation and slightly reduced SWECS costs will create immediate cost-effectiveness for this case. Similar results were obtained for the case where a hot water load is also added to the lighting and appliance load.

- The levelized energy cost of the SWECS system (and the diesel-only system) are substantially reduced (factor of two) when the hot water load is added. This suggests that it is unrealistic to consider SWECS for simply lighting and appliance loads, without using it for hot water.

- The primary reason for the SWECS-plus-diesel system looking favorable, as compared to the diesel only (same diesel), is that the SWECS-battery subsystem allows the diesel to always run at its optimum efficiency, since it is only used for battery charging. This results in significant fuel savings and becomes more important the higher the escalation of fuel costs above inflation.

- An "economy" SWECS system using a small gasoline generator and inexpensive batteries resulted in almost the same economics as the high quality SWECS-diesel system. This is due to the fixed costs being lower but variable costs higher, since the gasoline generator does not operate as efficiently as the diesel engine. The "economy" system is also less desirable since it is basically a manual system whereas the SWECS-diesel system is automatic (e.g., battery charging).

In summary, the remote residential case for SWECS is very close to, or is, cost-effective for non-utility served sites, and the cost-effectiveness will improve for higher fuel escalation rates above inflation. The SWECS size should be small (about 1 kW for the specific load that was examined) and a hot water load should be included to utilize excess energy.

Residential User with Utility Backup

This case was investigated for a residential user with a lighting and appliance load, plus a hot water load, plus space heating. Utility sellback of excess energy was investigated. It was assumed that the residential cost of electricity was 5¢/kWh (1980 dollars) and cases of zero and 1.5% escalation above inflation were examined. Wind speeds examined were 4.5, 5.5, and 6.5 m/s (mean annual). It was further assumed that the SWECS user would not be penalized, or rewarded, by changes in his rate structure. The following results were obtained:

- In no case examined, without utility sellback of excess energy, is the residential SWECS with utility backup cost-effective by 1990 for todays's off-the-shelf SWECS. However, with a slight improvement in costs, it would be.

- For the sellback case examined (1980 installation), it would take a sellback rate of 5¢/kWh for the SWECS to reach payback during its lifetime (20 years).

- The first SWECS application that most probably will reach economic viability will be the all-electric home, followed by one using electricity for lighting, appliances, and hot water. Residential SWECS for lighting and appliance loads only are unrealistic due to the excess energy, unless the energy can be sold to the utility at rates of 5¢/kWh, or greater, for the cases examined.

- Based upon the preliminary analyses done in this report, there seems to be an optimum size SWECS for each load condition. For lighting and appliance loads, it is about 1 kW. When the hot water load is also added, it is about 3-4 kW. When space heating is also added, it is 8 kW, or greater. Sellback rates can alter this somewhat, as could wind regime and SWECS characteristics.

- For the machines, loads, and winds used in this study, once a machine is selected there is substantially less change in economics with wind speed than might be expected from the wind cubed law. Little change was experienced from 4.5 to 6.5 m/s, and this is due to the additional wind energy being introduced into a system that already is generating excess energy. Unless the excess can be sold back at a large sellback rate (5¢/kWh, or greater), the incremental value of the extra wind energy is small. However, this result could be modified by selection of an optimum SWECS for each load. This study was limited to using existing, representative SWECS.

ECONOMIC ANALYSIS - CENTRAL STATION WIND SYSTEMS

INTRODUCTION

This section presents the results of a synthesis of wind energy systems' value to utilities (JBF, 1979), based upon several studies that have been conducted in the past three years. The studies, shown in Table 5 and funded by DOE and EPRI, were conducted by several different companies, each using different analytical models. The values assumed for model inputs also varied from study to study, as did the year in which the analyses were done, and the wind system characteristics.

The objective of this paper is to synthesize the results into a composite picture of WECS value to utilities, to the extent possible, by scaling the study results to a common set of parameter values. The parameters selected for scaling were: cost of capital, fuel escalation, and WECS lifetime. Using sensitivity analyses curves from the studies, scaling factors were developed for these parameters and the WECS value then scaled to a common set of assumptions (e.g., 11% cost of capital, 3% fuel escalation above inflation, WECS lifetime of 30 years). The scaled values were then used to synthesize a picture of the value of WECS to utilities in the time period 1980-2000. The synthesis addressed the cases of fuel savings and capacity credit.

The value of WECS to each utility is different, depending upon its equipment mix, loads, reliability requirements, financial parameters, wind resources, etc. This difference is amplified in the synthesis due to different generation expansion models being used, different approaches to reliability and capacity credit, different wind machines used in the analysis, etc. In spite of these differences, a number of conclusions can be drawn at this time relative to WECS value to utilities. These are:

1. Prime early utility markets could support \$1400/kW WECS (high winds-oil fired).

2. \$800/kW WECS first units could find utility markets in the late 1980s without capacity credit and immediately with capacity credit if fuels escalate at 3%/year above inflation.

3. WECS value is very sensitive to fuel escalations above inflation — this report assumed 3%/year escalation.

4. The spread in WECS values due to analysis differences and utility differences cannot be separated at this time.

5. WECS value seems to be directly related to the percent of oil and coal in the mix.

6. Critical factors affecting the value are fuel escalation, cost of capital, O&M assumed, the manner in which capacity credit is taken, WECS lifetime, and WECS availability.

TABLE 5

STUDIES EXAMINED

TITLE	SPONSOR	CONTRACTOR
REQUIREMENTS ASSESSMENT OF WIND POWER PLANTS IN ELECTRIC UTILITY SYSTEMS	EPRI	GENERAL ELECTRIC
WIND ENERGY SYSTEMS APPLICATION TO REGIONAL UTILITIES	DOE	JBF SCIENTIFIC CORPORATION
ELECTRIC UTILITY APPLICATION OF WIND ENERGY CONVERSION SYSTEMS ON THE ISLAND OF OAHU	DOE	AEROSPACE CORPORATION
SOUTHWEST PROJECT -RESOURCE/INSTITUTIONAL REQUIREMENTS ANALYSIS	DOE	STONE AND WEBSTER ENGINEERING CORPORATION
NORTHEAST REGIONAL ASSESSMENT STUDY FOR SOLAR ELECTRIC 1980-2000	DOE	JBF SCIENTIFIC CORPORATION

7. The values in this report were based upon the assumptions that utility fuels will escalate at 3%/year above inflation and that no significant operational problems (load following, reliability reduction) will result due to the WECS being introduced into the conventional mix. Changes in these assumptions will change the values obtained.

DATA ANALYSIS

Methodology

The basic methodology used for establishing the value of WECS to a utility is shown in Figure 58. A standard utility generation expansion model is run, without WECS present, to determine the fixed and variable costs that a utility would incur to meet its load and reliability requirements, using conventional sources. Wind systems are then added to the mix, at no cost to the utility, and hourly wind power used as a negative load to cancel out some of the hourly load that the utility would experience without WECS present. The remaining load is then run through the standard utility generation model and the fixed and variable costs associated with meeting this reduced load are calculated. The difference between the costs of meeting the total load, and the costs of meeting the reduced load (due to WECS) are an indication of the "savings" in conventional source costs due to the WECS being present. When these "Savings" have been adjusted to account for WECS siting and grid interface costs and any additional operating reserve required, the remaining savings can be used to calculate the value of WECS to the utility. This methodology has been described in detail in the reports from which this synthesis is derived and will not be discussed further.

Results of the Synthesis

The studies covered an interesting combination of utility situations, as shown in Figure 59. Table 6 is a summary of the utility parameters used in the original studies. It should be noted that the values calculated for WECS values are affected by the utilities' future mix of equipment and the future mixes used were those that the utility plans to have. If other mixes are actually installed, the value of WECS will be different than reported here. Note in Table 6 that the cost of capital and fuel escalation rates varied from study to study.

Table 7 shows the WECS parameter values used in the original studies. The wind speeds and WECS characteristics varied substantially from study to study as did the assumed land requirements and O&M costs. One study assumed a 20-year life for the WECS.

The approach used for normalizing the values was to scale the calculated WEC values ($/kW) by adjusting the cost of capital and WECS lifetime to 11% and 30 years, respectively. To the extent possible, sensitivity results from each study were used to scale its values. Figure 60 illustrated the results for one case, Southern California Edison (SCE). The SWRAS report values were extrapolated back to the first unit value. This value was then scaled up to reflect a 30-year WECS life and 11% cost of capital,resulting in the

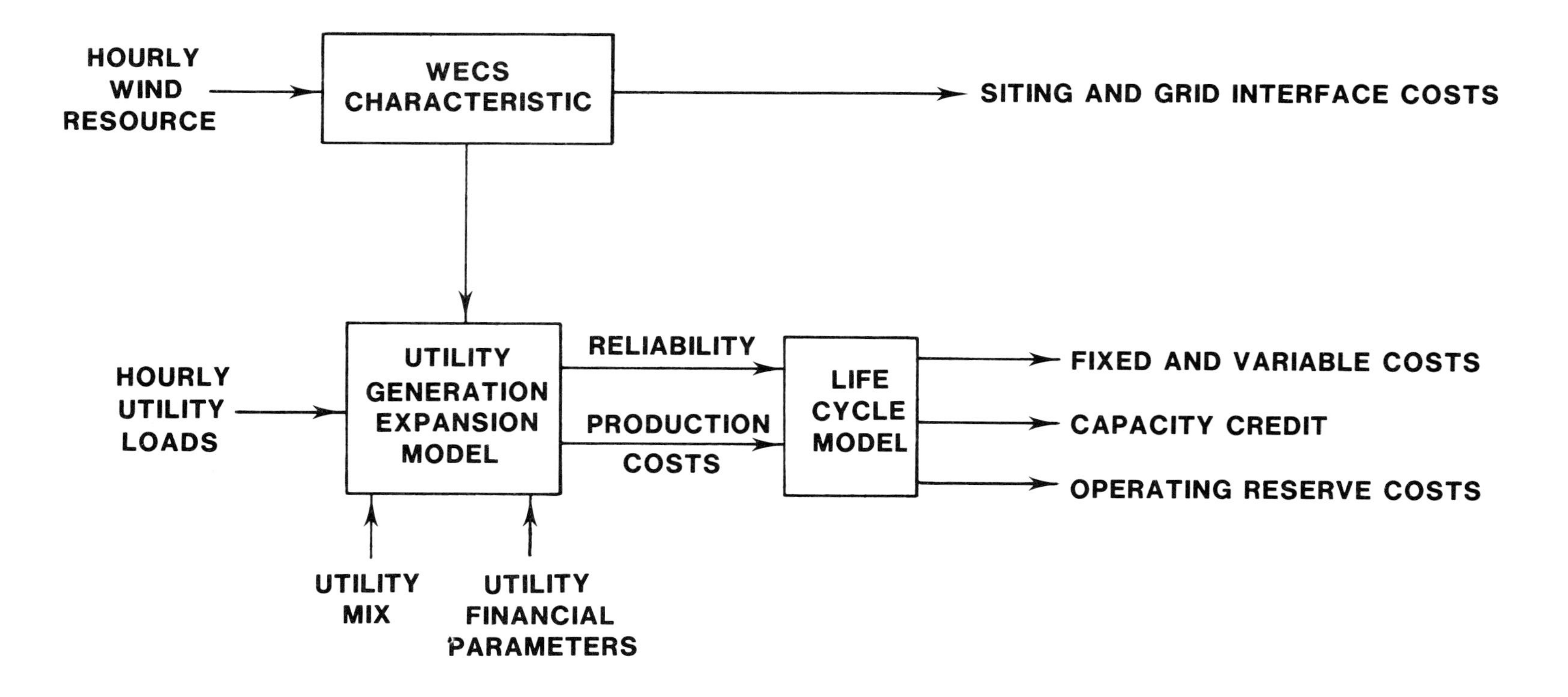

Fig. 58. Methodology for calculating the value of WECS to a utility

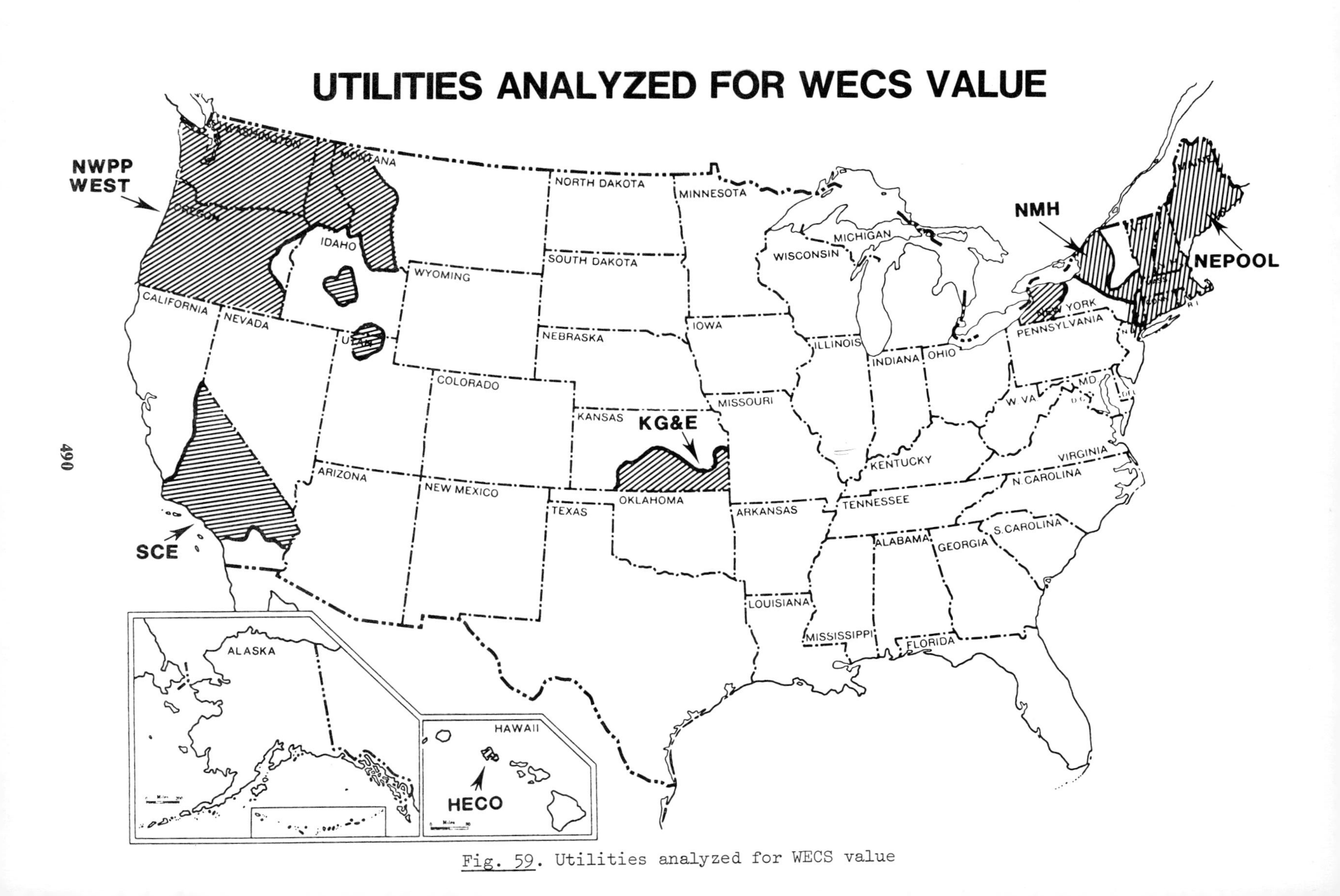

Fig. 59. Utilities analyzed for WECS value

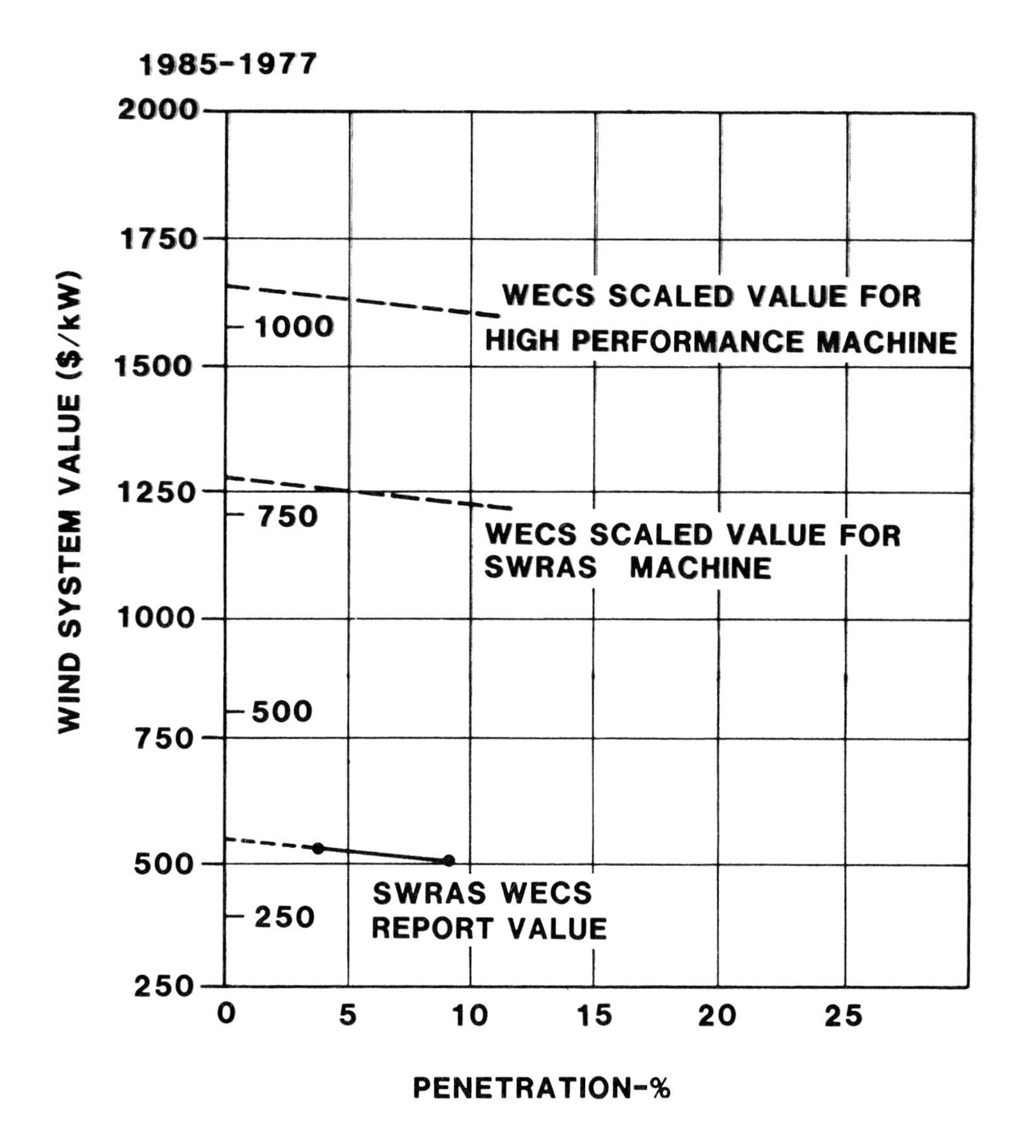

Fig. 60. Year 2000 WECS value for Southern California Edison Company normalized for economic parameters and for a high performance WECS

TABLE 6

UTILITY PARAMETERS USED IN THE ORIGINAL STUDIES

UTILITY	CODE	STUDY	FUTURE[1] EQUIPMENT MIX (%) Oil/Coal/Nuclear/Other	COST OF CAPITAL (%)	INFLATION (%/YEAR)	FUEL ESCALATION ABOVE INFLATION (%/YEAR)
NEW ENGLAND POWER POOL	NEPOOL	JBF	34/2/55/9	11	6	3
NIAGARA MOHAWK	NMH	GE	30/31.5/13.5/25	10	6	0
KANSAS GAS AND ELECTRIC	KGE	GE	28/57/15/0	10	6	0
NORTHWEST POWER POOL (COAST SITE)	$NWPP_c$	GE	2/13/31/53	10	6	0
NORTHWEST POWER POOL (GORGE SITE)	$NWPP_g$	GE	2/13/31/53	10	6	0
HAWAIIAN ELECTRIC	HECO	AERO-SPACE	100/0/0/0	11.7	6	3
SOUTHERN CALIFORNIA EDISON	SCE	STONE AND WEBSTER	53/17/25/5	9	6	0.7

(1) THESE REPRESENT THE UTILITIES PLANNED FUTURE MIX

TABLE 7

WECS PARAMETER VALUES USED IN THE ORIGINAL STUDIES

PARAMETER	UTILITY						
	NEPOOL	NMH	KGE	$NWPP_c$	$NWPP_g$	HECO	SCE
WIND SPEED- MPH (MEAN ANNUAL- 30FT)	11.2	10.1	12.1	17.9	16.3	18	(1)
WECS CHARACTERISTIC							
RATED POWER (MW)	2.5	1.5	1.5	2.0	2.0	1.5	1.5
V cut-in/V rated (MPH)	14/28	7.9/23	11.3/23	13/26	9.3/26	11.4/27.7	11/30
LIFETIME (YEARS)	30	30	30	30	30	30	20
LAND REQUIREMENTS (ACRES/$ PER ACRE)	9/$1250	(1)	(1)	(1)	(1)	16/$300	112/$200
CAPACITY FACTOR	0.3	0.36	0.446	0.512	0.45	0.595	0.26
O&M (%)	3.0 (2)	1.7 (3)	1.7 (3)	1.7 (3)	1.7 (3)	3.3 (5)	4 (4)

(1) NOT SPECIFIED
(2) PERCENT OF WECS VALUE
(3) BASED ON $10/KW/YEAR-$1976-EXPRESSED AS A PERCENTAGE OF CAPITAL COSTS
(4) BASED UPON AN ESTIMATED O&M COST-EXPRESSED AS A PERCENTAGE OF CAPITAL COSTS
(5) BASED ON $13/KW/YEAR-$1976-EXPRESSED AS A PERCENTAGE OF VALUE

line "WECS scaled value for SWRAS machine". As shown in Table 7, the SCE capacity factor from the study was only 0.26. This capacity factor was derived using an early version of the DOE MOD-1, apparently. However, the wind resource in San Gorginia Pass would probably justify a higher performance machine; thus, a capacity factor of 0.4 was assumed for this next generation WECS and the normalized economic value was then scaled up by the ratio of the study capacity factor and the assumed value of 0.4. There is no intent on the part of the author to persuade the reader that this is rigorous — it is not. However, it probably more closely reflects the value of WECS to Southern California than the original study values. The intent here is to synthesize existing studies, whereas to be rigorous the analysis should actually be rerun.

Figures 61-64 show the value of the first WECS to each of the utilities in the fuel displacement mode and in the fuel displacement plus capacity credit mode, for the economic factors normalization as well as for capacity factor scaling. The method used for capacity factor scaling was to use the studies' actual capacity factor if it was 0.4 or greater. If it was less than 0.4, the capacity factor was scaled up to a value of 0.4. The reason for this is that an attempt was made to find the value of the first WECS to each utility and, undoubtedly, the first unit will be installed in a good wind site. Furthermore, it is assumed it will be a high performance machine. A second reason for doing the capacity factor scaling was to attempt to reduce the spread in the WECS values, caused by the varying wind regimes and different WECS characteristics assumed, so that insight could be obtained on the spread due to non-wind related factors.

Figure 65 shows the impact of capacity credit on the study values. A number of utilities have indicated that they would not be able to accept the capacity credit concept until the WECS reliability and performance had been verified. Furthermore, the capacity credit values shown probably would not be obtained today, in some utilities, because of the current cost of peaking fuels.

Figures 66-67 show the values obtained after normalization for economics and capacity factor, for the cases with and without capacity credit. It should be stressed that these are optimistic because they reflect:

- a 3%/year escalation of fuel above inflation;
- a capacity factor of 0.4;
- the value of the first WECS.

However, in the case of oil-fired utilities, the 3%/year may be too conservative, based upon the 1970s experience.

Table 8 shows the "average" value from Figures 66 and 67 for the mainland and Hawaii. It indicates that mainland utilities might experience a value as much as 50% lower than Hawaii, due to the utility analyzed there having 100% oil-fired capacity and excellent wind resources. The values in Table 8 may be "optimistic" for the reasons stated earlier. It should also be

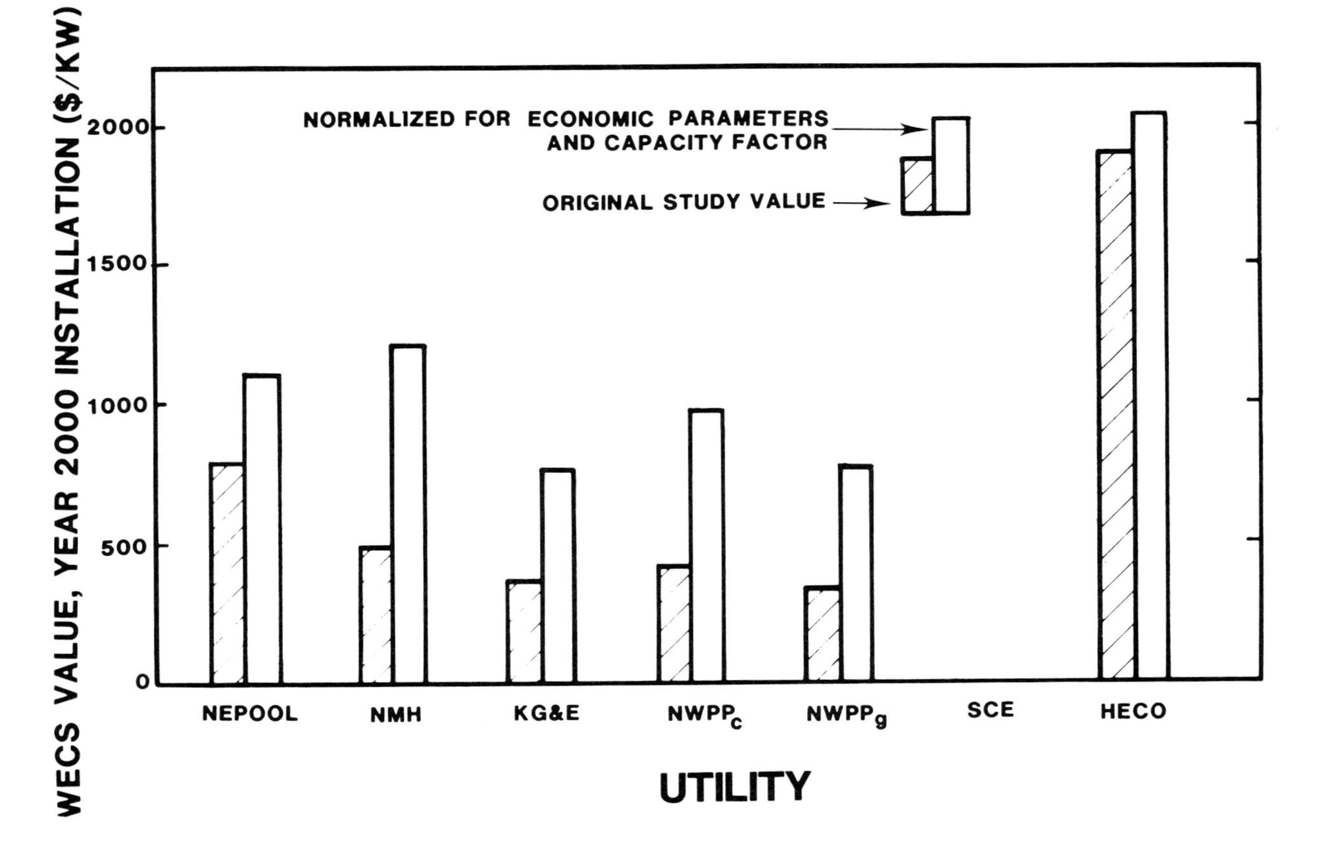

Fig. 61. Value of first WECS to utility in year 2000—fuel displacement only (1977$)

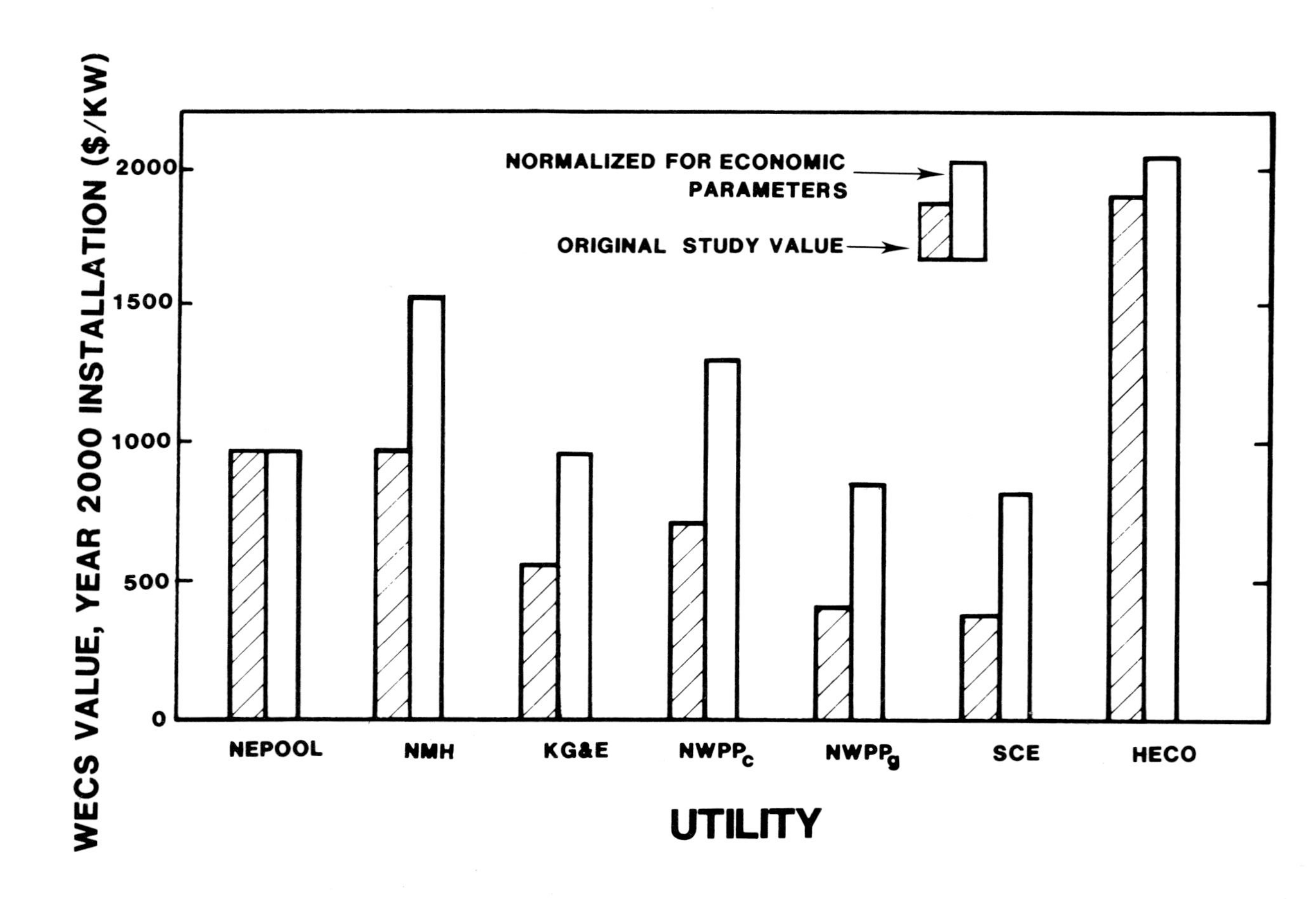

Fig. 62. Same as Fig.61 but without capacity credit (1977$)

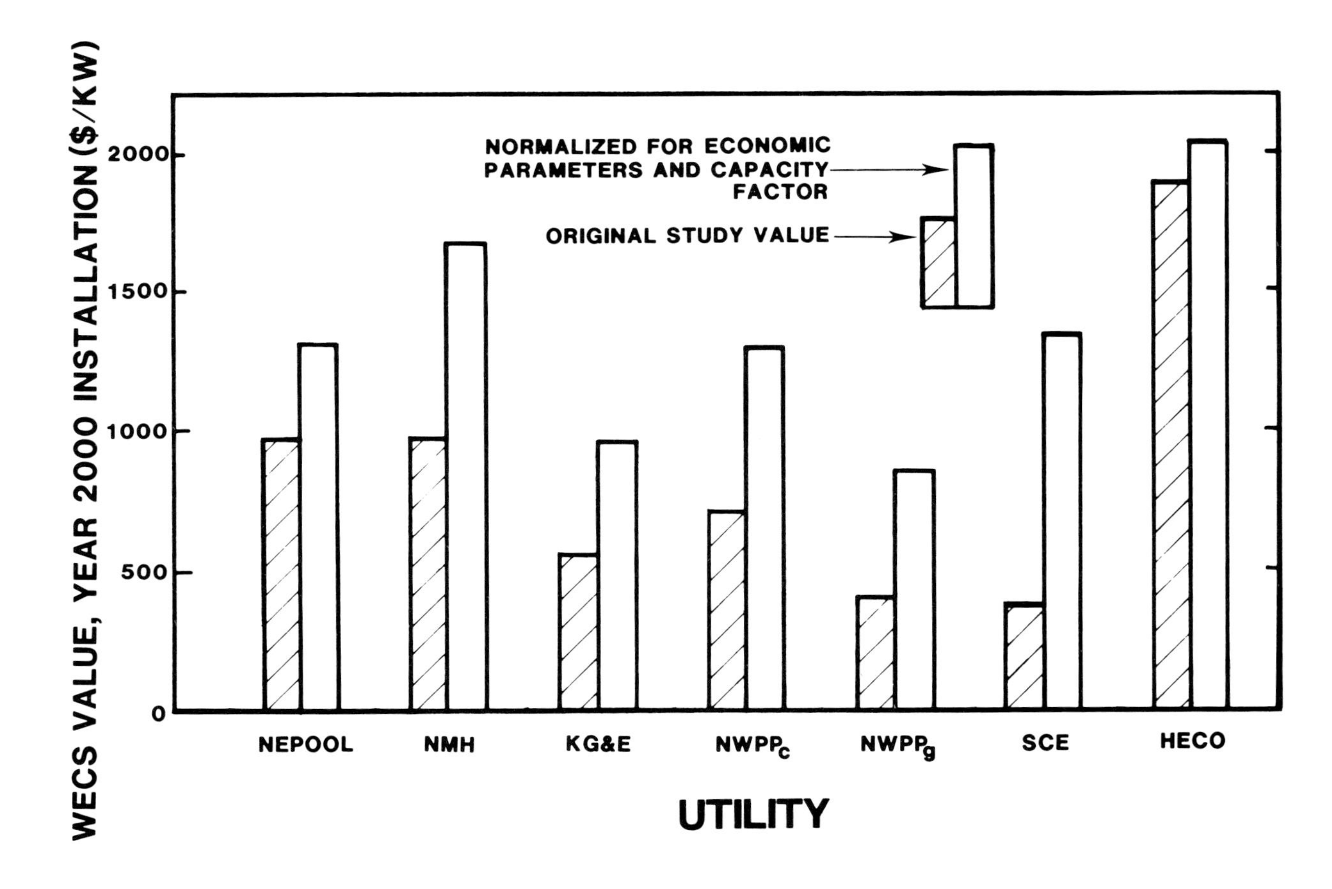

Fig. 63. Same as Fig. 61, but including fuel displacement and capacity credit (1977$)

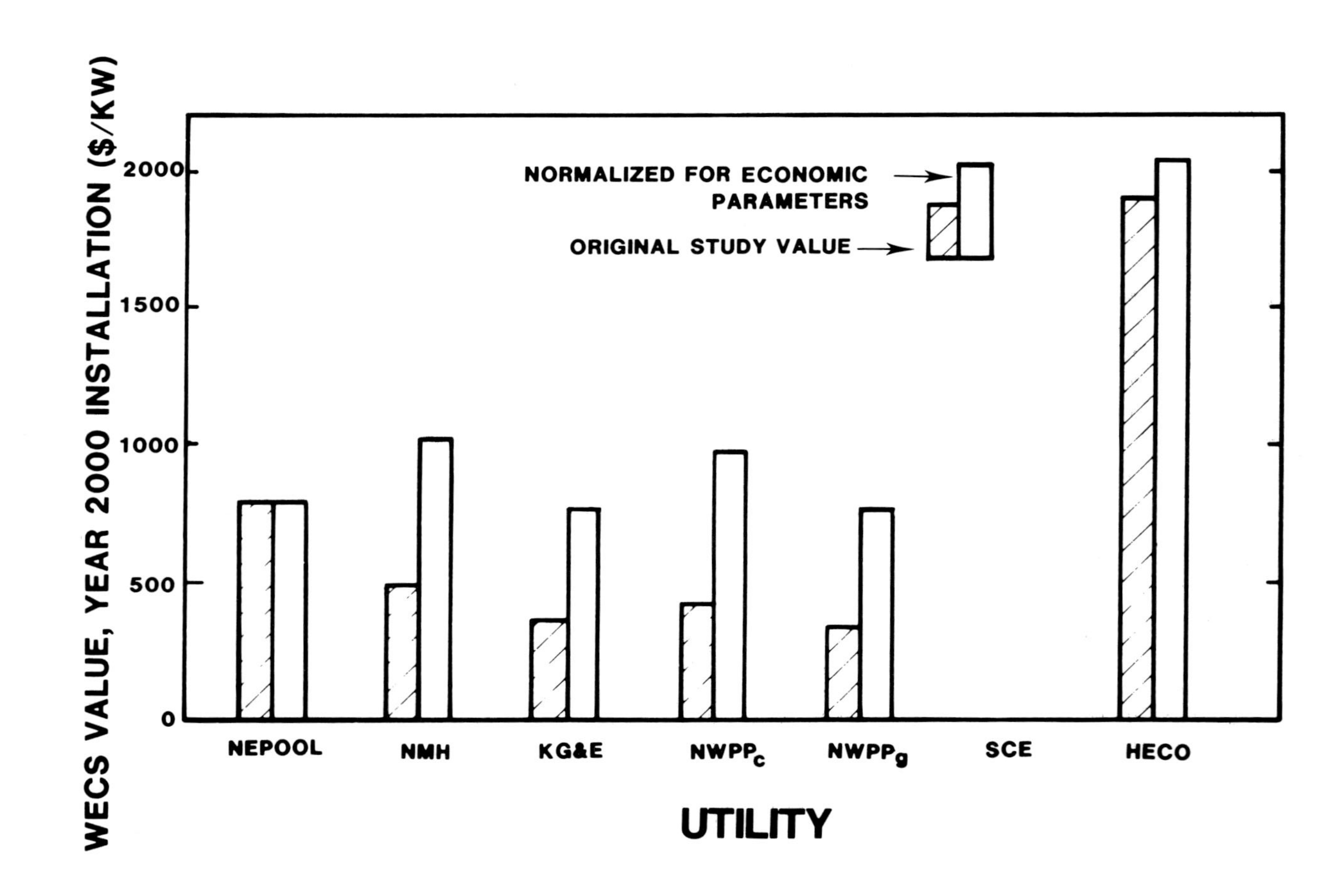

Fig. 64. Same as Fig. 61 including fuel displacement but without capacity credit (1977$)

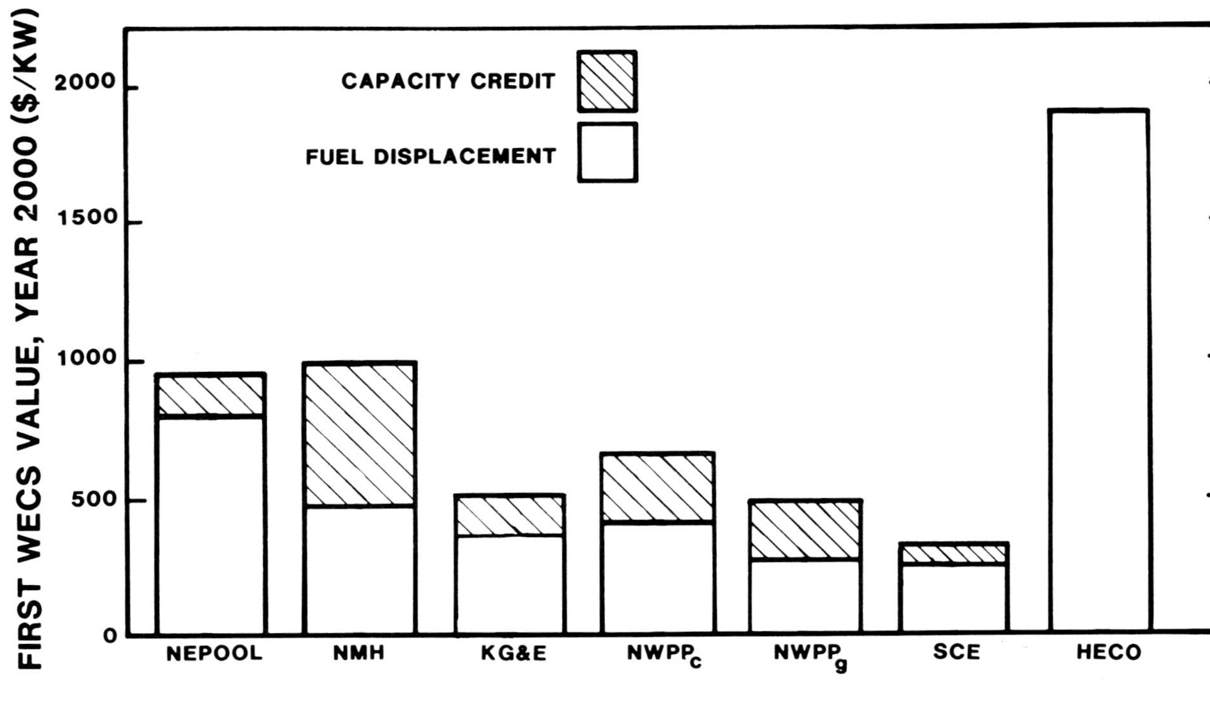

Fig.65. First WECS unit fuel displacement and capacity credit value—year 2000 (1977$)

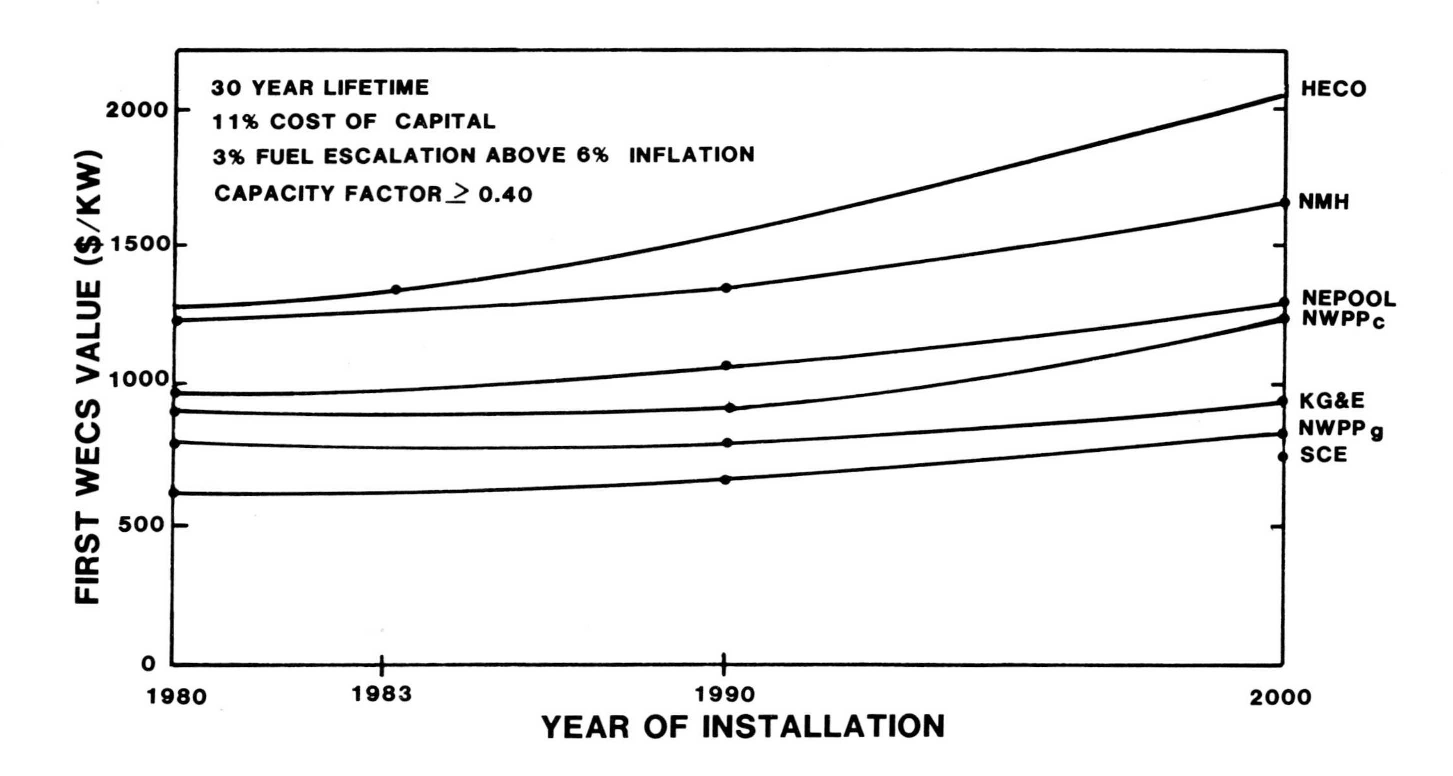

Fig. 66. First WECS value for different installation years normalized for economic parameters and capacity factor (fuel displacement value only) (1977$)

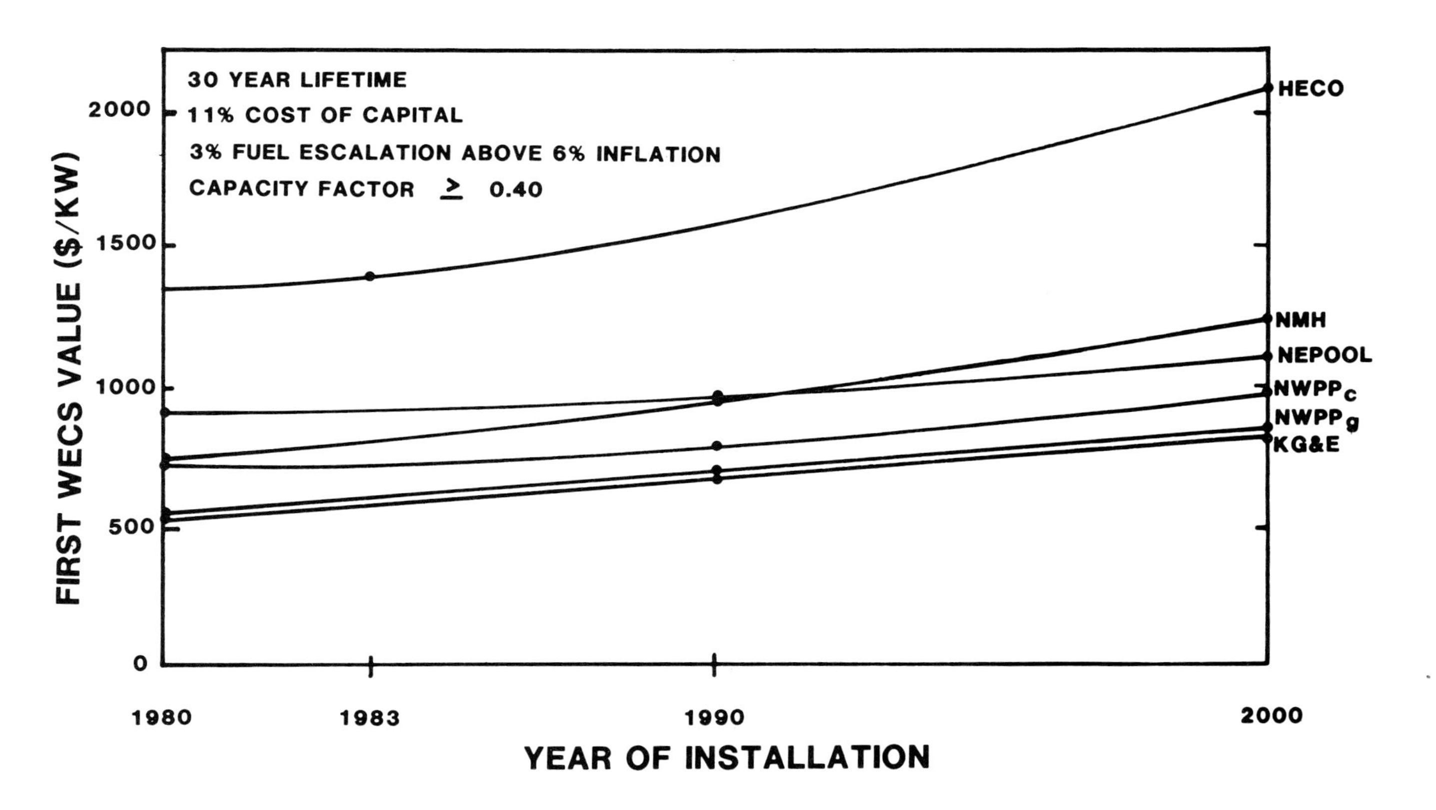

Fig. 67. Same as Fig. 66 but with fuel displacement and capacity credit (1977$)

TABLE 8

AVERAGE* VALUE OF FIRST WECS UNIT IN $/kW ($-1977)

	1980		1990		2000	
	MAINLAND	HAWAII	MAINLAND	HAWAII	MAINLAND	HAWAII
FUEL ONLY	710	1380	810	1600	1000	2060
FUEL AND C.C.	915	—	1020	—	1270	—

*NORMALIZED FOR ECONOMICS AND CAPACITY FACTOR

noted that the "averages" presented are the averages for those utilities analyzed and are not averages for all mainland utilities.

The results presented here are indicative of the range of values that utilities will experience for WECS and suggest that it will be difficult to generalize results from one utility to another.

ENVIRONMENTAL ISSUES ASSESSMENT

INTRODUCTION

The development and commercialization of wind turbine systems present a number of environmental problems that must be solved before widespread implementation of these systems can occur. In some cases, the barriers are similar to those encountered by other alternative energy technologies; but, for the most part, the barriers to wind turbine system use are unique and require detailed study. The major areas that have been investigated (ERDA, 1977 and DOE, 1978) include:

Interference with Electromagnetic Transmissions (Air Navigation Systems, Microwave Communications Links, Radio and Television Broadcasts)

Sound Intensity

Safety

Siting Problems

Ecological Concerns

Public Acceptance

Legal and Institutional Issues

This presentation discusses each of these potential problem areas, identifies the completed and active research projects in each area, summarizes the conclusions of each study, and provides an overview of the current status of environmental barriers to the commercialization of large-scale wind turbine systems (Kornreich and Kottler, 1979).

INTERFERENCE WITH ELECTROMAGNETIC TRANSMISSIONS

The University of Michigan Radiation Laboratory (Senior and Sengupta, 1978 and Sengupta and Senior, 1978) has carried out an extensive two-year investigation to determine the effects of large scale horizontal-axis wind turbine systems on the electromagnetic environment. The investigation included theoretical analyses, computer and laboratory simulations, and field tests to ascertain whether wind turbine systems could produce undesirable interference with the signals emanating from:

Air navigation systems;
Microwave communications links; and
Television and radio broadcasts.

Electromagnetic interference may be produced when a direct radio wave signal strikes the rotating blades of a wind system, producing a reflection and scattering of a secondary interference signal. It has been found that, under certain conditions, sufficiently strong, repetitive pulses are produced leading to significant interference with electromagnetic signals. In general, the interference level depends on the wind turbine blade geometry and material of construction, the rotation speed of the blades and, most importantly, on the directions and distances of the transmitter and receiver with respect to the wind turbine. Experimental scattering measurements were made using the MOD-0 system at the NASA/Plum Brook test site and small scale models of the MOD-0, MOD-0A, and MOD-1 blades.

The range over which interference can exist is proportional to the scattering area of the blades. The scattering area depends on the projected (planar) area of the blades and their scattering efficiency. Experimental results indicate that the MOD-0 (aluminium) and MOD-1 (steel) blades have much greater potential for affecting electromagnetic signals because their scattering efficiency is more than twice that of the MOD-0A (composite) blades (65% vs. 28%). The size of the MOD-1 blades in conjunction with their high scattering efficiency make them a substantially larger threat than the MOD-0 and MOD-0A in terms of electromagnetic interference.

Conventional and Doppler VOR (Very High Frequency Omnidirectional Range) systems are an integral part of the air traffic control network and provide navigation (bearing) data to flying aircraft. The Federal Aviation Administration (FAA) has established regulations concerning the maximum tolerable interference with VOR signals and has indicated minimum allowable distances of scattering source locations from VOR sites. A theoretical analysis has indicated that, if wind turbine siting is carried out in accordance with existing FAA guidelines, large horizontal-axis wind systems should not produce any significant interference with air navigation signals.

Microwave communication links are used for telephonic communications as well as for television and data transmissions. Based on the specification of a minimum level of interference that would adversely affect the link's performance, a "forbidden" zone for wind turbine siting was established around the microwave link receiver (see Figure 68). The shape of the zone is primarily determined by the radiation pattern of the receiving antenna, and its size is proportional to the wind turbine blade scattering area and to certain microwave-link characteristics. Outside the forbidden zone there are no significant interference effects of the wind turbine.

Laboratory simulations and a series of field tests using the MOD-0 wind turbine at the Plum Brook, Ohio, site were carried out to investigate and record the video and audio distortion effects produced by a large horizontal-axis wind turbine. Commercial television receivers were used in the tests in conjunction with the television signals transmitted by UHF and VHF stations in Cleveland and Toledo, Ohio.

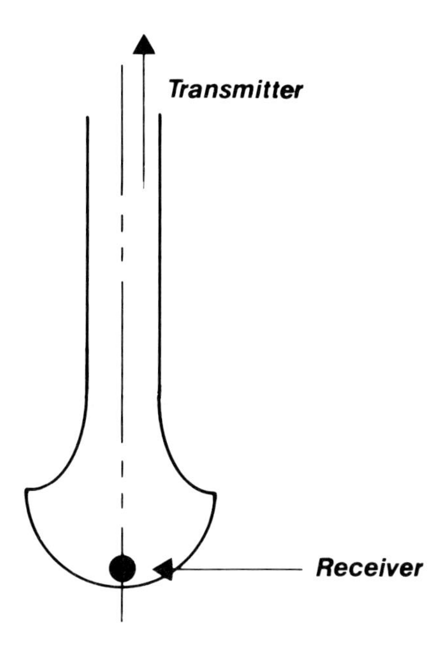

Fig. 68. Characteristic forbidden zone for microwave systems

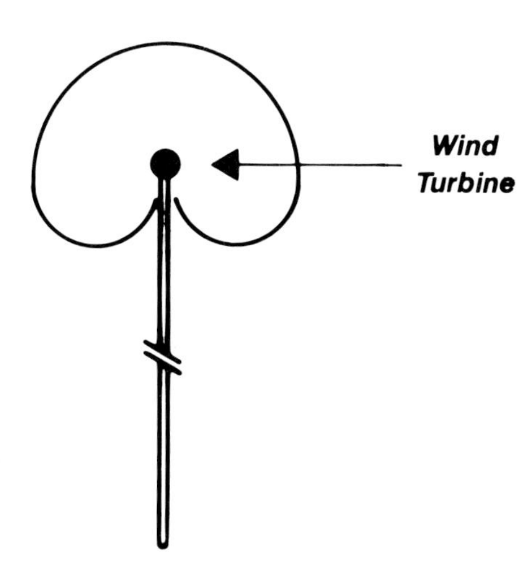

Fig. 69. Characteristic shape of TV interference zone

The level of observed interference increased with increasing television transmission frequency (or channel number) and decreased with increasing distance of the TV receiver from the wind turbine. Also, the interference radius increased as the distance of the transmitter from the turbine increased. In the worst case, objectionable video distortion was noted at distances up to a few km from the wind turbine on the upper UHF channels. Laboratory simulations of field test conditions substantiated these findings. In neither field tests nor in laboratory simulations was audio distortion observed.

The video interference observed was explained as follows. When the wind turbine blades are stationary the scattered signal may appear on the TV screen as a "ghost." A rotation of the blades causes the "ghost" to fluctuate and, if the ghost is sufficiently strong, resulting interference can be objectionable. In such cases, the received picture displays a horizontal jitter in synchronism with the blade rotation.

As the interference increases, the entire picture shows a pulsed brightening; still greater interference can disrupt the entire vertical synchronism of the TV receiver, leading to complete breakup of the picture.

As a result of the potential existence of television interference, a systematic procedure was developed for establishing the interference zone around a wind turbine system. This procedure requires input data on the television receiver, wind turbine blades, television transmitter, signal strengths, and geometric relationships between the various constituents.

The shape of the interference zone is a cardioid (see Figure 69) and is similar for all horizontal-axis wind systems and for all television channels. The size of the zone increases with increasing blade scatter and with increasing TV channel number.

At the higher frequencies, the MOD-0 produced interference with TV reception at distances up to several km from the wind turbine, based on the use of an omnidirectional antenna. Use of a directional antenna, which could discriminate against the interference signal coming from the wind turbine blades, could reduce this distance. Also, the availability of cable TV transmission or the use of careful wind turbine siting practices also offer a solution to the problem.

In practice, the interference observed will only be severe when the blades are positioned to direct the maximum scattered signal to the receiving antenna. Depending on the prevailing winds, the video interference would not be felt uniformly throughout the interference zone; some parts of the zone could suffer interference for only a small fraction of the total time.

The television interference phenomenon is site-specific. The effect of the local terrain on the interference problem, which was not considered in the University of Michigan study, needs to be carefully evaluated on a site by site basis. Since most television signals are horizontally polarized, interference by large vertical-axis wind systems would not likely be as severe as that produced by horizontal-axis systems.

SOUND INTENSITY

A recent study (Keast, 1978) was carried out to analyze the potential noise control problems that might be faced by wind energy systems. Potential problem areas were identified in both the audible sound range and at frequencies below the audible range (called infrasound).

Since there was no published information available (prior to that study) on the noise levels or noise design criteria for wind turbines, theoretical predictions of the noise spectra that might be generated by a typical wind machine (MOD-0) were developed. The calculated audible sound level spectrum corresponded to about 62 decibels of audible sound [dB(A)] which is 9 db(A) above the noise level produced by modern 765 kV transmission lines. Since the latter has elicited community-noise reactions from rural residents and sometimes contributed to lengthy project delays, it was felt that wind turbine noise considerations could possibly become a significant environmental consideration during the site approval process for large wind turbines.

The Keast study recommended that Wind Energy Conversion System (WECS) noise measurements be made to check the theoretical estimates and, if necessary, improvements in blade configuration and control which would alleviate potential noise problems should be developed. In response to these recommendations, the NASA/Lewis Research Center conducted a test program to determine the MOD-0 noise levels.

In the audible sound range (frequencies between 15 and 20,000 Hz) NASA/Lewis measurements indicated a maximum audible sould level of 64 dB(A), which was in good agreement with the theoretical estimates. With measured background noise levels of about 52 dB(A), NASA/Lewis estimated that the sound produced by the wind turbine would be indistinguishable from the background sound at about 250m from the wind turbine.

Measurements of infrasound, i.e., the accoustic energy generated at frequencies below the lower limit of human hearing (15 Hz), were also made by NASA/Lewis. Infrasound is generated when vortices are produced by the air flowing past obstructions, such as, supporting tower or turbine blades; it is measured in units of sound pressure level decibels (dB-SPL). The effects of infrasound depend on intensity, frequency, and exposure time, and may adversely affect the human respiratory system. Infrasound can cause discomfort at levels above 100dB-SPL but apparently produces no damaging effects until levels of at least 120 dB-SPL are reached. Peak infrasonic levels were experimentally measured at 78dB-SPL, therby indicating that the MOD-0 does not produce significant infrasound problems.

The Darrieus vertical-axis wind turbine has been investigated for sound level intensity by Sandia Laboratories. The 5 meter diameter wind turbine was found to produce a sound level intensity indistinguishable from the ambient sound level at a distance of 50 meters from the turbine. No data are yet available on the 17 meter Darrieus wind turbine sound levels, but measurements of a comprehensive nature are expected to be made in the future.

The results of the tests conducted to date indicate that the sound levels produced by large wind turbine systems are expected to be within tolerable limits even at distances relatively close to the wind turbine. In none of the cases examined did the wind turbine sound level exceed background sound levels at distances over 250m from the wind turbine. Mesurements of sound levels of newly designed systems should be routinely made to insure that they do not present a significant environmental problem.

SAFETY

Because the wind turbine system is composed of large blade and tower structural elements, there is a possibility of accidental structural failure even though these systems are being designed to withstand forces resulting from extreme environmental conditions. Therefore, a major concern of wind system manufacturers and users is safety.

Large wind turbines are currently being designed to withstand wind speeds up to 60 m/s. The major system failures that might occur during severe wind conditions are tower collapse and blade failure. In the event of a tower collapse, components may fall in any direction, thereby defining a danger zone encompassing a circle whose radius would be the maximum horizontal extension of the system at the time of failure. Blade throw is not expected to accompany this type of failure because the rotor should be feathered and braked before the wind force exceeds the design limits of the tower.

There are two hazards associated with blade failure, i.e., shedding of the total blade and blade break-up. The large wind turbine systems being developed under the auspices of the Federal Wind Program incorporate certain design features to minimize these hazards. For example, blade loading in the MOD-CA has been reduced through tower redesign and increased torsional stiffness in the yaw drive system. The rotating components have been redesigned to reduce stress. An automatic monitoring system which feathers and brakes the blades when ambient windspeed or gusts exceed 16 m/s has also been included. The blades themselves have been designed to withstand 70 m/s winds in the feathered position. The MOD-2 is incorporating an automatic crack detection system for the blades as an added safety feature.

The NASA/Lewis Research Center has also conducted an analysis of potential blade failure which indicated that the maximum blade throw distance for large scale systems would be approximately 150m at a blade rotation rate of 40 rpm and an optimum blade throw angle. If pieces of the blade were to break off the tip, this distance could be increased. However, if the projected angle were suboptimal, the blade throw distance would be significantly reduced.

Each wind turbine system installation is expected to be inspected regularly for defects (as part of a preventive maintenance program) to assist in the prevention of system failure. The risk to human life is expected to be very low because the potential failures would likely occur during extreme environmental conditions when only a remote chance exists that anyone would be in an exposed area near the wind turbine.

There are some additional safety problems that could occur because of unauthorized persons gaining access to wind turbine sites. The elimination of foothold and cables at ground level, the construction of security fences, and the shielding of ground level equipment are expected to minimize these problems. Safety problems associated with low-flying aircraft are also expected to be minimal.

SITING IMPACTS

The construction of a wind turbine system (or any other structure of significant size) will adversely impact the surrounding environment to some degree. The nature of these problems (and their potential solutions) will be site-specific, but some generic impacts may be discussed.

The most significant adverse effects will be the land grading and leveling, as well as the pouring of the tower foundation. For a machine such as the MOD-OA, these activities will adversely affect the total faunal and floral habitat within an area of about 0.5 acre. The potential destructive effects of these operations can be minimized by observance of careful work practices.

The siting of wind turbines in forested land areas might require clearcutting of trees so that the wind resource available to the turbine would not be restricted. This could lead to soil erosion, water pollution, and faunal and total floral habitat destruction. The latter could be mitigated by the planting of appropriate ground cover. Other environmental problems may be related to construction of access roads and limited-access safety zones surrounding the wind turbine.

In general, the effects of wind turbine construction on the surrounding environment appear to be minimal and have been encountered in many other construction activities. The corrective measures that have previously been applied in these activities will be used in the case of wind turbine construction to minimize these impacts.

ECOLOGICAL CONCERNS

The compatibility of wind turbines with their surrounding biophysical environment was investigated analytically and in field tests conducted at the MOD-0 Plum Brook Facility (Rogers, et al., 1977). The primary ecological concerns include the effects of wind turbine operation on the microclimate, bird migration, land dwelling mammals, and airborne insects.

The key findings of the microclimatic investigation indicated that fluctuations in rainfall, wind speed, temperature, and carbon dioxide level caused by the wind turbine, fall well within the natural variability of the environment, so that plants and animals in the area of the wind turbine wake are not likely to be significantly affected.

The only impact on animal life considered to be of significant magnitude was the potential for collisions of night migrating birds with a wind turbine. Therefore, a field study was carried out to gather data on this impact. This field study was conducted over a period of four migratory seasons. During this time only one nocturnal migratory bird kill was noted. However, some reservations as to the realism of these results were expressed because:

The MOD-0 was in operation only for 10% of the total nighttime hours of the migratory seasons, representing only 114 hours during peak migration periods.

Major bird kills at other locations have been known to be separated by periods of several years, so that four migratory seasons may not be a sufficiently long test interval to draw the necessary conclusions.

Despite these limitations, it was concluded that the MOD-0 wind turbine system does not appear to constitute a significant hazard to nocturnal migratory birds even during periods of high migration traffic, low cloud ceiling, and fog. Test observations indicated that migrating birds approaching the wind turbine at night tended to take evasive action to avoid the blades. However, if future wind turbine designs have blade tips reaching above the minimum altitude of most nocturnal migration (150m) or if wind turbines are sighted in locations where birds fly closer to the ground, bird collisions could present a greater risk on inclement nights.

The effects of wind turbine systems on land dwelling mammals should be quite limited where the basic habitat remains intact. In the case of extreme habitat alteration (e.g., deforestation) the impacts will likely be significant and a site-specific analysis would be required to determine the extent of the problems.

The impact of wind turbine operations on airborne insects and small invertebrates was also investigated by releasing large numbers of three species and observing their behavior when confronted with the wind turbine. These tests were designed to demonstrate a worst case example at a wind turbine site and involved the release of 50,000 insects (ladybird beetles, blow flies, and honey bees) in mid-air directly in the front of the operating MOD-0. Little evidence of injuries to insects was observed as they passed through the plane of the rotating wind turbine blades, leading to the conclusion that little impact of wind turbine operation on the insect population would be expected.

Based on the results obtained, wind turbines do not appear to pose a significant ecological threat. However, specific site conditions may be present which could alter these conclusions and therfore lead to significant ecological impacts. For example, the siting of wind turbines in relatively high altitude areas may increase the potential for bird kills.

Consideration of the potential wind turbine interactions with migratory birds during site selection can minimize these problems at future wind turbine installations. Thus, the range of prospective site-specific ecological impacts need to be carefully considered during the siting process.

It should also be noted that the results obtained relate to present wind turbine designs. The potentially adverse effects of using larger wind turbine rotors or higher blade rotation rates need to be carefully analyzed.

PUBLIC ACCEPTANCE

Public interest in alternative energy sources has grown rapidly since the oil embargo of 1973-74. While there is a realization that the United States' dependence on foreign energy sources needs to be reduced, there is still a question as to the public acceptability of alternative energy systems (such as wind turbines) as a replacement for conventional fossil-fueled systems.

A survey of public reactions to wind energy systems has been conducted (University of Illinois, 1977). This study indicated that over 80% of the people surveyed were favorably disposed to the use of wind energy as a means of generating electric power, though their knowledge of wind energy was very limited. Those who were familiar with the area of wind energy also heavily favored its use as an alternative energy source.

All of the possible classes of sites for wind turbines were viewed favorably by the survey sample. The use of wind turbines along shorelines met the strongest opposition but, even in that case, only 24% of those surveyed indicated opposition to wind turbine siting in coastal areas. The results of this survey indicate that public acceptance does not appear to constitute a barrier to wind turbine implementation. However, before firm conclusions on this topic can be formulated, information on public reaction to actual installed wind turbines should be gathered.

LEGAL AND INSTITUTIONAL ISSUES

A comprehensive investigation of the potential legal and institutional impediments to wind turbine utilization has been conducted (The George Washington University, 1977). While many of the issues considered in that study are outside the scope of this paper, there are a number of legal and institutional problems which fall in the realm of environmental considerations and will be discussed here. These problems include:

- Zoning;
- Wind "rights";
- State and local building, safety, and housing codes;
- Land requirements for large-sized systems;
- State-mandated requirements for environmental impact statements.

The effect of zoning regulations, which are likely to be mainly felt in urban or suburban areas generally classified as residential, can often lead to a prohibition on wind turbine use. Zoning restrictions often include limitations on height, setback, and use, as well as "aesthetic" zoning. It is not likely that large-scale wind systems for utility applications would be located in these areas and, in any event, federal utilities, municipal utilities, or other entities (e.g., investor-owned utilities) regulated by state public service commissions or subject to state power plant siting statutes may not be subject to these zoning regulations.

Depending on wind turbine siting, upwind obstructions could seriously impede the wind flow to the system and therefore seriously limit the energy that can be obtained. At present there is no specific legal recourse available for remedying problems of obstructions to wind flow. Existing statutes regarding rights to light or air, apportionment of water rights, or recent "solar rights" legislation, do not specifically consider wind rights. As matters now stand, affected wind system users would have to purchase preclusionary interests (title or a "negative" easement) in the surrounding land to ensure available wind flow. Again, this impediment will be of primary concern in urban or suburban areas.

Building, safety, and housing codes which typically control matters, such as, electrical wiring, structural loads, stress capacity, and enclosures of dangerous items attractive to children, are likely to totally preclude wind turbine use. However, compliance with these regulations could impose substantial burdens on the wind turbine user. Wind turbine activities by electrical utilities (municipal, cooperative, investor-owned, and federal) will sometimes be able to avoid these codes if regulated by the state public service commission or if subject to power plant siting statutes.

Land requirements for significantly-sized wind system operation could be substantial since adequate spacing (up to 10 rotor diameters) between individual units is necessary to allow for replenishment of the wind flow. However, the land area between the wind turbines appears to be suitable for compatible uses, such as, agricultural applications.

The greatest degree of uncertainty in interpretation concerns the item dealing with licenses or permits. Because of the requirements for land-use zoning or building permits for wind turbine installations, the local government agency may require the preparation of an environmental impact statement. This requirement would be contingent upon explicit language in state National Environmental Protection Agency (NEPA) regulations which impose local government participation in the Environmental Impact Statement (EIS) process. While the state may mandate such participation at the local level, it is often left unclear which party (public or private) is required to write and submit the EIS.

In summary, it appears that electric utilities are not expected to encounter significant institutional problems in incorporating wind turbine systems into their equipment mix. However, requirements for the preparation of state-mandated environmental impact statements are often unclear and vary from state to state.

SUMMARY

Electromagnetic interference, sound intensity, safety, siting impacts, ecological concerns, public acceptance, and legal and institutional issues represent potential environmental constraints to the wide-scale implentation of wind turbine systems. Of these potential problems, electromagnetic interference and certain legal and institutional issues are of most concern. The other potential barriers are not believed to constitute significant impediments to wind turbine commercialization at this time. However, the severity of environmental impacts is often site-specific and could present significant problems in certain situations. Therfore, the site selection process needs to incorporate a careful evaluation of all the prospective environmental problems.

The major environmental problem (interference with television reception) has been studied in detail for large-scale horizontal-axis wind systems. A method has been developed to define the interference zone of a wind turbine sytem so that wind turbine siting criteria can be established. No significant interference to FM broadcast reception has been found. Legal and institutional considerations (such as, zoning, building, and safety codes) may pose a signigicant barrier to prospective wind turbine use in urban and suburban areas but are not likely to affect rural wind turbine installations to as great a degree.

Overall, it is believed that there are no major environmental barriers to the implementation of wind turbine systems. In order to validate the research results obtained up to this point, experimental measurements and assessments will be made at the operational wind turbine sites being used in the Federal Wind Energy Program.

Acknowledgements

All the material presented in the various sections of this Wind Energy Summary is based on research and development carried out within the U.S. Department of Energy, and on reports, reviews, presentations, and papers, published by its laboratories, subcontractors, universities, industry, and other organizations and research institutions. The reader is provided with a detailed reference list. All the referenced material is easily available through U.S. government printing offices or other specified outlets.

The individuals who are responsible for the monitoring of the various sections within the wind program of the Department of Energy deserve a special mention, they are: George Tennyson, Dan Ancona, Terry Healy, Richard Williams, Irwin Vas, Charles Elderkin, Larry Wendell, Richard Braasch, Emil Kadleck, William Robbins and Ronald Thomas.

However, the backbone, the leader, the scientist and engineer behind it all is Dr. Louis Divone, the Chief of the Wind Systems Program. With enthusiasm and an untiring persistence, he has brought this program forward on a steady and unwavering course from its infancy to where it is today. His leadership in this field has not only benefitted the U.S. Wind Program, but has also had a considerable influence and impact on wind system developments in other countries.

REFERENCES

Elliott, D. L., 1979, "Meteorological and Topographical Indicators of Wind Energy for Regional Assessment," Presented at Conference and Workshop on Wind Energy Characteristics and Wind Energy Siting 1979, held June 19-21, Portland, Oregon. To be published in Proceedings of the Conference and Workshop on Wind Energy Characteristics and Wind Energy Siting 1979, PNL-3214, Pacific Northwest Laboratories, Richland, Washington, January 1980.

Elliott, D. L. and W. R. Barchet, 1979, "Northwest Regional Wind Energy Assessment," In Proceedings of Solar 79 Northwest, Held August 10-12, 1979, in Seattle, Washington, Available from Pacific Northwest Solar Energy Association, Seattle, Washington.

Elliott, D. L., 1977, Synthesis of National Wind Energy Assessments, PNL-2220 WIND-5, Pacific Northwest Laboratories, Richland, Washington.

Foreman, K.M., and Gilbert, B. L., "Further Investigations of Diffuser-Augmented Wind Turbines", Grumman Aerospace Corporation, COO-2612-2, July 1979.

Foreman, K. M., Gilbert, B., and Oman, R. A., "Diffuser Augmentation of Wind Turbines", Solar Energy, Vol. 20, No. 4, April 1978, pp. 305-311.

George Washington University, "Legal-Institutional Implications of Wind Energy Conversion Systems, Final Report," Program of Policy Studies in Science and Technology, Report No. NSF/RA-770203, September 1977.

Glasgow, J., and Robbins, W., "Utility Operation Experience on the NASA/DOE 200 kW Wind Turbine", DOE/NASA/1004-79/1, NASA TM-79084, Paper presented at Energy Technology VI Conference, Washington, D.C., Feb. 26-28, 1979.

Gilbert, B. L., Oman, R. A., and Foreman, K. M., "Fluid Dynamics of Diffuser Augmented Wind Turbines", Proceedings 12th Intersociety Energy Conversion Engineering Conference, Vol. 2, 1977, pp. 1651-1659.

Hewson, E. W., J. E. Wade, and R. W. Baker, 1979, A Handbook on the Use of Trees as Indicators of Wind Power Potential, RLO/2227-T24-79/3, available from National Technical Information Service, Springfield, Virginia.

Hewson, E. W., and J. Wade, 1977, Vegetation as an Indicator of High Wind Velocity, RLO/2227-T24-77/2, available from National Technical Information Service, Springfield, Virginia.

Keast, D. N., Noise Control Needs in the Developing Energy Technologies, Bolt, Beranek and Newman, Inc., Report No. COO-4389-1, prepared for the U.S. Department of Energy under Contract No. EE-77-C-02-4389, March 1978.

Kolm, K. E., and R. W. Marrs, 1977, Predicting the Wind Surface Characteristics of Southern Wyoming from Remote Sensing and Eolian Geomorphology, RLO/2343-78/1, available from National Technical Information Service, Springfield, Virginia.

Kornrich, T. R., ed., 1977, Proceedings of the Third Biennial Conference and Workshop on Wind Energy Conversion Systems, CONF-770921, held September 19-21, 1977, Washington, D.C. Available from National Technical Information Service, Springfield, Virginia.

Lowe, J., "Status and Outlook of Megawatt Size Wind Turbines for Utility Applications", Paper presented at Energy Technology VI Conference, Washington, D.C., Feb. 26-28, 1979.

Marrs, R. W., and S. Kopriva, 1978, Regions of the United States Susceptible to Eolian Action, RLO/2343-78/2, Available from National Technical Information Service, Springfield, Virginia.

Meroney, R. N., et al., 1978, Wind Characteristics Over Complex Terrain: Laboratory Simulation and Field Measurement at Rakaia Gorge, New Zealand, RLO/2438-77/2, available from National Technical Information Service, Springfield, Virginia.

"Mod-1 Wind Turbine Generator Analysis and Design Report — Executive Summary." General Electric Company, Philadelphia, Pennsylvania, under NASA Contract NAS3-20058, Report No. DOE/NASA/0058-79/3, NASA CR-159497, March 1979.

Oman, R. A., and Foreman, K. M., "Advantages of the Diffuser Augmented Wind Turbine", Proceedings of the NSF-NASA Workshop on Wind Driven Generator Systems, Report NSF/RA/W-73-006, Dec. 1973, pp. 103-106.

Oman, R. A., Foreman, K. M., and Gilbert, B. L., "Investigation of Diffuser Augmented Wind Turbines. Part II - Technical Report". ERDA Report COO-2612-2, Jan. 1977; Grumman Aerospace Corporation Research Department Report RE-534, Jan. 1977.

Powell, D. C. and J. R. Connell, 1980, One-Dimensional Gust Modeling for Wind Turbine Design: Literature Review and Analysis, PNL-3138, Pacific Northwest Laboratories, Richland, Washington.

Ramler, J. and Donovan, R., "Wind Turbines for Electric Utilities: Development Status and Economics", DOE/- NASA/1028-79, NASA TM-79170.

Reed, J. W., 1975, Wind Power Climatology of the U.S., SAND 74-0348, Sandia National Laboratory, Albuquerque, New Mexico.

Renne, D. S., and D. L. Elliott, 1978, "Overview of Techniques for Analyzing the Wind Energy Potential Over Large Areas," In Proceedings of Solar 78 Northwest Conference, DOE/IR/0114-1, held July 14-16, 1978, in Portland, Oregon. Available from Pacific Northwest Solar Energy Association, Seattle, Washington.

Rogers, S. E., et al., "Environmental Studies Related to the Operation of Wind Energy Conversion Systems," Battelle Columbus Laboratories, Report No. COO-0092-77/2, prepared for the U.S. Department of Energy under Contract W-7405-ENG-92-091, December 1977.

Sengupta, D. L., and T. B. A. Senior, "Electromagnetic Interference by Wind Turbine Generators," The University of Michigan Radiation Laboratory Report No. TID-28828, prepared for the U.S. Department of Energy under Contract DY-76-S-02-2846, January 1978.

Senior, T. B. A., and D. L. Sengupta, Wind Turbine Generator Siting and TV Reception Handbook, The University of Michigan Radiation Laboratory Technical Report COO-2846-1, prepared for the U.S. Department of Energy under Contract DY-76-S-02-2846, January 1978.

Traci, R. M., G. T. Phillips, and P. C. Patnaik, 1978, Wind Energy Site Selection Methodology Development, RLO/2440-78/2, available from National Technical Information Service, Springfield, Virginia.

University of Illinois, Public Reactions to Wind Energy Devices, Final Report, Survey Research Laboratory, Report No. NSF/RA-77-0026, October 1977.

U.S. Department of Energy, Draft Environmental Impact Statement for a Wind Turbine Generator System on Block Island, Rhode Island, Report No. DOE/EIS-0006-D, March 1978.

U.S. Department of the Interior, 1970, The National Atlas of the United States of America, U.S. Department of the Interior, Geological Survey, Washington, D.C.

U.S. Energy Research & Development Administration, Solar Program Assessment: Environmental Factors, Wind Energy Conversion, Report No. ERDA 77-47/6, March 1977.

Verholek, M. G., and P. A. Eckstrom, 1978, Remote Wind Measurements With a New Microprocessor-Based Accumulator Device, PNL-2515, Pacific Northwest Laboratories, Richland, Washington.

Walker, S. N., and T. G. Zambrano, 1979, "Wind Energy Assessment of the San Gorginia Pass Region," Presented at the Conference and Workshop on Wind Energy Characteristics and Wind Energy Siting 1979, held June 19-21, Portland, Oregon. To be published in Proceedings of the Conference and Workshop on Wind Energy Characteristics and Wind Energy Siting 1979, PNL-3214, Pacific Northwest Laboratories, Richland, Washington, January 1980.

Wegley, H. L., M. M. Orgill, and R. L. Drake, 1978, A Siting Handbook for Small Wind Energy Conversion Systems, PNL-2521, Pacific Northwest Laboratories, Richland, Washington.

Wendell, L. L., H. L. Wegley, and M. G. Verholek, 1978, Report From a Working Group Meeting on Wind Forecasts for WECS Operation, PNL-2513, Pacific Northwest Laboratories, Richland, Washington.

Whitford, D. H., Minardi, J. E., West, B. S., and Dominic, R. J., "An Analysis of the Madaras Rotor Power Plant — An Alternative Method of Extracting Large Amounts of Power from the Wind", University of Dayton Research Institute, HQS-2554-78/2, June 1979.